Texts in Applied Mathematics 13

T0156102

Springer
New York
Berlin
Heidelberg
Hong Kong
London
Milan
Paris
Tokyo

Texts in Applied Mathematics

(continued after index)

Michael Renardy   Robert C. Rogers

An Introduction to Partial Differential Equations

Second Edition

With 41 Illustrations

 Springer

Michael Renardy
Robert C. Rogers
Department of Mathematics
460 McBryde Hall
Virginia Polytechnic Institute
and State University
Blacksburg, VA 24061
USA
renardym@math.vt.edu
rogers@math.vt.edu

Series Editors

J.E. Marsden
Control and Dynamical Systems, 107–81
California Institute of Technology
Pasadena, CA 91125
USA
marsden@cds.caltech.edu

L. Sirovich
Division of Applied Mathematics
Brown University
Providence, RI 02912
USA
chico@camelot.mssm.edu

S.S. Antman
Department of Mathematics
and
Institute for Physical Science
and Technology
University of Maryland
College Park, MD 20742-4015
USA
ssa@math.umd.edu

Mathematics Subject Classification (2000): 35-01, 46-01, 47-01, 47-05

Library of Congress Cataloging-in-Publication Data
Renardy, Michael
    An introduction to partial differential equations / Michael Renardy, Robert C. Rogers.–
2nd ed.
        p. cm. – (Texts in applied mathematics ; 13)
    Includes bibliographical references and index.

    1. Differential equations, Partial. I. Rogers, Robert C. II. Title. III. Series.
QA374.R4244   2003
515′.353—dc21   2003042471

ISBN 978-1-4419-1820-8   e-ISBN 978-0-387-21687-4

Printed on acid-free paper.

9 8 7 6 5 4 3 2 1

www.springer-ny.com

Springer-Verlag   New York Berlin Heidelberg
A member of BertelsmannSpringer Science+Business Media GmbH

Series Preface

Mathematics is playing an ever more important role in the physical and biological sciences, provoking a blurring of boundaries between scientific disciplines and a resurgence of interest in the modern as well as the classical techniques of applied mathematics. This renewal of interest, both in research and teaching, has led to the establishment of the series Texts in Applied Mathematics (TAM).

The development of new courses is a natural consequence of a high level of excitement on the research frontier as newer techniques, such as numerical and symbolic computer systems, dynamical systems, and chaos, mix with and reinforce the traditional methods of applied mathematics. Thus, the purpose of this textbook series is to meet the current and future needs of these advances and to encourage the teaching of new courses.

TAM will publish textbooks suitable for use in advanced undergraduate and beginning graduate courses, and will complement the Applied Mathematical Sciences (AMS) series, which will focus on advanced textbooks and research-level monographs.

Pasadena, California   J.E. Marsden
Providence, Rhode Island   L. Sirovich
College Park, Maryland   S.S. Antman

Preface

Partial differential equations are fundamental to the modeling of natural phenomena; they arise in every field of science. Consequently, the desire to understand the solutions of these equations has always had a prominent place in the efforts of mathematicians; it has inspired such diverse fields as complex function theory, functional analysis and algebraic topology. Like algebra, topology and rational mechanics, partial differential equations are a core area of mathematics.

Unfortunately, in the standard graduate curriculum, the subject is seldom taught with the same thoroughness as, say, algebra or integration theory. The present book is aimed at rectifying this situation. The goal of this course was to provide the background which is necessary to initiate work on a Ph.D. thesis in PDEs. The level of the book is aimed at beginning graduate students. Prerequisites include a truly advanced calculus course and basic complex variables. Lebesgue integration is needed only in Chapter 10, and the necessary tools from functional analysis are developed within the course.

The book can be used to teach a variety of different courses. Here at Virginia Tech, we have used it to teach a four-semester sequence, but (more often) for shorter courses covering specific topics. Students with some undergraduate exposure to PDEs can probably skip Chapter 1. Chapters 2–4 are essentially independent of the rest and can be omitted or postponed if the goal is to learn functional analytic methods as quickly as possible. Only the basic definitions at the beginning of Chapter 2, the Weierstraß approximation theorem and the Arzela-Ascoli theorem are necessary for subsequent chapters. Chapters 10, 11 and 12 are independent of each other (except that Chapter 12 uses some definitions from the beginning of Chapter 11) and can be covered in any order desired.

We would like to thank the many friends and colleagues who gave us suggestions, advice and support. In particular, we wish to thank Pavel Bochev, Guowei Huang, Wei Huang, Addison Jump, Kyehong Kang, Michael Keane, Hong-Chul Kim, Mark Mundt and Ken Mulzet for their help. Special thanks is due to Bill Hrusa, who read a good deal of the manuscript, some of it with great care and made a number of helpful suggestions for corrections and improvements.

Notes on the second edition

We would like to thank the many readers of the first edition who provided comments and criticism. In writing the second edition we have, of course, taken the opportunity to make many corrections and small additions. We have also made the following more substantial changes.

- We have added new problems and tried to arrange the problems in each section with the easiest problems first.

- We have added several new examples in the sections on distributions and elliptic systems.

- The material on Sobolev spaces has been rearranged, expanded, and placed in a separate chapter. Basic definitions, examples, and theorems appear at the beginning while technical lemmas are put off until the end. New examples and problems have been added.

- We have added a new section on nonlinear variational problems with "Young-measure" solutions.

- We have added an expanded reference section.

Contents

1
Introduction

This book is intended to introduce its readers to the mathematical theory
of partial differential equations. But to suggest that there is a "theory"
of partial differential equations (in the same sense that there is a theory
of ordinary differential equations or a theory of functions of a single complex variable) is misleading. PDEs is a much larger subject than the two
mentioned above (it includes both of them as special cases) and a less well
developed one. However, although a casual observer may decide the subject
is simply a grab bag of unrelated techniques used to handle different types
of problems, there are in fact certain themes that run throughout.

In order to illustrate these themes we take two approaches. The first is
to pose a group of questions that arise in many problems in PDEs (existence, multiplicity, etc.). As examples of different methods of attacking
these problems, we examine some results from the theories of ODEs, advanced calculus and complex variables (with which the reader is assumed
to have some familiarity). The second approach is to examine three partial
differential equations (Laplace's equation, the heat equation and the wave
equation) in a very elementary fashion (again, this will probably be a review for most readers). We will see that even the most elementary methods
foreshadow deeper results found in the later chapters of this book.

1.1 Basic Mathematical Questions

1.1.1 Existence

Questions of existence occur naturally throughout mathematics. The question of whether a solution exists *should* pop into a mathematician's head any time he or she writes an equation down. Appropriately, the problem of existence of solutions of partial differential equations occupies a large portion of this text. In this section we consider precursors of the PDE theorems to come.

Initial-value problems in ODEs

The prototype existence result in differential equations is for initial-value problems in ODEs.

Theorem 1.1 (ODE existence, Picard-Lindelöf). *Let $D \subseteq \mathbb{R} \times \mathbb{R}^n$ be an open set, and let $\mathbf{F} : D \to \mathbb{R}^n$ be continuous in its first variable and* **uniformly Lipschitz** *in its second; i.e., for $(t, \mathbf{y}) \in D$, $\mathbf{F}(t, \mathbf{y})$ is continuous as a function of t, and there exists a constant γ such that for any $(t, \mathbf{y}_1)$ and $(t, \mathbf{y}_2)$ in D we have*

$$|\mathbf{F}(t, \mathbf{y}_1) - \mathbf{F}(t, \mathbf{y}_2)| \leq \gamma |\mathbf{y}_1 - \mathbf{y}_2|. \qquad (1.1)$$

Then, for any $(t_0, \mathbf{y}_0) \in D$, there exists an interval $I := (t^-, t^+)$ containing t_0, and at least one solution $\mathbf{y} \in C^1(I)$ of the **initial-value problem**

$$\frac{d\mathbf{y}}{dt}(t) = \mathbf{F}(t, \mathbf{y}(t)), \qquad (1.2)$$

$$\mathbf{y}(t_0) = \mathbf{y}_0. \qquad (1.3)$$

The proof of this can be found in almost any text on ODEs. We make note of one version of the proof that is the source of many techniques in PDEs: the construction of an equivalent integral equation. In this proof, one shows that there is a continuous function $\mathbf{y}$ that satisfies

$$\mathbf{y}(t) = \mathbf{y}_0 + \int_{t_0}^{t} \mathbf{F}(s, \mathbf{y}(s))\, ds. \qquad (1.4)$$

Then the fundamental theorem of calculus implies that $\mathbf{y}$ is differentiable and satisfies (1.2), (1.3) (cf. the results on smoothness below). The solution of (1.4) is obtained from an iterative procedure; i.e., we begin with an initial guess for the solution (usually the constant function $\mathbf{y}_0$) and proceed to calculate

$$
\begin{aligned}
\mathbf{y}_1(t) &= \mathbf{y}_0 + \int_{t_0}^{t} \mathbf{F}(s, \mathbf{y}_0)\, ds, \\
\mathbf{y}_2(t) &= \mathbf{y}_0 + \int_{t_0}^{t} \mathbf{F}(s, \mathbf{y}_1(s))\, ds, \\
&\vdots \\
\mathbf{y}_{k+1}(t) &= \mathbf{y}_0 + \int_{t_0}^{t} \mathbf{F}(s, \mathbf{y}_k(s))\, ds, \\
&\vdots
\end{aligned}
\tag{1.5}
$$

Of course, to complete the proof one must show that this sequence converges to a solution.

We will see generalizations of this procedure used to solve PDEs in later chapters.

Existence theorems of advanced calculus

The following theorems from advanced calculus give information on the solution of algebraic equations. The first, the inverse function theorem, considers the problem of n equations in n unknowns.

Theorem 1.2 (Inverse function theorem). *Suppose the function*

$$
\mathbf{F} : \mathbb{R}^n \ni \mathbf{x} := (x_1, \ldots, x_n) \mapsto \mathbf{F}(\mathbf{x}) := (F_1(\mathbf{x}), \ldots, F_n(\mathbf{x})) \in \mathbb{R}^n
$$

is C^1 in a neighborhood of a point $\mathbf{x}_0$. Further assume that

$$
\mathbf{F}(\mathbf{x}_0) = \mathbf{p}_0
$$

and

$$
\frac{\partial \mathbf{F}}{\partial \mathbf{x}}(\mathbf{x}_0) :=
\begin{pmatrix}
\frac{\partial F_1}{\partial x_1}(\mathbf{x}_0) & \cdots & \frac{\partial F_1}{\partial x_n}(\mathbf{x}_0) \\
\vdots & \ddots & \vdots \\
\frac{\partial F_n}{\partial x_1}(\mathbf{x}_0) & \cdots & \frac{\partial F_n}{\partial x_n}(\mathbf{x}_0)
\end{pmatrix}
$$

is nonsingular. Then there is a neighborhood N_x of $\mathbf{x}_0$ and a neighborhood N_p of $\mathbf{p}_0$ such that $F : N_x \to N_p$ is one-to-one and onto; i.e., for every $\mathbf{p} \in N_p$ the equation

$$
\mathbf{F}(\mathbf{x}) = \mathbf{p}
$$

has a unique solution in N_x.

Our second result, the implicit function theorem, concerns solving a system of p equations in $q + p$ unknowns.

Theorem 1.3 (Implicit function theorem). *Suppose the function*

$$
\mathbf{F} : \mathbb{R}^q \times \mathbb{R}^p \ni (\mathbf{x}, \mathbf{y}) \mapsto \mathbf{F}(\mathbf{x}, \mathbf{y}) \in \mathbb{R}^p
$$

is C^1 in a neighborhood of a point $(\mathbf{x}_0, \mathbf{y}_0)$. Further assume that

$$
\mathbf{F}(\mathbf{x}_0, \mathbf{y}_0) = \mathbf{0},
$$

and that the $p \times p$ matrix

$$\frac{\partial \mathbf{F}}{\partial \mathbf{y}}(\mathbf{x}_0, \mathbf{y}_0) := \begin{pmatrix} \frac{\partial F_1}{\partial y_1}(\mathbf{x}_0, \mathbf{y}_0) & \cdots & \frac{\partial F_1}{\partial y_p}(\mathbf{x}_0, \mathbf{y}_0) \\ \vdots & \ddots & \vdots \\ \frac{\partial F_p}{\partial y_1}(\mathbf{x}_0, \mathbf{y}_0) & \cdots & \frac{\partial F_p}{\partial y_p}(\mathbf{x}_0, \mathbf{y}_0) \end{pmatrix}$$

is nonsingular. Then there is a neighborhood $N_x \subset \mathbb{R}^q$ of $\mathbf{x}_0$ and a function $\hat{\mathbf{y}} : N_x \to \mathbb{R}^p$ such that

$$\hat{\mathbf{y}}(\mathbf{x}_0) = \mathbf{y}_0,$$

and for every $\mathbf{x} \in N_x$

$$\mathbf{F}(\mathbf{x}, \hat{\mathbf{y}}(\mathbf{x})) = \mathbf{0}.$$

The two theorems illustrate the idea that a nonlinear system of equations behaves essentially like its linearization as long as the linear terms dominate the nonlinear ones. Results of this nature are of considerable importance in differential equations.

1.1.2   Multiplicity

Once we have asked the question of whether a solution to a given problem exists, it is natural to consider the question of how many solutions there are.

Uniqueness for initial-value problems in ODEs

The prototype for uniqueness results is for initial-value problems in ODEs.

Theorem 1.4 (ODE uniqueness). *Let the function $\mathbf{F}$ satisfy the hypotheses of Theorem 1.1. Then the initial-value problem (1.2), (1.3) has at most one solution.*

A proof of this based on Gronwall's inequality is given below.

It should be noted that although this result covers a very wide range of initial-value problems, there are some standard, simple examples for which uniqueness fails. For instance, the problem

$$\frac{dy}{dt} = y^{1/3},$$
$$y(0) = 0$$

has an entire family of solutions parameterized by $\gamma \in [0, 1]$:

$$y_\gamma(t) := \begin{cases} 0, & 0 \leq t \leq \gamma \\ \left[\frac{2}{3}(t - \gamma)\right]^{3/2}, & \gamma < t \leq 1. \end{cases}$$

Nonuniqueness for linear and nonlinear boundary-value problems

While uniqueness is often a desirable property for a solution of a problem (often for physical reasons), there are situations in which multiple solutions are desirable. A common mathematical problem involving multiple solutions is an eigenvalue problem. The reader should, of course, be familiar with the various existence and multiplicity results from finite-dimensional linear algebra, but let us consider a few problems from ordinary differential equations. We consider the following second-order ODE depending on the parameter λ:

$$u'' + \lambda u = 0. \tag{1.6}$$

Of course, if we imposed two initial conditions (at one point in space) Theorem 1.4 would imply that we would have a unique solution. (To apply the theorem directly we need to convert the problem from a second-order equation to a first-order system.) However, if we impose the two-point boundary conditions

$$u(0) = 0, \tag{1.7}$$
$$u'(1) = 0, \tag{1.8}$$

the uniqueness theorem does not apply. Instead we get the following result.

Theorem 1.5. *There are two alternatives for the solutions of the boundary-value problem (1.6), (1.7), (1.8).*

1. *If $\lambda = \lambda_n := ((2n+1)^2 \pi^2)/4$, $n = 0, 1, 2, \ldots$, then the boundary-value problem has a family of solutions parameterized by $A \in (-\infty, \infty)$:*

   $$u_n(x) = A \sin \frac{(2n+1)\pi}{2} x.$$

   In this case we say λ is an eigenvalue.

2. *For all other values of λ the only solution of the boundary-value problem is the trivial solution*

   $$u(x) \equiv 0.$$

This characteristic of having either a unique (trivial) solution or an infinite linear family of solutions is typical of linear problems. More interesting multiplicity results are available for nonlinear problems and are the main subject of modern *bifurcation theory*. For example, consider the following nonlinear boundary-value problem, which was derived by Euler to describe the deflection of a thin, uniform, inextensible, vertical, elastic beam under a load λ:

$$\theta''(x) + \lambda \sin \theta(x) = 0, \tag{1.9}$$

$$\theta(0) = 0, \tag{1.10}$$
$$\theta'(1) = 0. \tag{1.11}$$

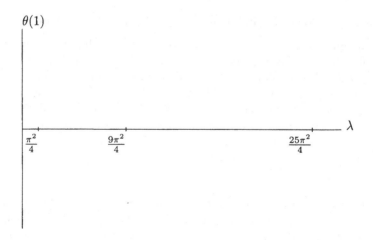

Figure 1.1. Bifurcation diagram for the nonlinear boundary-value problem

(Note that the linear ODE (1.6) is an approximation of (1.9) for small θ.) Solutions of this nonlinear boundary-value problem have been computed in closed form (in terms of Jacobi elliptic functions) and are probably best displayed by a *bifurcation diagram* such as Figure 1.1. This figure displays the amplitude of a solution θ as a function of the value of λ at which the solution occurs. The λ axis denotes the trivial solution $\theta \equiv 0$ (which holds for every λ). Note that a *branch* of nontrivial solutions emanates from each of the eigenvalues of the linear problem above. Thus for $\lambda \in (\lambda_{n-1}, \lambda_n)$, $n = 1, 2, 3, \ldots$, there are precisely $2n$ nontrivial solutions of the boundary-value problem.

1.1.3 Stability

The term stability is one that has a variety of different meanings within mathematics. One often says that a problem is stable if it is "continuous with respect to the data"; i.e., a problem is stable if when we change the problem "slightly," the solution changes only slightly. We make this precise below in the context of initial-value problems for ODEs. Another notion of stability is that of "asymptotic stability." Here we say a problem is stable if all of its solutions get close to some "nice" solution as time goes to infinity. We make this notion precise with a result on linear systems of ODEs with constant coefficients.

Stability with respect to initial conditions

In this section we assume that **F** satisfies the hypotheses of Theorem 1.1, and we define $\hat{\mathbf{y}}(t, t_0, \mathbf{y}_0)$ to be the unique solution of (1.2), (1.3). We then have the following standard result.

Theorem 1.6 (Continuity with respect to initial conditions). *The function $\hat{\mathbf{y}}$ is well defined on an open set*

$$U \subset \mathbb{R} \times D.$$

Furthermore, at every $(t, t_0, \mathbf{y}_0) \in U$ the function

$$(t_0, \mathbf{y}_0) \mapsto \hat{\mathbf{y}}(t, t_0, \mathbf{y}_0)$$

is continuous; i.e., for any $\epsilon > 0$ there exists δ (depending on $(t, t_0, \mathbf{y}_0)$ and ϵ) such that if

$$|(t_0, \mathbf{y}_0) - (\tilde{t}_0, \tilde{\mathbf{y}}_0)| < \delta,$$

then $\hat{\mathbf{y}}(t, \tilde{t}_0, \tilde{\mathbf{y}}_0)$ is well defined and

$$|\hat{\mathbf{y}}(t, t_0, \mathbf{y}_0) - \hat{\mathbf{y}}(t, \tilde{t}_0, \tilde{\mathbf{y}}_0)| < \epsilon. \tag{1.12}$$

Thus, we see that small changes in the initial conditions result in small changes in the solutions of the initial-value problem.

1.1.4 Linear Systems of ODEs and Asymptotic Stability

We now examine a concept called asymptotic stability in the context of linear system of ODEs. We consider the problem of finding a function $\mathbf{y} : \mathbb{R} \to \mathbb{R}^n$ that satisfies

$$\frac{d\mathbf{y}}{dt}(t) = \mathbf{A}(t)\mathbf{y}(t) + \mathbf{f}(t), \tag{1.13}$$

$$\mathbf{y}(t_0) = \mathbf{y}_0, \tag{1.14}$$

where $t_0 \in \mathbb{R}$, $\mathbf{y}_0 \in \mathbb{R}^n$, the vector valued function $\mathbf{f} : \mathbb{R} \to \mathbb{R}^n$ and the matrix valued function $\mathbf{A} : \mathbb{R} \to \mathbb{R}^{n \times n}$ are given.

Asymptotic stability describes the behavior of solutions of homogeneous systems as t goes to infinity.

Definition 1.7. The linear homogeneous system

$$\mathbf{y}' = \mathbf{A}(t)\mathbf{y} \tag{1.15}$$

is

1. **asymptotically stable** if every solution of (1.15) satisfies

$$\lim_{t \to \infty} |\mathbf{y}(t)| = 0, \tag{1.16}$$

2. **completely unstable** if every nonzero solution of (1.15) satisfies

$$\lim_{t \to \infty} |\mathbf{y}(t)| = \infty. \tag{1.17}$$

The following fundamental result applies to constant coefficient systems.

Theorem 1.8. *Let* $\mathbf{A} \in \mathbb{R}^{n \times n}$ *be a constant matrix with eigenvalues*

$$\lambda_1, \lambda_2, \ldots, \lambda_n.$$

Then the linear homogeneous system of ODEs

$$\mathbf{y}' = \mathbf{A}\mathbf{y} \tag{1.18}$$

is

1. *asymptotically stable if and only if all the eigenvalues of* $\mathbf{A}$ *have negative real parts; and*

2. *completely unstable if and only if all the eigenvalues of* $\mathbf{A}$ *have positive real parts.*

The proof of this theorem is based on a diagonalization procedure for the matrix $\mathbf{A}$ and the following formula for all solutions of the initial-value problem associated with (1.18)

$$\mathbf{y}(t) := e^{\mathbf{A}(t-t_0)}\mathbf{y}_0. \tag{1.19}$$

Here the matrix $e^{\mathbf{A}t}$ is defined by the uniformly convergent power series

$$e^{\mathbf{A}t} := \sum_{n=0}^{\infty} \frac{\mathbf{A}^n t^n}{n!}. \tag{1.20}$$

Formula 1.19 is the precursor of formulas in semigroup theory that we encounter in Chapter 12.

1.1.5   Well-Posed Problems

We say that a problem is *well-posed (in the sense of Hadamard)* if

1. there exists a solution,

2. the solution is unique,

3. the solution depends continuously on the data.

If these conditions do not hold, a problem is said to be *ill-posed.* Of course, the meaning of the term *continuity with respect to the data* has to be made more precise by a choice of norms in the context of each problem considered.

In the course of this book we classify most of the problems we encounter as either well-posed or ill-posed, but the reader should avoid the assumption that well-posed problems are always "better" or more "physically realistic" than ill-posed problems. As we saw in the problem of buckling of a beam mentioned above, there are times when the conditions of a well-posed problem (uniqueness in this case) are physically unrealistic. The importance of ill-posedness in nature was stressed long ago by Maxwell [Max]:

For example, the rock loosed by frost and balanced on a singular point of the mountain-side, the little spark which kindles the great forest, the little word which sets the world afighting, the little scruple which prevents a man from doing his will, the little spore which blights all the potatoes, the little gemmule which makes us philosophers or idiots. Every existence above a certain rank has its singular points: the higher the rank, the more of them. At these points, influences whose physical magnitude is too small to be taken account of by a finite being may produce results of the greatest importance. All great results produced by human endeavour depend on taking advantage of these singular states when they occur.

We draw attention to the fact that this statement was made a full century before people "discovered" all the marvelous things that can be done with cubic surfaces in $\mathbb{R}^3$.

1.1.6 Representations

There is one way of proving existence of a solution to a problem that is more satisfactory than all others: writing the solution explicitly. In addition to the aesthetic advantages provided by a representation for a solution there are many practical advantages. One can compute, graph, observe, estimate, manipulate and modify the solution by using the formula. We examine below some representations for solutions that are often useful in the study of PDEs.

Variation of parameters

Variation of parameters is a formula giving the solution of a nonhomogeneous linear system of ODEs (1.13) in terms of solutions of the homogeneous problem (1.15). Although this representation has at least some utility in terms of actually computing solutions, its primary use is analytical.

The key to the variations of constants formula is the construction of a *fundamental solution matrix* $\Phi(t, \tau) \in \mathbb{R}^{n \times n}$ for the linear homogeneous system. This solution matrix satisfies

$$\frac{d}{dt}\Phi(t, \tau) = \mathbf{A}(t)\Phi(t, \tau), \qquad (1.21)$$

$$\Phi(\tau, \tau) = \mathbf{I}, \qquad (1.22)$$

where $\mathbf{I}$ is the $n \times n$ identity matrix. The proof of existence of the fundamental matrix is standard and is left as an exercise. Note that the unique solution of the initial-value problem (1.15), (1.14) for the homogeneous system is given by

$$\mathbf{y}(t) := \Phi(t, t_0)\mathbf{y}_0. \qquad (1.23)$$

The use of Leibniz' formula reveals that the *variation of parameters formula*

$$\mathbf{y}(t) := \Phi(t, t_0)\mathbf{y}_0 + \int_{t_0}^{t} \Phi(t, s)\mathbf{f}(s)ds \qquad (1.24)$$

gives the solution of the initial-value problem (1.13), (1.14) for the nonhomogeneous system.

Cauchy's integral formula

Cauchy's integral formula is the most important result in the theory of complex variables. It provides a representation for analytic functions in terms of its values at distant points. Note that this representation is rarely used to actually compute the values of an analytic function; rather it is used to deduce a variety of theoretical results.

Theorem 1.9 (Cauchy's integral formula). *Let f be analytic in a simply connected domain $D \subset \mathbb{C}$ and let C be a simple closed positively oriented curve in D. Then for any point z_0 in the interior of C*

$$f(z_0) = \frac{1}{2\pi i} \oint_C \frac{f(z)}{z - z_0} \, dz. \qquad (1.25)$$

1.1.7   Estimation

When we speak of an *estimate* for a solution we refer to a relation that gives an indication of the solution's size or character. Most often these are inequalities involving norms of the solution. We distinguish between the following two types of estimate. An *a posteriori* estimate depends on knowledge of the existence of a solution. This knowledge is usually obtained through some sort of construction or explicit representation. An *a priori* estimate is one that is conditional on the existence of the solution; i.e., a result of the form, "If a solution of the problem exists, then it satisfies ..." We present here an example of each type of estimate.

Gronwall's inequality and energy estimates

In this section we derive an *a priori* estimate for solutions of ODEs that is related to the energy estimates for PDEs that we examine in later chapters. The uniqueness theorem 1.4 is an immediate consequence of this result. To derive our estimate we need a fundamental inequality called Gronwall's inequality.

Lemma 1.10 (Gronwall's inequality). *Let*

$$u : [a, b] \to [0, \infty),$$
$$v : [a, b] \to \mathbb{R},$$

be continuous functions and let C be a constant. Then if

$$v(t) \leq C + \int_a^t v(s)u(s)ds \tag{1.26}$$

for $t \in [a, b]$, it follows that

$$v(t) \leq C \exp\left(\int_a^t u(s)ds\right) \tag{1.27}$$

for $t \in [a, b]$.

The proof of this is left as an exercise.

Lemma 1.11 (Energy estimate for ODEs). *Let $\mathbf{F} : \mathbb{R} \times \mathbb{R}^n \to \mathbb{R}^n$ satisfy the hypotheses of Theorem 1.1, in particular let it be uniformly Lipschitz in its second variable with Lipschitz constant γ (cf. (1.1)). Let $\mathbf{y}_1$ and $\mathbf{y}_2$ be solutions of (1.2) on the interval $[t_0, T]$; i.e.,*

$$\mathbf{y}_i'(t) = \mathbf{F}(t, \mathbf{y}_i(t))$$

for $i = 1, 2$ and $t \in [t_0, T]$. Then

$$|\mathbf{y}_1(t) - \mathbf{y}_2(t)|^2 \leq |\mathbf{y}_1(t_0) - \mathbf{y}_2(t_0)|^2 e^{2\gamma(t-t_0)}. \tag{1.28}$$

Proof. We begin by using the differential equation, the Cauchy-Schwarz inequality and the Lipschitz condition to derive the following inequality.

$$
\begin{aligned}
&|\mathbf{y}_1(t) - \mathbf{y}_2(t)|^2 \\
&= |\mathbf{y}_1(t_0) - \mathbf{y}_2(t_0)|^2 + \int_{t_0}^t \frac{d}{ds}|\mathbf{y}_1(s) - \mathbf{y}_2(s)|^2 \, ds \\
&= |\mathbf{y}_1(t_0) - \mathbf{y}_2(t_0)|^2 \\
&\quad + \int_{t_0}^t 2(\mathbf{y}_1(s) - \mathbf{y}_2(s)) \cdot (\mathbf{F}(s, \mathbf{y}_1(s)) - \mathbf{F}(s, \mathbf{y}_2(s))) \, ds \\
&\leq |\mathbf{y}_1(t_0) - \mathbf{y}_2(t_0)|^2 + \int_{t_0}^t 2|\mathbf{y}_1(s) - \mathbf{y}_2(s)||\mathbf{F}(s, \mathbf{y}_1(s)) - \mathbf{F}(s, \mathbf{y}_2(s))| \, ds \\
&\leq |\mathbf{y}_1(t_0) - \mathbf{y}_2(t_0)|^2 + \int_{t_0}^t 2\gamma|\mathbf{y}_1(s) - \mathbf{y}_2(s)|^2 \, ds.
\end{aligned}
$$

Now (1.28) follows directly from Gronwall's inequality. $\qquad\square$

Note we can derive the uniqueness result for ODEs (Theorem 1.4) by simply setting $\mathbf{y}_1(t_0) = \mathbf{y}_2(t_0)$ and using (1.28). Also observe that these results are indeed obtained *a priori*: nothing we did depended on the existence of a solution, only on the equations that a solution would satisfy if it *did* exist.

Maximum principle for analytic functions

As an example of an *a posteriori* result we consider the following theorem.

Theorem 1.12 (Maximum modulus principle).   *Let $D \subset \mathbb{C}$ be a bounded domain and let f be analytic on D and continuous on the closure of D. Then $|f|$ achieves its maximum on the boundary of D; i.e., there exists $z_0 \in \partial D$ such that*

$$|f(z_0)| = \sup_{z \in \overline{D}} |f(z)|. \tag{1.29}$$

The reader is encouraged to prove this using Cauchy's integral formula (cf. Problem 1.10). Such a proof, based on an explicit representation for the function f, is *a posteriori*. We note, however, that it is possible to give an *a priori* proof of the result; and Chapter 4 is dedicated to finding *a priori* maximum principles for PDEs.

1.1.8 Smoothness

One of the most important modern techniques for proving the existence of a solution to a partial differential equation is the following process.

1. Convert the original PDE into a "weak" form that might conceivably have very rough solutions.

2. Show that the weak problem has a solution.

3. Show that the solution of the weak equation actually has more smoothness than one would have at first expected.

4. Show that a "smooth" solution of the weak problem is a solution of the original problem.

We give a preview of parts one, two, and four of this process in Section 1.2.1 below, but in this section let us consider precursors of the methods for part three: showing smoothness.

Smoothness of solutions of ODEs

The following is an example of a "bootstrap" proof of regularity in which we use the fact that $\mathbf{y} \in C^0$ to show that $\mathbf{y} \in C^1$, etc. Note that this result can be used to prove the regularity portion of Theorem 1.1 (which asserted the existence of a C^1 solution).

Theorem 1.13. *If $\mathbf{F} : \mathbb{R} \times \mathbb{R}^n \to \mathbb{R}^n$ is in $C^{m-1}(\mathbb{R} \times \mathbb{R}^n)$ for some integer $m \geq 1$, and $\mathbf{y} \in C^0(\mathbb{R})$ satisfies the integral equation*

$$\mathbf{y}(t) = \mathbf{y}(t_0) + \int_{t_0}^{t} \mathbf{F}(s, \mathbf{y}(s)) \, ds, \tag{1.30}$$

then in fact $\mathbf{y} \in C^m(\mathbb{R})$.

Proof. Since $\mathbf{F}(s, \mathbf{y}(s))$ is continuous, we can use the Fundamental Theorem of Calculus to deduce that the right-hand side of (8.173) is continuously differentiable, so the left-hand side must be as well, and

$$\mathbf{y}'(t) = \mathbf{F}(t, \mathbf{y}(t)). \tag{1.31}$$

Thus, $\mathbf{y} \in C^1(\mathbb{R})$. If $\mathbf{F}$ is in C^1, we can repeat this process by noting that the right-hand side of (1.31) is differentiable (so the left-hand side is as well) and

$$\mathbf{y}''(t) = \mathbf{F}_{\mathbf{y}}(t, \mathbf{y}(t)) \cdot \mathbf{y}'(t) + \mathbf{F}_t(t, \mathbf{y}(t)),$$

so $\mathbf{y} \in C^2(\mathbb{R})$. This can be repeated as long as we can take further continuous derivatives of $\mathbf{F}$. We conclude that, in general, $\mathbf{y}$ has one order of differentiablity more than $\mathbf{F}$. □

Smoothness of analytic functions

A stronger result can be obtained for analytic functions by using Cauchy's integral formula.

Theorem 1.14. *If a function $f : \mathbb{C} \to \mathbb{C}$ is analytic at $z_0 \in \mathbb{C}$ (i.e., if it has at least one complex derivative in a neighborhood of z_0), then it has complex derivatives of arbitrary order. In fact,*

$$f^{(n)}(z_0) = \frac{n!}{2\pi i} \oint_C \frac{f(z)}{(z - z_0)^{n+1}} \, dz \tag{1.32}$$

for any simple, closed, positively oriented curve C lying in a simply connected domain in which f is analytic and having z_0 in its interior.

The proof can be obtained by differentiating Cauchy's integral formula (1.25) under the integral sign. This is a common technique in PDEs, and one with which the reader should be familiar (cf. Problem 1.11).

Problems

1.1. Let $\mathbf{y}_k$ be the sequence defined by (1.5). Show that

$$|\mathbf{y}_{k+1}(t) - \mathbf{y}_k(t)| \le \gamma \, (t - t_0) \max_{\tau \in [t_0, t]} |\mathbf{y}_k(\tau) - \mathbf{y}_{k-1}(\tau)|.$$

Use this to show that the sequence converges uniformly for $t_0 \le t \le T$ for any $T < t_0 + 1/\gamma$.

1.2. Use the implicit function theorem to determine when the equation

$$x^2 + y^2 + z^2 = 1$$

defines implicitly a function $\hat{x}(y, z)$. Give a geometric interpretation of this result.

1.3. Show that if $\mathbf{F}$ as described in Theorem 1.3 is C^k, then $\hat{\mathbf{y}}$ is C^k as well. Hint: First consider difference quotients to show $\hat{\mathbf{y}}$ is C^1.

1.4. Show that there is an infinite family of minimizers of

$$\mathcal{E}(u) := \int_0^1 (1 - (u'(t))^2)^2 \, dt$$

over the set of all piecewise C^1 functions satisfying $u(0) = u(1) = 0$.

1.5. Show that there is no piecewise C^1 minimizer of

$$\mathcal{E}(u) := \int_0^1 u(t)^2 + (1 - (u'(t))^2)^2 \, dt$$

satisfying $u(0) = u(1) = 0$. Hint: Use a sequence of the solutions of the previous problem to show that a minimizer $\bar{u}$ would have to satisfy $\mathcal{E}(\bar{u}) = 0$. Remark: Minimization problems with features like these arise in the modeling of phase transitions.

1.6. Give an example that shows that δ in Theorem 1.6 cannot be chosen independent of t as $t \to \infty$.

1.7. Prove Theorem 1.8 in the case where the eigenvalues of $\mathbf{A}$ are distinct.

1.8. Prove the existence and uniqueness of the solution of (1.21), (1.22).

1.9. Prove Gronwall's inequality.

1.10. Prove Theorem 1.12 using Cauchy's integral formula.

1.11. Suppose $g : \mathbb{R}^2 \to \mathbb{R}$ is C^1. Define

$$f(x) := \int_a^b g(x, y) \, dy.$$

Use difference quotients to show that one can differentiate f "under the integral sign."

1.2   Elementary Partial Differential Equations

In the last section we discussed the basic types of mathematical questions that are considered throughout the rest of this book, and we looked at how those questions had been answered by two subdisciplines of PDEs: ODEs and complex variables. We now look at how these questions are often approached in elementary courses on partial differential equations. To do this, we consider three basic PDEs (Laplace's equation, the heat equation and the wave equation). Although we sometimes use an analytical approach to investigate their character, our basic technique is the explicit calculation of solutions. At this point we are not terribly concerned with either rigor or generality but rather with foreshadowing material to come; all of the methods and observations presented here are generalized later on.

1.2.1 Laplace's Equation

Perhaps the most important of all partial differential equations is

$$\Delta u := u_{x_1 x_1} + u_{x_2 x_2} + \cdots + u_{x_n x_n} = 0, \qquad (1.33)$$

known as Laplace's equation. You will find applications of it to problems in gravitation, elastic membranes, electrostatics, fluid flow, steady-state heat conduction and many other topics in both pure and applied mathematics.

As the remarks of the last section on ODEs indicated, the choice of boundary conditions is of paramount importance in determining the well-posedness of a given problem. The following two common types of boundary conditions on a bounded domain $\Omega \subset \mathbb{R}^n$ yield well-posed problems and will be studied in a more general context in later chapters.

Dirichlet conditions. Given a function $f : \partial\Omega \to \mathbb{R}$, we require

$$u(\mathbf{x}) = f(\mathbf{x}), \qquad \mathbf{x} \in \partial\Omega. \qquad (1.34)$$

In the context of elasticity, u denotes a change of position, so Dirichlet boundary conditions are often referred to as *displacement* conditions.

Neumann conditions. Given a function $f : \partial\Omega \to \mathbb{R}$, we require

$$\frac{\partial u}{\partial n}(\mathbf{x}) = f(\mathbf{x}), \qquad \mathbf{x} \in \partial\Omega. \qquad (1.35)$$

Here $\frac{\partial u}{\partial n}$ is the partial derivative of u with respect to the unit outward normal of $\partial\Omega$, $\mathbf{n}$. In linear elasticity $\frac{\partial u}{\partial n}(\mathbf{x}) = \nabla u(\mathbf{x}) \cdot \mathbf{n}(\mathbf{x})$ can be interpreted as a force, so Neumann boundary conditions are often referred to as *traction* boundary conditions.

We have been intentionally vague about the smoothness required of $\partial\Omega$ and f, and the function space in which we wish u to lie. These are central areas of concern in later chapters.

Solution by separation of variables

The first method we present for solving Laplace's equation is the most widely used technique for solving partial differential equations: separation of variables. The technique involves reducing a partial differential equation to a system of ordinary differential equations and expressing the solution of the PDE as a sum or infinite series.

Let us consider the following Dirichlet problem on a square in the plane. Let

$$\Omega = \{(x, y) \in \mathbb{R}^2 \mid 0 < x < 1, \ \ 0 < y < 1\}.$$

We wish to find a function $u : \overline{\Omega} \to \mathbb{R}$ satisfying Laplace's equation

$$u_{xx} + u_{yy} = 0 \qquad (1.36)$$

at each point in Ω and satisfying the boundary conditions

$$u(0, y) = 0, \tag{1.37}$$
$$u(1, y) = 0, \tag{1.38}$$
$$u(x, 0) = 0, \tag{1.39}$$
$$u(x, 1) = f(x). \tag{1.40}$$

The key to separation of variables is to look for solutions of (1.36) of the form

$$u(x, y) = X(x)Y(y). \tag{1.41}$$

When we put a function of this form into (1.36), the partial derivatives in the differential equation appear as ordinary derivatives on the functions X and Y; i.e., (1.36) becomes

$$X''(x)Y(y) + X(x)Y''(y) = 0. \tag{1.42}$$

At any point (x, y) at which u is nonzero we can divide this equation by u and rearrange to get

$$\frac{X''(x)}{X(x)} = -\frac{Y''(y)}{Y(y)}. \tag{1.43}$$

We now argue as follows: Since the right side of the equation does not depend on the variable x, neither can the left side; likewise, since the left side does not depend on y, neither does the right side. The only function on the plane that is independent of both x and y is a constant, so we must have

$$\frac{X''(x)}{X(x)} = -\frac{Y''(y)}{Y(y)} = \lambda. \tag{1.44}$$

This gives us

$$X'' = \lambda X, \tag{1.45}$$
$$Y'' = -\lambda Y. \tag{1.46}$$

Solving these equations and using (1.41), we get the following four-parameter family of solutions of the differential equation (1.36):

$$u(x, y) = A \left(e^{\sqrt{\lambda}x} + Be^{-\sqrt{\lambda}x} \right) \left(e^{\sqrt{-\lambda}y} + Ce^{-\sqrt{-\lambda}y} \right). \tag{1.47}$$

(Since we can verify directly that each of these functions is indeed a solution of the differential equation (1.36), there is no need to make the formal argument used to derive (1.45) and (1.46) rigorous.)

The more interesting aspect of separation of variables involves finding a combination of the solutions in (1.47) that satisfies given boundary conditions (and justifying this combination rigorously). In the rather simple set of boundary conditions chosen above, enforcing the three conditions

(1.37), (1.38) and (1.39) reduces the family (1.47) to the following infinite collection.

$$u(x,y) = A \sin n\pi x \, \sinh n\pi y, \quad n = 1, 2, 3, \ldots \tag{1.48}$$

The final condition (1.40) presents a problem. Of course, if the function f is rigged to be a finite linear combination of sine functions,

$$f(x) := \sum_{n=1}^{N} \alpha_n \sin n\pi x, \tag{1.49}$$

then we can simply take

$$A_n := \frac{\alpha_n}{\sinh n\pi}, \quad n = 1, \ldots, N, \tag{1.50}$$

and define

$$u(x,y) := \sum_{n=1}^{N} A_n \sin n\pi x \sinh n\pi y. \tag{1.51}$$

Since this is a finite sum, we can differentiate term by term; so u satisfies the differential equation (1.36). The boundary conditions can be confirmed simply by plugging in the boundary points.

However, the question remains: What is to be done about more general functions f? The answer was deduced by Joseph Fourier in his 1807 paper on heat conduction. Fourier claimed, in effect, that "any" function f could be "represented" by an infinite trigonometric series (now referred to as a Fourier sine series):

$$f(x) := \sum_{n=1}^{\infty} \alpha_n \sin n\pi x. \tag{1.52}$$

The removal of the quotation marks from the sentence above was one of the more important mathematical projects of the nineteenth century. Specifically, mathematicians needed to specify the type of convergence implied by the representation (1.52) and then identify the class of functions that can be achieved by that type of convergence.[1] In later chapters we describe some of the main results in this area, but for the moment let us just accept Fourier's assertion and try to deduce its consequences.

The first question we need to consider is the determination of the Fourier coefficients α_n. The key here is the mutual orthogonality of the sequence

---

[1] Anyone interested in the history of mathematics or the philosophy of science will find the history of Fourier's work fascinating. In the early nineteenth century the entire notion of convergence and the meaning of infinite series was not well formulated. Lagrange and his cohorts in the Academy of Sciences in Paris criticized Fourier for his lack of rigor. Although they were technically correct, they were essentially castigating Fourier for not having produced a body of mathematics that it took generations of mathematicians (including the likes of Cauchy) to finish.

of sine functions making up our series. That is,

$$\int_0^1 \sin(i\pi x)\sin(j\pi x)\,dx = \frac{\delta_{ij}}{2}. \tag{1.53}$$

Here δ_{ij} is the Kronecker delta:

$$\delta_{ij} := \begin{cases} 0, & i \neq j \\ 1, & i = j. \end{cases} \tag{1.54}$$

Thus, if we proceed formally and multiply (1.52) by $\sin j\pi x$ and integrate, we get

$$\int_0^1 f(x)\sin(j\pi x)\,dx = \sum_{n=1}^{\infty} \alpha_n \int_0^1 \sin(n\pi x)\sin(j\pi x)\,dx = \alpha_j/2. \tag{1.55}$$

As before, we postpone the justification for taking the integral under the summation until later chapters and proceed to examine the consequences of this result.

Of course, the main consequence from our point of view is that we can use the formulas above to write down a formal solution of the boundary-value problem (1.36)-(1.40). Namely, we write

$$u(x,y) := \sum_{n=1}^{\infty} A_n \sin n\pi x \sinh n\pi y, \tag{1.56}$$

where

$$A_n := \frac{2}{\sinh n\pi} \int_0^1 f(x)\sin n\pi x\,dx. \tag{1.57}$$

It remains to answer the following questions:

- Does the series (1.56) converge and if so in what sense?

- Is the limit of the series differentiable, and if so, does it satisfy (1.36)? That is, can we take the derivatives under the summation sign?

- In what sense are the boundary conditions met?

- Is the separation of variables solution the only solution of the problem? More generally, is the problem well-posed?

All of these questions will be answered in a more general context in later chapters.

Example 1.15. Let us ignore for the moment the theoretical questions that remain to be answered and do a calculation for a specific problem. We wish to solve the Dirichlet problem (1.36)-(1.40) with data

$$f(x) := \begin{cases} x, & 0 \leq x \leq 1/2 \\ 1-x, & 1/2 < x \leq 1. \end{cases} \tag{1.58}$$

We begin by calculating the Fourier coeficients of f using (1.55);

$$\alpha_n := 2 \int_0^1 f(x) \sin n\pi x \, dx = \frac{4}{n^2\pi^2} \sin \frac{n\pi}{2}. \tag{1.59}$$

Note that the even coefficients vanish. Thus, we can modify (1.56) to get the following separation of variables solution of our Dirichlet problem

$$u(x,y) := 4 \sum_{k=0}^{\infty} \frac{(-1)^k \sin(2k+1)\pi x \sinh(2k+1)\pi y}{(2k+1)^2\pi^2 \sinh(2k+1)\pi}. \tag{1.60}$$

Poisson's integral formula in the upper half-plane

In this section we describe *Poisson's integral formula in the upper half-plane*. This formula gives the solution of Dirichlet's problem in the upper half-plane. It is often derived in elementary complex variables courses.

For a suitable class of functions $g : \mathbb{R} \to \mathbb{R}$ (which we do not make precise here) it can be shown that the function $u : (-\infty, \infty) \times (0, \infty)$ defined by

$$u(x,y) := \frac{y}{\pi} \int_{-\infty}^{\infty} \frac{g(\xi)}{(x-\xi)^2 + y^2} \, d\xi \tag{1.61}$$

satisfies Laplace's equation (1.36) in the upper half-plane and and that it can be extended continuously to the x axis so that it satisfies the Dirichlet boundary conditions

$$u(x,0) = g(x), \tag{1.62}$$

for $x \in \mathbb{R}$.

Poisson's integral formula is an example of the use of *integral operators* to solve boundary-value problems. In later chapters we will generalize the technique through the use of *Green's functions*.

Variational formulations

In this section we give a demonstration of a *variational* technique for proving the existence of solutions of Dirichlet's problem on a "general" domain $\Omega \subset \mathbb{R}^n$. The technique should probably not be considered elementary (since as we shall see in later chapters its rigorous application requires some rather heavy machinery), but it is presented in many elementary courses (particularly in Physics and Engineering) using the formal arguments we sketch here.

We begin by defining an energy functional

$$\mathcal{E}(u) := \int_{\Omega} |\nabla u(\mathbf{x})|^2 \, d\mathbf{x} \tag{1.63}$$

and a class of admissible functions

$$\mathcal{A} := \{u : \Omega \to \mathbb{R} \mid u(\mathbf{x}) = f(\mathbf{x}) \text{ for } \mathbf{x} \in \partial\Omega, \ \mathcal{E}(u) < \infty\}. \tag{1.64}$$

We can now show the following.

Theorem 1.16. *If $\mathcal{A}$ is nonempty, and if there exists $\bar{u} \in \mathcal{A}$ that minimizes $\mathcal{E}$ over $\mathcal{A}$; i.e.,*

$$\mathcal{E}(\bar{u}) \leq \mathcal{E}(u) \quad \forall \; u \in \mathcal{A}, \tag{1.65}$$

then $\bar{u}$ is a solution of the Dirichlet problem.

Before giving the proof we note that there are some serious questions to be answered before this theorem can be applied.

1. Is $\mathcal{A}$ nonempty? More specifically, what properties do the boundary of a domain Ω and the boundary data function f defined on $\partial\Omega$ need to satisfy so that f can be extended into Ω using a function of finite energy?

2. Does there exist a minimizer $\bar{u} \in \mathcal{A}$?

These questions are often ignored (either explicitly or tacitly) in elementary presentations, but we shall see that they are far from easy to answer.

Proof. We give only a sketch of the proof and that will contain a number of holes to be filled later on. Let us define

$$\mathcal{A}_0 := \{v : \Omega \to \mathbb{R} \mid v(\mathbf{x}) = 0 \text{ for } \mathbf{x} \in \partial\Omega, \; \mathcal{E}(v) < \infty\}. \tag{1.66}$$

Note that using elementary inequalities one can show that if $u \in \mathcal{A}$ and $v \in \mathcal{A}_0$, then $(u + \epsilon v) \in \mathcal{A}$ for any $\epsilon \in \mathbb{R}$. We take any $v \in \mathcal{A}_0$ and define a function $\alpha : \mathbb{R} \to \mathbb{R}$ by

$$\begin{aligned} \alpha(\epsilon) \quad &:= \quad \mathcal{E}(\bar{u} + \epsilon v) \\ &= \quad \int_\Omega \{|\nabla\bar{u}|^2 + 2\epsilon\nabla\bar{u}\nabla v + \epsilon^2|\nabla v|^2\} \, d\mathbf{x} \tag{1.67} \\ &= \quad \mathcal{E}(\bar{u}) + 2\epsilon \int_\Omega \nabla\bar{u}\nabla v \, d\mathbf{x} + \epsilon^2 \mathcal{E}(v). \end{aligned}$$

Inequality (1.65) and the calculations above imply that $\epsilon \mapsto \alpha(\epsilon)$ is a quadratic function that is minimized when $\epsilon = 0$. Taking its first derivative at $\epsilon = 0$ yields

$$\int_\Omega \nabla\bar{u} \cdot \nabla v \, d\mathbf{x} = 0, \tag{1.68}$$

and this holds for every $v \in \mathcal{A}_0$.

The result that allows us to use (1.68) to deduce that $\bar{u}$ satisfies Laplace's equation is a version of the *fundamental lemma of the calculus of variations*. (This name has been given to a wide range of results that allow one to deduce that a function that satisfies a variational equation such as (1.68) also satisfies a pointwise differential equation. Another name commonly used in the same way is the DuBois-Reymond lemma in honor of the first versions of such a result.) We now prove a very weak version of this result.

Lemma 1.17. *Let* $\mathbf{F} : \Omega \to \mathbb{R}^n$ *be in* $C^1(\overline{\Omega})$ *and satisfy the variational equation*

$$\int \mathbf{F} \cdot \nabla v \, d\mathbf{x} = 0, \tag{1.69}$$

for every $v \in \mathcal{A}_0$ *with compact support. Then*

$$\operatorname{div} \mathbf{F} = 0 \tag{1.70}$$

in Ω.

Proof. We have assumed sufficient smoothness on $\mathbf{F}$ so that we can use the divergence theorem to get

$$0 = \int_\Omega \mathbf{F} \cdot \nabla v \, d\mathbf{x} = -\int_\Omega (\operatorname{div} \mathbf{F}) v \, d\mathbf{x} + \int_{\partial\Omega} v\mathbf{F} \cdot \mathbf{n} \, dS, \tag{1.71}$$

where $\mathbf{n}$ is the unit outward normal to $\partial\Omega$. Since any $v \in \mathcal{A}_0$ is zero on $\partial\Omega$ this implies

$$\int_\Omega (\operatorname{div} \mathbf{F}) v \, d\mathbf{x} = 0 \quad \forall \ v \in \mathcal{A}_0. \tag{1.72}$$

Since $\operatorname{div} \mathbf{F}$ is continuous, if there is a point x_0 at which it is nonzero (without loss of generality let us assume it is positive there) there is a ball B around x_0 contained in Ω such that $\operatorname{div} \mathbf{F} > \delta > 0$. We can then use a function $\bar{v}$ whose graph is a positive "blip" inside of B and zero outside of B (such a function is easy to construct, and the task is left to the reader) to obtain

$$\int_\Omega (\operatorname{div} \mathbf{F})\bar{v} \, d\mathbf{x} = \int_B (\operatorname{div} \mathbf{F})\bar{v} \, d\mathbf{x} > \delta \int_B \bar{v} \, d\mathbf{x} > 0. \tag{1.73}$$

This is a contradiction, and the proof of the lemma is complete. $\qquad\square$

Now to complete the proof of the theorem, we note that if $\bar{u}$ is in $C^2(\overline{\Omega})$ we can use Lemma 1.17 and (1.68) to deduce

$$\Delta\bar{u} := \operatorname{div} \nabla\bar{u} = 0. \tag{1.74}$$

However, at this point all we know is that $\bar{u} \in \mathcal{A}$. We know nothing more about its smoothness. Thus, the completion of this proof awaits the results on elliptic regularity of later chapters. $\qquad\square$

Equation (1.68) is known as the *weak form* of Laplace's equation. We refer to a solution of (1.33) as a *strong solution* and a solution of (1.68) as a *weak solution* of Laplace's equation. We will generalize these notions to many other types of equations in later chapters.

Note that every strong solution of Laplace's equation is also a weak solution. To see this, we simply multiply (1.33) by an arbitrary function

$v \in \mathcal{A}_0$, integrate by parts (use Green's identity) and use the fact that $v \equiv 0$ on $\partial\Omega$. This gives

$$0 = \int_\Omega (\Delta u)v \ d\mathbf{x} = -\int_\Omega \nabla u \cdot \nabla v \ d\mathbf{x} + \int_{\partial\Omega} v\nabla u \cdot \mathbf{n} \ dS = -\int_\Omega \nabla u \cdot \nabla v \ d\mathbf{x}.$$
$$(1.75)$$

However, as we noted above when we showed that a solution of the minimum energy problem was a weak solution of Laplace's equation, unless we know more about the continuity of a weak solution we cannot show it is a strong solution. This is a common theme in the modern theory of PDEs. It is often easy to find some sort of weak solution to an equation, but relatively hard to show that the weak solution is in fact a strong solution.

Problems

1.12. Compute the Fourier sine series coefficients for the following functions defined on the interval $[0, 1]$.

(a)
$$f(x) = x^2 - x.$$

(b)
$$f(x) = \cos\frac{\pi x}{4}.$$

(c)
$$f(x) = \begin{cases} 3x, & x \in [0, 1/4) \\ 1-x, & x \in [1/4, 1]. \end{cases}$$

1.13. Write a computer program that calculates partial sums of the series defined above and displays them graphically superimposed on the limiting function.

1.14. A function on the interval $[0, 1]$ can also be expanded in a *Fourier cosine series* of the form

$$f(x) = \sum_{n=0}^{\infty} \beta_n \cos n\pi x. \qquad (1.76)$$

Derive a formula for the cosine coefficients.

1.15. Compute the Fourier cosine coefficients for the functions given in Problem 1.12. Use a modification of the computer program developed in Problem 1.13 to display partial sums of the cosine series.

1.16. Both the Fourier sine and cosine series given above converge not only in the interval $[0, 1]$, but on the entire real line. If one computed both the sine and cosine series for the functions graphed below, what would you expect the respective graphs of the limits of the series to be on the whole real line.

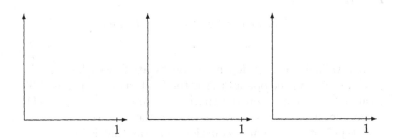

1.17. Solve Laplace's equation on the square $[0,1] \times [0,1]$ for the following boundary conditions:

(a)
$$
\begin{aligned}
u_y(x,0) &= 0, \\
u(x,1) &= x^2 - x, \\
u(0,y) &= 0, \\
u(1,y) &= 0.
\end{aligned}
$$

(b)
$$
\begin{aligned}
u(x,0) &= 0, \\
u_y(x,1) &= \sin \pi x, \\
u_x(0,y) &= 0, \\
u(1,y) &= 0.
\end{aligned}
$$

(c)
$$
\begin{aligned}
u(x,0) &= \begin{cases} x, & x \in [0,1/2) \\ 1-x, & x \in [1/2,1], \end{cases} \\
u(x,1) &= 0, \\
u_x(0,y) &= 0, \\
u(1,y) &= 0.
\end{aligned}
$$

1.18. Verify that the Laplacian takes the following form in polar coordinates in $\mathbb{R}^2$:

$$
\Delta u := \frac{1}{r} \frac{\partial}{\partial r} \left(r \frac{\partial u}{\partial r} \right) + \frac{1}{r^2} \frac{\partial^2 u}{\partial \theta^2}.
$$

1.19. Use the method of separation of variables to find solutions of Laplace's equation of the form

$$
u(r,\theta) = R(r)\Theta(\theta).
$$

1.20. Use the divergence theorem to derive Green's identity

$$
\int_\Omega (\Delta u)v \, d\mathbf{x} = -\int_\Omega \nabla u \cdot \nabla v \, d\mathbf{x} + \int_{\partial \Omega} v \nabla u \cdot \mathbf{n} \, dS.
$$

1.2.2 The Heat Equation

The next elementary problem we examine is the *heat equation*:

$$u_t = \Delta u. \tag{1.77}$$

Here u is a real-valued function depending on "spatial" variables $\mathbf{x} \in \mathbb{R}^n$ and on "time" $t \in \mathbb{R}$, and the operator Δ is the Laplacian defined in (1.33) which is assumed to act only on the spatial variables $(x_1, \ldots, x_n)$. (The reason for the quotation marks above is that in the next section we will describe the "type" of a differential equation in a way that is independent of any particular interpretation of the independent variables as spatial or temporal. However, even after we have done this, we will often lapse back to the terminology of space and time in order to draw analogies to the elementary Laplace, wave and heat equation described in this chapter.) As the name suggests, (1.77) describes the conduction of heat (with the dependent variable u usually interpreted as temperature), but more generally it governs a range of physical phenomena described as *diffusive*.

In discussing typical boundary conditions we confine ourselves to problems posed on a cylinder in space-time: $\Omega_t^+ := \{(\mathbf{x}, t) \in \Omega \times (0, \infty)\}$ where Ω is a bounded domain in $\mathbb{R}^n$. Since the heat equation is first order in time we place one *initial condition* on the solution. We let $\theta : \Omega \to \mathbb{R}$ be a given function and require

$$u(\mathbf{x}, 0) = \theta(\mathbf{x}). \tag{1.78}$$

There are a variety of conditions typically posed on the boundary of the body.

Temperature conditions. Here we fix the dependent variable u on some portion of the boundary.

$$u(\mathbf{x}, t) = f(\mathbf{x}) \tag{1.79}$$

for $\mathbf{x} \in \partial\Omega$ and $t \in (0, \infty)$. In problems of heat conduction, this corresponds to placing a portion of the boundary in contact with a constant temperature source (an ice bath, etc.). Of course, such conditions can be identified with Dirichlet conditions for Laplace's equation.

Heat flux conditions. Here we fix the normal derivative of u on some portion of the boundary.

$$\frac{\partial u}{\partial n}(\mathbf{x}, t) = g(\mathbf{x}) \tag{1.80}$$

for $\mathbf{x} \in \partial\Omega$ and $t \in (0, \infty)$, where $\mathbf{n}$ is the unit outward normal to $\partial\Omega$. A simplified version of Fourier's law of heat conduction says that the *heat flux vector* $\mathbf{q}$ at a point $\mathbf{x}$ at time t is given by

$$\mathbf{q}(\mathbf{x}, t) = -\kappa \nabla u(\mathbf{x}, t), \tag{1.81}$$

where κ is a positive constant called the *thermal conductivity*. Thus, condition (1.80) can be thought of as fixing the flow of heat through a portion of the boundary. (If $g = 0$, we say that portion of the boundary is insulated.) The connection between heat flux conditions and Neumann conditions for Laplace's equation should be obvious.

Linear radiation conditions. Here we require

$$-\frac{\partial u}{\partial n}(\mathbf{x}, t) = \alpha u(\mathbf{x}, t) \tag{1.82}$$

for $\mathbf{x} \in \partial \Omega$ and $t \in (0, \infty)$, where α is a positive constant. This can be thought of as the linearization of *Stefan's radiation law*

$$\mathbf{q}(\mathbf{x}, t) \cdot \mathbf{n}(\mathbf{x}) = \beta u^4(\mathbf{x}, t) \tag{1.83}$$

about a steady-state solution of the boundary-value problem. Stefan's law describes the loss of heat energy of a body through radiation into its surroundings.

Solution by separation of variables

As part of our review of elementary solution methods we now examine the solution of a one-dimensional heat conduction problem by the method of separation of variables. We consider the following *initial/boundary-value problem*. Let

$$D^+ := \{(x, t) \in \mathbb{R}^2 \mid 0 < x < 1, \ 0 < t < \infty\}.$$

Find a function $u : \overline{D^+} \to \mathbb{R}$ that satisfies the differential equation

$$u_t = u_{xx} \tag{1.84}$$

for $(x, t) \in D^+$, the initial condition

$$u(x, 0) = f(x) \tag{1.85}$$

for $x \in (0, 1)$, and the boundary conditions

$$u(0, t) = 0, \tag{1.86}$$
$$u(1, t) = 0 \tag{1.87}$$

for $t > 0$.

As before, we seek solutions of the form

$$u(x, t) := X(x)T(t). \tag{1.88}$$

Plugging this into the differential equation (1.84) gives us

$$XT' = X''T. \tag{1.89}$$

When u is nonzero we get

$$\frac{T'(t)}{T(t)} = \frac{X''(x)}{X(x)}. \tag{1.90}$$

Again, we make the argument that since the right side of the equation is independent of t and the left side is independent of x, each side must be independent of both variables and hence must be equal to a constant. We then get the two ordinary differential equations

$$T' = \lambda T, \tag{1.91}$$
$$X'' = \lambda X. \tag{1.92}$$

Solving these equations using (1.88) gives us the following three-parameter family of solutions of (1.84):

$$u(x,t) = e^{\lambda t}(Ae^{\sqrt{\lambda}x} + Be^{-\sqrt{\lambda}x}). \tag{1.93}$$

Members of this family satisfy the boundary conditions (1.86) and (1.87) only when

$$\lambda = -n^2\pi^2, \quad n = 1, 2, 3, \ldots, \tag{1.94}$$

in which case the collection of solutions reduces to

$$u(x,t) = Ae^{-n^2\pi^2 t} \sin n\pi x. \tag{1.95}$$

To satisfy the initial condition, we use the Fourier expansion of the initial function f via (1.52), (1.55) to define

$$u(x,t) := \sum_{n=1}^{\infty} A_n e^{-n^2\pi^2 t} \sin n\pi x, \tag{1.96}$$

where A_n are equal to the Fourier coefficients of f. The questions that we asked at the end of section 1.2.1 concerning the series solution of Laplace's equation can be asked again about this solution of the heat equation. (Does the series converge? Does its limit satisfy the differential equation and the boundary conditions?) Once again, general answers await later chapters.

Example 1.18. We now formally compute a series solution for the initial boundary-value problem (1.84)-(1.87). In order to compare and contrast the solution we get here to the one obtained for Laplace's equation on page 19 we use as our initial data the function f defined in (1.58). Since we have already computed the Fourier coefficients in (1.59) we can use these in (1.96) to define our solution.

$$u(x,t) := \sum_{k=0}^{\infty} \frac{4(-1)^k}{(2k+1)^2\pi^2} e^{-(2k+1)^2\pi^2 t} \sin(2k+1)\pi x. \tag{1.97}$$

Instability of backwards heat equation

In this section we consider the following problem. Let

$$D^- := \{(x,t) \in \mathbb{R}^2 \mid 0 < x < 1, \ -\infty < t < 0\}.$$

We wish to find a function $u : \overline{D^-} \to \mathbb{R}$ that satisfies the heat equation (1.84) in D^-, the boundary conditions (1.86) and (1.87) for $t < 0$, and the *terminal condition*

$$u(x,0) = f(x) \tag{1.98}$$

for $x \in (0,1)$.

The problem is often transformed using the change of variables

$$\bar{t} = -t. \tag{1.99}$$

Under this transformation we seek to solve the differential equation

$$u_t = -u_{xx} \tag{1.100}$$

for $(x,t) \in D^+$. The boundary conditions (1.86) and (1.87) are now applied for positive t, and the terminal condition (1.98) remains unchanged but is now thought of as an initial condition. This version of the problem is known as the *backwards heat equation*. The name stems, of course, from the fact that in the original formulation we are trying to solve the heat equation for times *before* the terminal condition (1.98).

The formal separation of variables solution for the backwards heat equation is

$$u(x,t) := \sum_{n=1}^{\infty} A_n e^{n^2\pi^2 t} \sin n\pi x, \tag{1.101}$$

where A_n are the Fourier coefficients of f. The derivation is almost exactly the same as before with the sole difference that we end up with exponentials that *grow rather than decay with time*. As a result we can get the following instability result that states that we can find initial data for which the solution of the backwards heat equation blows up as quickly as desired.

Lemma 1.19. *For any $T > 0$, $M > 0$, and $\epsilon > 0$ there exists an initial function f satisfying*

$$\|f\|_{C([0,1])} = \epsilon \tag{1.102}$$

such that the backwards heat conduction problem has a separation of variables solution $u(x,t)$ defined by (1.101) that satisfies

$$\|u(\cdot,T)\|_{C([0,1])} \geq M. \tag{1.103}$$

Here $\|f\|_{C([0,1])}$ denotes the max of $|f|$ on $[0,1]$.

Proof. Choose n sufficiently large so that

$$n^2\pi^2 T \geq \ln\left(\frac{M}{\epsilon}\right). \tag{1.104}$$

Then if

$$f(x) = \epsilon \sin n\pi x, \tag{1.105}$$

the solution

$$u(x,t) = \epsilon e^{n^2\pi^2 t} \sin n\pi x \tag{1.106}$$

satisfies the requirement.   □

Energy inequality

We end our discussion of the heat equation by proving an *energy estimate*. The simple estimate we derive in this section should act as a prototype for estimates that we will derive in later chapters. We will show the following.

Lemma 1.20. *Let* $u : \overline{D^+} \to \mathbb{R}$ *be a* C^2 *solution of the heat equation (1.84) satisfying the boundary conditions (1.86) and (1.87). Then for any* $t_1 \geq t_0 \geq 0$, *the solution* u *satisfies*

$$\int_0^1 u^2(x, t_1)\, dx \leq \int_0^1 u^2(x, t_0)\, dx. \qquad (1.107)$$

In the language of Chapter 6, for any solution of the heat equation satisfying the given boundary conditions, the L^2 norm (in space) decreases with time.

Proof. We first use the heat equation to derive the following differential identity for u.

$$\begin{aligned} \frac{d}{dt} u^2 &= 2uu_t \\ &= 2uu_{xx} \qquad (1.108) \\ &= 2(uu_x)_x - 2u_x^2. \end{aligned}$$

Integrating both sides of this identity with respect to x gives us

$$\frac{1}{2} \int_0^1 (u^2(x,t))_t\, dx = u(1,t)u_x(1,t) - u(0,t)u_x(0,t) - \int_0^1 u_x^2(x,t)\, dx. \qquad (1.109)$$

We now use the boundary conditions to eliminate the boundary terms in the equation above and integrate the result with respect to time. After changing the order of integration on the left side we get

$$\int_0^1 u^2(x,t_1)\, dx - \int_0^1 u^2(x,t_0)\, dx = \int_0^1 \int_{t_0}^{t_1} (u^2)_t\, dt\, dx \qquad (1.110)$$

$$= -2 \int_0^1 \int_{t_0}^{t_1} u_x^2\, dt\, dx \leq 0.$$

This completes the proof.   □

Problems

1.21. Solve the one-dimensional heat equation via separation of variables for the following boundary conditions:

(a)
$$\begin{array}{rcl} u(x,0) & = & x^2 - x, \\ u_x(0,t) & = & 0, \\ u_x(1,t) & = & 0. \end{array}$$

(b)
$$\begin{array}{rcl} u(x,0) & = & \sin \pi x, \\ u(0,t) & = & 0, \\ u_x(1,t) & = & 0. \end{array}$$

(c)
$$\begin{array}{rcl} u(x,0) & = & \begin{cases} x, & x \in [0,1/2) \\ 1 - x, & x \in [1/2, 1], \end{cases} \\ u_x(0,t) & = & 0, \\ u(1,t) & = & 0. \end{array}$$

1.22. In a typical physical problem in heat conduction, one studies the differential equation

$$c\rho u_t = \kappa \Delta u$$

where c is the *specific heat*, ρ is the *density*, and κ is the thermal conductivity of the medium under consideration. If c, ρ, and κ are constants, show that there is a linear change in time scale $\bar{t} = \gamma t$ that transforms the differential equation above into (1.77).

1.23. Suppose $f : \overline{D^+} \to \mathbb{R}$ is continuous and $u : \overline{D^+} \to \mathbb{R}$ is a solution of the following nonhomogeneous initial/boundary-value problem:

$$u_t(x,t) - u_{xx}(x,t) = f(x,t), \quad (x,t) \in D^+,$$

$$u(x,0) = 0, \quad x \in [0,1],$$

$$u(0,t) = u(1,t) = 0, \quad t \in [0,\infty).$$

Now, for each $\tau \in [0,\infty)$, let $w(x,t,\tau)$ be the solution of the following *pulse problem:*

$$w_t - w_{xx} = 0, \quad (x,t) \in (0,1) \times (\tau, \infty),$$

$$w(x,\tau,\tau) = f(x,\tau), \quad x \in [0,1]$$

$$w(0,t,\tau) = w(1,t,\tau) = 0, \quad t \in [\tau, \infty).$$

Show that u and w satisfy the relation

$$u(x,t) = \int_0^t w(x,t,\tau) d\tau.$$

This and similar methods of relating nonhomogeneous PDEs with homogeneous initial conditions to homogeneous PDEs with nonhomogeneous initial conditions are known as Duhamel's principle.

1.24. Solve the Cauchy problem

$$u_t = u_{xx},$$
$$u(x,0) = \begin{cases} 0, & x < 0 \\ 1, & x > 0. \end{cases}$$

Hint: Seek a solution in the form $u(x,t) = \phi(x/\sqrt{t})$.

1.2.3   The Wave Equation

Our next elementary equation is the *wave equation*. Here we seek a real-valued function u depending on spatial variables $\mathbf{x} \in \mathbb{R}^n$ and a time variable $t \in \mathbb{R}$ satisfying

$$u_{tt} = \Delta u. \tag{1.111}$$

Once again the Laplacian acts only on the spatial variables. This equation describes many types of elastic and electromagnetic waves.

We once again describe some typical boundary conditions on the space-time cylinder $\Omega_t^+ := \{(\mathbf{x},t) \in \Omega \times (0,\infty)\}$, where Ω is a bounded domain in $\mathbb{R}^n$. Since the wave equation is second order in time one usually specifies two initial conditions

$$u(\mathbf{x},0) = f(\mathbf{x}), \tag{1.112}$$
$$u_t(\mathbf{x},0) = g(\mathbf{x}). \tag{1.113}$$

In problems in elasticity this amounts to specifying the position and velocity at time zero. Dirichlet or Neumann conditions are usually prescribed on various parts of the boundary. In elasticity applications these are usually interpreted as displacement and traction conditions, respectively.

Solution of an initial/boundary-value problem by separation of variables

The first initial/boundary-value problem we consider describes a string of unit length fixed at each end and given an initial position and velocity. The problem is described as follows. Let D^+ be the (x,t) domain defined in the previous subsection. We seek a function $u : \overline{D^+} \to \mathbb{R}$ satisfying the one-dimensional (in space) wave equation

$$u_{tt} = u_{xx} \tag{1.114}$$

for $(x,t) \in D^+$, the initial conditions

$$u(x,0) = f(x), \tag{1.115}$$
$$u_t(x,0) = g(x) \tag{1.116}$$

for $x \in (0,1)$ and the Dirichlet boundary conditions

$$u(0,t) = 0, \tag{1.117}$$
$$u(1,t) = 0 \tag{1.118}$$

for $t > 0$.

If we carry out the method of separation as before, we get the following family of solutions to both the wave equation (1.114) and the boundary conditions.

$$u_n(x, t) = (\alpha_n \cos n\pi t + \beta_n \sin n\pi t) \sin n\pi x. \qquad (1.119)$$

If our initial conditions have Fourier expansions of the form

$$f(x) \quad = \quad \sum_{n=1}^{\infty} A_n \sin n\pi x, \qquad (1.120)$$

$$g(x) \quad = \quad \sum_{n=1}^{\infty} B_n \sin n\pi x, \qquad (1.121)$$

then the formal series solution for the initial/boundary-value problem is

$$u(x, t) := \sum_{n=1}^{\infty} \left(A_n \cos n\pi t + \frac{B_n}{n\pi} \sin n\pi t \right) \sin n\pi x. \qquad (1.122)$$

D'Alembert's solution for the Cauchy problem

In this section we consider the *Cauchy problem* for the one-dimensional wave equation. Specifically, we wish to find a real-valued function u that satisfies the wave equation (1.114) in the half-plane $(x, t) \in (-\infty, \infty) \times (0, \infty)$ and the initial conditions

$$u(x, 0) \quad = \quad f(x), \qquad (1.123)$$
$$u_t(x, 0) \quad = \quad g(x) \qquad (1.124)$$

for $x \in (-\infty, \infty)$.

To derive a solution to this problem we first examine two special *traveling wave* solutions of the wave equation. Suppose F and G are real-valued functions in $C^2(\mathbb{R})$. We observe that

$$u_1(x, t) := F(x + t) \qquad (1.125)$$

and

$$u_2(x, t) := G(x - t) \qquad (1.126)$$

each solve the wave equation. Note that u_1 is simply a translation of the function F to the left with speed one, whereas u_2 is a translation of G to the right. In fact, we can show that any solution of the wave equation has the form

$$u(x, t) = F(x + t) + G(x - t). \qquad (1.127)$$

To see this we simply make the change of variables

$$\xi \quad = \quad x + t, \qquad (1.128)$$
$$\tau \quad = \quad x - t, \qquad (1.129)$$

so that

$$\bar{u}(\xi, \tau) := u\left(\frac{\xi + \tau}{2}, \frac{\xi - \tau}{2}\right). \tag{1.130}$$

Using the chain rule we see that if u satisfies the wave equation then $\bar{u}$ satisfies

$$\bar{u}_{\xi\tau} = 0. \tag{1.131}$$

This implies

$$\bar{u}(\xi, \tau) = F(\xi) + G(\tau). \tag{1.132}$$

Changing back to the independent variables (x, t) gives us (1.127).

We now apply this general form for solutions to the Cauchy problem by plugging in the initial conditions (1.123) and (1.124) to get the following equations for the unknown functions F and G:

$$f(x) = F(x) + G(x), \tag{1.133}$$
$$g(x) = F'(x) - G'(x). \tag{1.134}$$

These yield

$$F'(x) = \frac{1}{2}(f'(x) + g(x)), \tag{1.135}$$

$$G'(x) = \frac{1}{2}(f'(x) - g(x)). \tag{1.136}$$

Integrating these equations and using the result in (1.127) gives us *D'Alembert's solution of the Cauchy problem*

$$u(x, t) := \frac{1}{2}[f(x + t) + f(x - t)] + \frac{1}{2}\int_{x-t}^{x+t} g(s)\, ds. \tag{1.137}$$

One of the most striking things about D'Alembert's solution (or more specifically, the form of the solution implied by (1.132)) is that the formula for the solution makes perfectly good sense even when f and g are discontinuous. Such a "solution" would consist of a "jump" in u moving to the left or right with unit speed. The existence of such solutions should not violate our intuition about the wave equation since physical wave-like phenomena that we would call discontinuous (such as breaking waves in the surf and shock waves from explosions) occur every day. But what about the mathematical nature of the solution? How can we say that a solution satisfies a differential equation at a point at which it is not differentiable? In later chapters we will examine this question more fully, and especially in the context of generalized wave equations we will get some fairly detailed answers.

Energy conservation

In this section we derive a result for solutions of the wave equation known as conservation of energy. We prove a version here that holds for the one-dimensional wave equation with fixed ends defined above and leave generalizations for later chapters.

Lemma 1.21. *Let* $u : \overline{D^+} \to \mathbb{R}$ *be a* C^2 *solution of the wave equation (1.114) satisfying the boundary conditions (1.117) and (1.118). Then for any* $t_1 \geq t_0 \geq 0$, *the solution* u *satisfies*

$$\int_0^1 u_t^2(x, t_1) + u_x^2(x, t_1) \, dx = \int_0^1 u_t^2(x, t_0) + u_x^2(x, t_0) \, dx. \qquad (1.138)$$

Proof. As we did in the proof of the energy inequality for the heat equation, we begin by deriving a differential identity. Let u satisfy the wave equation. Then

$$0 = (u_{tt} - u_{xx})u_t \qquad (1.139)$$

$$= \frac{d}{dt}(u_t^2 + u_x^2)/2 - \frac{d}{dx}(u_x u_t). \qquad (1.140)$$

We now use this in an integration over the rectangle $(x, t) \in [0, 1] \times [t_0, t_1]$, in which we change the order of integration at will, and we obtain the following:

$$\int_0^1 u_t^2(x, t_1) + u_x^2(x, t_1) \, dx - \int_0^1 u_t^2(x, t_0) + u_x^2(x, t_0) \, dx$$

$$= \int_0^1 \int_{t_0}^{t_1} \frac{d}{dt}(u_t^2 + u_x^2) \, dt \, dx$$

$$= \int_{t_0}^{t_1} \int_0^1 2 \frac{d}{dx}(u_x u_t) \, dx \, dt$$

$$= \int_{t_0}^{t_1} 2 u_x(1, t) u_t(1, t) \, dt - \int_{t_0}^{t_1} 2 u_x(0, t) u_t(0, t) \, dt.$$

However, the boundary conditions (1.117) and (1.118) imply

$$u_t(0, t) = u_t(1, t) \equiv 0, \qquad (1.141)$$

so this gives us (1.138). ☐

Note that the quantity we call the energy for solutions of the wave equation and the quantity we call the energy for solutions of the heat equation seem very different mathematically. However, the mathematical techniques that we use to study the quantities (multiplication of the differential equation by the solution or its derivative and (essentially) integrating by parts in order to obtain an estimate) are common to both. This technique of obtaining estimates on solutions of PDEs is extremely useful and is generalized in later chapters.

Problems

1.25. Solve the one-dimensional wave equation via separation of variables for the following boundary conditions:

(a)
$$\begin{aligned}
u(x,0) &= x^2 - x, \\
u_t(x,0) &= 0, \\
u_x(0,t) &= 0, \\
u_x(1,t) &= 0.
\end{aligned}$$

(b)
$$\begin{aligned}
u(x,0) &= 0, \\
u_t(x,0) &= \sin \pi x, \\
u(0,t) &= 0, \\
u_x(1,t) &= 0.
\end{aligned}$$

(c)
$$\begin{aligned}
u(x,0) &= \begin{cases} x, & x \in [0,1/2) \\ 1-x, & x \in [1/2,1], \end{cases} \\
u_t(x,0) &= 0, \\
u_x(0,t) &= 0, \\
u(1,t) &= 0.
\end{aligned}$$

1.26. Give a specific definition of well-posedness (in particular, make precise in what sense the problem is continuous with respect to the data) for the Cauchy problem (1.114), (1.123), (1.124) on the domain $(x,t) \in (-\infty, \infty) \times (0, \infty)$. Derive conditions on the initial data under which the problem is well-posed. How do your results differ if the domain under consideration is $(x,t) \in (-\infty, \infty) \times (0,T)$ for some $0 < T < \infty$. Hint: If $u(x,0) = 0$ and $u_t(x,0) = \epsilon > 0$ for $x \in (-\infty, \infty)$, then u grows arbitrarily large with time. Figure out conditions on the initial data that assure that u stays bounded.

1.27. Suppose f and g are identically zero outside the interval $[-1,1]$. In what region in $(-\infty, \infty) \times [0, \infty)$ can you ensure that the solution u of the Cauchy problem is identically zero.

1.28. Is there a similar result to the previous problem for the heat equation? Hint: Use

$$\chi_{[-1,1]}(x) := \begin{cases} 1, & x \in [-1,1] \\ 0, & x \notin [-1,1] \end{cases}$$

as initial datum. Use Problem 1.24 to obtain a solution.

1.29. We define a weak solution of the one-dimensional wave equation to be a function $u(x,t)$ such that

$$\int_{-\infty}^{\infty}\int_{-\infty}^{\infty} u(x,t)(\phi_{tt}(x,t) - \phi_{xx}(x,t))\ dx\ dt = 0 \tag{1.142}$$

for every $\phi \in C_0^2(\mathbb{R}^2)$. Here $C_0^2(\mathbb{R}^2)$ is the set of functions in $C^2(\mathbb{R}^2)$ that have compact support; i.e., that are identically zero outside of some bounded set.

(a) Show that any strong (classical C^2) solution of the wave equation is also a weak solution.

(b) Show that discontinuous functions of the form

$$u(x,t) := H(x - t) \tag{1.143}$$

and

$$u(x,t) := H(x + t) \tag{1.144}$$

are weak solutions of the wave equation. Here H is the Heaviside function:

$$H(x) := \begin{cases} 0, & x < 0 \\ 1, & x \geq 0. \end{cases} \tag{1.145}$$

2
Characteristics

2.1   Classification and Characteristics

The typical problem in partial differential equations consists of finding the solution of a PDE (or a system of PDEs) subject to certain boundary and/or initial conditions. The nature of boundary and initial conditions which lead to well-posed problems depends in a very essential way on the specific PDE under consideration. For example, we saw in the examples in the Introduction that a natural choice of conditions for Laplace's equation,

$$u_{xx} + u_{yy} = f(x, y), \quad 0 < x < 1, \quad 0 < y < 1, \tag{2.1}$$

consists of prescribing u on the boundary,

$$u(x, 0) = \phi_0(x), \quad u(x, 1) = \phi_1(x), \quad u(0, y) = \psi_0(y), \quad u(1, y) = \psi_1(y). \tag{2.2}$$

For the wave equation,

$$u_{xx} - u_{yy} = f(x, y), \tag{2.3}$$

posed on the same domain (with y taking the role of time) a natural choice of conditions is, for example,

$$u(0, y) = \phi_0(y), \quad u(1, y) = \phi_1(y), \quad u(x, 0) = \psi_0(x), \quad u_y(x, 0) = \psi_1(x). \tag{2.4}$$

Ill-posed problems result if one tries to impose the conditions (2.2) on the wave equation or the conditions (2.4) on Laplace's equation.

Laplace's equation and the wave equation differ in other important aspects. For example, solutions of (2.1) will always be smooth in the interior of the domain as long as f is smooth. On the other hand, solutions of (2.3) may have discontinuities even for $f = 0$. Indeed, as we mentioned in the previous chapter, any twice differentiable function of the form $u = F(x-y)+G(x+y)$ is a solution of (2.3), and we shall later introduce "generalized" solutions which dispense with the requirement that F and G have to be twice differentiable.

An important ingredient of a systematic theory of partial differential equations is a classification scheme which identifies classes of equations with common properties. The "type" of an equation determines the nature of boundary and initial conditions which may be imposed, the nature of singularities which solutions may have and the nature of methods which can be used to approximate a solution. In this section, we shall provide the basic definitions underlying the classification of PDEs.

2.1.1 The Symbol of a Differential Expression

The notation of multi-indices is very convenient in avoiding excessively cumbersome notations in PDEs. A multi-index is a vector

$$\alpha = (\alpha_1, \alpha_2, \dots, \alpha_n)$$

whose components are non-negative integers. The notation $\alpha \geq \beta$ indicates that $\alpha_i \geq \beta_i$ for each i. For any multi-index α, we make the following definitions:

$$|\alpha| = \alpha_1 + \alpha_2 + \cdots + \alpha_n, \quad \alpha! = \alpha_1!\alpha_2!\cdots\alpha_n!; \tag{2.5}$$

moreover, for any vector $\mathbf{x} = (x_1, x_2, \dots, x_n) \in \mathbb{R}^n$, we set

$$\mathbf{x}^\alpha = x_1^{\alpha_1} x_2^{\alpha_2} \cdots x_n^{\alpha_n}. \tag{2.6}$$

The following notation for partial derivatives is extremely convenient in writing partial differential equations:

$$D^\alpha = \frac{\partial^{|\alpha|}}{\partial x_1^{\alpha_1} \partial x_2^{\alpha_2} \cdots \partial x_n^{\alpha_n}}. \tag{2.7}$$

For example, if $\alpha = (1, 2)$, then

$$D^\alpha u = \frac{\partial^3 u}{\partial x_1 \partial x_2^2}. \tag{2.8}$$

We now consider a linear differential expression of the form

$$L(\mathbf{x}, D)u = \sum_{|\alpha| \leq m} a_\alpha(\mathbf{x}) D^\alpha u, \tag{2.9}$$

where $u : \mathbb{R}^n \to \mathbb{R}$. With this analytic operation on functions we associate an algebraic operation called the symbol.

Definition 2.1. The **symbol** of the expression $L(\mathbf{x}, D)$ as given by (2.9) is

$$L(\mathbf{x}, i\boldsymbol{\xi}) := \sum_{|\alpha| \le m} a_\alpha(\mathbf{x})(i\boldsymbol{\xi})^\alpha. \qquad (2.10)$$

The **principal part** of the symbol is

$$L^P(\mathbf{x}, i\boldsymbol{\xi}) := \sum_{|\alpha| = m} a_\alpha(\mathbf{x})(i\boldsymbol{\xi})^\alpha. \qquad (2.11)$$

Example 2.2. The symbol of Laplace's operator $\partial^2/\partial x_1^2 + \partial^2/\partial x_2^2$ is $-\xi_1^2 - \xi_2^2$, the symbol of the heat operator $\partial/\partial x_1 - \partial^2/\partial x_2^2$ is $i\xi_1 + \xi_2^2$, and the symbol of the wave operator $\partial^2/\partial x_1^2 - \partial^2/\partial x_2^2$ is $-\xi_1^2 + \xi_2^2$. For the Laplace and wave operator, the symbols are equal to their principal parts; the principal part for the heat operator is ξ_2^2.

In an analogous fashion, we can associate a matrix-valued symbol with a system of partial differential equations. The definition of the principal part in this case is more involved; we shall address this issue in a later subsection.

If the coefficients of the partial differential equation are constant, and we are looking at solutions on all of space, the symbol is easily interpreted in terms of the Fourier transform: If

$$\hat{u}(\boldsymbol{\xi}) := (2\pi)^{-n/2} \int_{\mathbb{R}^n} u(\mathbf{x}) \exp(-i\boldsymbol{\xi} \cdot \mathbf{x}) \, d\mathbf{x} \qquad (2.12)$$

is the Fourier transform of $u(\mathbf{x})$, then $L(i\boldsymbol{\xi})\hat{u}(\boldsymbol{\xi})$ is the Fourier transform of $L(D)u(\mathbf{x})$. (We examine Fourier transforms in detail in Chapter 5.)

In general, the symbol tells us how a differential expression acts on functions which have their support contained in a small neighborhood of a given point $\mathbf{x}$. If the coefficients are smooth, they are approximately constant in such a small neighborhood. Moreover, if u varies very rapidly, the highest-order derivatives are dominant over lower-order derivatives, and the principal part therefore contains the most important terms. How the differential expression acts on rapidly varying functions of small support is of crucial importance for many basic properties of PDEs. The classification into types is based on the principal part of the symbol.

2.1.2 Scalar Equations of Second Order

Let us consider a second-order PDE in two space dimensions,

$$
\begin{aligned}
Lu =& a(x,y)u_{xx} + b(x,y)u_{xy} + c(x,y)u_{yy} \\
& + d(x,y)u_x + e(x,y)u_y + f(x,y)u \\
=& g(x,y).
\end{aligned}
\qquad (2.13)
$$

The principal part of the symbol of L is

$$L^p(x,y;i\xi,i\eta) = -a(x,y)\xi^2 - b(x,y)\xi\eta - c(x,y)\eta^2. \qquad (2.14)$$

Second-order PDEs are classified according to the behavior of L^p, viewed as a quadratic form in ξ and η. The quadratic form given by (2.14) can be represented in matrix form as

$$L^p(x,y;i\xi,i\eta) = (\xi,\eta)\begin{pmatrix} -a(x,y) & -\frac{1}{2}b(x,y) \\ -\frac{1}{2}b(x,y) & -c(x,y) \end{pmatrix}\begin{pmatrix} \xi \\ \eta \end{pmatrix}. \qquad (2.15)$$

Recall that a quadratic form is called definite if the associated symmetric matrix is (positive or negative) definite, it is called indefinite if the matrix has eigenvalues of both signs, and it is called degenerate if the matrix is singular.

Definition 2.3. The differential equation (2.13) is called **elliptic** if the quadratic form given by (2.14) is strictly definite, **hyperbolic** if it is indefinite and **parabolic** if it is degenerate.

The terms elliptic, parabolic and hyperbolic are motivated by the analogy with the classification of conic sections.

Example 2.4. Laplace's equation is elliptic, the heat equation is parabolic and the wave equation is hyperbolic. For these three cases, the matrices associated with the principal part of the symbol are

$$\begin{pmatrix} -1 & 0 \\ 0 & -1 \end{pmatrix}, \quad \begin{pmatrix} 0 & 0 \\ 0 & 1 \end{pmatrix} \quad \text{and} \quad \begin{pmatrix} -1 & 0 \\ 0 & 1 \end{pmatrix}, \qquad (2.16)$$

respectively.

Example 2.5. In general, equations may have different type in different parts of the region in which they are to be solved. A typical example of this is the Tricomi equation

$$yu_{xx} + u_{yy} = 0. \qquad (2.17)$$

The symbol is $-y\xi_1^2 - \xi_2^2$; hence the equation is elliptic for $y > 0$, parabolic for $y = 0$ and hyperbolic for $y < 0$. Equations which change type arise in some physical applications, for example the study of steady transonic flow. Such problems are generally very difficult to analyze.

Consider now a second-order PDE in n space dimensions:

$$Lu = a_{ij}(\mathbf{x})\frac{\partial^2 u}{\partial x_i \partial x_j} + b_i(\mathbf{x})\frac{\partial u}{\partial x_i} + c(\mathbf{x})u = 0. \qquad (2.18)$$

Because the matrix of second partials of u is symmetric, we may assume without loss of generality that $a_{ij} = a_{ji}$. The principal symbol of this second-order PDE is still a quadratic form in $\boldsymbol{\xi}$; we can represent this quadratic form as $\boldsymbol{\xi}^T \mathbf{A}(\mathbf{x})\boldsymbol{\xi}$, where $\mathbf{A}$ is the $n \times n$ matrix with components $-a_{ij}$.

Definition 2.6. Equation (2.18) is called **elliptic** if all eigenvalues of $\mathbf{A}$ have the same sign, **parabolic** if $\mathbf{A}$ is singular and **hyperbolic** if all but one of the eigenvalues of $\mathbf{A}$ have the same sign and one has the opposite sign. If $\mathbf{A}$ is nonsingular and there is more than one eigenvalue of each sign, the equation is called **ultrahyperbolic**.

In this definition, it is understood that eigenvalues are counted according to their multiplicities.

The notion of characteristic surfaces is closely related to that of type. We make the following definition:

Definition 2.7. The surface described by $\phi(x_1, x_2, \ldots, x_n) = 0$ is **characteristic** at the point $\hat{\mathbf{x}}$, if $\phi(\hat{\mathbf{x}}) = 0$ and, in addition,

$$a_{ij}(\hat{\mathbf{x}}) \frac{\partial \phi}{\partial x_i}(\hat{\mathbf{x}}) \frac{\partial \phi}{\partial x_j}(\hat{\mathbf{x}}) = 0. \qquad (2.19)$$

A surface is called characteristic if it is characteristic at each of its points.

In matrix form, condition (2.19) reads $(\nabla \phi)^T \mathbf{A} (\nabla \phi) = 0$. The matrix $\mathbf{A}$ is strictly definite, i.e., (2.18) is elliptic if and only if there are no nonzero real vectors with this property. We can therefore characterize elliptic equations as those without (real) characteristic surfaces.

For hyperbolic equations, on the other hand, all but one of the eigenvalues of $\mathbf{A}$ have the same sign, say one eigenvalue is negative and the rest positive. Let $\mathbf{n}$ be a unit eigenvector corresponding to the negative eigenvalue. The span of $\mathbf{n}$ and its orthogonal complement are both invariant subspaces of $\mathbf{A}$, and, utilizing the decomposition

$$\nabla \phi = (\mathbf{n} \cdot \nabla \phi) \mathbf{n} + (\nabla \phi - (\mathbf{n} \cdot \nabla \phi) \mathbf{n}), \qquad (2.20)$$

we find

$$(\nabla \phi)^T \mathbf{A} (\nabla \phi) = -\lambda (\mathbf{n} \cdot \nabla \phi)^2 + [\nabla \phi - (\mathbf{n} \cdot \nabla \phi) \mathbf{n}]^T \mathbf{B} [\nabla \phi - (\mathbf{n} \cdot \nabla \phi) \mathbf{n}] = 0, \qquad (2.21)$$

where $-\lambda$ is the negative eigenvalue of $\mathbf{A}$ and $\mathbf{B}$ is positive definite on the $(n-1)$–dimensional subspace perpendicular to $\mathbf{n}$. Let us now regard $\nabla \phi - (\mathbf{n} \cdot \nabla \phi) \mathbf{n}$, i.e., the part of $\nabla \phi$ that is perpendicular to $\mathbf{n}$, as given. Then $\mathbf{n} \cdot \nabla \phi$ can be determined from (2.21). For any nonzero choice of the perpendicular part of $\nabla \phi$, we get two real and distinct solutions for $\mathbf{n} \cdot \nabla \phi$.

Note that if we take any C^2 function $u : \mathbb{R} \to \mathbb{R}$ and compose it with ϕ, the resulting function satisfies

$$L^p u(\phi) = a_{ij}(\mathbf{x}) \frac{\partial^2 u}{\partial x_i \partial x_j} = u''(\phi) \left[a_{ij}(\mathbf{x}) \frac{\partial \phi}{\partial x_i} \frac{\partial \phi}{\partial x_j} \right] + u'(\phi) a_{ij}(\mathbf{x}) \frac{\partial^2 \phi}{\partial x_i \partial x_j}, \qquad (2.22)$$

and if the surfaces $\phi = \text{const.}$ are characteristic, the coefficient of $u''(\phi)$ on the right-hand side vanishes. That is, the function $u(\phi)$ satisfies the equation $Lu = 0$ "to leading order." Because of this property, characteristics are important in the study of singularities of solutions of partial differential equations. As we shall see in Chapter 3, partial differential equations can have solutions that are (or whose derivatives are) discontinuous across a characteristic surface. For example, we can guess from the results of Problem 1.29 that $F(x-t)+G(x+t)$ satisfies the weak form of the wave equation $u_{tt} = u_{xx}$ even when F and G are discontinuous. The lines $x \pm t = \text{const.}$ are the characteristics of this equation.

A related property will be important in connection with the Cauchy-Kovalevskaya theorem. Suppose u and its normal derivative $\nabla u \cdot \nabla \phi$ are prescribed on a surface given by $\phi(\mathbf{x}) = 0$. (Note that this implies that tangential derivatives of all orders are automatically specified.) Can we use (2.18), in conjunction with the given data, to find the second derivative of u in the direction of $\nabla \phi$? To decide this, let $\mathbf{q}_1, \mathbf{q}_2, \mathbf{q}_3, \ldots, \mathbf{q}_n$ be an orthonormal basis such that $\mathbf{q}_1$ is in the direction of $\nabla \phi$. To simplify notation, we shall write $\mathbf{q}$ for $\mathbf{q}_1$. We have

$$\mathbf{A} = (\mathbf{q}^T \mathbf{A} \mathbf{q}) \mathbf{q} \mathbf{q}^T + \mathbf{B}, \tag{2.23}$$

where

$$\mathbf{q}^T \mathbf{B} \mathbf{q} = 0. \tag{2.24}$$

The matrix $\mathbf{B}$ can be represented as

$$\mathbf{B} = \sum_{i=2}^{n} (\mathbf{q}^T \mathbf{A} \mathbf{q}_i) \mathbf{q} \mathbf{q}_i^T + (\mathbf{q}_i^T \mathbf{A} \mathbf{q}) \mathbf{q}_i \mathbf{q}^T \tag{2.25}$$

$$+ \sum_{i,j=2}^{n} (\mathbf{q}_i^T \mathbf{A} \mathbf{q}_j) \mathbf{q}_i \mathbf{q}_j^T. \tag{2.26}$$

Let $D^2 u$ denote the matrix of the second derivatives $\partial^2 u / \partial x_i \partial x_j$. From (2.23), we find

$$\mathbf{A} : D^2 u := -a_{ij} \frac{\partial^2 u}{\partial x_i \partial x_j} = (\mathbf{q}^T \mathbf{A} \mathbf{q}) \mathbf{q}^T (D^2 u) \mathbf{q} + \cdots, \tag{2.27}$$

where the second derivatives of u indicated by the dots involve at least one differentiation in a direction perpendicular to $\mathbf{q}$ (this is clear from (2.26)). If u and its normal derivative are prescribed, these terms can therefore be considered known. The condition for being able to determine the second normal derivative is therefore that $\mathbf{q}^T \mathbf{A} \mathbf{q} \neq 0$, i.e., that the surface $\phi = 0$ is noncharacteristic.

2.1.3 Higher-Order Equations and Systems

The generalization of the definitions above to equations of higher order than second is straightforward.

Definition 2.8. Let L be the mth-order operator defined in (2.9). **Characteristic surfaces** are defined by the equation

$$L^p(\mathbf{x}, \nabla\phi) = 0. \tag{2.28}$$

An equation is called **elliptic** at $\mathbf{x}$ if there are no real characteristics at $\mathbf{x}$ or, equivalently, if

$$L^p(\mathbf{x}, i\boldsymbol{\xi}) \neq 0, \quad \forall \boldsymbol{\xi} \neq 0. \tag{2.29}$$

An equation is called **strictly hyperbolic**[1] in the direction $\mathbf{n}$ if

1. $L^p(\mathbf{x}, i\mathbf{n}) \neq 0$, and

2. all the roots ω of the equation

$$L^p(\mathbf{x}, i\boldsymbol{\xi} + i\omega\mathbf{n}) = 0 \tag{2.30}$$

are real and distinct for every $\boldsymbol{\xi} \in \mathbb{R}^n$ which is not collinear with $\mathbf{n}$.

In applications, $\mathbf{n}$ is usually a coordinate direction associated with time. In this case, let us set $\mathbf{x} = (x_1, x_2, \ldots, x_{n-1}, t)$ and let $\boldsymbol{\xi} = (\xi_1, \ldots, \xi_{n-1}, 0)$ be a spatial vector.

For rapidly oscillating functions of small support, we may think of the coefficients of L^p as approximately constant; let us assume they are constant. If ω is a root of (2.30), then $u = \exp(i(\boldsymbol{\xi} \cdot \mathbf{x}) + i\omega t)$ is a solution of $L^p u = 0$. If ω has negative imaginary part, then this solution grows exponentially in time. Moreover, since L^p is homogeneous of degree m, i.e., $L^p(\mathbf{x}, \lambda(i\boldsymbol{\xi} + i\omega\mathbf{n})) = \lambda^m L^p(\mathbf{x}, i\boldsymbol{\xi} + i\omega\mathbf{n})$ for any scalar λ, there are always roots with negative imaginary parts if there are any roots which are not real (if we change the sign of $\boldsymbol{\xi}$, we also change the sign of ω). Moreover, if we multiply $\boldsymbol{\xi}$ by a scalar factor λ, then ω is multiplied by the same factor, and hence solutions would grow more and more rapidly the faster they oscillate in space. The condition that the roots in (2.30) are real is therefore a necessary condition for well-posedness of initial-value problems.

We now turn our attention to systems of k partial differential equations involving k unknowns u_j, $j = 1, 2, \ldots, k$:

$$L_{ij}(\mathbf{x}, D)u_j = 0, \quad i = 1, 2, \ldots, k. \tag{2.31}$$

As for systems of algebraic equations, well-posed problems require equal numbers of equations and unknowns, so we shall assume that the operators L_{ij} form a square matrix $\mathbf{L}$. The generalization of the notions above is in principle quite straightforward.

Definition 2.9. Characteristic surfaces are defined by the equation

$$\det \mathbf{L}^p(\mathbf{x}, \nabla\phi) = 0, \tag{2.32}$$

---

[1] We shall not give a general definition of what it means to be nonstrictly hyperbolic, although such definitions exist. Below we shall define nonstrict hyperbolicity for first-order systems.

and equations without real characteristic surfaces are called **elliptic**. **Strict hyperbolicity** is also defined as above, with L^p in the definition replaced by det L^p. A system in which all components of $\mathbf{L}^p$ are operators of first order is called hyperbolic (not necessarily strictly) in the direction $\mathbf{n}$ if

1. det $\mathbf{L}^p(\mathbf{x}, \mathbf{n}) \neq 0$, and

2. for $\boldsymbol{\xi}$ not collinear with $\mathbf{n}$, all eigenvalues ω of the problem

$$\det \mathbf{L}^p(\mathbf{x}, i\boldsymbol{\xi} + i\omega\mathbf{n}) = 0$$

are real and there is a complete set of eigenvectors.

Note that, since we assumed that the components of $\mathbf{L}^p$ are of first order, we have $\mathbf{L}^p(\mathbf{x}, i\boldsymbol{\xi} + i\omega\mathbf{n}) = \mathbf{L}^p(\mathbf{x}, i\boldsymbol{\xi}) + \omega\mathbf{L}^p(\mathbf{x}, i\mathbf{n})$; hence the problem det $\mathbf{L}^p = 0$ is a matrix eigenvalue problem for ω. If the eigenvalues are distinct, there is always a complete set of eigenvectors; hence strict hyperbolicity implies hyperbolicity.

In general, we need to be careful about defining the "principal part" of a system. A naive approach of simply taking the "terms of highest order" turns out to be unsatisfactory.

Example 2.10. To see the problem, let us consider Laplace's equation in two dimensions

$$u_{xx} + u_{yy} = 0, \tag{2.33}$$

and rewrite it as a system of first-order equations by setting $v = u_x$, $w = u_y$. The resulting system is

$$u_x = v, \quad u_y = w, \quad v_x + w_y = 0. \tag{2.34}$$

If we define $\mathbf{L}^p$ to be the part involving first-order terms, it is easy to see that det $\mathbf{L}^p$ turns out to be identically zero. On the other hand, since Laplace's equation is the standard example of an elliptic equation, it would be desirable to have the equivalent first-order system also defined as "elliptic." Obviously, we then cannot throw away the terms v and w in the first two equations of (2.34).

The difficulty is resolved by assigning "weights" s_i to each equation and t_j to each dependent variable in such a way that the order of each operator L_{ij} does not exceed $s_i + t_j$. The principal part L_{ij}^p is then defined to consist of those terms which have order exactly equal to $s_i + t_j$. We assume that the weights can be assigned in such a way that det $\mathbf{L}^p$ does not vanish identically[2]; in this case det $\mathbf{L}^p$ consists of all the terms of order $\sum_{i,j} s_i + t_j$

---

[2]There are examples where this assumption fails, e.g., the system $u_x + v_y = 0$, $u_x + v_y + v = 0$. The difficulty here disappears if we use the equivalent form $u_x + v_y = 0$, $v = 0$. It is known that weights with the desired properties always exist for nondegenerate systems [Vo]. Here nondegenerate means the following: If det $\mathbf{L}(i\boldsymbol{\xi})$ is expressed in the

which appear in det **L**. In (2.34), for example, we would set $s_1 = s_2 = t_2 = t_3 = 0$ and $t_1 = s_3 = 1$. (Here, it is understood that the ordering of the variables is u, v, w.) With these weights, the principal part of (2.34) is actually identical to (2.34), and we compute

$$\det \mathbf{L}^p(i\xi) = \det \begin{pmatrix} i\xi_1 & -1 & 0 \\ i\xi_2 & 0 & -1 \\ 0 & i\xi_1 & i\xi_2 \end{pmatrix} = -\xi_1^2 - \xi_2^2, \tag{2.35}$$

which is equal to the symbol of Laplace's equation.

The term "order" for systems is used with two different meanings. Such terms as "first-order" or "second-order" systems usually refer to the equations of which the system is composed. However, it is also possible to assign an order to the system as a whole. This order is defined as the degree of the symbol of det $\mathbf{L}^p$, which is equal to $\sum_{i,j} s_i + t_j$.

Remark 2.11. We note that weights need not be unique. The following is a simple example of a system which is elliptic under more than one choice of weights. Both choices are useful, and lead to different elliptic regularity results. Consider the system

$$\Delta u - v = 0, \quad \Delta v = 0. \tag{2.36}$$

We can choose the weights $s_1 = s_2 = 2$, $t_1 = t_2 = 0$. With these choices, the principal part of the symbol is

$$\mathbf{L}^p(i\xi) = \begin{pmatrix} -|\xi|^2 & 0 \\ 0 & -|\xi|^2 \end{pmatrix}. \tag{2.37}$$

But we can also choose the weights $t_1 = 2$, $t_2 = 0$, $s_1 = 0$, $s_2 = 2$. Now the principal part is

$$\mathbf{L}^p(i\xi) = \begin{pmatrix} -|\xi|^2 & -1 \\ 0 & -|\xi|^2 \end{pmatrix}. \tag{2.38}$$

In both cases, we have $\det \mathbf{L}^p(i\xi) = |\xi|^4$.

2.1.4 Nonlinear Equations

For nonlinear equations and systems, the type can depend not only on the point in space but on the solution itself. We simply linearize the equation at a given solution and define the type to be that of the linearized equation. Characteristic surfaces are similarly defined as the characteristic surfaces of the linearized equation.

For future use we give the definition of quasilinear and semilinear:

---

usual way as a sum of products, then the degree of each of these products as a polynomial in ξ does not exceed the degree of the determinant.

Definition 2.12. A system is called **quasilinear** if derivatives of principal order occur only linearly (with coefficients which may depend on derivatives of lower order). It is called **semilinear** if it is quasilinear and the coefficients of the terms of principal order depend only on $\mathbf{x}$, but not on the solution.

Example 2.13. The equation

$$\alpha(u_x)u_{xx} + u_{yy} = 0 \tag{2.39}$$

is quasilinear; it is elliptic if $\alpha(u_x) > 0$ and hyperbolic if $\alpha(u_x) < 0$. The equation

$$\phi(u_{xx}) + u_{yy} = 0 \tag{2.40}$$

is not quasilinear; it is elliptic if $\phi'(u_{xx}) > 0$ and hyperbolic if $\phi'(u_{xx}) < 0$. The equation

$$(x^2 + 1)u_{xx} + u_{yy} + \phi(u_x, u_y) = 0 \tag{2.41}$$

is semilinear elliptic.

Problems

2.1. Determine the type of the following equations:
(a) $u_{xy} + 2u_x = 0$,
(b) $u_{xxxx} - u_{xxyy} + u_{yyyy} = 0$.
(c) $u_{tt} + u_{xxxx} = 0$.

2.2. Find the characteristics of the following equations.
(a) $xu_{xx} + u_{yy} = 0$. (b) $yu_{xx} + u_{yy} = 0$.

2.3. Consider the first-order equation $yu_x + (x^2 + 1)u_y = 0$.
(a) Determine the characteristics.
(b) Show that the most general solution is any function which is constant along characteristics and use this fact to give a formula for the general solution.

2.4. The Stokes system in $\mathbb{R}^3$ consists of the equations

$$\Delta \mathbf{u} - \nabla p = \mathbf{0}, \quad \text{div } \mathbf{u} = 0. \tag{2.42}$$

Here the unknowns are the vector function $\mathbf{u} : \mathbb{R}^3 \to \mathbb{R}^3$ and the scalar function $p : \mathbb{R}^3 \to \mathbb{R}$. Show that this system is elliptic.

2.5. The Euler equations in $\mathbb{R}^3$ are

$$(\mathbf{u} \cdot \nabla)\mathbf{u} - \nabla p = \mathbf{0}, \quad \text{div } \mathbf{u} = 0. \tag{2.43}$$

Here the unknowns are the vector function $\mathbf{u} : \mathbb{R}^3 \to \mathbb{R}^3$ and the scalar function $p : \mathbb{R}^3 \to \mathbb{R}$. Show that this system is neither elliptic nor hyperbolic.

2.6. Consider the system

$$\mathbf{u}_t + \mathbf{A}\mathbf{u}_x + \mathbf{B}\mathbf{u}_y = 0. \tag{2.44}$$

What condition must $\mathbf{A}$ and $\mathbf{B}$ satisfy for the system to be hyperbolic in the t direction? The condition which you will find is, in general, difficult to verify. Can you give a simple special case? Consider now the special case where $\mathbf{A}$ and $\mathbf{B}$ are diagonal. Under what conditions is the system strictly hyperbolic?

2.2   The Cauchy-Kovalevskaya Theorem

The theorem of Cauchy and Kovalevskaya quite generally asserts the local existence of solutions to a system of partial differential equations with initial conditions on a noncharacteristic surface. The coefficients in the equations, the initial data and the surface on which they are prescribed are required to be analytic. This is a severe restriction which, in general, cannot be removed. Moreover, we shall see that the theorem does not distinguish between well-posed and ill-posed problems; it covers situations where a small change in the data leads to a large change in the solution. For these reasons, the theorem has little practical importance. Historically, however, it is the first existence theorem for a general class of PDEs and it is one of very few such theorems which can be proved without the tools of functional analysis.

We shall state and prove the theorem for quasilinear first-order systems of the form

$$\frac{\partial u_i}{\partial x_n} = \sum_{k=1}^{n-1}\sum_{j=1}^{N} a_{ij}^k(\mathbf{p})\frac{\partial u_j}{\partial x_k} + b_i(\mathbf{p}), \quad i = 1, \ldots, N, \tag{2.45}$$

where $\mathbf{p}$ stands for the vector $(x_1, \ldots, x_{n-1}, u_1, \ldots, u_N)$, and the functions a_{ij}^k and b_i are assumed analytic. The initial conditions are

$$u_i = 0 \quad \text{on } x_n = 0, \quad i = 1, \ldots, N. \tag{2.46}$$

A general noncharacteristic initial-value problem for a system of PDEs can always be reduced to the form (2.45), (2.46); below we shall discuss the reduction algorithm in detail. We shall start the section by reviewing some basic facts about real analytic functions.

2.2.1   Real Analytic Functions

Analytic functions are functions which can be represented locally by power series. We shall use the multi-index notation introduced in the previous section and write the power series of a function of n variables in the form

$$f(\mathbf{x}) = \sum_\alpha c_\alpha \mathbf{x}^\alpha, \tag{2.47}$$

where $\alpha = (\alpha_1, \dots, \alpha_n)$ is a multi-index and $\mathbf{x}^\alpha$ has the meaning introduced in equation (2.6).

We note the following facts about power series:

1. Suppose that (2.47) converges absolutely for $\mathbf{x} = \mathbf{y}$, where all components of $\mathbf{y}$ are different from zero. Then it converges absolutely in the domain $D = \{\mathbf{x} \in \mathbb{R}^n \mid |x_i| < |y_i|, \ i = 1, \dots, n\}$ and it converges uniformly absolutely in any compact subset of D.

2. In D, the power series (2.47) can be differentiated term by term. We shall obtain an estimate for the derivatives. Let $|x_i| \le q|y_i|$ for $i = 1, \dots, n$, where $0 \le q < 1$. We compute

$$D^\beta f(\mathbf{x}) = \sum_{\alpha \ge \beta} c_\alpha D^\beta \mathbf{x}^\alpha = \sum_{\alpha \ge \beta} c_\alpha \frac{\alpha!}{(\alpha - \beta)!} \mathbf{x}^{\alpha - \beta}, \tag{2.48}$$

and hence

$$\begin{aligned}
|D^\beta f(\mathbf{x})| &\le \sum_{\alpha \ge \beta} \frac{\alpha!}{(\alpha - \beta)!} |c_\alpha| q^{|\alpha - \beta|} |\mathbf{y}^{\alpha - \beta}| \\
&\le \frac{1}{|\mathbf{y}^\beta|} \sup_\alpha (|c_\alpha| |\mathbf{y}^\alpha|) \sum_{\alpha \ge \beta} \frac{\alpha!}{(\alpha - \beta)!} q^{|\alpha - \beta|}.
\end{aligned}$$

We have (see Problem 2.7)

$$\sum_{\alpha \ge \beta} \frac{\alpha!}{(\alpha - \beta)!} q^{|\alpha - \beta|} = \frac{\beta!}{(1 - q)^{n + |\beta|}}, \tag{2.49}$$

and with

$$M = (1 - q)^{-n} \sup_\alpha (|c_\alpha| |\mathbf{y}^\alpha|), \quad r = (1 - q) \min_i |y_i|, \tag{2.50}$$

we finally obtain

$$|D^\beta f(\mathbf{x})| \le M |\beta|! r^{-|\beta|}. \tag{2.51}$$

3. We have

$$c_\alpha = \frac{1}{\alpha!} D^\alpha f(0). \tag{2.52}$$

With these preliminaries we are ready to define real analytic functions.

Definition 2.14. Let f be a real-valued function defined on the open set $\Omega \subseteq \mathbb{R}^n$. We call f **real analytic** at $\mathbf{y}$ if there is a neighborhood of $\mathbf{y}$ within which f can be represented as a Taylor series

$$f(\mathbf{x}) = \sum_\alpha c_\alpha (\mathbf{x} - \mathbf{y})^\alpha. \tag{2.53}$$

We say f is real analytic in Ω if it is analytic at every point in Ω.

Vector- or matrix-valued functions will be called analytic if their components are analytic. The symbol $C^\omega(\Omega)$ is used to denote the class of functions analytic in Ω, whereas $C^\infty(\Omega)$ denotes functions which have derivatives of all orders. Obviously $C^\omega(\Omega) \subset C^\infty(\Omega)$. Like holomorphic functions of a single complex variable, analytic functions have a unique continuation property.

Theorem 2.15. *Let Ω be a domain (i.e., an open connected set), and let f and g be analytic in Ω. If, for some point $\mathbf{x}_0 \in \Omega$, we have $D^\alpha f(\mathbf{x}_0) = D^\alpha g(\mathbf{x}_0)$ for every α, then $f = g$ in Ω.*

Proof. Let

$$S = \{\mathbf{x} \in \Omega \mid D^\alpha f(\mathbf{x}) = D^\alpha g(\mathbf{x}) \; \forall \alpha\}. \tag{2.54}$$

Then S is the intersection of sets which are relatively closed in Ω; hence S is itself relatively closed. On the other hand S is also open, because if $\mathbf{y} \in S$, then the Taylor coefficients of f and g agree at the point $\mathbf{y}$, and hence $f = g$ in a neighborhood of $\mathbf{y}$. Since Ω is connected and, by assumption, $S \neq \emptyset$, we must have $S = \Omega$. $\square$

If f is analytic at a point $\mathbf{y}$, then the derivatives of f satisfy a bound of the form (2.51) in some neighborhood of $\mathbf{y}$. It turns out that this property characterizes real analytic functions.

Definition 2.16. Let f be defined in a neighborhood of the point $\mathbf{y}$. For given positive numbers M and r, we say that $f \in C_{M,r}(\mathbf{y})$ if f is of class C^∞ in a neighborhood of $\mathbf{y}$ and

$$|D^\beta f(\mathbf{y})| \leq M|\beta|! r^{-|\beta|} \quad \forall \beta. \tag{2.55}$$

The following equivalence holds.

Theorem 2.17. *Let Ω be an open set and let $f \in C^\infty(\Omega)$. Then $f \in C^\omega(\Omega)$ if and only if the following holds: For every compact set $S \subset \Omega$ there exist positive numbers M and r with $f \in C_{M,r}(\mathbf{y})$ for every $\mathbf{y} \in S$.*

Proof. If $f \in C^\omega(\Omega)$, then for every $\mathbf{y} \in \Omega$ we find a neighborhood $N(\mathbf{y})$ and numbers $M(\mathbf{y})$ and $r(\mathbf{y})$ such that (2.55) holds in $N(\mathbf{y})$. A finite number of these neighborhoods covers S and it suffices to take the maximum of the M's and the minimum of the r's.

For the converse, choose $\mathbf{x} \in \Omega$ and let S be a closed ball of radius s centered at $\mathbf{x}$, with s chosen small enough so that $S \subset \Omega$. Let M, r be the values for which $f \in C_{M,r}(\mathbf{y})$ for all $\mathbf{y} \in S$. We shall prove that

$$f(\mathbf{y}) = \sum_\alpha \frac{1}{\alpha!} D^\alpha f(\mathbf{x})(\mathbf{y} - \mathbf{x})^\alpha, \tag{2.56}$$

whenever $d := \sum_{i=1}^{n} |y_i - x_i| < \min(r, s)$. For such a $\mathbf{y}$, we introduce the scalar function

$$\phi(t) := f(\mathbf{x} + t(\mathbf{y} - \mathbf{x})). \tag{2.57}$$

We compute, for any integer $j \geq 0$ and $0 \leq t \leq 1$,

$$\left| \frac{1}{j!} \frac{d^j}{dt^j} \phi(t) \right| = \left| \sum_{|\alpha|=j} \frac{1}{\alpha!} D^\alpha f(\mathbf{x} + t(\mathbf{y} - \mathbf{x}))(\mathbf{y} - \mathbf{x})^\alpha \right|$$

$$\leq \sum_{|\alpha|=j} M \frac{|\alpha|!}{\alpha!} r^{-|\alpha|} |(\mathbf{y} - \mathbf{x})^\alpha|$$

$$= M r^{-j} d^j.$$

From Taylor's theorem, we find

$$f(\mathbf{y}) = \phi(1) = \sum_{k=0}^{j-1} \frac{1}{k!} \phi^{(k)}(0) + \frac{1}{j!} \phi^{(j)}(\tau_j), \tag{2.58}$$

where $0 \leq \tau_j \leq 1$. The remainder term in (2.58) is bounded by $M r^{-j} d^j$ and tends to zero for $d < r$. Hence (2.56) follows.   □

Real analytic functions can also be characterized as restrictions of complex analytic functions. One direction is clear. If the Taylor series (2.53) converges absolutely, say for $|\mathbf{x} - \mathbf{y}| < R$, it will still converge if the components of $\mathbf{x}$ are allowed to be complex. Thus every real analytic function can be extended into a subset of the complex plane, and since power series can be differentiated term by term, the extended function is differentiable.

On the other hand, consider a differentiable function $f(\mathbf{z})$ of n complex variables, defined in the neighborhood of the point $\mathbf{x}$, say in a domain including the set $S = \{\mathbf{z} \in \mathbb{C}^n \mid |z_i - x_i| \leq r, \ i = 1, \dots, n\}$. Choose $\mathbf{y}$ in the interior of S. By repeated application of Cauchy's formula for functions of a single complex variable, we find that

$$f(\mathbf{y}) = (2\pi i)^{-n} \int_{\Gamma_1} \frac{1}{z_1 - y_1} \int_{\Gamma_2} \frac{1}{z_2 - y_2} \cdots \int_{\Gamma_n} \frac{1}{z_n - y_n} f(\mathbf{z}) \, dz_n \, dz_{n-1} \cdots dz_1, \tag{2.59}$$

where Γ_i is the positively oriented circle $z_i - x_i = r$. We now write

$$\frac{1}{z_i - y_i} = \frac{1}{(z_i - x_i)\left(1 - \frac{y_i - x_i}{z_i - x_i}\right)}, \tag{2.60}$$

and expand in a geometric series with respect to powers of $(y_i - x_i)/(z_i - x_i)$. By inserting this series in (2.59), we obtain

$$f(\mathbf{y}) = \sum_\alpha c_\alpha (\mathbf{y} - \mathbf{x})^\alpha, \tag{2.61}$$

where

$$c_\alpha = (2\pi i)^{-n} \int_{\Gamma_1} \frac{1}{(z_1 - x_1)^{1+\alpha_1}} \int_{\Gamma_2} \frac{1}{(z_2 - x_2)^{1+\alpha_2}} \cdots$$
$$\int_{\Gamma_n} \frac{1}{(z_n - x_n)^{1+\alpha_n}} f(\mathbf{z})\, dz_n\, dz_{n-1} \cdots dz_1.$$

$$(2.62)$$

The characterization of real analytic functions as restrictions of complex differentiable functions allows a simple proof of the implicit function theorem for real analytic functions.

Theorem 2.18. *Let the functions* $F_i(x_1, \ldots, x_n, y_1, \ldots, y_m)$, $i = 1, \ldots, m$, *be real analytic at the point* $(\mathbf{x}^0, \mathbf{y}^0) \in \mathbb{R}^n \times \mathbb{R}^m$. *Assume that*

$$\mathbf{F}(\mathbf{x}^0, \mathbf{y}^0) = \mathbf{0}, \quad \det\left(\frac{\partial F_i}{\partial y_j}\right)(\mathbf{x}^0, \mathbf{y}^0) \neq 0. \qquad (2.63)$$

Then, in a neighborhood of the point $(\mathbf{x}^0, \mathbf{y}^0)$, *the system* $\mathbf{F}(\mathbf{x}, \mathbf{y}) = \mathbf{0}$ *has a unique solution* $\hat{\mathbf{y}}(\mathbf{x})$ *and the function* $\hat{\mathbf{y}}(\mathbf{x})$ *is real analytic at* $\mathbf{x}^0$.

By extending $\mathbf{x}$ and $\mathbf{y}$ to $\mathbb{C}^n$, the theorem is reduced to the implicit function theorem for differentiable functions, a generalization of Theorem 1.3.

The characterization of analytic functions as differentiable functions of complex variables also implies that the composite of two analytic functions is analytic.

2.2.2 Majorization

The proof of the Cauchy-Kovalevskaya theorem uses the method of majorization, which consists of comparing analytic functions with other functions which have larger Taylor coefficients, but can be given explicitly. We make the following definition.

Definition 2.19. Let f, F be real analytic functions defined in a neighborhood of the origin of $\mathbb{R}^n$. Then we say that f is **majorized** by F, $f \preceq F$ if $|D^\alpha f(\mathbf{0})| \leq D^\alpha F(\mathbf{0})$ for every α.

Example 2.20. If $f \in C_{M,r}(\mathbf{0})$, then f is majorized by the function

$$\phi(\mathbf{x}) = \frac{Mr}{r - x_1 - x_2 - \cdots - x_n}. \qquad (2.64)$$

This is clear, since $D^\alpha \phi(\mathbf{0}) = M|\alpha|! r^{-|\alpha|}$.

We need the following result concerning composite functions.

Theorem 2.21. *Let* $\mathbf{f}$, $\mathbf{F}$ *be vector-valued real analytic functions from a neighborhood of the origin in* $\mathbb{R}^n$ *into* $\mathbb{R}^m$ *such that* $\mathbf{f}(\mathbf{0}) = \mathbf{F}(\mathbf{0}) = \mathbf{0}$. *Let* g, G *be real analytic functions from a neighborhood of the origin in* $\mathbb{R}^m$ *into* $\mathbb{R}$. *Assume that* $f_i \preceq F_i$ *for* $i = 1, \ldots, m$ *and* $g \preceq G$. *Then* $g \circ \mathbf{f} \preceq G \circ \mathbf{F}$.

The proof follows by noting that all derivatives of $g \circ \mathbf{f}$ can be expressed as polynomials involving derivatives of g and $\mathbf{f}$. All these polynomials have positive coefficients. Hence they can be estimated by the corresponding polynomials involving derivatives of G and $\mathbf{F}$.

2.2.3 Statement and Proof of the Theorem

Consider the system of partial differential equations

$$\frac{\partial u_i}{\partial x_n} = \sum_{k=1}^{n-1} \sum_{j=1}^{N} a_{ij}^k(\mathbf{p}) \frac{\partial u_j}{\partial x_k} + b_i(\mathbf{p}), \quad i = 1, \ldots, N, \tag{2.65}$$

where $\mathbf{p}$ stands for the vector $(x_1, \ldots, x_{n-1}, u_1, \ldots, u_N)$, with the initial conditions

$$u_i = 0 \quad \text{on } x_n = 0, \quad i = 1, \ldots, N. \tag{2.66}$$

We shall establish the following local existence result.

Theorem 2.22. *Let the functions a_{ij}^k and b_i be real analytic at the origin of $\mathbb{R}^{N+n-1}$. Then the system (2.65) with initial conditions (2.66) has a unique (among real analytic functions) system of solutions u_i that is real analytic at the origin.*

Proof. We begin by formally computing all derivatives of the u_i at the origin. From (2.66) it follows that all tangential derivatives of all orders are zero. Hence the only nonzero first derivative is in the x_n direction, and (2.65) leads to $\partial u_i / \partial x_n(\mathbf{0}) = b_i(\mathbf{0})$. Next, we can differentiate (2.65) to obtain second derivatives of u_i. We find

$$\frac{\partial^2 u_i}{\partial x_n \partial x_l}(\mathbf{0}) = \sum_{k=1}^{n-1} \sum_{j=1}^{N} a_{ij}^k(\mathbf{0}) \frac{\partial^2 u_j}{\partial x_k \partial x_l}(\mathbf{0}) + \sum_{j=1}^{N+n-1} \frac{\partial b_i}{\partial p_j}(\mathbf{0}) \frac{\partial p_j}{\partial x_l}(\mathbf{0}), \tag{2.67}$$

which yields, for $l = 1, \ldots, n-1$,

$$\frac{\partial^2 u_i}{\partial x_n \partial x_l} = \frac{\partial b_i}{\partial p_l}, \tag{2.68}$$

and, for $l = n$, $i = 1, \ldots, N$,

$$\begin{aligned}
\frac{\partial^2 u_i}{\partial x_n^2} &= \sum_{k=1}^{n-1} \sum_{j=1}^{N} a_{ij}^k(\mathbf{0}) \frac{\partial^2 u_j}{\partial x_k \partial x_n}(\mathbf{0}) + \sum_{j=1}^{N} \frac{\partial b_i}{\partial p_{n-1+j}}(\mathbf{0}) \frac{\partial u_j}{\partial x_n}(\mathbf{0}) \\
&= \sum_{k=1}^{n-1} \sum_{j=1}^{N} a_{ij}^k(\mathbf{0}) \frac{\partial b_j}{\partial p_k}(\mathbf{0}) + \sum_{j=1}^{N} \frac{\partial b_i}{\partial p_{n-1+j}}(\mathbf{0}) b_j(\mathbf{0}).
\end{aligned} \tag{2.69}$$

Proceeding in a similar fashion, we can compute derivatives of all orders. The resulting expression for any derivative of u_i at the origin is a polynomial with positive coefficients involving derivatives of a_{ij}^k and b_i at the origin.

We can thus construct a formal Taylor series for the u_i. To show that this Taylor series converges, we use the method of majorants. Let

$$a_{ij}^k(\mathbf{p}) \preceq A_{ij}^k(\mathbf{p}), \quad b_i(\mathbf{p}) \preceq B_i(\mathbf{p}), \tag{2.70}$$

and let U_i be the solution of the problem

$$\frac{\partial U_i}{\partial x_n} = \sum_{k=1}^{n-1} \sum_{j=1}^{N} A_{ij}^k(\mathbf{p}) \frac{\partial U_j}{\partial x_k} + B_i(\mathbf{p}), \quad i = 1, \dots, N, \tag{2.71}$$

with the initial conditions

$$U_i = 0 \quad \text{on } x_n = 0, \quad i = 1, \dots, N. \tag{2.72}$$

Then clearly $|D^\alpha u_i(\mathbf{0})| \leq D^\alpha U_i(\mathbf{0})$ for any α; hence the formal power series for u_i converges if U_i is analytic.

It remains to construct appropriate functions A_{ij}^k and B_i. Let us assume that a_{ij}^k and b_i are in $C_{M,r}(\mathbf{0})$. Then we can find a majorant from (2.64), i.e.,

$$A_{ij}^k = B_i = \frac{Mr}{r - x_1 - \cdots - x_{n-1} - U_1 - \cdots - U_N}. \tag{2.73}$$

That is, we have to consider the initial-value problem

$$\frac{\partial U_i}{\partial x_n} = \frac{Mr}{r - x_1 - \cdots - x_{n-1} - U_1 - \cdots - U_N} \left(1 + \sum_{k=1}^{n-1} \sum_{j=1}^{N} \frac{\partial U_j}{\partial x_k}\right), \tag{2.74}$$

with initial conditions (2.72). A solution of this problem can be found in the form $U_i(\mathbf{x}) = V(x_1 + \cdots + x_{n-1}, x_n)$; the resulting equation for $V(s, t)$ is

$$V_t = \frac{Mr}{r - s - NV}(1 + N(n-1)V_s), \quad V(s, 0) = 0. \tag{2.75}$$

This equation can be solved explicitly. We first note that the characteristic lines $s = s(t)$ are given by the equation

$$\frac{ds}{dt} = -\frac{(n-1)NMr}{r - s - NV(s(t), t)}. \tag{2.76}$$

Along a characteristic, we have

$$\frac{d}{dt}V(s(t), t) = \frac{Mr}{r - s(t) - NV(s(t), t)} = -\frac{1}{(n-1)N}\frac{ds}{dt}. \tag{2.77}$$

Integration with respect to t yields

$$V(s(t), t) = -\frac{1}{(n-1)N}s(t) + \frac{1}{(n-1)N}s(0). \tag{2.78}$$

By inserting (2.78) into (2.76), we find

$$\frac{ds}{dt} = -\frac{(n-1)MNr}{r - s + \dfrac{s}{n-1} - \dfrac{s(0)}{n-1}}, \tag{2.79}$$

which can be integrated to obtain

$$s(0) = \frac{n-1}{n}\left[r + \frac{1}{n-1}s - \sqrt{(r-s)^2 - 2nNMrt}\right]. \qquad (2.80)$$

We insert this into (2.78) and finally find

$$V(s,t) = \frac{1}{Nn}\left(r - s - \sqrt{(r-s)^2 - 2nNMrt}\right). \qquad (2.81)$$

This expression is analytic at $s = t = 0$. This concludes the proof.   □

The theorem guarantees the existence of a solution in a neighborhood of each point on the initial surface. By taking the union of all these neighborhoods, we obtain the existence of a solution in an open set containing the initial surface. We also note that the Taylor series for the solution is guaranteed to converge in a ball whose radius depends only on n, N, M, and r.

We emphasize that the theorem is strictly local in character, that is, the solution may cease to exist or at least lose analyticity at some finite value of x_n. We shall see examples of this later when we study development of shocks. Also, the theorem does not rule out the existence of nonanalytic solutions even in a neighborhood of the initial surface. Uniqueness is only guaranteed within the class of analytic functions. Holmgren's theorem, proved in the next section, asserts that for linear equations the analytic solution is unique in a larger function class.

2.2.4 Reduction of General Systems

We now consider a general system of partial differential equations, for which we shall prescribe initial data on a noncharacteristic surface. We shall demonstrate how such a problem can be reduced to the form given by (2.45), (2.46). First, we reduce the initial surface to a plane.

A surface is usually described in one of three ways:

1. As a level set of a function: $\phi(x_1, \ldots, x_n) = \lambda$. Here it is assumed that $\nabla\phi \neq 0$.

2. As a graph: $x_i = \psi(x_1, \ldots, x_{i-1}, x_{i+1}, \ldots, x_n)$.

3. By a parametrization, $\mathbf{x} = \mathbf{F}(\mathbf{y})$, where $\mathbf{y} \in \mathbb{R}^{n-1}$. Here it is assumed that the matrix $\partial F_i/\partial y_j$ has maximal rank.

We say that a surface is of class C^k, $k \geq 1$ (analytic) if, respectively, the functions ϕ, ψ, $\mathbf{F}$ are of class C^k (analytic). The implicit function theorem implies that locally all three definitions are equivalent. After possible renumbering of coordinates, we may therefore assume that an analytic surface is given in the form $x_n = \psi(x_1, \ldots, x_{n-1})$, where ψ is analytic. We now set $y_i = x_i$ for $i \leq n-1$ and $y_n = x_n - \psi(x_1, \ldots, x_{n-1})$. In the new coordinates the initial surface is $y_n = 0$. Henceforth we shall therefore assume that the initial surface is the plane $x_n = 0$.

In reducing the equations, we shall sometimes have to differentiate equations with respect to x_n. We note that after such a differentiation, the plane $x_n = 0$ remains a noncharacteristic surface of the new system. Moreover, if the differentiated equation is satisfied everywhere and the original equation is satisfied at $x_n = 0$, then the original equation is satisfied everywhere. That is, whenever we differentiate an equation with respect to x_n, we shall view the original equation as a constraint which has to be satisfied by the initial data for the new system.

We now note that any system of PDEs can be made quasilinear by differentiating one or more of the equations with respect to x_n. Hence we can restrict our attention to quasilinear systems. Consider a system of the form

$$\sum_{j=1}^{N} \sum_{|\alpha|=s_i+t_j} a_{ij}^{\alpha}(\mathbf{p}_i)D^{\alpha}u_j + b_i(\mathbf{p}_i) = 0, \quad i = 1, \ldots, N, \qquad (2.82)$$

where s_i and t_j are the weights assigned to the equations and dependent variables, respectively (cf. the discussion in Example 2.10), and where the vector $\mathbf{p}_i$ consists of all the independent variables and all derivatives of the dependent variables u_j of orders less than $s_i + t_j$. Since only the sums $s_i + t_j$ are relevant, we can assume that all the s_i are non-positive and all the t_j are non-negative. By differentiating the ith equation $-s_i$ times with respect to x_n, we can then make s_i for the new system equal to zero. After doing so, a natural set of initial conditions is obtained by prescribing $\partial^l u^j / \partial x_n^l$ for $l = 0, \ldots, t_j - 1$. These initial conditions must be subjected to constraints arising from any equations which have been differentiated with respect to x_n.

At this point, we have arrived at a system of the form

$$\sum_{j=1}^{N} \sum_{|\alpha|=t_j} a_{ij}^{\alpha}(\mathbf{p})D^{\alpha}u_j + b_i(\mathbf{p}) = 0, \quad i = 1, \ldots, N, \qquad (2.83)$$

where $\mathbf{p}$ contains all the independent variables and all derivatives of all dependent variables u_j of order less than t_j. The initial conditions are

$$\frac{\partial^l u_i}{\partial x_n^l} = f_{il}(x_1, \ldots, x_{n-1}), \quad i = 1, \ldots, N; \; l = 0, \ldots, t_i - 1. \qquad (2.84)$$

Note that the vector $\mathbf{p}$ is determined by the initial data. If (for the given data) the surface $x_n = 0$ is noncharacteristic, we can solve (2.83) for the derivatives

$$\frac{\partial^{t_i} u_i}{(\partial x_n)^{t_i}}, \qquad (2.85)$$

and obtain a system of the form

$$\frac{\partial^{t_i} u_i}{(\partial x_n)^{t_i}} = \sum_{j=1}^{N} \sum_{\substack{|\alpha|=t_j \\ \alpha \neq (0,\ldots,t_j)}} c_{ij}^{\alpha}(\mathbf{p}) D^{\alpha} u_j + d_i(\mathbf{p}). \tag{2.86}$$

We can now reduce this system to first-order by introducing all derivatives of u_j which are of order less than t_j as new dependent variables. We then add new equations accordingly, exploiting (2.86) and the equality of mixed partial derivatives. For example if $t_1 = 2$, we would introduce the new variables $v_{1k} = \partial u_1 / \partial x_k$ and add the equations

$$\frac{\partial u_1}{\partial x_n} = v_{1n}, \quad \frac{\partial v_{1k}}{\partial x_n} = \frac{\partial v_{1n}}{\partial x_k}, \quad k = 1, \ldots, n-1, \tag{2.87}$$

and the equation for $\partial v_{1n}/\partial x_n$ would come from (2.86). In this fashion, we arrive at an initial-value problem for a first-order system. The only differences from (2.45), (2.46) which are left are that the initial data may not be zero and that the coefficients may depend on x_n. The former can be taken care of by a substitution. If the initial condition for u_i is $u_i = f_i(x_1, \ldots, x_{n-1})$, we simply introduce $v_i = u_i - f_i(x_1, \ldots, x_{n-1})$ as new variables. Finally, we can remove the dependence of the coefficients on the independent variable x_n by introducing an additional dependent variable w which satisfies the equation

$$\frac{\partial w}{\partial x_n} = 1, \quad w = 0 \quad \text{on } x_n = 0. \tag{2.88}$$

In all coefficients which depend on x_n, we then replace x_n by w.

Example 2.23. Let us consider the scalar equation

$$\phi(u_{xx}, u_{xy}, u_{yy}, u_x, u_y, u, x, y) = 0 \tag{2.89}$$

with initial conditions

$$u(x,0) = f(x), \quad u_y(x,0) = g(x). \tag{2.90}$$

We assume that ϕ, f and g are analytic functions of their arguments and that, at least in a neighborhood of the origin, the plane $y = 0$ is noncharacteristic, i.e., the equation

$$\phi(f''(x), g'(x), z, f'(x), g(x), f(x), x, 0) = 0 \tag{2.91}$$

has a solution $z(x)$, and

$$\phi_{,3}(f''(x), g'(x), z(x), f'(x), g(x), f(x), x, 0) \neq 0. \tag{2.92}$$

Here and in the following, $\phi_{,i}$ denotes the derivative of ϕ with respect to the ith argument.

We first make the equation quasilinear by differentiating with respect to y. This leads to the problem

$$
\begin{aligned}
0 = & \phi_{,1}(\cdots)u_{xxy} + \phi_{,2}(\cdots)u_{xyy} + \phi_{,3}(\cdots)u_{yyy} \\
& + \phi_{,4}(\cdots)u_{xy} + \phi_{,5}(\cdots)u_{yy} \\
& + \phi_{,6}(\cdots)u_{y} + \phi_{,8}(\cdots),
\end{aligned}
\tag{2.93}
$$

(where the dots $(\cdots)$ denote the arguments $(u_{xx}, u_{xy}, u_{yy}, u_x, u_y, u, x, y)$) and the initial conditions are

$$
u(x,0) = f(x), \quad u_y(x,0) = g(x), \quad u_{yy}(x,0) = z(x), \tag{2.94}
$$

with $z(x)$ as above. We now introduce new variables $p = u_x$, $q = u_y$, $a = u_{xx}$, $b = u_{xy}$, $c = u_{yy}$, $y = w$ and obtain the system

$$
\begin{aligned}
& u_y = q, \quad p_y = b, \quad q_y = c, \quad a_y = b_x, \quad b_y = c_x, \\
& c_y = -\frac{1}{\phi_{,3}(\cdots)}\Big[\phi_{,1}(\cdots)b_x + \phi_{,2}(\cdots)c_x \\
& \qquad + \phi_{,4}(\cdots)b + \phi_{,5}(\cdots)c + \phi_{,6}(\cdots)q + \phi_{,8}(\cdots)\Big], \\
& w_y = 1.
\end{aligned}
\tag{2.95}
$$

Here the dots $(\cdots)$ indicate the argument (a, b, c, p, q, u, x, w). The initial conditions are

$$
\begin{aligned}
u(x,0) = f(x), \quad p(x,0) = f'(x), \quad q(x,0) = g(x), \quad a(x,0) = f''(x), \\
b(x,0) = g'(x), \quad c(x,0) = z(x), \quad w(x,0) = 0.
\end{aligned}
\tag{2.96}
$$

Example 2.24. As an example for a system, consider the Stokes system in two space dimensions,

$$
\begin{aligned}
u_{xx} + u_{yy} - p_x &= 0, \\
v_{xx} + v_{yy} - p_y &= 0, \\
u_x + v_y &= 0,
\end{aligned}
\tag{2.97}
$$

with initial conditions on $y = 0$. With the understanding that the ordering of variables is (u, v, p), we can assign the weights $s_1 = s_2 = 0$, $s_3 = -1$, $t_1 = t_2 = 2$, $t_3 = 1$. According to the above algorithm, we must differentiate the last equation in (2.97) with respect to y, leading to

$$
u_{xy} + v_{yy} = 0. \tag{2.98}
$$

The system can then be restated as

$$
\begin{aligned}
u_{yy} &= -u_{xx} + p_x, \\
v_{yy} &= -u_{xy}, \\
p_y &= v_{xx} - u_{xy}.
\end{aligned}
\tag{2.99}
$$

Appropriate initial conditions are

$$u(x,0) = f_1(x), \ u_y(x,0) = f_2(x),$$
$$v(x,0) = g_1(x), \ v_y(x,0) = g_2(x),$$
$$p(x,0) = h(x);$$

(2.100)

however, in order to obtain a solution of the original system, these initial data must satisfy the constraint given by the last equation of (2.97), namely,

$$f_1'(x) + g_2(x) = 0.$$

(2.101)

To obtain a first-order system, we finally set $u_x = a$, $u_y = b$, $v_x = c$, $v_y = d$. We obtain

$$u_y = b, \quad a_y = b_x, \quad b_y = -a_x + p_x, \quad v_y = d,$$
$$c_y = d_x, \quad d_y = -b_x, \quad p_y = c_x - b_x.$$

(2.102)

2.2.5 A PDE without Solutions

Every now and then a paper appears with a title like "A method to solve all partial differential equations." The content of such papers is always very far from satisfying the claims made in the title. It is rumored that a paper of this kind inspired Lewy to construct his famous example of a linear PDE which has no solutions at all. This example also highlights the importance of analyticity in the Cauchy-Kovalevskaya result.

Theorem 2.25. *For a complex-valued function $u(x,y,z)$, let*

$$Lu = -u_x - iu_y + 2i(x + iy)u_z.$$

(2.103)

Then there is a real-valued function $f(x,y,z)$, of class $C^\infty(\mathbb{R}^3)$, such that the equation

$$Lu = f(x,y,z)$$

(2.104)

has no solutions of class $C^1(\Omega)$ in any open subset $\Omega \subset \mathbb{R}^3$.

We note that when f is analytic, the Cauchy-Kovalevskaya theorem applies and noncharacteristic initial-value problems for (2.104) have local solutions. In contrast, for nonanalytic f there may be no solutions, even if no initial conditions are prescribed.

We shall not give a full proof of the theorem, but outline some of the main ideas. First, we shall prove the following lemma.

Lemma 2.26. *Let $\psi \in C^\infty(\mathbb{R})$ be real-valued and such that ψ is not real analytic at z_0. Then the equation*

$$Lu = \psi'(z)$$

(2.105)

has no solution of class C^1 in any neighborhood of the point $(0, 0, z_0)$.

Proof. Assume the contrary and let u be a solution in a neighborhood of $(0, 0, z_0)$, say for $x^2 + y^2 < \epsilon$, $|z - z_0| < \epsilon$. We set

$$v(r, \theta, z) = e^{i\theta} \sqrt{r} u(\sqrt{r} \cos \theta, \sqrt{r} \sin \theta, z). \qquad (2.106)$$

After some algebra, we find that v satisfies the equation

$$-2v_r - \frac{i}{r} v_\theta + 2iv_z = \psi'(z). \qquad (2.107)$$

We set

$$V(z, r) = \int_0^{2\pi} v(r, \theta, z) \, d\theta, \qquad (2.108)$$

and integrate (2.107) with respect to θ. This yields

$$V_z + iV_r = -\pi i \psi'(z). \qquad (2.109)$$

Clearly V is of class C^1 for $0 < r < \epsilon$, $|z - z_0| < \epsilon$ and continuous for $0 \le r < \epsilon$, $|z - z_0| < \epsilon$. Moreover, $V(z, 0) = 0$. Consider now $W = V(z, r) + \pi i \psi(z)$. Then (2.109) yields $W_z + iW_r = 0$, which makes W a holomorphic function of the complex variable $z + ir$ in the domain $|z - z_0| < \epsilon$, $0 < r < \epsilon$. Moreover, W is continuous up to the boundary $r = 0$ and it is purely imaginary there. By the Schwarz reflection principle (see Problem 2.15), we can extend W to a holomorphic function on $|r| < \epsilon$, $|z - z_0| < \epsilon$ by defining $W(z, -r) = -\overline{W(z, r)}$. But this implies that $\pi \psi(z)$, the imaginary part of $W(z, 0)$, is real analytic. $\qquad \square$

The equation

$$Lu = \psi'(z - 2y_0 x + 2x_0 y) \qquad (2.110)$$

is transformed into (2.105) by the simple substitution

$$U(x, y, z) = u(x + x_0, y + y_0, z + 2y_0 x - 2x_0 y). \qquad (2.111)$$

Hence, given any point (x_0, y_0, z_0), we can find a function $f(x, y, z)$ such that (2.104) has no solutions in any neighborhood of (x_0, y_0, z_0).

Consider now a finite number of points (x_i, y_i, z_i), $i = 1, \ldots, k$. We shall construct a function f such that (2.104) has no solutions in the neighborhood of any of these points. To do so, we choose ψ to be a real-valued function in $C^\infty(\mathbb{R})$ which is not analytic anywhere (see Problem 2.17) and make the ansatz

$$f(x, y, z) = \sum_{i=1}^{k} c_i \psi'(z - 2y_i x + 2x_i y), \qquad (2.112)$$

with real coefficients c_i. Then, for any choice of $(c_2, c_3, \ldots, c_k)$, there is at most one value $c_1 = c_1(c_2, \ldots, c_k)$ for which (2.104) has solutions in any neighborhood of (x_1, y_1, z_1). Assume the contrary, i.e., that there are two such values c_1 and $\tilde{c}_1$. Then we conclude that the equation

$$Lu = (c_1 - \tilde{c}_1)\psi'(z - 2y_1 x + 2x_1 y) \qquad (2.113)$$

has a solution in a neighborhood of (x_1, y_1, z_1), in contradiction to the result proved above. Likewise, for given $(c_1, c_3, \ldots, c_k)$ there is at most one value $c_2 = c_2(c_1, c_3, \ldots, c_k)$ for which (2.104) can have a solution in any neighborhood of (x_2, y_2, z_2). Now restrict c_i to a set of l elements, $l > k$. There are l^k possible choices of $(c_1, \ldots, c_k)$. However, there are at most kl^{k-1} choices for which any of the relations $c_1 = c_1(c_2, \ldots, c_k)$, $c_2 = c_2(c_1, c_3, \ldots, c_k)$, etc. hold. Hence there are choices for which (2.104) has no solutions in any neighborhood of any of the points (x_i, y_i, z_i). In fact, if l is large relative to k, this will be the case for "most" choices of the c_i.

To complete the proof of the theorem, one needs to extend this argument to a countable number of points. Let (x_i, y_i, z_i), $i \in \mathbb{N}$, be a sequence of points that is dense in $\mathbb{R}^3$. Then we make the ansatz

$$f(x, y, z) = \sum_{i=1}^{\infty} c_i \psi'(z - 2y_i x + 2x_i y). \tag{2.114}$$

If c_i converges to zero sufficiently rapidly as $i \to \infty$, then f is of class C^{∞}. It can still be shown that, for "most" choices of the c_i, (2.104) has no solutions in any neighborhood of any point (x_i, y_i, z_i); hence it does not have solutions anywhere. Carrying out this argument is not easy and requires the methods of functional analysis. We shall not pursue this point here and instead refer to the literature, see e.g., [Jo].

Problems

2.7. If $|x_i| < 1$ for $i = 1, \ldots, n$, show that

$$\sum_{\alpha} \mathbf{x}^{\alpha} = \frac{1}{(1 - x_1)(1 - x_2) \cdots (1 - x_n)}. \tag{2.115}$$

Apply D^{β} to both sides and compare. Then set $\mathbf{x} = (q, q, \ldots, q)$. This will yield (2.49).

2.8. Fill in the remaining details in the proof of claims 1-3 about power series (proceed analogously as for power series in one variable).

2.9. Let the Taylor series

$$\sum_{\alpha} c_{\alpha} \mathbf{x}^{\alpha} \tag{2.116}$$

converge in $D = \{\mathbf{x} \in \mathbb{R}^n \mid |\mathbf{x}| < r\}$. Show that the function represented by the Taylor series is real analytic in D.

2.10. Let f be real analytic in a convex domain D which contains the origin. Let

$$\sum_{\alpha} c_{\alpha} \mathbf{x}^{\alpha} \tag{2.117}$$

be the associated Taylor series. Assume the series converges absolutely for some $\mathbf{x} \in D$. Show that it converges to $f(\mathbf{x})$. Hint: Consider $f(\lambda \mathbf{x})$ for $0 < \lambda < 1$ and pass to the limit $\lambda \to 1$.

2.11. Find the solution of the initial-value problem

$$u_{xx} + u_{yy} = 0, \quad u(x,0) = 0, \quad u_y(x,0) = \frac{\sin nx}{n}. \tag{2.118}$$

Discuss what happens as $n \to \infty$. Compare with the analogous initial-value problem for $u_{xx} - u_{yy} = 0$.

2.12. Consider the initial-value problem

$$u_y = u u_x, \quad u(x,0) = f(x), \tag{2.119}$$

where f is analytic. Show that unless f is monotone decreasing, there cannot be an analytic (or even continuous) solution for all positive values of y. Hint: u is constant along characteristics given by $dx/dy = -u$. If f is not monotone decreasing, characteristics must intersect.

2.13. Show in detail that (2.95) with initial conditions (2.96) is equivalent to (2.93) with initial conditions (2.94). Is (2.95) taken by itself equivalent to (2.93)? If not, how do they differ?

2.14. The system (2.34) is not of the form (2.45). Discuss how an initial-value problem for (2.34) can be reduced to the standard form (2.45), (2.46).

2.15. Let Ω be a domain in $\mathbb{C}$, symmetric with respect to the real axis. Let f be holomorphic in $\Omega \cap \{\operatorname{Im} z > 0\}$ and continuous on $\Omega \cap \{\operatorname{Im} z \geq 0\}$. Moreover, assume that f takes real values on $\Omega \cap \{\operatorname{Im} z = 0\}$. Show that f can be extended to a function that is holomorphic in all of Ω by setting $f(z) = \overline{f(\bar{z})}$ for $\operatorname{Im} z < 0$. Hint: Show that $\int_C f(z)\, dz = 0$ for any closed rectifiable curve C such that C and its interior lie in Ω. It suffices to show this when C is a triangle.

2.16. Show that the three definitions of a C^k (or analytic) surface are indeed equivalent.

2.17. Show that the function

$$f(x) = \sum_{n=1}^{\infty} \frac{\cos(n!x)}{(n!)^n} \tag{2.120}$$

is of class C^∞ on $\mathbb{R}$, but is not real analytic anywhere. Hint: Show first that f is not in $C_{M,r}(0)$ for any M and r. Next show that $f(x) - f(x + 2\pi q)$ is analytic for any rational number q.

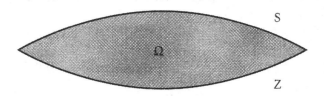

Figure 2.1. A lens-shaped region

2.3   Holmgren's Uniqueness Theorem

The theorem in the previous section shows existence and uniqueness of solutions for a noncharacteristic initial-value problem. However, uniqueness
was only guaranteed within the class of analytic functions; the existence of
other, nonanalytic solutions was not ruled out. Holmgren's theorem shows
that this cannot happen for linear equations; we shall prove uniqueness
assuming only that the solution is smooth enough so that all derivatives
appearing in the partial differential equation are continuous (using the concept of "generalized" solutions, defined later in this book, this assumption
can be relaxed further). The proof of uniqueness is achieved by proving
existence of solutions for an "adjoint" system of differential equations. To
obtain this existence, we shall use the Cauchy-Kovalevskaya theorem; this
requires us to assume analytic coefficients in the equations. If, however, we
had an existence theory which works without analyticity of the coefficients,
this assumption would be unnecessary.

2.3.1   An Outline of the Main Idea

Consider a system of linear equations

$$a_{ij}^k(\mathbf{x})\frac{\partial u_j}{\partial x_k} + b_{ij}(\mathbf{x})u_j = 0, \quad i = 1, \ldots, N. \tag{2.121}$$

Let $\mathbf{u} = (u_1, \ldots, u_N)$ be a solution in a "lens-shaped" domain $\Omega \subset \mathbb{R}^n$
bounded by two surfaces S and Z. Assume that $\mathbf{u} = 0$ on Z and that S
is noncharacteristic and analytic. We also assume that the coefficients in
(2.121) are analytic.

Let v_i, $i = 1, \ldots, N$ be arbitrary functions in $C^1(\overline{\Omega})$. We multiply the ith equation of (2.121) by v_i, sum over i, and integrate over Ω. This yields

$$
\begin{aligned}
0 &= \int_\Omega v_i(\mathbf{x}) a_{ij}^k(\mathbf{x}) \frac{\partial u_j}{\partial x_k}(\mathbf{x}) + v_i(\mathbf{x}) b_{ij}(\mathbf{x}) u_j(\mathbf{x}) \, d\mathbf{x} \\
&= \int_\Omega -\frac{\partial}{\partial x_k} \left[v_i(\mathbf{x}) a_{ij}^k(\mathbf{x}) \right] u_j(\mathbf{x}) + v_i(\mathbf{x}) b_{ij}(\mathbf{x}) u_j(\mathbf{x}) \, d\mathbf{x} \qquad (2.122) \\
&\quad + \int_{\partial\Omega} a_{ij}^k(\mathbf{x}) v_i(\mathbf{x}) u_j(\mathbf{x}) n_k \, dS,
\end{aligned}
$$

where $\mathbf{n}$ is the outer normal to $\partial\Omega$.

Assume now that $\mathbf{v}$ satisfies the "adjoint" system of PDEs,

$$
-\frac{\partial}{\partial x_k}(a_{ij}^k v_i) + b_{ij} v_i = 0, \quad j = 1, \ldots, N, \qquad (2.123)
$$

with initial conditions

$$
v_i = f_i \qquad (2.124)
$$

on S. Then (2.122) reduces to

$$
0 = \int_S a_{ij}^k(\mathbf{x}) f_i(\mathbf{x}) u_j(\mathbf{x}) n_k \, dS. \qquad (2.125)
$$

If this holds for arbitrary continuous functions f_i on S, then we conclude that $a_{ij}^k u_j n_k = 0$ on S, and since $\det a_{ij}^k n_k \neq 0$ (S is noncharacteristic), we conclude that $\mathbf{u} = \mathbf{0}$ on S.

The Cauchy-Kovalevskaya theorem guarantees that (2.123) has a solution in a neighborhood of S if the f_i are analytic. Unfortunately, we can in general not claim that this neighborhood includes all of Ω. If it did, we would obtain (2.125) for analytic $\mathbf{f}$. The Weierstraß approximation theorem states that any continuous function on a compact subset of $\mathbb{R}^n$ can be approximated uniformly by polynomials. Therefore, if (2.125) holds for $\mathbf{f}$ whose components are polynomials, it also holds for continuous $\mathbf{f}$.

2.3.2 Statement and Proof of the Theorem

In order to overcome the difficulty that we cannot guarantee a solution of (2.123) throughout all of Ω, we shall replace the surface S by a one-parameter family of surfaces S_λ and then take "small steps" in λ. More precisely, we shall presume the following situation.

Let D be a bounded domain in $\mathbb{R}^n$, such that the coefficients of (2.121) are analytic on D. Let $Z = D \cap \{x_n = 0\}$ and assume that Z is nonempty and noncharacteristic. Let $\Phi(\mathbf{x})$ be an analytic function defined on D such that $\nabla\Phi \neq 0$ and let $S_\lambda = \{\Phi = \lambda\} \cap \{x_n \geq 0\} \cap D$. We assume there are real numbers a and b, $a < b$, such that the following hold:

1. The set $\bigcup_{\lambda \in [a,b]} S_\lambda$ is compact.

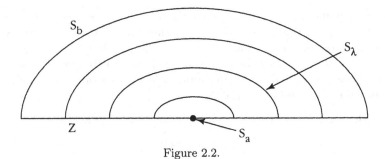

Figure 2.2.

2. S_a consists of a single point located on Z.

3. For $a < \lambda \leq b$, S_λ is a regular surface intersecting Z transversally (the intersection of two surfaces is called transverse if their normals are not collinear). The intersection of S_λ and Z is then a regular (analytic) $(n-2)$-dimensional surface. Moreover, we assume that S_λ is noncharacteristic.

We shall establish the following result:

Theorem 2.27. *Let* $\Omega = \{\mathbf{x} \in D \mid x_n > 0,\ a < \Phi(\mathbf{x}) < b\}$. *Let* $\mathbf{u} \in C^1(\overline{\Omega})$ *be a solution of (2.121) such that* $\mathbf{u} = \mathbf{0}$ *on* $\partial\Omega \cap Z$. *Then* $\mathbf{u} = \mathbf{0}$ *in* $\overline{\Omega}$.

Proof. Let $\Lambda = \{\lambda \in [a,b] \mid \mathbf{u} = \mathbf{0} \text{ on } S_\lambda\}$. We know that $a \in \Lambda$, and it follows from the continuity of $\mathbf{u}$ that Λ is closed. We shall show that Λ is also open in $[a,b]$. This implies $\Lambda = [a,b]$ and hence the theorem.

We note that $\overline{\Omega}$ is compact, and hence there is M, ρ, independent of $\mathbf{x}$, such that a_{ij}^k, b_{ij} and Φ are in $C_{M,\rho}(\mathbf{x})$ for every $\mathbf{x} \in \overline{\Omega}$. Consequently, if Cauchy data of class $C_{M,\rho}$ are prescribed on S_μ, a solution of (2.123) exists in an ϵ-neighborhood of S_μ, with ϵ independent of $\mu \in (a,b]$. We note that any polynomial lies in some $C_{M,\rho}$, where we can choose ρ as large as we wish, at the expense of making M large. However, (2.123) is linear, and hence the domain on which the solution exists does not change if the Cauchy data are multiplied by a constant factor. Hence the class of Cauchy data for which solutions to (2.123) exist in an ϵ-neighborhood of S_μ includes all polynomials.

We claim that, for given $\lambda \in [a,b]$ and $\epsilon > 0$, there is a $\delta > 0$ such that S_λ is contained in the ϵ-neighborhood of S_μ whenever $\mu \in [a,b]$ and $|\mu - \lambda| < \delta$. To see this, we first note that, in the neighborhood of any point $\mathbf{x} \in S_\lambda$, the equation $\Phi(\mathbf{x}) = \mu$ can be solved for one of the coordinates $x_i = x_i(x_1, \ldots, x_{i-1}, x_{i+1}, \ldots, x_n, \mu)$, and if $\mathbf{x} \in Z$ and $\lambda \neq a$ we can choose $i \neq n$. If $\lambda = a$, we have to choose $i = n$, and x_n is an increasing function of μ. In all cases, an immediate consequence is that if $\delta(\mathbf{x})$ is chosen sufficiently small, then for every $\mu \in [a,b]$ with $|\mu - \lambda| < \delta(\mathbf{x})$, there is a point $\mathbf{y} \in S_\mu$ with $|\mathbf{y} - \mathbf{x}| < \epsilon/2$. Since S_λ is compact, there is a finite number of points $\mathbf{x}_k$, $k = 1, \ldots, K$, such that S_λ is covered by the balls centered at $\mathbf{x}_k$ with radius $\epsilon/2$. The claim then follows with $\delta = \min \delta(\mathbf{x}_k)$.

Assume now that $\lambda \in \Lambda$ and $\mu \in (a, b]$ with $|\mu - \lambda| < \delta$, where δ is as above. We can then apply the argument explained in the previous section to the domain bounded by S_λ, S_μ and Z. We thus reach the conclusion that $\mathbf{u} = \mathbf{0}$ on S_μ, and hence $\mu \in \Lambda$.   □

Example 2.28. Consider the wave equation in two dimensions

$$u_{yy} - u_{xx} = 0, \tag{2.126}$$

with Cauchy data prescribed for $y = 0$, $-1 < x < 1$. Let $\Phi(x, y) = (x - y + 1)(x + y - 1)$, and let $D = (-1, 1) \times (-1, 1)$. Then S_λ, $-1 \le \lambda < 0$ is the arc of the hyperbola $(x - y + 1)(x + y - 1) = \lambda$ that lies within the triangle with corners $(-1, 0)$, $(1, 0)$ and $(0, 1)$. It is easy to show that all the hypotheses of the theorem are satisfied with $a = -1$ and any $b \in (-1, 0)$. Since the S_λ fill the interior of the triangle, u is determined within the whole triangle by its prescribed Cauchy data. In general, if u is determined in Ω by its Cauchy data on Z, we call Ω a *domain of determinacy* for Z.

2.3.3 The Weierstraß Approximation Theorem

In the above proof, we have used the following theorem, known as the Weierstraß approximation theorem.

Theorem 2.29. *Let S be a compact subset of $\mathbb{R}^n$ and let f be a continuous function $S \to \mathbb{R}$. Then there is a sequence of polynomials $p_m(\mathbf{x})$ such that $p_m(\mathbf{x}) \to f(\mathbf{x})$ uniformly on S.*

Proof. Let a_i, b_i, $i = 1, \ldots, n$, be such that $S \subset P := \prod_{i=1}^{n}(a_i, b_i)$. By the extension theorem of Tietze-Urysohn, f can be extended to a continuous function on all of $\mathbb{R}^n$, which agrees with the given function on S and vanishes outside of P. (Most good topology texts give a proof of this result.) We shall again call the extended function f and show that f can be approximated by polynomials uniformly on P. Without loss of generality, we may assume that $0 < a_i < b_i < 1$.

We first consider the one-dimensional case, i.e., let $f \in C(\mathbb{R})$ be such that $f = 0$ outside the interval (a, b), where $0 < a < b < 1$. We shall construct a sequence p_m of polynomials such that $p_m \to f$ uniformly on $[a, b]$.

First, we define

$$I_m = \int_0^1 (1 - v^2)^m \, dv, \quad I_{m,\delta} = \int_\delta^1 (1 - v^2)^m \, dv. \tag{2.127}$$

From the elementary inequalities

$$I_m > \int_0^1 (1 - v)^m \, dv = \frac{1}{m + 1}, \quad I_{m,\delta} < (1 - \delta^2)^m, \tag{2.128}$$

we conclude that, for any positive δ,

$$\lim_{m \to \infty} \frac{I_{m,\delta}}{I_m} = 0. \tag{2.129}$$

We now choose α and β such that $0 < \alpha < a < b < \beta < 1$. We set

$$p_m(x) = \frac{\int_\alpha^\beta f(y)\left[1 - (y - x)^2\right]^m dy}{\int_{-1}^1 (1 - y^2)^m \, dy}. \tag{2.130}$$

Obviously, p_m is a polynomial of degree $2m$. We shall now show that p_m converges to f, uniformly on $[a, b]$. In the numerator, we set $y = v + x$, and obtain

$$\int_{\alpha-x}^{\beta-x} f(v + x)(1 - v^2)^m \, dv = \int_{\alpha-x}^{-\delta} \cdots + \int_{-\delta}^{\delta} \cdots + \int_{\delta}^{\beta-x} \cdots = I_1 + I_2 + I_3. \tag{2.131}$$

Evidently, $|I_1|$ and $|I_3|$ are bounded by $MI_{m,\delta}$, where M is the maximum of $|f|$. Let now $\epsilon > 0$ be given and choose δ such that $|f(v + x) - f(x)| \leq \epsilon$ for $|v| \leq \delta$. Then we compute

$$I_2 = f(x)\int_{-\delta}^{\delta} (1 - v^2)^m \, dv + \int_{-\delta}^{\delta} (f(x + v) - f(x))(1 - v^2)^m \, dv. \tag{2.132}$$

The first term on the right equals $2f(x)(I_m - I_{m,\delta})$ and the second term can be estimated by $2\epsilon I_m$. By combining the estimates obtained, we find

$$|p_m(x) - f(x)| \leq 2M\frac{I_{m,\delta}}{I_m} + \epsilon. \tag{2.133}$$

We can make this quantity as small as we wish by first choosing ϵ small, then choosing δ accordingly and then choosing m sufficiently large.

In several space dimensions, we proceed in an analogous fashion with

$$p_m(\mathbf{x}) = \frac{\int_{\alpha_1}^{\beta_1} \cdots \int_{\alpha_n}^{\beta_n} f(\mathbf{y})[1-(y_1-x_1)^2]^m \cdots [1-(y_n-x_n)^2]^m \, dy_n \cdots dy_1}{\left[\int_{-1}^1 (1-\xi^2)^m \, d\xi\right]^n}. \tag{2.134}$$

$\square$

Problems

2.18. Let u be a solution of Laplace's equation in $\mathbb{R}^2$ and assume that $u = u_y = 0$ for $y = 0$ and $-1 < x < 1$. Show that $u = 0$ everywhere.

2.19. Consider the wave equation $u_{yy} - u_{xx} = 0$ with Cauchy data on $(-1, 1) \times \{0\}$. Show that no domain of determinacy extends beyond the square with corners $(-1, 0)$, $(0, 1)$, $(1, 0)$ and $(0, -1)$.

2.20. Consider a system of linear homogeneous first-order PDEs with constant coefficients such that the planes $x_n = $ const. are noncharacteristic. Show that any solution which vanishes on $x_n = 0$ vanishes everywhere.

2.21. Verify that

$$u(x,t) = \sum_{m=0}^{\infty} \frac{x^{2m}}{(2m)!} \frac{d^m}{dt^m} \exp\left(\frac{-1}{t^2}\right) \qquad (2.135)$$

is a solution of $u_t = u_{xx}$ with initial condition $u(x,0) = 0$. Why does Holmgren's theorem not apply? Hint: Use Cauchy's formula to estimate derivatives of $\exp(-1/t^2)$. For the contour, choose a circle centered at t with radius $t/2$.

2.22. Let f be a continuous function on $\mathbb{R}^n$. Show that there is a sequence of polynomials p_m such that $p_m \to f$ on all of $\mathbb{R}^n$, uniformly on any bounded set. Give an example showing that in general p_m cannot converge to f uniformly on all of $\mathbb{R}^n$.

3

Conservation Laws and Shocks

Recall that in Problem 1.29 we defined a weak solution of the one-dimensional wave equation to be a function $u(x,t)$ such that

$$\int_{-\infty}^{\infty} \int_{-\infty}^{\infty} u(x,t)(\phi_{tt}(x,t) - \phi_{xx}(x,t)) \, dx \, dt = 0 \qquad (3.1)$$

for every $\phi \in C_0^2(\mathbb{R}^2)$. In the problem, one was asked to show the following:

1. that any strong (classical C^2) solution of the wave equation is also a weak solution,

2. that discontinuous functions of the form

$$u(x,t) := H(x-t), \qquad (3.2)$$

and

$$u(x,t) := H(x+t), \qquad (3.3)$$

where H is the Heaviside function,

$$H(x) := \begin{cases} 0, & x < 0 \\ 1, & x \geq 0 \end{cases} \qquad (3.4)$$

are weak solutions of the wave equation.

In this section, we extend the notion of weak solution to strictly hyperbolic systems of conservation laws in one space dimension. As in the case of the wave equation, we will be able to observe discontinuous solutions. The use of the term *shock wave* to describe a discontinuous weak solution originated in the study of gas dynamics, but now it is used in connection with any discontinuous solution of a hyperbolic problem.

3.1 Systems in One Space Dimension

We consider the Cauchy problem for the following first-order quasilinear system of equations

$$\mathbf{u}_t + \mathbf{f}(\mathbf{u})_x = 0. \tag{3.5}$$

Here (x, t) lies in the upper half-plane $\mathbb{R}^{2+} := (-\infty, \infty) \times [0, \infty)$; the unknown is $\mathbf{u} : \mathbb{R}^{2+} \to \mathbb{R}^n$; and $\mathbf{f} : \mathbb{R}^n \to \mathbb{R}^n$ is a given function assumed to be sufficiently smooth (usually at least $C^2(D)$, for some open set $D \subseteq \mathbb{R}^n$). We refer to the space $\mathbb{R}^n$ in which the dependent variables of the unknown lie as *state space*. Our initial condition is

$$\mathbf{u}(x, 0) = \mathbf{u}_0(x) \tag{3.6}$$

for $x \in \mathbb{R}$, where the function $\mathbf{u}_0 : \mathbb{R} \to \mathbb{R}^n$ is given.

   The following examples will be mentioned repeatedly throughout this chapter.

Example 3.1. Burgers' equation. The following equation is the most commonly used example of a single ($n = 1$) nonlinear conservation law.

$$u_t + \left(\frac{u^2}{2} \right)_x = u_t + u u_x = 0. \tag{3.7}$$

Here $u : \mathbb{R}^{2+} \to \mathbb{R}$. The equation was studied by Burgers and Hopf, and was considered to be a crude model for turbulence.

Example 3.2. The wave equation. The linear one dimensional wave equation (1.114) can be converted from a second-order single equation into a first-order system by setting $v = u_t$ and $w = u_x$. The resulting system is

$$w_t = v_x, \tag{3.8}$$
$$v_t = w_x. \tag{3.9}$$

Of course the first equation is simply the equality of mixed partials of u, whereas the second is the original wave equation.

Example 3.3. The p system. This is simply the first-order system derived from the nonlinear wave equation

$$u_{tt} = \phi(u_x)_x,$$

where $\phi : \mathbb{R} \to \mathbb{R}$ is a given function. Setting $v = u_t$ and $w = u_x$ as we did for the wave equation yields

$$w_t = v_x, \tag{3.10}$$
$$v_t = \phi(w)_x. \tag{3.11}$$

It is very common to define $p(w) := -\phi(w)$ and write the system in the form

$$w_t - v_x = 0, \qquad (3.12)$$
$$v_t + p(w)_x = 0. \qquad (3.13)$$

A common model for isentropic gas dynamics uses this system with

$$p(w) := kw^{-\gamma}, \qquad (3.14)$$

and with the restriction $w > 0$. Here $\gamma \geq 1$ and $k > 0$ are constants.

Example 3.4. Gas dynamics in Lagrangian coordinates. The following equations describe the motion of an inviscid gas that does not conduct heat:

$$v_t - u_x = 0, \qquad (3.15)$$
$$u_t + p_x = 0, \qquad (3.16)$$
$$E_t + (up)_x = 0. \qquad (3.17)$$

Here v is the *specific volume*, u is the *velocity*, p is the *pressure* and E is the *specific energy per unit mass*. The specific volume is defined to be the reciprocal of the *density* ρ

$$v := 1/\rho. \qquad (3.18)$$

Equation (3.15) represents conservation of mass, (3.16) represents conservation of linear momentum and (3.17) represents conservation of energy.

In order to make this system of three equations in four unknowns well-posed, we must add a *constitutive equation* or *equation of state* that describes one of the variables as a given function of the other three. This is done here with the pressure, which is usually given by

$$p := \hat{p}(e, v), \qquad (3.19)$$

where $e := E - u^2/2$ is the *internal energy*.

Example 3.5. Gas dynamics in Eulerian coordinates. In the *Lagrangian* description of gas dynamics above, the variable x describes a fixed particle of gas. In the *Eulerian* description, x describes a fixed point in space. When the equations are derived using such a model, the following system of equations results:

$$\rho_t + (\rho u)_x = 0, \qquad (3.20)$$
$$(\rho u)_t + (\rho u^2 + p)_x = 0, \qquad (3.21)$$
$$\left[\rho\left(\frac{u^2}{2} + e\right)\right]_t + \left[\rho u\left(\frac{u^2}{2} + i\right)\right]_x = 0. \qquad (3.22)$$

Here ρ, u, p and e are defined as above; and $i := e + p/\rho$ is the *specific enthalpy*. Similarly to the equations above, (3.20) represents conservation of mass, (3.21) represents conservation of linear momentum and (3.22) represents conservation of energy.

3.2 Basic Definitions and Hypotheses

We begin our study of conservation laws by computing their characteristics and giving conditions under which the systems are strictly hyperbolic.

Lemma 3.6. *A curve* $t \mapsto \hat{x}(t)$ *is a characteristic curve for the conservation law (3.5) with solution* $\mathbf{u}(x,t)$ *if the matrix*

$$\hat{x}'(t)\mathbf{I} - \nabla\mathbf{f}(\mathbf{u}(\hat{x}(t),t)) \tag{3.23}$$

is singular. Furthermore, the system is strictly hyperbolic at a solution $\mathbf{u}$ *if the eigenvalues of* $\nabla\mathbf{f}(\mathbf{u})$ *are real and distinct.*

The proof is left to the reader. All that is involved is interpreting the definition of a characteristic curve and strict hyperbolicity for a nonlinear system in the case where the curve is described by a graph rather than a level set. (Recall the comments about different representations for surfaces in Section 2.2.)

Of course, the slopes of characteristic curves are nothing more than the eigenvalues of the matrix $\nabla\mathbf{f}(\mathbf{u})$. In light of this we introduce some notation describing eigenvalues and eigenvectors of $\nabla\mathbf{f}$. We assume that our system is strictly hyperbolic so that there are n real distinct eigenvalues $\lambda_1(\mathbf{u}) < \cdots < \lambda_n(\mathbf{u})$ with corresponding right and left eigenvectors $\mathbf{r}_k(\mathbf{u})$ and $\mathbf{l}_k(\mathbf{u})$ satisfying

$$\nabla\mathbf{f}(\mathbf{u})\mathbf{r}_k(\mathbf{u}) = \lambda_k(\mathbf{u})\mathbf{r}_k(\mathbf{u}), \tag{3.24}$$

$$\mathbf{l}_k(\mathbf{u})^T\nabla\mathbf{f}(\mathbf{u}) = \lambda_k(\mathbf{u})\mathbf{l}_k(\mathbf{u})^T. \tag{3.25}$$

Recall that since the eigenvectors are distinct, each of the sets of right and left eigenvectors $\{\mathbf{r}_1(\mathbf{u}), \ldots, \mathbf{r}_n(\mathbf{u})\}$ and $\{\mathbf{l}_1(\mathbf{u}), \ldots, \mathbf{l}_n(\mathbf{u})\}$ forms a basis for the state space.

We now define some functions on the state space, called *Riemann invariants*, that are instrumental in finding solutions to problems with discontinuous initial conditions. These functions are defined locally in a neighborhood $U \subset \mathbb{R}^n$.

Definition 3.7. A k-**Riemann invariant** is a smooth function $w : U \to \mathbb{R}$ such that for every $\mathbf{u} \in U$

$$\mathbf{r}_k(\mathbf{u}) \cdot \nabla w(\mathbf{u}) = 0. \tag{3.26}$$

The following lemma gives an existence result for an appropriate system of Riemann invariants.

Lemma 3.8. *For every* $\bar{\mathbf{u}} \in \mathbb{R}^n$ *there is a neighborhood* $\bar{U} \subset \mathbb{R}^n$ *of* $\bar{\mathbf{u}}$ *on which there are* $n - 1$ *k-Riemann invariants whose gradients are linearly independent at each point* $\mathbf{u} \in \bar{U}$.

Proof. Let S be a smooth surface through the point $\bar{\mathbf{u}}$, transversal to the vector $\mathbf{r}_k(\bar{\mathbf{u}})$. In a neighborhood of $\bar{\mathbf{u}}$, we now consider a system the ODEs $d\mathbf{u}/dt = \mathbf{r}_k(\mathbf{u})$. Then $w(\mathbf{u})$ is a k-Riemann invarient if it is constant along every trajectory of this system of ODEs. Now every trajectory that passes through a sufficiently small neighborhood of $\bar{\mathbf{u}}$ intersects S exactly once. The coordinates of this point of intersection (in a suitably chosen coordinate system on S) will serve as our Riemann invariants. $\square$

Example 3.9. The p system. We now consider the p system

$$\begin{pmatrix} w \\ v \end{pmatrix}_t + \begin{pmatrix} -v \\ p(w) \end{pmatrix}_x = \begin{pmatrix} 0 \\ 0 \end{pmatrix}. \tag{3.27}$$

Here we have

$$\nabla f(w, v) = \begin{pmatrix} 0 & -1 \\ p'(w) & 0 \end{pmatrix}. \tag{3.28}$$

To ensure strict hyperbolicity we assume

$$p' < 0. \tag{3.29}$$

We now have eigenvalues $\lambda_1(w) := -\sqrt{-p'(w)}$ and $\lambda_2(w) := \sqrt{-p'(w)}$ with corresponding right eigenvectors

$$\mathbf{r}_1(w) := \begin{pmatrix} 1 \\ \sqrt{-p'(w)} \end{pmatrix}, \tag{3.30}$$

$$\mathbf{r}_2(w) := \begin{pmatrix} 1 \\ -\sqrt{-p'(w)} \end{pmatrix}. \tag{3.31}$$

As indicated by the lemma above, there is one Riemann invariant corresponding to each eigenvalue; they are given as follows:

$$\rho_1(w, v) := v - \Psi(w), \tag{3.32}$$
$$\rho_2(w, v) := v + \Psi(w), \tag{3.33}$$

where

$$\Psi(w) := \int^w \sqrt{-p'(\xi)} d\xi. \tag{3.34}$$

The relationship between the Riemann invariants and the characteristic curves for this system is given by the following result.

Theorem 3.10. *Let* $(w(x,t), v(x,t))$ *be a* C^1 *solution of the p system given above. Then the Riemann invariant* $\rho_i(w(x,t), v(x,t))$ *is constant along characteristic curves satisfying* $\hat{x}'(t) = -\lambda_i(w(\hat{x}(t), t))$.

Proof. We do only the calculation for ρ_1.

$$\frac{d}{dt}\rho_1(w(\hat{x}(t),t),v(\hat{x}(t),t))$$

$$= v_x\hat{x}' + v_t - \sqrt{-p'(w)}(w_x\hat{x}' + w_t)$$

$$= -\sqrt{-p'(w)}(w_t - v_x) + (v_t + p'(w)w_x)$$

$$= 0.$$

The calculation for ρ_2 is identical.   □

One of the nice things about the p system is that we can use the Riemann invariants as a convenient change of coordinates in state space; i.e., since the system is strictly hyperbolic, $\Psi'(w) = \sqrt{-p'(w)} > 0$; hence Ψ and the map $(w,v) \rightarrow (\rho_1,\rho_2)$ are invertible. If we rewrite the p system in terms of ρ_1, ρ_2, we get the diagonal system

$$\rho_{1,t} + \hat{\lambda}(\rho_1 - \rho_2)\rho_{1,x} = 0, \tag{3.35}$$

$$\rho_{2,t} - \hat{\lambda}(\rho_1 - \rho_2)\rho_{2,x} = 0. \tag{3.36}$$

Here

$$\hat{\lambda}(s) := \sqrt{-p'\left(\Psi^{-1}\left(-\frac{s}{2}\right)\right)}. \tag{3.37}$$

Both the Theorem 3.10 and the diagonalization procedure above can be generalized to any system of two strictly hyperbolic conservation laws.

We can also use Riemann invariants to describe a hypothesis that often holds for systems of conservation laws coming from physics.

Definition 3.11. A system of conservation laws (3.5) is said to be **genuinely nonlinear** in a region $D \subseteq \mathbb{R}^n$ if

$$\nabla\lambda_k \cdot \mathbf{r}_k \neq 0, \quad k = 1,2,\ldots,n \tag{3.38}$$

in D.

Example 3.12. In the case of a single conservation law (3.41) we have $\lambda(u) = f'(u)$ and $r = 1$, so $\nabla\lambda(u) \cdot r = f''(u)$. We refer to a function satisfying $f'' > 0$ (< 0) as *strongly convex (concave)*. In conservation laws, such a function is sometimes refered to as *strictly* convex (concave). This is (strictly speaking) incorrect. Thus, genuine nonlinearity is implied by either strong convexity or concavity of f. Strong convexity is often assumed for physical reasons. (Variational problems that represent the steady state of conservation laws are usually stated as minimization rather than maximization problems.)

Example 3.13. For the p system we have

$$\nabla\lambda_1 \cdot \mathbf{r}_1 = -\nabla\lambda_2 \cdot \mathbf{r}_2 = \frac{p''}{2\sqrt{-p'}}. \tag{3.39}$$

Once again, strong convexity or strong concavity of p is sufficient to ensure genuine nonlinearity. In typical applications in gas dynamics one assumes p to be strongly convex.

We should note that there are interesting physical problems that are not genuinely nonlinear. In particular, in the p system the function p is sometimes assumed to have an inflection point. We do not address such problems in detail in this book, but we should introduce the reader to the following terminology.

Definition 3.14. We say that the kth characteristic field is **linearly degenerate** at $\mathbf{u}$ if

$$\nabla \lambda_k(\mathbf{u}) \cdot \mathbf{r}_k(\mathbf{u}) = 0. \tag{3.40}$$

3.3 Blowup of Smooth Solutions

As we noted above, the main purpose of this chapter is to study PDEs with discontinuous solutions. We are now prepared to show how discontinuous solutions of conservation laws can develop from continuous ones.

3.3.1 Single Conservation Laws

We consider a single conservation law of the form

$$u_t + f'(u)u_x = 0. \tag{3.41}$$

Here f is assumed sufficiently smooth. Characteristic curves for (3.41) must satisfy

$$\hat{x}'(t) = f'(u(\hat{x}(t), t)). \tag{3.42}$$

As a result of this relation we get the following very strong result for single conservation laws.

Theorem 3.15. *Any C^1 solution of the single conservation law (3.41) is constant along characteristics. Accordingly, characteristic curves for (3.41) are straight lines.*

Proof. Using (3.41) and (3.42), we get

$$\frac{d}{dt}u(\hat{x}(t), t) = u_x\hat{x}' + u_t = u_x f'(u) + u_t = 0. \tag{3.43}$$

Thus

$$u(\hat{x}(t), t) \equiv C \tag{3.44}$$

where C is a constant. So (3.42) implies

$$\hat{x}(t) = kt + \hat{x}(0), \tag{3.45}$$

where k is the constant $k := f'(C)$. □

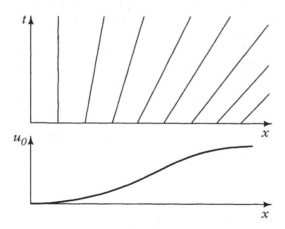

Figure 3.1. Defining a solution by characteristics.

To see what this implies in general about solutions of the Cauchy problem let us focus on Burgers' equation

$$u_t + uu_x = 0 \qquad (3.46)$$

with the initial condition

$$u(x,0) = u_0(x). \qquad (3.47)$$

Note that the equation for characteristics reduces to

$$\hat{x}'(t) = u(\hat{x}(t), t). \qquad (3.48)$$

Thus, the initial data give us the slopes of the characteristic rays emanating from the x axis. For certain initial data this gives us a method for "solving" the Cauchy problem. We simply go along the x axis, drawing characteristic rays with slope depending on the initial data, and let the solution take the value of the corresponding initial data along the characteristic (cf. Figure 3.1).

Unfortunately, some simple examples of discontinuous initial data show us just how easily the procedure falls apart. In Figure 3.2 we see that for an initial condition corresponding to a step function, there is a region that is untouched by any characteristics from the initial data; the procedure above does not identify a solution in this region. As we shall see below, in this case we will be able to identify a continuous solution called a rarefaction or fan wave.

However, in Figure 3.3 we have a more difficult problem. For a decreasing step function the characteristics overlap. Since our solution cannot be multivalued, we must conclude (in light of Theorem 3.15) that it cannot

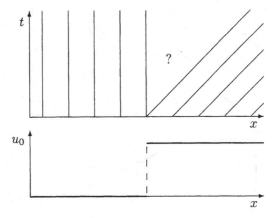

Figure 3.2. Characteristics do not specify the solution in the blank region.

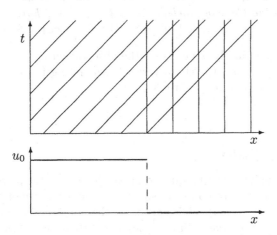

Figure 3.3. Characteristics overlap.

be smooth. For this type of initial data we will have to develop a theory of discontinuous solutions, or "shock waves."

Note that smoothing out the data does not help matters in this case; it merely delays the problem. In fact, the following theorem shows that the problem of overlapping characteristics and the development of singularities is a generic problem.

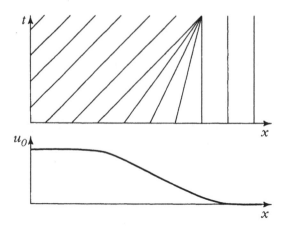

Figure 3.4. Intersecting characteristics from continuous initial data.

Theorem 3.16. *If $f'' > 0$ and the initial data u_0 is not monotone increasing, then the Cauchy problem for (3.41) does not have a C^1 solution defined on the entire upper half-plane $(x,t) \in (-\infty, \infty) \times [0, \infty)$.*

Proof. The proof simply depends on the observation that if $f'(u_0(x_1)) > f'(u_0(x_2))$ for $x_1 < x_2$, the characteristics emanating from x_1 and x_2 will intersect in finite time (cf. Figure 3.4). $\qquad\square$

3.3.2   The p System

We use more analytical techniques to prove blowup for the p system, where characteristic curves are no longer so simple. We make use of the diagonal form of the system (3.35), (3.36) given by changing to Riemann invariant coordinates in state space. Since Theorem 3.10 implies that ρ_1 and ρ_2 are constant along their respective characteristic curves, we cannot expect them to become unbounded as long as the solution stays C^1. However, if we examine the evolution of the slopes $\rho_{1,x}$ and $\rho_{2,x}$, we can expect something to go wrong. Thus, we differentiate (3.35) and (3.36) with respect to x to obtain

$$\rho_{1,xt} + \hat\lambda(\rho_1 - \rho_2)\rho_{1,xx} \;=\; -\hat\lambda'(\rho_1 - \rho_2)(\rho_{1,x}^2 - \rho_{1,x}\rho_{2,x}), \quad (3.49)$$

$$\rho_{2,xt} - \hat\lambda(\rho_1 - \rho_2)\rho_{2,xx} \;=\; -\hat\lambda'(\rho_1 - \rho_2)(\rho_{2,x}^2 - \rho_{1,x}\rho_{2,x}). \quad (3.50)$$

The product terms $\rho_{1,x}\rho_{2,x}$ cause an inconvenient coupling, but we can get rid of them by using the change of variables $r := \hat\lambda^{1/2}\rho_{1,x}$, $s := \hat\lambda^{1/2}\rho_{2,x}$.

Under this change our system becomes

$$r_t + \hat{\lambda}(\rho_1 - \rho_2)r_x = -\hat{\lambda}^{-1/2}(\rho_1 - \rho_2)\hat{\lambda}'(\rho_1 - \rho_2)r^2, \qquad (3.51)$$
$$s_t - \hat{\lambda}(\rho_1 - \rho_2)s_x = -\hat{\lambda}^{-1/2}(\rho_1 - \rho_2)\hat{\lambda}'(\rho_1 - \rho_2)s^2. \qquad (3.52)$$

Hence, the derivatives of r and s along characteristics are proportional to r^2 and s^2, respectively. With this in mind, consider the following lemma.

Lemma 3.17. *Let z be the solution of the ODE initial-value problem*

$$z'(t) = a(t)z^2(t), \qquad (3.53)$$
$$z(0) = m > 0, \qquad (3.54)$$

where $0 < B \le a(t) \le A$. Then z becomes infinite (has a vertical asymptote) at time $t_c \in ((mA)^{-1}, (mB)^{-1})$. Furthermore, for $t < t_c$ we have the estimate

$$\frac{1}{A(t_c - t)} < z(t) < \frac{1}{B(t_c - t)}. \qquad (3.55)$$

Using this and the calculations above, we can show

Theorem 3.18. *Assume that*

$$\sup_{x_1, x_2 \in \mathbb{R}} \hat{\lambda}'(\rho_1(x_1, 0) - \rho_2(x_2, 0)) < 0$$

and that

$$m := \max[\sup_x r(x, 0), \sup_x s(x, 0)] > 0.$$

Let

$$A := - \inf_{x_1, x_2 \in \mathbb{R}} \hat{\lambda}^{-1/2}\hat{\lambda}'(\rho_1(x_1, 0) - \rho_2(x_2, 0)),$$

$$B := - \sup_{x_1, x_2 \in \mathbb{R}} \hat{\lambda}^{-1/2}\hat{\lambda}'(\rho_1(x_1, 0) - \rho_2(x_2, 0)).$$

Then at least one of r and s becomes unbounded at a time between $(mA)^{-1}$ and $(mB)^{-1}$.

The proof of this theorem and Lemma 3.17 are left to the reader as exercises.

3.4 Weak Solutions

As we observed in the previous section, smooth solutions of hyperbolic conservation laws can blow up (develop discontinuities or singularities) in finite time. But this is not simply a mathematical oddity. It was observed in the nineteenth century that there were types of physical wave motion that were essentially discontinuous in nature, and which were not predicted by linear wave equations. In such a case one could not follow the practice of accepting the solution of a differential equation even when the equation itself failed to make sense (as we were able to do in the case of D'Alembert's solution of the wave equation) because closed form solutions of the nonlinear problems could not be computed. In order to understand (and compute) discontinuous solutions, one needed to extend the notion of solution itself.

Definition 3.19. A **weak solution** of (3.5), (3.6) is a function $\mathbf{u} : \mathbb{R}^{2+} \to \mathbb{R}^n$ such that

$$\int_0^\infty \int_{-\infty}^\infty [\mathbf{u}(x,t) \cdot \boldsymbol{\phi}_t(x,t) + \mathbf{f}(\mathbf{u}(x,t)) \cdot \boldsymbol{\phi}_x(x,t)] \, dx \, dt$$

$$\tag{3.56}$$

$$+ \int_{-\infty}^\infty \mathbf{u}_0(x)\phi(x,0) \, dx = 0$$

for every $\phi \in C_0^1(\mathbb{R}^{2+})$. Here

$$C_0^1(\mathbb{R}^{2+}) := \{\phi \in C^1(\mathbb{R}^{2+}) \mid \exists r > 0 \text{ s.t. supp } \phi \subset B_r((0,0)) \cap \mathbb{R}^{2+}\}.$$
$$\tag{3.57}$$

We begin our study of weak solutions by noting that the definition is indeed an extension of the classical notion of solution.

Theorem 3.20. *Suppose* $\mathbf{u} \in C^1(\mathbb{R}^{2+})$ *is a classical solution of (3.5), (3.6). Then* $\mathbf{u}$ *is also a weak solution.*

Proof. The proof is a simple application of Green's theorem in the plane. Take any $\phi \in C_0^1(\mathbb{R}^{2+})$ and let r be large enough so that supp $\phi \subseteq S$ where $S = B_r((0,0)) \cap \mathbb{R}^{2+}$. Note that since $\mathbf{u}$ satisfies (3.5) classically we have

$$(\mathbf{u} \cdot \boldsymbol{\phi})_t + (\mathbf{f}(\mathbf{u}) \cdot \boldsymbol{\phi})_x = \mathbf{u} \cdot \boldsymbol{\phi}_t + \mathbf{f}(\mathbf{u}) \cdot \boldsymbol{\phi}_x. \tag{3.58}$$

Thus, using this with Green's theorem and our information about the support of ϕ we have

$$\int\int_{\mathbb{R}^{2+}} \mathbf{u} \cdot \boldsymbol{\phi}_t + \mathbf{f}(\mathbf{u}) \cdot \boldsymbol{\phi}_x \, dx \, dt$$

$$= \int\int_S (\mathbf{u} \cdot \boldsymbol{\phi})_t + (\mathbf{f}(\mathbf{u}) \cdot \boldsymbol{\phi})_x \, dx \, dt$$

$$= -\int_{\partial S} \mathbf{u} \cdot \boldsymbol{\phi} \, dx - \mathbf{f}(\mathbf{u}) \cdot \boldsymbol{\phi} \, dt$$

$$= -\int_{-\infty}^\infty \mathbf{u}(x,0) \cdot \boldsymbol{\phi}(x,0) \, dx$$

$$= -\int_{-\infty}^\infty \mathbf{u}_0(x) \cdot \boldsymbol{\phi}(x,0) \, dx.$$

Here the next to the last equality was obtained from the fact that $\phi = 0$ on the half circle $t = (r^2 - x^2)^{1/2}$. The final equation derived from the fact that $\mathbf{u}$ satisfies the initial condition (3.6). $\qquad\square$

3.4.1 The Rankine-Hugoniot Condition

Now that we have defined a weak solution, let us find necessary conditions
for a discontinuous weak solution.

The following necessary condition (3.59) on piecewise smooth weak
solutions is known as the Rankine-Hugoniot condition.

Theorem 3.21 (Rankine-Hugoniot). *Let N be an open neighborhood
in the open upper half-plane, and suppose a curve $C : (\alpha, \beta) \ni t \mapsto \hat{x}(t)$
divides N into two pieces, N^l and N^r, lying to the left and right of the
curve, respectively. Let $\mathbf{u}$ be a weak solution of (3.5) (the initial conditions
do not matter here) such that*

1. $\mathbf{u}$ is a classical solution of (3.5) in both N^l and N^r,

2. $\mathbf{u}$ undergoes a jump discontinuity $[\mathbf{u}]$ at the curve C, and

3. the jump $[\mathbf{u}]$ is continuous along C.

*For any $\mathbf{p} \in C$, let $s := \hat{x}'(\mathbf{p})$ be the slope of C at $\mathbf{p}$. Then the following
relation holds between the curve and the jumps:*

$$s[\mathbf{u}] = [\mathbf{f}(\mathbf{u})]. \tag{3.59}$$

Here, for any $\mathbf{p} = (x_0, t_0) \in C$, we define

$$[\mathbf{u}](\mathbf{p}) := \mathbf{u}^r(\mathbf{p}) - \mathbf{u}^l(\mathbf{p}) := \lim_{(x^r,t^r)\overset{r}{\to}\mathbf{p}} \mathbf{u}(x^r, t^r) - \lim_{(x^l,t^l)\overset{l}{\to}\mathbf{p}} \mathbf{u}(x^l, t^l), \tag{3.60}$$

where the symbol $\overset{r}{\to} \mathbf{p}$ indicates the limit of points $(x^r, t^r) \in N^r$ converging
to $\mathbf{p}$ and $\overset{l}{\to} \mathbf{p}$ indicates a limit of points $(x^l, t^l) \in N^l$ converging to $\mathbf{p}$.

Proof. Let $\phi \in C_0^1(\mathbb{R}^{2+})$ be any test function with support in N. Since $\mathbf{u}$
is a weak solution we can write

$$\begin{aligned} 0 &= \int_N \mathbf{u} \cdot \phi_t + \mathbf{f}(\mathbf{u}) \cdot \phi_x \, dx \, dt \\ &= \int_{N^r} \mathbf{u} \cdot \phi_t + \mathbf{f}(\mathbf{u}) \cdot \phi_x \, dx \, dt + \int_{N^l} \mathbf{u} \cdot \phi_t + \mathbf{f}(\mathbf{u}) \cdot \phi_x \, dx \, dt. \end{aligned}$$

We now use Green's theorem in the plane, the fact that $\phi \equiv \mathbf{0}$ outside of
a compact set contained in N and the fact that $\mathbf{u}$ is a classical solution of

(3.5) in N^r and N^l to get the following:

$$
\begin{aligned}
0 &= \int_{N^r} \mathbf{u} \cdot \boldsymbol{\phi}_t + \mathbf{f}(\mathbf{u}) \cdot \boldsymbol{\phi}_x \, dx \, dt + \int_{N^l} \mathbf{u} \cdot \boldsymbol{\phi}_t + \mathbf{f}(\mathbf{u}) \cdot \boldsymbol{\phi}_x \, dx \, dt \\
&= -\int_{N^r} (\mathbf{u}_t + \mathbf{f}(\mathbf{u})_x) \cdot \boldsymbol{\phi} \, dx \, dt - \int_C \boldsymbol{\phi} \cdot (-\mathbf{u}^r \, dx + \mathbf{f}(\mathbf{u}^r) \, dt) \\
&\quad - \int_{N^l} (\mathbf{u}_t + \mathbf{f}(\mathbf{u})_x) \cdot \boldsymbol{\phi} \, dx \, dt + \int_C \boldsymbol{\phi} \cdot (-\mathbf{u}^l \, dx + \mathbf{f}(\mathbf{u}^l) \, dt) \\
&= -\int_\alpha^\beta (-[\mathbf{u}]\hat{x}' + [\mathbf{f}(\mathbf{u})]) \cdot \boldsymbol{\phi} \, dt.
\end{aligned}
$$

Since ϕ was an arbitrary test function we can conclude that the Rankine-Hugoniot condition (3.59) holds for every point $\mathbf{p} \in C$.   □

Example 3.22. Let us consider Burgers' equation with the initial data

$$
u_0(x) := \begin{cases} 1, & x < 0 \\ 0, & x \geq 0. \end{cases} \tag{3.61}
$$

This is the case examined in Figure 3.3 in which the characteristics intersect. If we seek a discontinuous solution that is identically one to the left of a "shock curve" of slope s and identically zero to the right, the Rankine-Hugoniot condition (3.59) gives us

$$
s(1 - 0) = \frac{1^2}{2} - \frac{0^2}{2}, \tag{3.62}
$$

or $s = 1/2$. Thus, the shock follows a straight line. Figure 3.5 illustrates this solution. (Note that since $s = dx/dt$, the slope s is the reciprocal of the slope for the usual orientation of the (x,t) axes.)

Remark 3.23. It is important to note that while smooth changes of the dependent variable $\mathbf{u}$ may transform *smooth* solutions of a conservation law to solutions of an "equivalent equation," the Rankine-Hugoniot conditions for the new equation may be very different from the old. Thus, the two "equivalent" equations may have very different discontinuous solutions. For example, if we multiply Burgers' equation by u, we get the equation

$$
uu_t + u^2 u_x = 0. \tag{3.63}
$$

Thus, if u is any smooth positive solution of Burgers' equation and we define $v := u^2$, then v satisfies the equation

$$
v_t + \left(\frac{2}{3} v^{3/2} \right)_x = 0. \tag{3.64}
$$

However, if we use the step function such as the one defined in (3.61) as initial data for this new equation, the shock induced has slope $s = 2/3$; where the slope of the shock for Burgers' equation was $s = 1/2$.

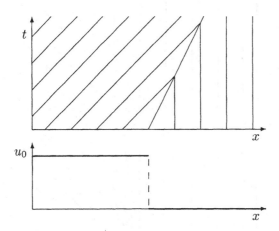

Figure 3.5. Shock wave solution.

3.4.2 Multiplicity

Let us again consider Burgers' equations, but this time with the initial data

$$u_0(x) := \begin{cases} 0, & x < 0 \\ 1, & x \geq 0. \end{cases} \tag{3.65}$$

This is the case examined in Figure 3.2 in which the method of character-istics leaves a blank patch in which the solution is undefined. How should we fill in the blank spot? If we seek a discontinuous solution that is iden-tically zero to the left of the shock and identically one to the right, the Rankine-Hugoniot condition (3.59) gives us

$$s(0 - 1) = \frac{0^2}{2} - \frac{1^2}{2}, \tag{3.66}$$

or, as before, $s = 1/2$. This solution is illustrated by the upper characteristic diagram in Figure 3.6.

However, there is another way of filling in the blank patch and coming up with a solution. The following continuous solution is called a *rarefaction wave* and is given further motivation in Section 3.5 below.

$$u(x, t) := \begin{cases} 0, & x < 0 \\ x/t, & 0 \leq x < t \\ 1, & t \leq x. \end{cases} \tag{3.67}$$

The reader should verify that any continuous function that is a piecewise classical solution of (3.5) is a weak solution (cf. the proof of Theorem 3.21).

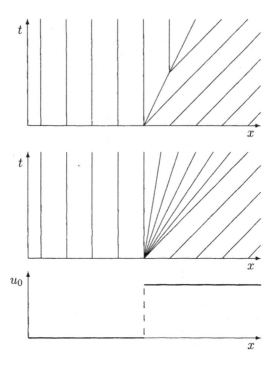

Figure 3.6. Two weak solutions for the same initial data: shock (top) and rarefaction or "fan" wave (bottom).

This solution is illustrated by the lower characteristic diagram in Figure 3.6.

In fact, we can create a whole continuum of solutions by combining shocks and rarefaction waves. For

$$u_0(x) := \begin{cases} 0, & x < 0 \\ 1, & x \geq 0, \end{cases} \tag{3.68}$$

where we have the following family of solutions parameterized by $\gamma \in [0, 1]$:

$$u_\gamma(x, t) := \begin{cases} 0, & -\infty < x \leq 0 \\ x/t, & 0 < x \leq \gamma t \\ \gamma, & \gamma t < x \leq \frac{\gamma+1}{2}t \\ 1, & \frac{\gamma+1}{2}t < x < \infty. \end{cases} \tag{3.69}$$

We ask the reader to sketch a characteristic diagram for this solution in Problem 3.6 below.

3.4.3   The Lax Shock Condition

Such a huge multiplicity of solutions is unacceptable. The physics associated with conservation laws, particularly with the example of gas dynamics, leads one to expect a unique solution. Thus, we need a "selection criterion" that picks out the physically reasonable solution from among the many possible weak solutions.

A number of such criteria have been proposed, but there is as yet no general agreement on which is the "right" one. Roughly speaking, all of the conditions agree for certain "easy" types of conservation laws, but there are pathological cases for which they disagree. Unfortunately (or fortunately if you plan to be doing research in this area) some of the pathological cases occur in important physical problems (such as oil recovery models), so the question of the best selection criterion is not simply mathematical sophistry.

The criterion we present in this section is one of the most widely used, primarily because it is rather easy to implement in the Riemann problems discussed below. It was introduced by Peter Lax in the 1950s [La]. We will examine some of the Lax condition's competitors in Section 3.6.

Definition 3.24 (Lax shock condition). Suppose $\mathbf{u}$ is a piecewise classical solution of a strictly hyperbolic system of conservation laws. Suppose that $\mathbf{u}$ has a jump discontinuity on a curve C, and that $\mathbf{u}^l$ and $\mathbf{u}^r$ are the respective left and right limiting values of $\mathbf{u}$ at a point $\mathbf{p} \in C$ (cf. (3.60)). Suppose further that C has slope s at $\mathbf{p}$. Then the discontinuous solution is deemed admissible, and is called a **k-shock**, if at each $\mathbf{p} \in C$ it satisfies the Rankine-Hugoniot condition (3.59); and if there is an integer k with $1 \leq k \leq n$ such that

$$\begin{aligned} \lambda_1(\mathbf{u}^l) < \lambda_{k-1}(\mathbf{u}^l) < s < \lambda_k(\mathbf{u}^l) < \lambda_n(\mathbf{u}^l), \\ \lambda_1(\mathbf{u}^r) < \lambda_k(\mathbf{u}^r) < s < \lambda_{k+1}(\mathbf{u}^r) < \lambda_n(\mathbf{u}^r). \end{aligned} \tag{3.70}$$

Thus, there are exactly $k-1$ characteristic speeds $\lambda_j(\mathbf{u}^l)$ on the left that are less than the slope s and exactly $n-k$ speeds $\lambda_j(\mathbf{u}^r)$ on the right that are greater than the slope s.

Example 3.25. (Single conservation laws.) For Burgers' equation the Lax shock condition (3.70) reduces to

$$u^l > s > u^r, \tag{3.71}$$

so that we can only "jump down" across a shock. This rules out the discontinuous weak solution described in Figure 3.6 in which the characteristics leave the discontinuity. More generally, for a single conservation law

$$u_t + f(u)_x = 0,$$

we have

$$f'(u^l) > s > f'(u^r). \tag{3.72}$$

Thus, as was the case in Burgers' equation, Lax's criterion requires characteristics to impinge on a shock for any single conservation law. This is often described as a "loss of information," and, as we shall see, the Lax condition is related to a version of the second law of thermodynamics.

The reader should also note the relation between convexity and the Lax shock condition for single conservation laws (cf. Problem 3.7).

Example 3.26. (The p system.) For a 2×2 system the Lax conditions require that a "1-shock" satisfy

$$s < \lambda_1(\mathbf{u}^l), \qquad \lambda_1(\mathbf{u}^r) < s < \lambda_2(\mathbf{u}^r), \tag{3.73}$$

and that a "2-shock" satisfy

$$\lambda_1(\mathbf{u}^l) < s < \lambda_2(\mathbf{u}^l), \qquad \lambda_2(\mathbf{u}^r) < s. \tag{3.74}$$

In the case of the p system, where we have

$$\lambda_1 = -\sqrt{-p'(w)} < 0 < \sqrt{-p'(w)} = \lambda_2; \tag{3.75}$$

this implies that 1-shocks have negative speed (they are often called "back-shocks") and satisfy

$$-\sqrt{-p'(w^r)} < s < -\sqrt{-p'(w^l)}, \tag{3.76}$$

whereas 2-shocks (also called "front-shocks") have positive speed and satisfy

$$\sqrt{-p'(w^r)} < s < \sqrt{-p'(w^l)}. \tag{3.77}$$

Note that if we assume $p'' > 0$ to ensure genuine nonlinearity $w \mapsto \sqrt{-p'(w)}$ is strictly decreasing. Thus, in this case condition (3.76) for a 1-shock implies

$$w^r < w^l, \tag{3.78}$$

whereas condition (3.77) for a 2-shock implies

$$w^l < w^r. \tag{3.79}$$

3.5   Riemann Problems

The "shock tube" experiment is one of the classic experiments of gas dynamics. To perform it one takes a long cylindrical tube separated into halves by a thin membrane. A gas is placed into each side, usually with both sides at rest, but with different pressures and densities. The membrane is then suddenly removed, and the evolution of the gas is observed.

The mathematical problem illustrated by the shock tube experiment was analyzed by Riemann, and this problem (and the analogous problem for other conservation laws) now bears his name. The problem consists in solving the Cauchy problem for the conservation law (3.5)

$$\mathbf{u}_t + \mathbf{f}(\mathbf{u})_x = 0$$

with the piecewise constant initial data

$$\mathbf{u}(x,0) = \begin{cases} \mathbf{u}^l, & x < 0 \\ \mathbf{u}^r, & x \geq 0. \end{cases} \tag{3.80}$$

The study of the Riemann problem is pedagogically important, in that it allows us to examine a variety of wave-like behavior that includes shocks in as simple a setting as possible. But the problem also has great practical importance in that some of the most useful numerical techniques for studying conservation laws are based on solving a succession of Riemann problems. Furthermore, these numerical techniques are the basis for general existence proofs.

We will limit our study to just two simple cases: the single conservation law and the p system. These cases, however, give only a tempting hint of the full breadth of this subject. The interested reader should consult the references given at the end of this section for further material.

3.5.1 Single Equations

The Riemann problem for a single conservation law (3.41) is exceedingly simple, at least in the case where f is strongly convex. We assume throughout that $f \in C^2(\mathbb{R})$. We need only consider three cases here.

1. The initial condition is a constant. When $u^l = u^r$ we get the trivial, classical, constant solution, $u(x,t) \equiv u^l$.

2. The initial condition jumps down. In this case, where $u^l > u^r$, we can use the shock solution

$$u(x,t) := \begin{cases} u^l, & x < st \\ u^r, & x \geq st, \end{cases} \tag{3.81}$$

where the shock speed is given by the Rankine-Hugoniot condition

$$s := \frac{f(u^l) - f(u^r)}{u^l - u^r}. \tag{3.82}$$

Note that because f is convex our shock meets the Lax shock criterion

$$f'(u^l) > s > f'(u^r). \tag{3.83}$$

Hence, the shock satisfies the entropy condition as well.

3. The initial condition jumps up. In this case we introduce a continuous *rarefaction wave* (the term, like so many others in the subject, comes from gas dynamics), which generalizes example (3.67) given above. To give

some mathematical motivation for the formula for rarefaction waves given
below, we note that since the jump in our initial data occurs at $x = 0$,
we can take any weak solution $u(x, t)$ of (3.41) and form a parameterized
family of solutions via the formula

$$u^\lambda(x, t) := u(\lambda x, \lambda t). \qquad (3.84)$$

If we expect our problem to have a unique solution, then u should have the
form

$$u(x, t) := \tilde{u}(x/t). \qquad (3.85)$$

Placing this in (3.41) gives us

$$-\tilde{u}' \frac{x}{t^2} + f'(\tilde{u})\tilde{u}' \frac{1}{t} = 0. \qquad (3.86)$$

Thus, either $\tilde{u}$ is constant or

$$f'(\tilde{u}(x/t)) = x/t. \qquad (3.87)$$

In this case, we use the fact that $f'' > 0$ to deduce that f' is invertible and
get

$$\tilde{u}(x/t) = f'(x/t)^{-1}. \qquad (3.88)$$

We thus justify the following formula for the classical rarefaction solution

$$u(x, t) := \begin{cases} u^l, & x < f'(u^l)t \\ f'(x/t)^{-1}, & f'(u^l)t \le x < f'(u^r)t \\ u^r, & f'(u^r)t \le x. \end{cases} \qquad (3.89)$$

3.5.2 Systems

In this section we state a collection of results that allow us to solve the
Riemann problem for systems of equations, but some of our proofs are only
for the special case of the p system. This allows us to keep our treatment
fairly brief and concrete while displaying most of the ideas involved in the
more general proofs.

For the single conservation law we were able to connect any pair of left
and right states using a single wave, either a shock or a rarefaction wave.
In higher dimensions, we will have to use intermediate states and several
different waves to make the connection. However, as a first step, we will
see what left and right states can be "hooked up" using a single shock or
rarefaction wave.

Shock waves

We begin by considering the possibility of using a single shock wave to
connect the left and right states. Thus, we have to ask the question: given
$\mathbf{u}^l$, what states $\mathbf{u}^r$ satisfy the Rankine-Hugoniot condition (3.59) and the

Lax shock condition (3.70)? The answer is that, emanating from each point $\mathbf{u}^l$ in state space, there are n *shock curves* that describe the possible right states that can be connected by a single shock. More specifically, we have the following theorem.

Theorem 3.27. *Suppose that (3.5) is a strictly hyperbolic system of conservation laws defined on a region $\Omega \subset \mathbb{R}^n$ of state space. Then for any $\mathbf{u}^l \in \Omega$ there exist n open intervals I_k containing 0 and n one-parameter families of states $\hat{\mathbf{u}}_k(\epsilon)$ and shock speeds $\hat{s}_k(\epsilon)$ defined on $\epsilon \in I_k$ such that*

$$\mathbf{u}_k(0) = \mathbf{u}^l \tag{3.90}$$

and such that for $\epsilon \in I_k$, $\hat{\mathbf{u}}_k(\epsilon)$ and $\hat{s}(\epsilon)$ satisfy the Rankine-Hugoniot condition

$$\hat{s}(\epsilon)[\mathbf{u}^l - \hat{\mathbf{u}}_k(\epsilon)] = \mathbf{f}(\mathbf{u}^l) - \mathbf{f}(\hat{\mathbf{u}}_k(\epsilon)).$$

Furthermore, if the kth characteristic field is genuinely nonlinear, then the parameterization can be chosen so that

$$\hat{\mathbf{u}}'_k(0) = \mathbf{r}_k(\mathbf{u}^l), \tag{3.91}$$
$$\hat{s}(0) = \lambda_k(\mathbf{u}^l), \tag{3.92}$$
$$\hat{s}'(0) = 1/2, \tag{3.93}$$

where the prime ' refers to differentiation with respect to ϵ. Moreover, with this parameterization, the Lax shock conditions hold if and only if $\epsilon < 0$.

We will not prove this theorem in general, but will instead calculate the shock curves explicitly for the p system. To ensure strict hyperbolicity and genuine nonlinearity we will assume $p' < 0$ and $p'' > 0$. We can also either assume that p is defined on all of $\mathbb{R}$ or make appropriate restrictions on the states chosen below. Thus, we take any admissible $\mathbf{u}^l := (w^l, v^l)^t$ and suppose $\hat{\mathbf{u}} := (\hat{w}, \hat{v})^t$ is connected to $\mathbf{u}^l$ by one of the two shock curves whose existence was asserted in the theorem. In this case the Rankine-Hugoniot condition reduces to

$$s(w^l - \hat{w}) = -(v^l - \hat{v}), \tag{3.94}$$
$$s(v^l - \hat{v}) = p(w^l) - p(\hat{w}). \tag{3.95}$$

By eliminating s from these equations we get

$$(v^l - \hat{v})^2 = (\hat{w} - w^l)(p(w^l) - p(\hat{w})). \tag{3.96}$$

Since $p' < 0$, there are two curves of solutions, defined for all ϵ in the domain of p.

$$S_1 : \quad \hat{\mathbf{u}}_1(\epsilon) = \begin{pmatrix} \hat{w}_1(\epsilon) \\ \hat{v}_1(\epsilon) \end{pmatrix}$$
$$:= \begin{pmatrix} \epsilon \\ v^l + \text{sgn}(\epsilon - w^l)\sqrt{(\epsilon - w^l)(p(w^l) - p(\epsilon))} \end{pmatrix}, \tag{3.97}$$

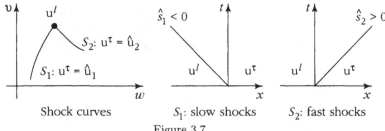

Shock curves S_1: slow shocks S_2: fast shocks

Figure 3.7.

$$S_2 : \quad \hat{\mathbf{u}}_2(\epsilon) = \begin{pmatrix} \hat{w}_2(\epsilon) \\ \hat{v}_2(\epsilon) \end{pmatrix}$$

$$:= \begin{pmatrix} \epsilon \\ v^l - \mathrm{sgn}(\epsilon - w^l)\sqrt{(\epsilon - w^l)(p(w^l) - p(\epsilon))} \end{pmatrix}. \tag{3.98}$$

The corresponding shock speeds are

$$\hat{s}_1(\epsilon) := -\sqrt{\frac{p(w^l) - p(\epsilon)}{\epsilon - w^l}}, \tag{3.99}$$

$$\hat{s}_2(\epsilon) := \sqrt{\frac{p(w^l) - p(\epsilon)}{\epsilon - w^l}}. \tag{3.100}$$

Only half of each curve satisfies the Lax shock conditions. Conditions (3.78) and (3.79) imply that any right state of a 1-shock would have to lie on the curve

$$\mathbf{u}^r = \hat{\mathbf{u}}_1(\epsilon), \quad \epsilon < w^l, \tag{3.101}$$

and any right state of a 2-shock would have to lie on the curve

$$\mathbf{u}^r = \hat{\mathbf{u}}_2(\epsilon), \quad w^l < \epsilon. \tag{3.102}$$

The reader is asked to verify that these curves can be reparameterized so that they satisfy the stated initial conditions; and more importantly, that these states and the corresponding shock speeds satisfy (3.76) and (3.77) (cf. Problem 3.10).

Pictorially, we see that emanating from each left state $\mathbf{u}^l$ we have the two shock curves S_1 and S_2. Shocks with negative speed (sometimes called slow shocks or back-shocks) lie along S_1; shocks with positive speed (fast shocks or front-shocks) lie along S_2.

Rarefaction waves

We now construct a family of continuous waves that generalize the rarefaction waves for the single conservation law. As in the case of shocks, we prove the existence of n curves emanating from a left state $\mathbf{u}^l$ giving the possible right states $\mathbf{u}^r$ that can be connected directly using a single rarefaction wave.

The general idea is based on the construction for the single conservation law. Suppose we have a situation where for some $k = 1, 2, \ldots, n$ we have

$$\lambda_k(\mathbf{u}^l) < \lambda_k(\mathbf{u}^r). \tag{3.103}$$

Note that the Lax condition immediately rules out connecting the two states with a single shock. We now mimic the procedure followed for the single equation case and draw characteristic lines $x = \lambda(\mathbf{u}^l)t$ and $x = \lambda(\mathbf{u}^r)t$ emanating from the left and right of the origin. (Note that in the case of systems it is not necessary that solutions be constant along characteristics, though in this case, such a "guess" will lead us to a solution.) Observe that this characteristic diagram is very similar to Figure 3.2: we have two regions in the upper half of the (x, t)-plane covered by characteristics with a wedge-shaped blank region in between. If we yield to temptation and define a solution to be the constant $\mathbf{u}^l$ in the left-hand shaded region and $\mathbf{u}^r$ in the right-hand region, how are we to fill in the blank region? The answer is that we can do so with the following type of wave.

Definition 3.28. Let $\mathbf{u}$ be a C^1 solution of conservation law (3.5) in a domain D. Then $\mathbf{u}$ is said to be a **k-rarefaction wave** (or a **k-simple wave**) if all k-Riemann invariants are constant in D.

As we might have hoped from observing the results for the single conservation law, if we can find a rarefaction wave that fills in the blank wedge, the characteristics associated with λ_k form a "fan."

Theorem 3.29. *Let $\mathbf{u}$ be a k-rarefaction wave in a domain D. Then the characteristic curves $\hat{x}'(t) = \lambda_k(\mathbf{u}(\hat{x}(t), t))$ are straight lines along which $\mathbf{u}$ is constant.*

Proof. We wish to show

$$0 = \frac{d}{dt}\mathbf{u}(\hat{x}(t), t) = \mathbf{u}_t + \lambda_k \mathbf{u}_x. \tag{3.104}$$

Lemma 3.8 asserts that there exist $n - 1$ k-Riemann invariants w_i, $i = 1, 2, \ldots, n - 1$, whose gradients are linearly independent. Since $\mathbf{u}$ is a k-rarefaction wave, $w_i(\mathbf{u}(x, t))$ is constant, and hence

$$\frac{d}{dt} w_i(\mathbf{u}(\hat{x}(t), t)) = \nabla w_i \cdot (\mathbf{u}_t + \lambda_k \mathbf{u}_x) = 0, \tag{3.105}$$

for $i = 1, 2, \ldots, n - 1$. We now use the fact that $\mathbf{u}$ solves (3.5) and the definition of $\mathbf{l}_k$ to deduce

$$\mathbf{l}_k \cdot (\mathbf{u}_t + \lambda_k \mathbf{u}_x) = \mathbf{l}_k \cdot (\mathbf{u}_t + \nabla f \mathbf{u}_x) = 0. \tag{3.106}$$

Thus, $\mathbf{u}_t + \lambda_k \mathbf{u}_x$ is orthogonal to every vector in the set

$$V := \{\mathbf{l}_k, \nabla w_1, \nabla w_2, \ldots, \nabla w_{n-1}\}. \tag{3.107}$$

Thus, all that remains to complete the proof is to show that V is a basis for $\mathbb{R}^n$. This is left to the reader (cf. Problem 3.9). □

We now state our basic theorem on the existence of rarefaction curves.

Theorem 3.30. *Suppose that the system of conservation laws (3.5) is genuinely nonlinear in an open region $\Omega \subset \mathbb{R}^n$ in state space, and let the right eigenvectors $\mathbf{r}_k$ be normalized so that $\nabla \lambda_k \cdot \mathbf{r}_k = 1$. Then for any left state $\mathbf{u}^l \in \Omega$ there exist n intervals $J_k = [0, a_k)$ and n smooth, one-parameter families of right states $\tilde{\mathbf{u}}_k(\gamma)$ defined for $\gamma \in J_k$ that can be connected to $\mathbf{u}^l$ by a k-simple wave using the procedure above. Moreover, these one-parameter families satisfy the following properties:*

$$\tilde{\mathbf{u}}_k(0) = \mathbf{u}^l, \tag{3.108}$$

$$\tilde{\mathbf{u}}'_k(0) = \mathbf{r}_k, \tag{3.109}$$

and for $0 < \gamma \in J_k$,

$$\lambda_k(\mathbf{u}^l) < \lambda_k(\tilde{\mathbf{u}}_k(\gamma)). \tag{3.110}$$

Proof. The rarefaction curves are simply solutions of the ODE initial-value problem

$$\frac{d\tilde{\mathbf{u}}_k}{d\gamma}(\gamma) = \mathbf{r}_k(\tilde{\mathbf{u}}_k(\gamma)), \tag{3.111}$$

$$\tilde{\mathbf{u}}_k(0) = \mathbf{u}^l. \tag{3.112}$$

Existence on an interval about 0 follows from Theorem 1.1.

Note that

$$\frac{d}{d\gamma}\lambda_k(\tilde{\mathbf{u}}_k(\gamma)) = \nabla \lambda_k \cdot \tilde{\mathbf{u}}'_k = \nabla \lambda_k \cdot \mathbf{r}_k = 1. \tag{3.113}$$

Thus, $\gamma \mapsto \lambda_k(\tilde{\mathbf{u}}_k(\gamma))$ is increasing so that (3.110) holds. Moreover, using the initial condition (3.108), we get

$$\lambda_k(\tilde{\mathbf{u}}_k(\gamma)) = \gamma + \lambda_k(\mathbf{u}^l). \tag{3.114}$$

To see that we can use this curve to "hook up" a left and right state using a k-rarefaction wave, we simply let

$$\mathbf{u}(x, t) := \tilde{\mathbf{u}}_k(x/t - \lambda_k(\mathbf{u}^l)). \tag{3.115}$$

Note that this is indeed a solution of (3.5) and that for any k-Riemann invariant w_k

$$\frac{\partial}{\partial t} w_k(\mathbf{u}(x, t)) = \nabla w_k \cdot \tilde{\mathbf{u}}'_k\left(-\frac{x}{t^2}\right) = \nabla w_k \cdot \mathbf{r}_k\left(-\frac{x}{t^2}\right) = 0. \tag{3.116}$$

A similar calculation for the derivative with respect to x shows that any k-Riemann invariant is constant in the "fan" region so that the solution is a k-rarefaction wave. □

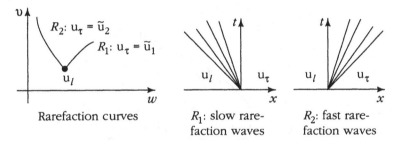

Figure 3.8.

In the case of the p system it is easier to solve (3.111) without normalizing the eigenvectors. We get the curves

$$R_1: \quad \tilde{\mathbf{u}}_1(\gamma) = \begin{pmatrix} \tilde{w}_1(\gamma) \\ \tilde{v}_1(\gamma) \end{pmatrix} := \begin{pmatrix} w^l + \gamma \\ v^l + \int_0^\gamma \sqrt{-p'(w^l + \xi)}\,d\xi \end{pmatrix} \qquad (3.117)$$

$$R_2: \quad \tilde{\mathbf{u}}_2(\gamma) = \begin{pmatrix} \tilde{w}_2(\gamma) \\ \tilde{v}_2(\gamma) \end{pmatrix} := \begin{pmatrix} w^l - \gamma \\ v^l + \int_0^\gamma \sqrt{-p'(w^l - \xi)}\,d\xi \end{pmatrix} \qquad (3.118)$$

Because we have not normalized the eigenvectors it is somewhat harder to determine $(w(x,t), v(x,t))$ from the rarefaction curves. To compute a 1-wave between $\mathbf{u}^l$ and $\mathbf{u}^r$ (with $\mathbf{u}^r$ on R_1) we take

$$\lambda_1(\mathbf{u}^l) < \frac{x}{t} < \lambda_1(\mathbf{u}^r), \qquad (3.119)$$

and solve the equation

$$\frac{x}{t} = \lambda_1 = -\sqrt{-p'(w(x/t))} \qquad (3.120)$$

for $w(x/t)$. Next, we let

$$\gamma = w(x/t) - w^l, \qquad (3.121)$$

and use this in (3.117) to determine $v(x/t)$.

The picture here is much the same as the one for the shock curves. The two rarefaction curves emanate from the left state $\mathbf{u}^l$; they share the tangent vectors $\mathbf{r}_k$ with the shock curves, but propagate in the opposite direction. The R_1 curve represents rarefaction waves in which both the left and right state have negative speed (sometimes called a slow wave or a back-wave) whereas the R_2 curve represents rarefaction waves in which both the left and right state have positive speed (fast waves or front-waves).

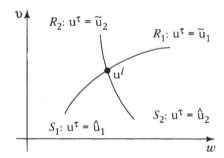

Figure 3.9. The slow curve $(S_1 \cup R_1)$ and the fast curve $(S_2 \cup R_2)$.

General solution

We now show how shocks and rarefaction waves can be put together to get a general solution for the Riemann problem. Our basic theorem is as follows.

Theorem 3.31. *Suppose that our system of conservation laws (3.5) is strictly hyperbolic and genuinely nonlinear in a region $\Omega \subset \mathbb{R}^n$ of state space. Then for any $\mathbf{u}^l \in \Omega$ there is a neighborhood $N \subset \Omega$ of $\mathbf{u}^l$ such that if $\mathbf{u}^r \in N$ there exists a weak solution of the Riemann problem (3.5), (3.80). This solution is composed of at most $n+1$ constant states separated by rarefaction waves and shocks satisfying the Lax shock condition.*

We will not prove this, but we show how the process works in the case of the p system. We start with the left state $\mathbf{u}^l$ and consider the two pairs of shock and rarefaction curves emanating from that point. It is best to think of these as being two C^1 curves: a "slow curve" consisting of the union of S_1 and R_1 (the slow shocks and the slow rarefaction waves) and a "fast curve" consisting of the union of S_2 and R_2. Of course, if the right state $\mathbf{u}^r$ in our Riemann problem lies on either of these curves, we can simply connect the left and right state with a single wave (slow or fast, shock or rarefaction), depending on which of the four original curves it lies on. The question remains, what happens if the right state lies in one of the four open regions cut out by our curves?

The solution is obtained by covering these four regions with a family of fast curves. Through each point $\bar{\mathbf{u}}$ on the slow curve through $\mathbf{u}^l$ we can construct the curves $\bar{S}_2$ and $\bar{R}_2$. These new curves will represent shocks and rarefaction waves, respectively, all with positive speed and all having $\bar{\mathbf{u}}$ as the left state. Taking the union of $\bar{S}_2$ and $\bar{R}_2$ gives us a family of curves (which we will call the "fast family") parameterized by the points $\bar{\mathbf{u}}$ on the original slow curve. It is left as an exercise (cf. Problem 3.11) to show that there is a neighborhood $\mathcal{N}$ of $\mathbf{u}^l$ that is covered univalently by the fast family; i.e., for any point $\mathbf{u}^r \in \mathcal{N}$ there is exactly one member of the fast family containing $\mathbf{u}^r$.

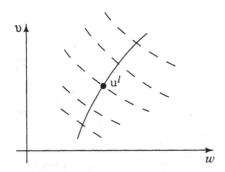

Figure 3.10. Slow curve and fast family.

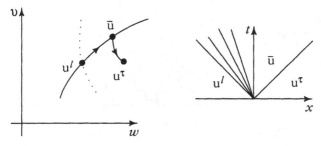

Figure 3.11. Slow rarefaction—fast shock.

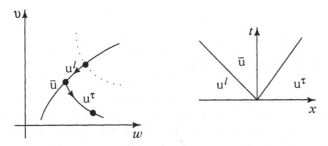

Figure 3.12. Slow shock—fast shock.

Now that we have used the left state to generate the slow curve and the fast family, the solution of the Riemann problem is simple. From any right state $\mathbf{u}^r \in \mathcal{N}$ we simply follow the appropriate member of the fast family back to a point $\bar{\mathbf{u}}$ on the slow curve. The point $\bar{\mathbf{u}}$ is now used as an intermediate state between two waves: a slow wave connecting $\mathbf{u}^l$ and $\bar{\mathbf{u}}$ and a fast wave connecting $\bar{\mathbf{u}}$ and $\mathbf{u}^r$. Of course, each of the two waves can be either a shock or rarefaction wave depending on which of the four regions defined by the original slow and fast curves the right state $\mathbf{u}^r$ lies in. The various possibilities are described in Figures 3.11-3.14.

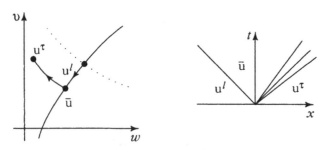

Figure 3.13. Slow shock—fast rarefaction.

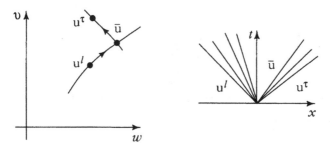

Figure 3.14. Slow rarefaction—fast rarefaction.

3.6   Other Selection Criteria

The Lax shock condition is not the only viable selection criterion used to pick the "physically reasonable" solution from among the possible weak solutions to systems of conservation laws. In this section we present some competing conditions and describe some of the relationships between them.

3.6.1   The Entropy Condition

The first alternative selection criterion we present is called the entropy condition. It is an outgrowth of the second law of thermodynamics, which is generalized in this situation to include physical systems other than mechanical and thermal. The key to the condition is the existence of an additional conservation law derived from (3.5).

Definition 3.32. An **entropy/entropy-flux pair**[1] is a pair of functions $(U, F) : \mathbb{R}^n \to \mathbb{R}^2$ satisfying

$$\nabla F = \nabla U \cdot \nabla \mathbf{f}. \tag{3.122}$$

It follows immediately from the definition and the chain rule that if $\mathbf{u}$ is a classical solution of (3.5), then

$$U(\mathbf{u})_t + F(\mathbf{u})_x = 0. \tag{3.123}$$

Of course, as we noted in Remark 3.23, a *weak* solution of (3.5) does not necessarily satisfy (3.123).

Definition 3.33. A weak solution of (3.5) is said to satisfy the **entropy condition** if there exists an entropy/entropy-flux pair with $\mathbf{u} \mapsto U(\mathbf{u})$ convex such that

$$-\int \int (U(\mathbf{u})\phi_t + F(\mathbf{u})\phi_x) \, dx \, dt \leq 0 \tag{3.124}$$

for every *non-negative* C^1 test function ϕ with compact support in the open upper half-plane.

Remark 3.34. An entropy/entropy-flux pair satisfying (3.124) is often described as satisfying the inequality

$$U(\mathbf{u})_t + F(\mathbf{u})_x \leq 0 \tag{3.125}$$

in the *sense of distributions*. Note that if an entropy/entropy-flux pair satisfies (3.125) in the classical sense, then we can multiply the inequality by a non-negative test function and derive (3.124) using Green's theorem in the plane.

Remark 3.35. If $\mathbf{u}$ satisfies (3.125) and has compact support, then by integrating (3.125) with respect to x we get

$$\int_{-\infty}^{\infty} U_t \, dx \leq 0. \tag{3.126}$$

Thus, the quantity

$$\int_{-\infty}^{\infty} U \, dx$$

(which, we repeat, is best thought of as a free energy, rather than an entropy) is a decreasing function of time. The idea that energy can only be lost can be thought of as a version of the second law of thermodynamics.

---

[1] This terminology is quite standard despite the fact that it contradicts any reasonable definitions from thermodynamics. Perhaps a better term would be free energy/energy-flux, but it seems hopeless to try to change things after so many years. Students of conservation laws should therefore resign themselves to forgetting everything they learned about thermodynamics while studying the subject. The culprit responsible for this terminology has not come forward to confess.

The following lemma translates the entropy condition into a jump condition for piecewise continuous weak solutions.

Lemma 3.36. *Suppose* **u** *is a piecewise continuous weak solution of (3.5) that satisfies the entropy condition. Suppose* **u** *has a jump discontinuity along a shock curve with slope* s. *Then*

$$s[U(\mathbf{u})] - [F(\mathbf{u})] \geq 0. \tag{3.127}$$

The proof of this is left to the reader (cf. Problem 3.8). The proof can be obtained by modifying the proof of the Rankine-Hugoniot condition.

The following theorem states that in the case of genuinely nonlinear systems, the entropy condition and the Lax shock condition are equivalent for weak shocks.

Theorem 3.37. *Suppose that the system of conservation laws (3.5) is strictly hyperbolic and genuinely nonlinear. In addition, suppose that there is an entropy/entropy-flux pair* (U, F) *with* U *strongly convex. Let* **u** *be a weak solution of (3.5), containing a shock with values* $\mathbf{u}^l$ *and* $\mathbf{u}^r$ *traveling with speed* s, *with* $\mathbf{u}^l - \mathbf{u}^r$ *sufficiently small. Then* **u** *satisfies the Lax shock condition at this shock if and only if the entropy jump condition (3.127) holds with strict inequality.*

Proof. Since **u** is a weak solution, $\mathbf{u}^l$, $\mathbf{u}^r$ and s must satisfy the Rankine-Hugoniot condition. Thus, by Theorem 3.27, $\mathbf{u}^r$ must lie along one of the n shock curves through $\mathbf{u}^l$; i.e., $\mathbf{u}^r = \hat{\mathbf{u}}_k(\epsilon)$ and $s = \hat{s}_k(\epsilon)$ for some $k = 1, 2, \ldots, n$ and some $\epsilon \in I_k$. Now, again by Theorem 3.27, the Lax shock condition is satisfied if and only if $\epsilon < 0$. Furthermore, the entropy jump condition holds if and only if

$$E(\epsilon) := \hat{s}_k(\epsilon)[U(\mathbf{u}^l) - U(\hat{\mathbf{u}}_k(\epsilon))] - [F(\mathbf{u}^l) - F(\hat{\mathbf{u}}_k(\epsilon))] < 0. \tag{3.128}$$

We now let a "prime" indicate differentiation with respect to ϵ and show $E(0) = E'(0) = E''(0) = 0$ and that $E'''(0) > 0$. This indicates that for ϵ sufficiently small, E is negative if and only if ϵ is negative and thus completes the proof. Using $\mathbf{u}^l = \hat{\mathbf{u}}_k(0)$ we get

$$E(0) = \hat{s}_k(0)[U(\mathbf{u}^l) - U(\hat{\mathbf{u}}_k(0))] - [F(\mathbf{u}^l) - F(\hat{\mathbf{u}}_k(0))] = 0. \tag{3.129}$$

We could use $\hat{s}_k(0) = \lambda_k$ and $\mathbf{u}_k'(0) = \mathbf{r}_k$ to calculate $E'(0)$ directly, but instead we differentiate the Rankine-Hugoniot condition to get

$$\hat{s}_k'(\epsilon)[\mathbf{u}^l - \hat{\mathbf{u}}_k(\epsilon)] - \hat{s}_k(\epsilon)\hat{\mathbf{u}}_k'(\epsilon) = -\mathbf{f}(\hat{\mathbf{u}}_k(\epsilon))'. \tag{3.130}$$

We now use $\nabla F = \nabla U \cdot \nabla \mathbf{f}$ to get

$$E'(\epsilon) = \hat{s}_k'(\epsilon)\{[U(\mathbf{u}^l) - U(\hat{\mathbf{u}}_k(\epsilon))] - \nabla U(\hat{\mathbf{u}}_k(\epsilon)) \cdot [\mathbf{u}^l - \hat{\mathbf{u}}_k(\epsilon)]\}, \tag{3.131}$$

from which it is easy to see that $E'(0) = 0$. Simply differentiating this gives us $E''(0) = 0$. The calculation of $E'''(0)$ contains many terms that go to zero in the same way as the terms of the preceding calculations, but one interesting term remains:

$$E'''(0) = \hat{s}'_k(0)[\hat{\mathbf{u}}'_k(0)^T \nabla^2 U(\hat{\mathbf{u}}_k(0))\hat{\mathbf{u}}'_k(0)] \qquad (3.132)$$

where $\nabla^2 U$ is the second gradient or Hessian matrix of U. Now, from Theorem 3.27 we have $\hat{s}'_k(0) = 1/2$ and since U is strictly convex its Hessian is positive definite. Thus $E''' > 0$, and the theorem is proved. $\qquad\square$

3.6.2 Viscosity Solutions

Another important selection criterion (whose physical significance is perhaps easier to understand) is the requirement that we accept only *viscosity solutions*.

Definition 3.38. We say that $\mathbf{u}$ is a **viscosity solution** of (3.5) if $\mathbf{u}$ can be obtained as the limit

$$\mathbf{u} = \lim_{\epsilon \to 0^+} \mathbf{u}^\epsilon \qquad (3.133)$$

of solutions of the parabolic system of differential equations

$$\mathbf{u}_t^\epsilon + \mathbf{f}(\mathbf{u}^\epsilon)_x = \epsilon \mathbf{A}\mathbf{u}_{xx}^\epsilon. \qquad (3.134)$$

for some positive definite matrix $\mathbf{A}$.

Remark 3.39. The reader should be wondering in what sense the limit in (3.133) is achieved. Well, we're not going to tell you yet. (All right, if you must know it's a weak-star limit in L^∞, but we're not going to explain this terminology until later chapters.) Suffice it to say that if $\mathbf{u}$ is a piecewise C^1 solution containing a single shock, the convergence is uniform off of any neighborhood containing the shock.

The rationale behind this choice of a selection criterion is that most conservation laws (again, gas dynamics being the system that we have foremost in mind) are simply approximate mathematical models of physical systems; and the "real" physical systems have some sort of dissipation effects like viscosity that are modeled by the $\mathbf{A}\mathbf{u}_{xx}$ term in (3.134). Of course, the question immediately comes up, "If (3.134) is the better model, why are we spending so much time solving the approximate conservation law (3.5)"? There are a few different answers to that question.

1. The viscosity effects embodied in the dissipation term are often very small and accordingly hard to measure. Thus, it is not easy to determine $\mathbf{A}$ or ϵ with any accuracy.

2. In a numerical implementation of (3.134) the small dissipation term is usually of no help in stabilizing the numerical algorithm.

3. There are reasonably efficient and accurate numerical methods of computing the solutions of (3.5), and there are analytical methods for determining simple discontinuous solutions.

Even if we accept the idea that we should continue to study hyperbolic conservation laws rather than parabolic systems, there are a few questions about viscosity solutions that remain unanswered.

1. Is there more than one viscosity solution? More precisely, how does the limit $\mathbf{u}$ depend on the choice of the matrix $\mathbf{A}$?

2. What is the relationship between the viscosity solution and the limit of other small higher order effects as the magnitude of the effect goes to zero? (For example, the third-order effect capillarity has been used in a manner similar to our use of viscosity.)

In short, should we question the notion that there should be a unique solution of a system of conservation laws? It seems that the current consensus is that uniqueness is required by the physics in most situations.

Our next theorem involves the relationship between viscosity solutions and solutions satisfying the entropy condition. Because of the vague nature of our definition of viscosity solutions, we will not be able to give a rigorous proof, but we do supply some formal justification.

Theorem 3.40. *For a system of conservation laws (3.5) for which there exists an entropy/entropy-flux pair* (U, F) *with convex entropy* U, *any viscosity solution also satisfies the entropy condition.*

Proof. We present here a plausibility argument rather than a proof. Although the arguments presented here cannot be justified without the tools of distribution theory and L^p spaces, they should give the reader an idea of why the theorem is true. In fact, a reader very familiar with the more advanced topics mentioned above would probably accept these arguments as sufficiently rigorous. For clarity, we consider only the case $\mathbf{A} = \mathbf{I}$; the generalization to other positive definite $\mathbf{A}$ is straightforward.

Multiplying (3.134) by ∇U and using (3.122) we get

$$
\begin{aligned}
U(\mathbf{u}^\epsilon)_t + F(\mathbf{u}^\epsilon)_x &= \nabla U \cdot \mathbf{u}_t^\epsilon + \nabla U^T \nabla \mathbf{f}\, \mathbf{u}_x^\epsilon \\
&= \epsilon \nabla U \cdot \mathbf{u}_{xx}^\epsilon \\
&= \epsilon (U_{xx} - (\mathbf{u}_x^\epsilon)^T \nabla^2 U \mathbf{u}_x^\epsilon)
\end{aligned}
$$

Using the convexity of U (which implies the positive definiteness of the Hessian matrix $\nabla^2 U$) we obtain

$$
U_t + F_x \leq \epsilon U_{xx}. \tag{3.135}
$$

The right-hand side goes to 0 (in the sense of distributions) as $\epsilon \to 0$, so we have (3.125). $\qquad \square$

3.6.3 Uniqueness

We have now discussed several selection criteria and noted some of the relationships between them. Our stated goal was to achieve some sort of uniqueness result. After all this work, are we in a position to do this? The answer, in general, is "no." Although the criteria we have suggested rule out the most obvious "physically unreasonable" weak solutions, the question of existence and uniqueness is, in general, open. At the time of this writing, this is a very active area of research. In the following, we summarize a number of results in special cases.

For the scalar conservation law with strongly convex f, the questions of existence and uniqueness are basically settled. For genuinely nonlinear systems, existence (but not uniqueness) is known for initial data of small total variation. For the p system, assuming strong convexity, much more is known. Solutions exist for arbitrary initial data, and uniqueness has been shown within the class of piecewise smooth solutions. We refer to [Sm] for a exposition.

There are many specialized results for other systems, e.g., those where genuine nonlinearity is violated in a specific fashion and for the system of gas dynamics. Existence results are usually based on finding estimates for approximated solutions and extracting convergent subsequences. Such approximate solutions usually come from finite difference schemes or, alternatively, from adding "viscosity" terms to the equations. Some of the main contributors to the field are Lax, Glimm, DiPerna, Tartar, Godunov, Liu, Smoller and Oleinik. Despite all of these efforts, general answers in this field have remained elusive. In fact, there are recent counterexamples where the usual admissibility conditions do not guarantee uniqueness [Se]. Of course, real world problems are usually in more than one space dimension. Almost everything is open for that situation.

Problems

3.1. Show that if $p' < 0$, then the p system is hyperbolic.

3.2. Give conditions on the constitutive functions ensuring that the two systems of gas dynamics equations are hyperbolic.

3.3. Prove Lemma 3.6.

3.4. Prove Lemma 3.17.

3.5. Prove Theorem 3.18.

3.6. Sketch a characteristic diagram and the wavefront for the set of solutions given by (3.68).

3.7. Let f be convex and let u be a piecewise smooth weak solution of (3.41) with a finite number of jumps. Show that if u is monotone decreasing as a function of x, then it satisfies the Lax shock condition at each discontinuity. Use an example to show that this is false if f is nonconvex.

3.8. Prove Lemma 3.36.

3.9. Show that the set V defined in (3.107) is a basis for $\mathbb{R}^n$. Hint: What is the relationship between $\mathbf{r}_i$ and $\mathbf{l}_j$ for $i \neq j$?

3.10. Show that the states defined by the curve defined in (3.101) and (3.102) with corresponding shock speeds satisfy (3.76) and (3.77), respectively. Hint: Use the convexity of p before taking square roots.

3.11. Show that the fast family covers a neighborhood of $\bar{\mathbf{u}}$ univalently.

3.12. Show that Eulerian and Lagrangian gas dynamics are equivalent for smooth solutions. What are the difficulties with weak solutions?

4

Maximum Principles

The maximum principle asserts that solutions of certain scalar elliptic equations of second order cannot have a maximum (or a minimum) in the interior of the domain where they are defined. The basic idea is quite simple. Consider, for simplicity, Laplace's equation $\Delta u = 0$. If u has a maximum at a point $\mathbf{x}$ and the second derivatives of u do not all vanish at $\mathbf{x}$, then Δu is negative at $\mathbf{x}$, in contradiction to the equation. The only case left to be ruled out is that of degenerate maxima where all second derivatives vanish. This is accomplished by an approximation argument which removes the degeneracy.

The maximum principle can be used to show that solutions of certain equations must be non-negative. This is important for quantities which have a physical interpretation as densities, concentrations, probabilities, etc. The maximum principle also leads to easy uniqueness results. In later chapters we shall see that in certain problems uniqueness also implies existence. The maximum principle itself can also be used to construct existence proofs. In the next section, we shall give Perron's existence proof for Dirichlet's problem. A very recent application of the maximum principle, too complicated to be discussed here, concerns "viscosity solutions" for Hamilton-Jacobi equations. In the third section of this chapter, we shall discuss a result of Gidas, Ni and Nirenberg [GNN], which asserts that positive solutions to certain elliptic boundary-value problems must be radially symmetric. The final section of the chapter is concerned with the extension of the maximum principle to parabolic equations.

4.1   Maximum Principles of Elliptic Problems

4.1.1   The Weak Maximum Principle

Throughout this section, we shall consider a second-order operator of the form

$$Lu = a_{ij}(\mathbf{x})\frac{\partial^2 u}{\partial x_i \partial x_j} + b_i(\mathbf{x})\frac{\partial u}{\partial x_i} + c(\mathbf{x})u. \tag{4.1}$$

The following assumptions are made throughout and will therefore not be stated with each theorem. Ω is a domain in $\mathbb{R}^n$. The coefficients a_{ij}, b_i and c are continuous on $\overline{\Omega}$, and u is in $C^2(\Omega) \cap C(\overline{\Omega})$. The matrix a_{ij} is symmetric and strictly positive definite at every point $\mathbf{x} \in \overline{\Omega}$, i.e., L is elliptic.

The weak maximum principle is expressed by the following theorem.

Theorem 4.1. *Assume that $Lu \geq 0$ (or, respectively, $Lu \leq 0$) in a bounded domain Ω and that $c(\mathbf{x}) = 0$ in Ω. Then the maximum (or, respectively, the minimum) of u is achieved on $\partial\Omega$.*

Proof. If $Lu > 0$ in Ω, then u cannot achieve its maximum anywhere in Ω. Suppose it did, say at the point $\mathbf{x}_0$. Then all first derivatives of u vanish at this point, and hence

$$Lu = a_{ij}\frac{\partial^2 u}{\partial x_i \partial x_j}. \tag{4.2}$$

But at a maximum the matrix of second partial derivatives is negative semidefinite and we conclude (see Problem 2) that $Lu(\mathbf{x}_0) \leq 0$, a contradiction.

For the general case, consider the function $u_\epsilon = u + \epsilon \exp(\gamma x_1)$. We find

$$Lu_\epsilon = Lu + \epsilon(\gamma^2 a_{11} + \gamma b_1)\exp(\gamma x_1). \tag{4.3}$$

We now choose γ large enough so that $\gamma^2 a_{11} + \gamma b_1 > 0$ throughout Ω (this is possible since a_{11} is positive and continuous on $\overline{\Omega}$). Then $Lu_\epsilon > 0$ for any positive ϵ. We conclude that

$$\max_{\overline{\Omega}} u_\epsilon = \max_{\partial\Omega} u_\epsilon. \tag{4.4}$$

The theorem follows by letting $\epsilon \to 0$. $\square$

Remark 4.2. For later use in connection with parabolic equations, we remark that the proof of Theorem 4.1 still works if the matrix a_{ij} is only positive semidefinite, as long as there is at least one vector $\boldsymbol{\xi}$ independent of $\mathbf{x} \in \overline{\Omega}$ such that $\xi_i a_{ij} \xi_j > 0$.

We have the following corollary of Theorem 4.1.

Corollary 4.3. *Let Ω be bounded and assume $c \leq 0$ in Ω. Let $Lu \geq 0$ (or, respectively, $Lu \leq 0$). Then*

$$\max_{\overline{\Omega}} u \leq \max_{\partial\Omega} u^+ \text{ (or, resp., } \min_{\overline{\Omega}} u \geq \min_{\partial\Omega} u^-). \tag{4.5}$$

Here, $u^+ = \max(u, 0)$, $u^- = \min(u, 0)$. In particular, if $Lu = 0$ in Ω, then

$$\max_{\overline{\Omega}} |u| = \max_{\partial\Omega} |u|. \tag{4.6}$$

Proof. If $u \leq 0$ throughout Ω, the corollary is trivially true. Hence we may assume that $\Omega^+ = \Omega \cap \{u > 0\} \neq \emptyset$. On Ω^+, we have $-cu \geq 0$, and hence

$$a_{ij} \frac{\partial^2 u}{\partial x_i \partial x_j} + b_i \frac{\partial u}{\partial x_i} \geq 0. \tag{4.7}$$

Hence the previous theorem implies that the maximum of u on the closure of Ω^+ is equal to its maximum on $\partial\Omega^+$. Since $u = 0$ on $\partial\Omega^+ \cap \Omega$, this maximum must be achieved on $\partial\Omega$. $\square$

The following corollary is typically used in applications. It yields a uniqueness result as well as a comparison principle.

Corollary 4.4. *Let Ω be bounded and $c \leq 0$. If $Lu = Lv$ in Ω and $u = v$ on $\partial\Omega$, then $u = v$ in Ω. If $Lu \leq Lv$ in Ω and $u \geq v$ on $\partial\Omega$, then $u \geq v$ in Ω.*

Remark 4.5. We draw the reader's attention to the particular case $v = 0$. The reader should also note the relationship between this result and the oscillation and comparison theorems of Sturm-Liouville theory in ODEs (cf. [In]).

We conclude this subsection with a definition.

Definition 4.6. Assume that $c \leq 0$. If $Lu \geq 0$ (or, resp., $Lu \leq 0$), then u is called a **subsolution (supersolution)** of the equation $Lu = 0$. Subsolutions of $\Delta u = 0$ are called **subharmonic**, and supersolutions are called **superharmonic**.

The terminology is motivated by Corollary 4.4. A subsolution is less than or equal to a solution with the same values on the boundary; a supersolution is greater than or equal to a solution.

4.1.2 The Strong Maximum Principle

Theorem 4.1 states that u assumes its maximum at the boundary. However, u may assume its maximum at many points, and therefore the theorem does not rule out the possibility that some of these points are interior. The strong maximum principle states that this is impossible, unless u is a constant. For the proof, we shall need the following lemma, which is interesting in its own right.

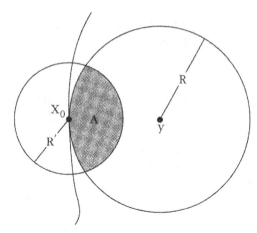

Figure 4.1.

Lemma 4.7. *Suppose that Ω lies on one side of $\partial\Omega$. Assume $Lu \geq 0$, and let $\mathbf{x}_0$ be a point on $\partial\Omega$ such that $u(\mathbf{x}_0) > u(\mathbf{x})$ for every $\mathbf{x} \in \Omega$. Also assume that, in a neighborhood of $\mathbf{x}_0$, $\partial\Omega$ is a C^2-surface and that u is differentiable at $\mathbf{x}_0$. Moreover, suppose that either*

1. $c = 0$,

2. $c \leq 0$ and $u(\mathbf{x}_0) \geq 0$, or

3. $u(\mathbf{x}_0) = 0$.

Then $\partial u/\partial n(\mathbf{x}_0) > 0$, where $\partial u/\partial n$ denotes the derivative in the direction of the outer normal to $\partial\Omega$.

Proof. Since $\partial\Omega$ was assumed C^2, we can choose (see Problem 4.5) a ball $B_R(\mathbf{y})$ such that $B_R(\mathbf{y}) \subset \Omega$ and $\mathbf{x}_0 \in \partial B_R(\mathbf{y})$. Here R and $\mathbf{y}$ denote the radius and center of the ball.

For $0 \leq r = |\mathbf{x} - \mathbf{y}| \leq R$, define

$$v(\mathbf{x}) = \exp(-\alpha r^2) - \exp(-\alpha R^2). \tag{4.8}$$

We find

$$Lv(\mathbf{x}) = \exp(-\alpha r^2)\left[4\alpha^2 a_{ij}(x_i - y_i)(x_j - y_j) - 2\alpha(a_{ii} + b_i(x_i - y_i))\right] + cv. \tag{4.9}$$

Now let $A = B_R(\mathbf{y}) \cap B_{R'}(\mathbf{x}_0)$, with R' chosen small. For large enough α, we have $Lv > 0$ in A. Moreover, if we choose $\epsilon > 0$ small enough, then $u - u(\mathbf{x}_0) + \epsilon v \leq 0$ on $\partial A \cap \partial B_{R'}(\mathbf{x}_0)$, and also on $\partial A \cap \partial B_R(\mathbf{y})$, where $v = 0$. Thus we find $L(u - u(\mathbf{x}_0) + \epsilon v) \geq -cu(\mathbf{x}_0) \geq 0$ in A and $u - u(\mathbf{x}_0) + \epsilon v \leq 0$ on ∂A.

If $c \leq 0$, the weak maximum principle (Corollary 4.3) implies that $u - u(\mathbf{x}_0) + \epsilon v \leq 0$ throughout A. We take the normal derivative at $\mathbf{x}_0$, and obtain

$$\frac{\partial u}{\partial n}(\mathbf{x_0}) \geq -\epsilon \frac{\partial v}{\partial n}(\mathbf{x_0}) = 2\alpha\epsilon R \exp(-\alpha R^2) > 0, \qquad (4.10)$$

which implies the lemma.

If $u(\mathbf{x_0}) = 0$, then, by assumption, u is negative in Ω. Now let $c^+(\mathbf{x}) = \max(0, c(\mathbf{x}))$. We find that $(L - c^+)u = Lu - c^+u \geq Lu \geq 0$, and hence we can apply the argument above with $L - c^+$ in place of L. □

Remark 4.8. Since Ω is assumed to be connected, it can be shown that Ω is on one side of $\partial\Omega$ if Ω is bounded and $\partial\Omega$ is globally smooth. (This is a multidimensional generalization of the Jordan curve theorem.) For a proof of this see, e.g., [Mas].

Remark 4.9. Lemma 4.7 still holds if the matrix a_{ij} is only positive semi-definite and $\mathbf{n}$ is not in the nullspace.

As a consequence of Lemma 4.7, we obtain the following strong maximum principle.

Theorem 4.10. *Assume $Lu \geq 0$ ($Lu \leq 0$) in Ω (not necessarily bounded) and assume that u is not constant. If $c = 0$, then u does not achieve its maximum (minimum) in the interior of Ω. If $c \leq 0$, u cannot achieve a non-negative maximum (non-positive minimum) in the interior. Regardless of the sign of c, u cannot be zero at an interior maximum (minimum).*

Proof. Assume that u assumes its maximum M at an interior point and let $\Omega^- = \Omega \cap \{u < M\}$. If Ω^- is not empty, then $\partial\Omega^- \cap \Omega$ is not empty. Let $\mathbf{y}$ be a point in Ω^- that is closer to $\partial\Omega^-$ than to $\partial\Omega$ and let B be the largest ball contained in Ω^- and centered at $\mathbf{y}$. Let $\mathbf{x_0}$ be a point on $\partial B \cap \partial\Omega^-$. Then we can apply the previous lemma to B. We conclude that ∇u is nonzero at $\mathbf{x_0}$, contradicting the assumption that u assumes its maximum there. □

4.1.3   A Priori Bounds

The maximum principle can be used to obtain pointwise estimates for solutions of $Lu = f$ in bounded domains. To state these bounds, we introduce the following quantities:

$$\lambda(\mathbf{x}) = \min_{\boldsymbol{\xi} \in \mathbb{R}^n \setminus \{0\}} \frac{a_{ij}(\mathbf{x})\xi_i\xi_j}{|\boldsymbol{\xi}|^2}, \qquad \beta = \max_{\mathbf{x} \in \overline{\Omega}} \frac{|\mathbf{b}(\mathbf{x})|}{\lambda(\mathbf{x})}. \qquad (4.11)$$

We have the following result.

Theorem 4.11. *Assume that Ω is bounded and contained in the strip between two parallel planes of distance d. Assume also that $c \leq 0$. If $Lu = f$, then*

$$\max_{\overline{\Omega}} |u| \leq \max_{\partial\Omega} |u| + C \max_{\overline{\Omega}} \frac{|f|}{\lambda}. \qquad (4.12)$$

Here $C = \exp((\beta + 1)d) - 1$.

Proof. We shall assume $Lu \geq f$ and prove that

$$\max_{\overline{\Omega}} u \leq \max_{\partial \Omega} u^+ + C \max_{\overline{\Omega}} \frac{|f^-|}{\lambda}. \qquad (4.13)$$

The theorem follows by applying this inequality to both u and $-u$. Since λ and β do not change under rotations of the coordinate system, we may assume that Ω is contained in the strip $0 < x_1 < d$. We set $L_0 = L - c$. For $\alpha \geq \beta + 1$, we have

$$L_0 \exp(\alpha x_1) = (\alpha^2 a_{11} + \alpha b_1) \exp(\alpha x_1) \geq \lambda(\alpha^2 - \alpha\beta) \exp(\alpha x_1) \geq \lambda. \quad (4.14)$$

Let

$$v = \max_{\partial \Omega} u^+ + (\exp(\alpha d) - \exp(\alpha x_1)) \max_{\overline{\Omega}} \frac{|f^-|}{\lambda}. \qquad (4.15)$$

Then obviously $v \geq u$ on $\partial \Omega$, and $v \geq 0$ on Ω. We compute

$$L(v - u) = L_0 v + cv - Lu \leq -\lambda \max_{\overline{\Omega}} \frac{|f^-|}{\lambda} - f \leq 0. \qquad (4.16)$$

Corollary 4.4 shows that $u \leq v$, which yields the desired result for $C = \exp(\alpha d) - 1$. $\qquad\qquad\qquad\qquad\qquad\qquad\qquad\qquad\qquad\qquad\qquad\square$

The following corollary shows that in certain cases it is possible to dispense with the requirement that $c \leq 0$.

Corollary 4.12. *Let* $Lu = f$ *in a bounded domain* Ω. *Let* C *be as in Theorem 4.11 and assume that*

$$C^* = 1 - C \max_{\overline{\Omega}} \frac{c^+}{\lambda} > 0. \qquad (4.17)$$

Then

$$\max_{\overline{\Omega}} |u| \leq \frac{1}{C^*} \left(\max_{\partial \Omega} |u| + C \max_{\overline{\Omega}} \frac{|f|}{\lambda} \right). \qquad (4.18)$$

The proof follows by applying the result of the previous theorem to the equation $(L_0 + c^-)u = f - c^+ u$.

Problems

4.1. Let u be a solution of $\Delta u = u^3 - u$ on a bounded domain Ω. Assume that $u = 0$ on $\partial \Omega$. Show that $u \in [-1, 1]$ throughout Ω. Can the values ± 1 be achieved?

4.2. Assume that the $n \times n$ matrices $\mathbf{A}$ and $\mathbf{B}$ are symmetric and positive semidefinite. Show that tr $(\mathbf{AB}) \geq 0$. Hint: $\mathbf{B} = \sum_k \lambda_k \mathbf{q}_k \mathbf{q}_k^T$, where the λ_k and $\mathbf{q}_k$ are the eigenvalues and eigenvectors of $\mathbf{B}$.

4.3. Give a counterexample showing that Corollary 4.3 does not hold if $c > 0$.

4.4. Show that Corollary 4.4 fails if Ω is unbounded. Hint: Consider the problem $\Delta u = 0$, $u = 0$ on $\partial\Omega$ when Ω is a strip bounded by parallel planes.

4.5. If $\partial\Omega$ is of class C^2 and $\mathbf{x}_0$ is on $\partial\Omega$, show that there is a ball lying entirely in Ω with $\mathbf{x}_0$ on its boundary.

4.6. (a) On the bounded domain Ω with smooth boundary, let u be a solution of the problem

$$\Delta u + a_i(\mathbf{x})\frac{\partial u}{\partial x_i} = f(\mathbf{x}), \quad \frac{\partial u}{\partial n} = 0 \text{ on } \partial\Omega. \tag{4.19}$$

Assume that $f \geq 0$ in Ω. Show that u is a constant and $f = 0$.
(b) Show that problem (4.19) can have a solution only if

$$\int_\Omega f(\mathbf{x})v(\mathbf{x})\,d\mathbf{x} = 0 \tag{4.20}$$

for every solution v of the "adjoint" equation

$$\Delta v - \frac{\partial}{\partial x_i}\left(a_i(\mathbf{x})v\right) = 0, \quad \frac{\partial v}{\partial n} - a_i(\mathbf{x})n_i v = 0 \text{ on } \partial\Omega. \tag{4.21}$$

(c) Using techniques to be developed in later chapters, one can show that the condition (4.20) is also sufficient and that the solution space of (4.21) is one-dimensional. Taking these facts for granted, show that solutions of (4.21) are either non-negative or non-positive. Equations of the form (4.21) are called Fokker-Planck equations and arise in statistical physics. Only non-negative solutions are physically meaningful.

4.7. Let Ω be a regular hexagon with side a. Let $\lambda \in \mathbb{R}$ be such that the equation $\Delta u + \lambda u = 0$ with boundary condition $u = 0$ has a nontrivial solution in Ω. Give a lower bound for λ.

4.2 An Existence Proof for the Dirichlet Problem

In this section, we shall establish existence of solutions for the Dirichlet problem. Specifically, we shall prove the following theorem:

Theorem 4.13. *Let Ω be a bounded domain in $\mathbb{R}^n$ with a C^2-boundary. Then, for any function $g \in C(\partial\Omega)$, there is a unique $u \in C^2(\Omega) \cap C(\overline{\Omega})$ satisfying $\Delta u = 0$ in Ω and $u = g$ on $\partial\Omega$.*

It will be evident from the proof that the assumption that $\partial\Omega$ is of class C^2 can be relaxed; for example, all convex domains are permissible.
The proof will be based on the ideas of Perron, which make use of the following notions. We call v a subsolution (supersolution) if $\Delta v \geq 0$ ($\Delta v \leq$

0) in Ω and $v \le g$ ($v \ge g$) on $\partial\Omega$. Obviously subsolutions exist, e.g., every sufficiently small constant is a subsolution. The maximum principle (weak form) shows that if u is a solution, then $v \le u$ for every subsolution (this is why we call them subsolutions). Thus, the pointwise supremum of all subsolutions (which is a well defined function) gives us an obvious candidate for a solution. If we can show that this function actually solves the problem, our existence proof will be complete. Before we prove this, we first need to develop a number of prerequisites.

4.2.1 The Dirichlet Problem on a Ball

The most old-fashioned way to prove existence of solutions is to give a formula for them. (We discussed some elementary methods for doing this in Section 1.2.1.) We now obtain an explicit solution for the Dirichlet problem on a ball.

Theorem 4.14. *Let B be the ball of radius R centered at the origin and let g be continuous on ∂B. Then the function*

$$u(\mathbf{x}) = \frac{R^2 - |\mathbf{x}|^2}{n\omega_n R} \int_{\partial B} \frac{g(\mathbf{y})}{|\mathbf{x} - \mathbf{y}|^n} \, dS_\mathbf{y} \tag{4.22}$$

is of class $C^2(B)$ and satisfies $\Delta u = 0$. Moreover, for every $\mathbf{y} \in \partial B$, we have

$$\lim_{\mathbf{x} \to \mathbf{y}} u(\mathbf{x}) = g(\mathbf{y}). \tag{4.23}$$

Here ω_n denotes the volume of the unit ball in $\mathbb{R}^n$, and the notation $dS_\mathbf{y}$ in the surface integral indicates the variable of integration.

Equation (4.22) is known as Poisson's formula. We note that the special case $\mathbf{x} = \mathbf{0}$ leads to the well known mean-value property: The value of a harmonic function at a point is equal to its average on any ball centered at that point.

Proof. Since we can differentiate under the integral, u is in fact of class C^∞ in B and it is a simple calculation to show that it is harmonic. It remains to establish (4.23). Let

$$K(\mathbf{x}, \mathbf{y}) = \frac{R^2 - |\mathbf{x}|^2}{n\omega_n R |\mathbf{x} - \mathbf{y}|^n}, \quad \mathbf{x} \in B, \ \mathbf{y} \in \partial B, \tag{4.24}$$

and

$$\psi(\mathbf{x}) = \int_{\partial B} K(\mathbf{x}, \mathbf{y}) \, dS_\mathbf{y}. \tag{4.25}$$

Obviously, (4.25) is just the special case $g = 1$ in (4.22); hence ψ is harmonic. It is also obvious that ψ is a radially symmetric function. For radially symmetric functions, Laplace's equation reads

$$u_{rr} + \frac{n-1}{r} u_r = 0, \tag{4.26}$$

and the only solutions of this equation which are regular at the origin are constants. Hence $\psi(\mathbf{x}) = \psi(\mathbf{0}) = 1$.

Now let $\mathbf{x}_0 \in \partial B$ and let $\epsilon > 0$ be given. Choose $\delta > 0$ so that $|g(\mathbf{y}) - g(\mathbf{x}_0)| < \epsilon$ for $|\mathbf{y} - \mathbf{x}_0| < \delta$ and let M be an upper bound for g on ∂B. For $|\mathbf{x} - \mathbf{x}_0| < \delta/2$, we have

$$
\begin{aligned}
|u(\mathbf{x}) - g(\mathbf{x}_0)| &= \left| \int_{\partial B} K(\mathbf{x}, \mathbf{y})(g(\mathbf{y}) - g(\mathbf{x}_0)) \, dS_{\mathbf{y}} \right| \\
&\leq \int_{|\mathbf{y} - \mathbf{x}_0| \leq \delta} K(\mathbf{x}, \mathbf{y}) |g(\mathbf{y}) - g(\mathbf{x}_0)| \, dS_{\mathbf{y}} \\
&\quad + \int_{|\mathbf{y} - \mathbf{x}_0| \geq \delta} K(\mathbf{x}, \mathbf{y}) |g(\mathbf{y}) - g(\mathbf{x}_0)| \, dS_{\mathbf{y}} \\
&\leq \epsilon + \frac{2M(R^2 - |\mathbf{x}|^2) R^{n-2}}{(\delta/2)^n}.
\end{aligned}
\tag{4.27}
$$

As $\mathbf{x} \to \mathbf{x}_0$, the last term on the right-hand side tends to zero and the theorem follows. $\qquad\square$

4.2.2 Subharmonic Functions

We shall need a notion of subsolutions to the Dirichlet problem which does not require them to be of class $C^2(\Omega)$. The definition is motivated by the maximum principle.

Definition 4.15. A function u in $C^0(\overline{\Omega})$ is called **subharmonic (superharmonic)**, if for every ball B with $\overline{B} \subset \Omega$ and every function $h \in C(\overline{B})$ with h harmonic in B and $u \leq h$ ($u \geq h$) on ∂B, we have $u \leq h$ ($u \geq h$) in B. A subsolution (supersolution) of the Dirichlet problem is a function $u \in C(\overline{\Omega})$ which is subharmonic (superharmonic) and such that $u \leq g$ ($u \geq g$) on $\partial \Omega$.

Clearly, if $\Delta u \geq 0$, then u is also subharmonic in the sense of the new definition. We note the following properties:

1. The strong maximum principle holds, i.e., if u is subharmonic and v is superharmonic with $v \geq u$ on $\partial \Omega$, then either $v > u$ in Ω or $v = u$ everywhere. We prove this by contradiction. Assume that $u - v$ assumes its maximum M at some point $\mathbf{x}_0 \in \Omega$, where $M \geq 0$. If $u - v = M$ throughout Ω, it follows that $u = v$; hence we may assume that there are points in Ω where $u - v \neq M$. In that case, we can choose $\mathbf{x}_0$ in such a way that there is a ball $B \subset \Omega$ centered at $\mathbf{x}_0$ such that $u - v$ does not equal M on all of ∂B. Let $\bar{u}$ and $\bar{v}$ denote

the harmonic functions on B which are equal to u and v, respectively, on ∂B. We find

$$M = (u - v)(\mathbf{x}_0) \leq (\bar{u} - \bar{v})(\mathbf{x}_0), \qquad (4.28)$$

and the right-hand side is strictly less than M by the strong maximum principle for harmonic functions. Hence we have a contradiction.

An immediate consequence is that every subsolution for the Dirichlet problem is less than or equal to every supersolution.

2. Let u be subharmonic in Ω and let B be a ball with $\overline{B} \subset \Omega$. Let $\bar{u}$ be the harmonic function on B satisfying $\bar{u} = u$ on ∂B. Then the function

$$U(\mathbf{x}) = \begin{cases} \bar{u}(\mathbf{x}), & x \in B \\ u(\mathbf{x}), & x \in \Omega \backslash B \end{cases} \qquad (4.29)$$

is also subharmonic in Ω (cf. Problem 4.9). U is called the harmonic lifting of u with respect to B.

3. If $u_1, u_2, \ldots, u_N$ are subharmonic, then $\max\{u_1, u_2, \ldots, u_N\}$ is also subharmonic.

4.2.3   The Arzela-Ascoli Theorem

The Arzela-Ascoli theorem states that sequences of functions on a compact set which satisfy certain conditions have uniformly convergent subsequences. Results of this nature are often useful in existence proofs; the thing which must be proved to exist is the limit of the convergent subsequence. To state the theorem, we need the following definition.

Definition 4.16. Let f_m be a sequence of real-valued functions defined in a subset D of $\mathbb{R}^n$. Let $\mathbf{x} \in D$. The sequence is called **equicontinuous** at $\mathbf{x}$ if, for every $\epsilon > 0$, there is a $\delta > 0$, independent of m, such that $|f_m(\mathbf{y}) - f_m(\mathbf{x})| < \epsilon$ for $\mathbf{y} \in D$ with $|\mathbf{y} - \mathbf{x}| < \delta$.

If the sequence f_m is equicontinuous at each point of a compact set S, it is uniformly equicontinuous, i.e., δ in the definition above can be chosen independently of $\mathbf{x} \in S$ (cf. Problem 4.11; it is not necessary that $D = S$). We note that a sequence of functions is equicontinuous at $\mathbf{x}$ if there exists a bound (independent of m) for the derivatives in some neighborhood of $\mathbf{x}$.

Theorem 4.17 (Arzela-Ascoli). *Let f_m be a sequence of real-valued functions defined on a compact subset S of $\mathbb{R}^n$. Assume that there is a constant M such that $|f_m(\mathbf{x})| \leq M$ for every $m \in \mathbb{N}$ and every $\mathbf{x} \in S$. Moreover, assume that the sequence f_m is equicontinuous at every point of S. Then there exists a subsequence which converges uniformly on S.*

We remark that (with rather obvious modifications) the theorem and proof can be extended to the case where S is a compact set in an abstract topological space and the values of f_m are in a complete metric space.

Proof. Let $\mathbf{x}_i$, $i \in \mathbb{N}$ be a sequence of points that is dense in S. The sequence $f_m(\mathbf{x}_1)$ is bounded; hence it has a convergent subsequence. That is, we can choose a sequence m_{1j} such that $f_{m_{1j}}(\mathbf{x}_1)$ converges as $j \to \infty$. Similarly, we can choose a subsequence m_{2j} of the sequence m_{1j} such that $f_{m_{2j}}(\mathbf{x}_2)$ converges. Since m_{2j} is a subsequence of m_{1j}, $f_{m_{2j}}(\mathbf{x}_1)$ converges as well. Next, we choose a subsequence m_{3j} of the sequence m_{2j} such that $f_{m_{3j}}$ converges also at $\mathbf{x}_3$. We proceed in this manner ad infinitum. Finally, consider the "diagonal" sequence $f_{m_{jj}}$. Except for the first $i-1$ terms, m_{jj} is a subsequence of m_{ij}; hence $f_{m_{jj}}(\mathbf{x}_i)$ converges for every $i \in \mathbb{N}$. To simplify notation, we shall set $g_j = f_{m_{jj}}$ in the following.

To conclude the proof, we show that the sequence g_m is uniformly Cauchy. Let $\epsilon > 0$ be given. The g_m, being a subsequence of the f_m, are uniformly equicontinuous on S; hence there is a $\delta > 0$ such that $|g_m(\mathbf{y}) - g_m(\mathbf{x})| < \epsilon/3$ whenever $|\mathbf{y} - \mathbf{x}| < \delta$. Since S is compact, there is a $K \in \mathbb{N}$ such that for every $\mathbf{x} \in S$ there exists $i \in \{1, \ldots, K\}$ with $|\mathbf{x}_i - \mathbf{x}| < \delta$. Now choose N large enough so that $|g_m(\mathbf{x}_i) - g_k(\mathbf{x}_i)| < \epsilon/3$ for $m, k > N$ and every $i \in \{1, \ldots, K\}$. For $m, k > N$ and arbitrary $\mathbf{x} \in S$, we now have

$$
\begin{aligned}
|g_m(\mathbf{x}) - g_k(\mathbf{x})| \leq & |g_m(\mathbf{x}) - g_m(\mathbf{x}_i)| \\
& + |g_m(\mathbf{x}_i) - g_k(\mathbf{x}_i)| + |g_k(\mathbf{x}_i) - g_k(\mathbf{x})| \qquad (4.30) \\
< & \epsilon,
\end{aligned}
$$

for some $i \in \{1, \ldots, K\}$. $\qquad\square$

Below, we shall have to apply the Arzela-Ascoli theorem to sequences of harmonic functions. In this particular case, we have the following result.

Theorem 4.18. *Let Ω be a domain in $\mathbb{R}^n$. Let f_m be a sequence of harmonic functions on Ω which is uniformly bounded, i.e., $|f_m(\mathbf{x})| \leq M$ for every $\mathbf{x} \in \Omega$ and $m \in \mathbb{N}$. Then f_m has a subsequence which converges to a harmonic function on Ω, uniformly on compact subsets of Ω.*

Proof. Let $\Omega_k = \{\mathbf{x} \in \Omega \mid |\mathbf{x}| \leq k,\ \text{dist}(\mathbf{x}, \partial\Omega) \geq 1/k\}$. Then Ω_k is compact and $\Omega = \cup_{k=1}^{\infty} \Omega_k$. If $\mathbf{x}$ is in Ω_k, then, in a neighborhood of $\mathbf{x}$, we can represent f_m by Poisson's formula, using a ball centered at $\mathbf{x}$ with radius $1/2k$. This yields uniform estimates for the derivatives of f_m on Ω_k. In particular, the f_m are equicontinuous on Ω_k. By the Arzela-Ascoli theorem, we can extract a uniformly convergent subsequence on each Ω_k, and using a diagonal argument similar to that used in the proof above, we obtain a subsequence converging on all of Ω, uniformly on each Ω_k.

It remains to show that the limit of this subsequence is harmonic. To see this, we simply note that harmonic functions are characterized by the property that they obey Poisson's formula on every ball. If Poisson's formula is obeyed by all functions in a uniformly convergent sequence, it also holds for the limit.   □

4.2.4 Proof of Theorem 4.13

We are now ready to prove our main result. Let S_g be the set of all subsolutions to the Dirichlet problem, and let

$$u(\mathbf{x}) = \sup_{v \in S_g} v(\mathbf{x}). \tag{4.31}$$

Since every subsolution is less than or equal to every supersolution and both subsolutions and supersolutions exist, the supremum in (4.31) is finite, and u is well defined.

Lemma 4.19. *The function u is harmonic in Ω.*

Proof. Let $\mathbf{x}$ be an arbitrary point in Ω and let v_m be a sequence in S_g such that $v_m(\mathbf{x}) \to u(\mathbf{x})$. Clearly v_m is bounded from above (by any supersolution), we may also assume it is bounded from below; if necessary, replace v_m by $\max(v_m, v_0)$, where v_0 is any subsolution. Choose R such that the closure of $B = B_R(\mathbf{x})$ is contained in Ω and let V_m be the harmonic lifting of v_m with respect to B. Then $V_m(\mathbf{x}) \to u(\mathbf{x})$ and, by Theorem 4.18, V_m has a subsequence V_{m_k} which converges on B to a harmonic function v; the convergence is uniform on every compact subset of B. Clearly $v \leq u$ in B. We shall now show that in fact $v = u$ in B. This clearly implies the lemma.

Assume on the contrary that $v(\mathbf{y}) < u(\mathbf{y})$ for some $\mathbf{y} \in B$. Then there exists a function $W \in S_g$ such that $v(\mathbf{y}) < W(\mathbf{y})$. Let $w_k = \max(W, V_{m_k})$ and let W_k be the harmonic lifting of w_k with respect to B. As before, a subsequence of W_k converges to a function w, harmonic in B. Then clearly $v \leq w$ on B and $v(\mathbf{x}) = w(\mathbf{x})$. By the maximum principle, this implies $v = w$ on B, contradicting the choice of W.   □

It remains to be shown that u is actually continuous up to $\partial\Omega$ and assumes the given boundary data. Let $\boldsymbol{\xi}$ be a point on $\partial\Omega$. Since $\partial\Omega$ was assumed of class C^2, there is a ball $B = B_R(\mathbf{y})$ in the exterior of Ω such that $\overline{B} \cap \overline{\Omega} = \{\boldsymbol{\xi}\}$. We now define

$$w(\mathbf{x}) = \begin{cases} R^{2-n} - |\mathbf{x} - \mathbf{y}|^{2-n} & n \geq 3 \\ \log \frac{|\mathbf{x}-\mathbf{y}|}{R} & n = 2. \end{cases} \tag{4.32}$$

Then

1. w is superharmonic (actually harmonic) in Ω.

2. $w > 0$ on $\overline{\Omega} \backslash \{\boldsymbol{\xi}\}$, $w(\boldsymbol{\xi}) = 0$.

A continuous function with properties (1) and (2) is called a *barrier* at $\boldsymbol{\xi}$ relative to Ω. We have the following result.

Lemma 4.20. *Let u be the harmonic function in Ω constructed above and let $\boldsymbol{\xi} \in \partial\Omega$. Then $u(\mathbf{x}) \to g(\boldsymbol{\xi})$ as $\mathbf{x} \to \boldsymbol{\xi}$.*

Proof. Choose $\epsilon > 0$ and let $M = \max_{\partial\Omega} |g|$. Let w be a barrier at $\boldsymbol{\xi}$ and let δ and k be such that $|g(\mathbf{x}) - g(\boldsymbol{\xi})| < \epsilon$ for $\mathbf{x} \in \partial\Omega$, $|\mathbf{x} - \boldsymbol{\xi}| < \delta$ and $kw(\mathbf{x}) \geq 2M$ for $\mathbf{x} \in \Omega$, $|\mathbf{x} - \boldsymbol{\xi}| \geq \delta$. The functions $g(\boldsymbol{\xi}) + \epsilon + kw(\mathbf{x})$ and $g(\boldsymbol{\xi}) - \epsilon - kw(\mathbf{x})$ are, respectively, a supersolution and subsolution, and hence

$$g(\boldsymbol{\xi}) - \epsilon - kw(\mathbf{x}) \leq u(\mathbf{x}) \leq g(\boldsymbol{\xi}) + \epsilon + kw(\mathbf{x}). \qquad (4.33)$$

Since $w(\mathbf{x}) \to 0$ as $\mathbf{x} \to \boldsymbol{\xi}$, the lemma is immediate. $\qquad\square$

It is clear from the proof that the smoothness assumption on $\partial\Omega$ can be relaxed as long as barriers can be constructed. The above construction works for all convex domains. In other cases, alternative barriers are often available, see, e.g., Problem 4.14.

Problems

4.8. Verify that the function given by (4.22) is harmonic.

4.9. Prove the second claim in the subsection on subharmonic functions.

4.10. Show that if u is of class C^2 and subharmonic in the sense defined in this section, then $\Delta u \geq 0$.

4.11. Let the sequence f_m be equicontinuous at each point of a compact set S. Show that it is uniformly equicontinuous on S.

4.12. Let $\Omega = \{(x, y) \in \mathbb{R}^2 \mid 0 < x^2 + y^2 < 1\}$. Prove that there is no solution to the Dirichlet problem $\Delta u = 0$ in Ω, $u(\mathbf{x}) = 1$ for $x^2 + y^2 = 1$, $u(0) = 0$. Hint: Show first that if there is a solution, then there is also a radially symmetric solution.

4.13. A function w is called a local barrier at $\boldsymbol{\xi}$ relative to Ω if the conditions (i) and (ii) in the definition hold only in some neighborhood N of $\boldsymbol{\xi}$. Show that if there is a local barrier, there is a barrier. Hint: Let B be a small ball centered at $\boldsymbol{\xi}$ and $m = \inf_{N \setminus B} w$. Define $\bar{w} = \min(m, w)$ on B and $\bar{w} = m$ otherwise.

4.14. Let Ω be a domain in $\mathbb{R}^2$ with the origin on its boundary. Let r and θ denote polar coordinates. Assume that a single-valued branch of θ exists in $\Omega \cap N$, where N is a neighborhood of the origin. Show that a local barrier exists at the origin. Hint: Consider the analytic function $1/\log z$.

4.15. Why does the proof of Theorem 4.13 not work for unbounded domains?

4.16. Show that $\omega_n = \pi^{n/2}/\Gamma(\frac{n}{2}+1)$. Hint: Evaluate $\int_{\mathbb{R}^n} \exp(-|\mathbf{x}|^2)\, d\mathbf{x}$ in both Cartesian and polar coordinates. By comparing the two expressions, you find a formula for $\Omega_n = n\omega_n$, the surface area of the unit sphere.

4.3 Radial Symmetry

In a 1979 paper, Gidas, Ni and Nirenberg [GNN] establish radial symmetry of positive solutions to certain nonlinear elliptic equations. The technique is based on the maximum principle. In this section, we shall consider the simplest case of their result, which is the following theorem.

Theorem 4.21. *Let* $f : \mathbb{R} \to \mathbb{R}$ *be of class* C^1. *In the ball* $B = B_R(0) \subset \mathbb{R}^n$, *let* $u > 0$ *be a solution (of class* $C^2(\overline{B})$*) of the equation*

$$\Delta u + f(u) = 0, \tag{4.34}$$

satisfying the boundary condition

$$u = 0 \quad \text{on } \partial B. \tag{4.35}$$

Then u *is radially symmetric and strictly monotone decreasing, i.e.,*

$$\frac{\partial u}{\partial r} < 0 \quad \text{for } 0 < r < R. \tag{4.36}$$

The remarkable fact is that the result holds regardless of what f is.

4.3.1 Two Auxiliary Lemmas

Throughout the rest of this section, let B and u be as given by the assumptions of Theorem 4.21.

Lemma 4.22. *Let* $\mathbf{x}_0 \in \partial B \cap \{x_1 > 0\}$. *Then, for some* $\delta > 0$, *we have* $\partial u/\partial x_1 < 0$ *in* $B \cap \{|\mathbf{x} - \mathbf{x}_0| < \delta\}$.

Proof. To simplify notation, we shall write u_1 for $\partial u/\partial x_1$, u_{11} for $\partial^2 u/\partial x_1^2$, etc.

If the lemma were false, there would be a sequence of points $\mathbf{x}_j \in B$ with $u_1(\mathbf{x}_j) \geq 0$ and $\mathbf{x}_j \to \mathbf{x}_0$. On the other hand, since $u > 0$ in B and $u = 0$ on the boundary, we must have $u_1(\mathbf{x}_0) \leq 0$. Hence, by continuity, we find $u_1(\mathbf{x}_0) = 0$. Note that since $u \equiv 0$ on ∂B the tangential derivatives are all zero. Since the x_1 direction is out of the tangent plane, we have shown that $\nabla u(\mathbf{x}_0) = \mathbf{0}$.

We claim that the second derivative $u_{11}(\mathbf{x}_0)$ is also zero. First, since $u(\mathbf{x}_0) = u_1(\mathbf{x}_0) = 0$, and $u > 0$ in B, we must have $u_{11}(\mathbf{x}_0) \geq 0$. Assume now that $u_{11}(\mathbf{x}_0) > 0$. Then u_{11} is also positive in a neighborhood N of $\mathbf{x}_0$. Consider now the straight line segment Γ in the positive x_1 direction connecting $\mathbf{x}_j$ to a point $\mathbf{y}_j$ on ∂B. For sufficiently large j, Γ lies completely in N so that

$$u_1(\mathbf{y}_j) - u_1(\mathbf{x}_j) = \int_\Gamma u_{11}(\mathbf{x}) \, dx_1 \qquad (4.37)$$

is positive. But this is a contradiction since $u_1(\mathbf{x}_j) \geq 0$ and $u_1(\mathbf{y}_j) \leq 0$.

Suppose now that $f(0) \geq 0$. Then we have

$$\Delta u + f(u) - f(0) \leq 0, \qquad (4.38)$$

and by the mean value theorem, we can find a function $c(\mathbf{x})$ so that $f(u) - f(0) = c(\mathbf{x})u$. Now Lemma 4.7, applied to $-u$, implies $u_1(\mathbf{x}_0) < 0$, a contradiction.

Hence suppose $f(0) < 0$. Then $\Delta u(\mathbf{x}_0) = -f(0) > 0$. On the other hand, since $u = 0$ on ∂B and $\nabla u(\mathbf{x}_0) = 0$, it is easy to check that $u_{11}(\mathbf{x}_0) = n_1^2 \Delta u(\mathbf{x}_0) \neq 0$, where $\mathbf{n}$ denotes the unit normal to ∂B (see Problem 4.19). Again we have a contradiction. □

For $\lambda \in \mathbb{R}$, let T_λ denote the plane $x_1 = \lambda$, and let $\Sigma(\lambda) = B \cap \{x_1 > \lambda\}$. Moreover, let $\Sigma'(\lambda)$ denote the reflection of $\Sigma(\lambda)$ across T_λ, and let $\mathbf{x}^\lambda$ denote the reflection of a point $\mathbf{x}$ across T_λ. We have the following lemma.

Lemma 4.23. *Assume that for some $\lambda \in [0, R)$ we have*

$$u_1(\mathbf{x}) \leq 0, \ u(\mathbf{x}) \leq u(\mathbf{x}^\lambda) \quad \forall \mathbf{x} \in \Sigma(\lambda), \qquad (4.39)$$

but that $u(\mathbf{x})$ does not identically equal $u(\mathbf{x}^\lambda)$ in $\Sigma(\lambda)$. Then $u(\mathbf{x}) < u(\mathbf{x}^\lambda)$ in all of $\Sigma(\lambda)$ and $u_1(\mathbf{x}) < 0$ on $B \cap T_\lambda$.

Proof. In $\Sigma'(\lambda)$, let $v(\mathbf{x}) = u(\mathbf{x}^\lambda)$. Then v satisfies the equation $\Delta v + f(v) = 0$. Let $w = v - u$, and let $c(\mathbf{x})$ be such that $f(v) - f(u) = c(\mathbf{x})w$ (mean value theorem). Then we have

$$\Delta w + c(\mathbf{x})w = 0 \qquad (4.40)$$

in $\Sigma'(\lambda)$. Moreover, by the assumptions of the lemma we have $w \leq 0$ in $\Sigma'(\lambda)$ and w is not identically zero in $\Sigma'(\lambda)$. Moreover, w vanishes on $T_\lambda \cap B$, which is part of the boundary of $\Sigma'(\lambda)$. It now follows from Theorem 4.10 and Lemma 4.7 that $w < 0$ in $\Sigma'(\lambda)$ and $w_1 > 0$ on $T_\lambda \cap B$. Since $w_1 = -2u_1$ on $T_\lambda \cap B$, the lemma follows. □

4.3.2 Proof of the Theorem

We shall use the two lemmas from the previous subsection to show that u is symmetric with respect to the plane $x_1 = 0$ and that $u_1 < 0$ for $\mathbf{x} \in B \cap \{x_1 > 0\}$. This obviously implies the theorem, because by the same argument u is symmetric with respect to any plane through the origin, and hence radially symmetric. We shall show the following.

Lemma 4.24. *For any* $\lambda \in (0, R)$, *we have*

$$u_1(\mathbf{x}) < 0, \; u(\mathbf{x}) < u(\mathbf{x}^\lambda) \quad \forall \mathbf{x} \in \Sigma(\lambda). \tag{4.41}$$

By continuity, we obtain $u(\mathbf{x}) \leq u(\mathbf{x}^0)$ in $\Sigma(0) = B \cap \{x_1 > 0\}$. Repeating the same argument with x_1 replaced by $-x_1$, we find that $u(\mathbf{x}) \geq u(\mathbf{x}^0)$ in $\Sigma(0)$. Hence u is symmetric across the plane $x_1 = 0$, and the theorem follows.

Proof. From Lemma 4.22, it follows that (4.41) holds for λ sufficiently close to R. Now let μ be a critical value such that (4.41) holds for $\lambda > \mu$, but not beyond. By continuity, we find that

$$u_1(\mathbf{x}) < 0, \; u(\mathbf{x}) \leq u(\mathbf{x}^\mu) \quad \forall \mathbf{x} \in \Sigma(\mu). \tag{4.42}$$

We need to show that $\mu = 0$. Assume the contrary, i.e., $\mu > 0$. For any point $\mathbf{x}_0 \in \partial\Sigma(\mu) \backslash T_\mu$, we have $\mathbf{x}_0^\mu \in B$, and hence $0 = u(\mathbf{x}_0) \neq u(\mathbf{x}_0^\mu)$. Hence $u(\mathbf{x})$ does not identically equal $u(\mathbf{x}^\mu)$ in $\Sigma(\mu)$, and Lemma 4.23 is applicable. Thus $u(\mathbf{x}) < u(\mathbf{x}^\mu)$ in $\Sigma(\mu)$ and $u_1 < 0$ on $B \cap T_\mu$. Hence (4.41) holds for $\lambda = \mu$. Moreover, by Lemma 4.22, we have $u_1 < 0$ in a neighborhood of any point on $T_\mu \cap \partial B$. Thus every point on $T_\mu \cap \overline{B}$ has a neighborhood where $u_1 < 0$, and since $T_\mu \cap \overline{B}$ is compact, we must have $u_1 < 0$ in $\Sigma(\mu - \epsilon)$ for ϵ sufficiently small.

Since we assumed that (4.41) does not hold beyond μ, there must be a sequence λ_j and $\mathbf{x}_j \in \Sigma(\lambda_j)$ such that $\lambda_j \to \mu$ and

$$u(\mathbf{x}_j) \geq u(\mathbf{x}_j^{\lambda_j}). \tag{4.43}$$

After taking a subsequence, we may assume that $\mathbf{x}_j$ converges; the limit $\mathbf{x}$ is necessarily in the closure of $\Sigma(\mu)$. Passing to the limit in (4.43), we find $u(\mathbf{x}) \geq u(\mathbf{x}^\mu)$. This cannot hold if $\mathbf{x} \in \partial B$, since in that case $u(\mathbf{x}) = 0$, $u(\mathbf{x}^\mu) > 0$. Since (4.41) holds for μ, we must have $\mathbf{x} \in T_\mu \cap B$; hence $\mathbf{x}^\mu = \mathbf{x}$. On the other hand, the straight line segment connecting $\mathbf{x}_j$ to $\mathbf{x}_j^{\lambda_j}$ belongs to B, and because of (4.43) and the mean value theorem, it must contain a point $\mathbf{y}_j$ where $u_1(\mathbf{y}_j) \geq 0$. Taking the limit $j \to \infty$, we find $u_1(\mathbf{x}) \geq 0$. This is a contradiction. $\qquad\square$

Problems

4.17. Show that, if $f(u) \geq 0$ for every u, then nontrivial solutions of (4.34), (4.35) are automatically positive. Also show that if $f(0) \geq 0$, then (4.36) holds also at $r = R$.

4.18. Show that if a non-negative solution u of (4.34), (4.35) exists and u is not identically zero, then $f(u(\mathbf{0}))$ must be strictly positive.

4.19. Let B be a ball centered at the origin. Let u be a C^2-function such that $u > 0$ in B and $u = 0$ on ∂B. Let $\mathbf{x}$ be a point on ∂B.

(a) Show that $\Delta u(\mathbf{x}) = u_{rr}(\mathbf{x}) + \frac{n-1}{r} u_r(\mathbf{x})$ where r is the radial direction.

(b) Assume that $u_r(\mathbf{x}) = 0$. Show that $\partial^2 u / \partial x_i \partial x_j(\mathbf{x}) = n_i n_j u_{rr}(\mathbf{x})$. Here $\mathbf{n}$ is the unit normal to ∂B. Hint: Show first that the matrix of second derivatives is positive semidefinite.

4.20. Let $u \in C^4(\overline{B})$ be radially symmetric and satisfy $u_r < 0$ for $0 < r \leq R$, and assume that $\Delta u(\mathbf{0}) \neq 0$. Show that there is a C^1-function f such that $\Delta u + f(u) = 0$.

4.21. Let $u > 0$ be a C^2-solution of (4.34) in the domain $R_1 < |\mathbf{x}| < R_2$. Assume that $u = 0$ on the outer boundary $|\mathbf{x}| = R_2$. Show that $u_r < 0$ for $(R_1 + R_2)/2 \leq |\mathbf{x}| < R_2$.

4.4 Maximum Principles for Parabolic Equations

In this section, we shall extend the maximum principle to parabolic equations of the form

$$Lu = -\frac{\partial u}{\partial t} + a_{ij}(\mathbf{x}, t)\frac{\partial^2 u}{\partial x_i \partial x_j} + b_i(\mathbf{x}, t)\frac{\partial u}{\partial x_i} + c(\mathbf{x}, t)u = f(\mathbf{x}, t). \quad (4.44)$$

We shall consider solutions defined on $\Omega \times (0, T)$, where Ω is an (open) domain in $\mathbb{R}^n$ (more general regions in $(\mathbf{x}, t)$ space can be considered). We set $D = \Omega \times (0, T]$, $Q = \Omega \times (0, T)$ and $\Sigma = (\partial \Omega \times [0, T]) \cup (\Omega \times \{0\})$. We shall assume that the coefficients are continuous on $\overline{D}$, and that the matrix a_{ij} is strictly positive definite on $\overline{D}$. Throughout we shall also assume that $u \in C^2(D) \cap C(\overline{D})$. The maximum principle for parabolic equations says that (under appropriate sign conditions on c and f) the maximum of u must be on Σ.

4.4.1 The Weak Maximum Principle

The analogue of Theorem 4.1 for parabolic equations is the following result.

Theorem 4.25. *Assume that Ω is bounded and $Lu \geq 0$ ($Lu \leq 0$). Moreover, let $c(\mathbf{x}, t) = 0$ in D. Then the maximum (minimum) of u is achieved on Σ.*

Proof. From Theorem 4.1 and Remark 4.2, we already know that the maximum is achieved on ∂D. Moreover, if $Lu > 0$, then the maximum cannot be achieved on $\Omega \times \{T\}$. For let $\mathbf{x} \in \Omega \times \{T\}$ be such that $u(\mathbf{x}, T) = \max_{\mathbf{y} \in \Omega \times \{T\}} u(\mathbf{y}, T)$. Then at the point $\mathbf{x}$ we have

$$a_{ij}\frac{\partial^2 u}{\partial x_i \partial x_j} + b_i\frac{\partial u}{\partial x_i} \leq 0, \quad (4.45)$$

and since $Lu > 0$, we conclude $u_t < 0$.

For the general case, let $u_\epsilon = u + \epsilon \exp(-t)$. This yields

$$Lu_\epsilon = Lu + \epsilon \exp(-t) > 0, \tag{4.46}$$

hence

$$\max_{\overline{D}} u_\epsilon = \max_{\Sigma} u_\epsilon. \tag{4.47}$$

The theorem follows by letting $\epsilon \to 0$. □

Obviously, the analogues of Corollaries 4.3 and 4.4 (with D in place of Ω and Σ in place of $\partial\Omega$) also hold. We note that by setting $u = v\exp(\gamma t)$, we find

$$Lu = \exp(\gamma t)(Lv - \gamma v). \tag{4.48}$$

Hence, if $Lu \geq 0$, then $(L - \gamma)v \geq 0$ and vice versa. By choosing γ large enough, we can always achieve that $c - \gamma \leq 0$. Hence the analogue of Corollary 4.4 holds without any sign condition on c.

4.4.2   The Strong Maximum Principle

The goal of this section is the following result.

Theorem 4.26. *Assume that $Lu \geq 0$ $(Lu \leq 0)$. Let $M = \sup_D u$ $(M = \inf_D u)$. Assume that $u = M$ at a point $(\mathbf{x}_0, t_0) \in D$ and that one of the following holds:*

1. *$c = 0$ and M is arbitrary,*

2. *$c \leq 0$ and $M \geq 0$ $(M \leq 0)$,*

3. *$M = 0$ and c is arbitrary.*

Then $u = M$ on $\overline{\Omega} \times [0, t_0]$.

The proof follows from three lemmas. From now on, we always assume $Lu \geq 0$ without explicitly stating so; the case $Lu \leq 0$ obviously follows by substituting $-u$ for u. The following lemma is immediate from Lemma 4.7 and Remark 4.9.

Lemma 4.27. *Let $B \subset \mathbb{R}^{n+1}$ be a ball with $\overline{B} \subset D$ and suppose $u < M$ in B and $u(\mathbf{x}_0, t_0) = M$, where $(\mathbf{x}_0, t_0) \in \partial B$. Then t_0 is either the smallest or the largest value which t assumes in $\overline{B}$.*

Next we show that if $u < M$ at any point $(\mathbf{x}_0, t_0)$ in Q, then $u < M$ on all of $\Omega \times \{t_0\}$.

Lemma 4.28. *Assume that $u(\mathbf{x}_0, t_0) < M$, where $\mathbf{x}_0 \in \Omega$ and $t_0 \in (0, T)$. Then $u(\mathbf{x}, t_0) < M$ for every $\mathbf{x} \in \Omega$.*

Proof. Assume that $\Omega \times \{t_0\}$ contains points where $u = M$. Then we can find points $\mathbf{x}_1$ and $\mathbf{x}_2$ in Ω such that $u(\mathbf{x}_1, t_0) = M$, $u(\mathbf{x}_2, t_0) < M$, and the line segment L connecting $(\mathbf{x}_1, t_0)$ and $(\mathbf{x}_2, t_0)$ is entirely in $\Omega \times \{t_0\}$. By moving the point $\mathbf{x}_1$, we can always achieve that $u < M$ along L. Now let

$$\delta = \min(|\mathbf{x}_1 - \mathbf{x}_2|, \operatorname{dist}(L, \partial D)). \tag{4.49}$$

For $\mathbf{x} \in L$ with $0 < |\mathbf{x} - \mathbf{x}_1| < \delta$, let

$$d(\mathbf{x}) = \operatorname{dist}((\mathbf{x}, t_0), Q \cap \{u = M\}). \tag{4.50}$$

Obviously, $d(\mathbf{x}) \le |\mathbf{x} - \mathbf{x}_1|$.

By Lemma 4.27, the point in $Q \cap \{u = M\}$ which is nearest to $(\mathbf{x}, t_0)$ is of the form $(\mathbf{x}, t)$, so that either $u(\mathbf{x}, t_0 + d(\mathbf{x})) = M$ or $u(\mathbf{x}, t_0 - d(\mathbf{x})) = M$. Let $\mathbf{n}$ be a unit vector in the direction of L. For sufficiently small $|\epsilon| > 0$, $d(\mathbf{x} + \epsilon \mathbf{n})$ is defined, and by the Pythagorean Theorem we have

$$d(\mathbf{x} + \epsilon \mathbf{n}) \le \sqrt{\epsilon^2 + d(\mathbf{x})^2}. \tag{4.51}$$

By the same argument,

$$d(\mathbf{x}) \le \sqrt{\epsilon^2 + d(\mathbf{x} + \epsilon \mathbf{n})^2}, \tag{4.52}$$

and hence

$$d(\mathbf{x} + \epsilon \mathbf{n}) \ge \sqrt{-\epsilon^2 + d(\mathbf{x})^2}. \tag{4.53}$$

From (4.52) and (4.53), it follows that the derivative of $d(\mathbf{x} + \epsilon \mathbf{n})$ at $\epsilon = 0$ exists and is zero. Hence $d(\mathbf{x})$ is constant along $L \cap \{0 < |\mathbf{x} - \mathbf{x}_1| < \delta\}$. But this is a contradiction since $d(\mathbf{x}) \ne 0$, but $d(\mathbf{x}) \to 0$ as $\mathbf{x} \to \mathbf{x}_1$. $\quad\square$

The final lemma which we require is the following.

Lemma 4.29. *Let $0 \le t_0 < t_1 \le T$ and assume that $u < M$ in $\Omega \times (t_0, t_1)$. Then $u < M$ on $\Omega \times \{t_1\}$.*

Proof. Assume on the contrary that there is $\mathbf{x}_1 \in \Omega$ with $u(\mathbf{x}_1, t_1) = M$. Define

$$v(\mathbf{x}, t) = \exp\left(-|\mathbf{x} - \mathbf{x}_1|^2 - \alpha(t - t_1)\right) - 1, \tag{4.54}$$

where $\alpha > 0$ is chosen large. We compute

$$
\begin{aligned}
Lv(\mathbf{x}, t) = {} & \exp\left(-|\mathbf{x} - \mathbf{x}_1|^2 - \alpha(t - t_1)\right) \\
& \times \left(4a_{ij}(x_i - x_{1i})(x_j - x_{1j}) - 2(a_{ii} + b_i(x_i - x_{1i})) + \alpha\right) + cv.
\end{aligned}
$$

If N is a small neighborhood of $(\mathbf{x}_1, t_1)$, we obtain $Lv > 0$ in $N \cap \{t \le t_1\}$ if we choose α large enough. Let now A be the domain bounded by the paraboloid $|\mathbf{x} - \mathbf{x}_1|^2 + \alpha(t - t_1) = 0$ and a sufficiently small sphere centered at $(\mathbf{x}_1, t_1)$. The boundary of A has two parts, one on the paraboloid, where $v = 0$, and one on the sphere and inside the paraboloid, where $u < M$.

If we choose $\epsilon > 0$ small, we therefore have $u + \epsilon v - M \leq 0$ on ∂A and $L(u + \epsilon v - M) \geq -cM \geq 0$ in A.

If $c \leq 0$, the weak maximum principle implies that $u + \epsilon v - M \leq 0$ throughout A. Taking the t derivative at the point $(\mathbf{x}_1, t_1)$, we conclude

$$u_t + \epsilon v_t = u_t - \epsilon \alpha \geq 0, \tag{4.55}$$

i.e., $u_t > 0$. But since $(\mathbf{x}_1, t_1)$ is a maximum with respect to $\mathbf{x}$, we find that at this point

$$Lu \leq -u_t + cu \leq -u_t, \tag{4.56}$$

and since $Lu \geq 0$ was assumed, we conclude $u_t \leq 0$, a contradiction.

If $M = 0$, we can, as in the proof of Lemma 4.7, apply the same argument with $(L - c^+)$ in place of L. $\qquad\qquad\square$

The proof of the theorem is now clear. For every $t > 0$, we have either $u(\mathbf{x}, t) = M$ or $u(\mathbf{x}, t) < M$ for all $\mathbf{x} \in \Omega$ (Lemma 4.28). The set Z of all t's for which $u < M$ is open and hence a countable union of intervals. By Lemma 4.29 none of these intervals can have an upper endpoint. The only possibility is then that $Z = (t_0, T]$ for some t_0, which is the claim of the theorem.

Finally we remark that if $\partial\Omega$ is of class C^2 and u is C^1 up to the boundary, then Lemma 4.7 applies; i.e., if $\mathbf{x}_0 \in \partial\Omega$, $t \in (0, T)$, $u(\mathbf{x}_0, t) = M$, but $u(\mathbf{x}, t) < M$ for $\mathbf{x} \in \Omega$, then $\partial u/\partial n(\mathbf{x}_0, t) > 0$.

Problems

4.22. Assume that Ω is bounded, $\partial\Omega$ is of class C^2 and that $u, v \in C^2(D) \cap C^1(\overline{D})$. Assume, moreover, that $Lu \leq Lv$ for $(\mathbf{x}, t) \in D$, that $u(\mathbf{x}, 0) \geq v(\mathbf{x}, 0)$ for $\mathbf{x} \in \Omega$ and that $\partial u/\partial n \geq \partial v/\partial n$ for $(\mathbf{x}, t) \in \partial\Omega \times (0, T)$. Show that $u \geq v$ in D.

4.23. Assume that Ω is bounded, $f \leq 0$ and that f and the coefficients of L are independent of t. We consider the equation $Lu = f$ with boundary conditions $u = 0$ on $\partial\Omega \times (0, \infty)$. An equilibrium solution is a solution which does not depend on time. An equilibrium solution v is called stable if, for every continuous function $u_0(\mathbf{x})$ satisfying $u_0 = 0$ on $\partial\Omega$, there is a solution u satisfying $Lu = f$ in $\Omega \times (0, \infty)$, $u = 0$ on $\partial\Omega \times (0, \infty)$ and $u(\mathbf{x}, 0) = u_0(\mathbf{x})$ for $\mathbf{x} \in \Omega$ and, moreover, $u(\mathbf{x}, t) \to v(\mathbf{x})$ as $t \to \infty$. Show that a stable equilibrium solution must be non-negative.

4.24. Let $D = \mathbb{R}^n \times (0, T]$. Let $u \in C^2(D) \cap C(\overline{D})$ be a bounded solution of the heat equation $u_t = \Delta u$. Show that $\sup_D u \leq \sup_{\mathbb{R}^n} u(\cdot, 0)$. Hint: Consider the function $v = u - \epsilon(2nt + |\mathbf{x}|^2)$.

4.25. Let Ω be bounded, and let $f : \mathbb{R} \to \mathbb{R}$ be of class C^1. Let $u_0 \in C(\overline{\Omega})$ be such that $u_0 = 0$ on $\partial\Omega$. Prove that the equation $u_t = \Delta u + f(u)$, subject to the boundary condition $u = 0$ on $\partial\Omega$ and the initial condition $u(\mathbf{x}, 0) = u_0(\mathbf{x})$, has at most one solution on $\Omega \times (0, T)$, for any $T > 0$.

4.26. Let Ω be bounded and let $f, g : \mathbb{R}^2 \to \mathbb{R}$ be of class C^1. Let u, v satisfy the equations $u_t = \Delta u + f(u, v)$, $v_t = \Delta v + g(u, v)$ on $\Omega \times (0, T)$, with boundary conditions $u = v = 0$ on $\partial\Omega$. Assume that $f(0, v) \geq 0$ for every v and that $g(u, 0) \geq 0$ for every u. Show that if $u(\mathbf{x}, 0) \geq 0$, $v(\mathbf{x}, 0) \geq 0$ for every $\mathbf{x} \in \Omega$, then $u \geq 0$, $v \geq 0$. Systems of the kind described here arise in reaction-diffusion problems and in mathematical biology, where u and v denote concentrations or populations of species.

5

Distributions

5.1 Test Functions and Distributions

5.1.1 Motivation

Many problems arising naturally in differential equations call for a generalized definition of functions, derivatives, convergence, integrals, etc. In this subsection, we discuss a number of such questions, which will be adequately answered below.

1. In Chapter 1, we noted that any twice differentiable function of the form $u(x,t) = F(x+t) + G(x-t)$ is a solution of the wave equation $u_{tt} = u_{xx}$. Clearly, it seems natural to call u a "generalized" solution even if F and G are not twice differentiable. A natural question is what meaning can be given to u_{tt} and u_{xx} in this case; obviously, they cannot be "functions" in the usual sense. The same question arises for the shock solutions of hyperbolic conservation laws which we discussed in Chapter 3.

2. Consider the ODE initial-value problem

$$u'(t) = f_\epsilon(t), \quad u(0) = 0, \tag{5.1}$$

where

$$f_\epsilon(t) = \begin{cases} 1/\epsilon & 1 < t < 1+\epsilon \\ 0 & \text{otherwise.} \end{cases} \tag{5.2}$$

Obviously, the solution is

$$u(t) = \begin{cases} 0 & 0 \le t \le 1 \\ (t-1)/\epsilon & 1 \le t \le 1 + \epsilon \\ 1 & t \ge 1 + \epsilon. \end{cases} \tag{5.3}$$

Note that the limit of u as $\epsilon \to 0$ exists; it is a step function. The function f_ϵ has unit integral; it is supported on shorter and shorter time intervals as ϵ tends to zero. It would be natural to regard the "limit" of f_ϵ as an instantaneous unit impulse. The question arises what meaning can be given to this limit and in what sense the differential equation holds in the limit. Similar questions arise in many physical problems involving idealized point singularities: the electric field of a point charge, light emitted by a point source, etc.

3. In Chapter 1, we outlined the solution of Dirichlet's problem by minimizing the integral $\int_\Omega |\nabla u|^2 \, d\mathbf{x}$. A fundamental ingredient in turning these ideas into a rigorous theory is obviously the definition of a class of functions for which the integral is finite; the square root of the integral naturally defines a norm on this space of functions. It turns out that $C^1(\overline{\Omega})$ is too restrictive; it is not a complete metric space in the norm defined by the integral. It is natural to consider the completion; this leads to functions for which ∇u does not exist in the sense of the classical definition as a pointwise limit of difference quotients.

4. The Fourier transform is a natural tool for dealing with PDEs with constant coefficients posed on all of space. However, the class of functions for which the Fourier integral exists in the conventional sense is rather restrictive; in particular, such functions must be integrable at infinity. Clearly, it would be useful to have a notion of the Fourier transform for functions which do not satisfy such a restriction, e.g., constant functions.

The idea behind generalized functions is roughly this: Given a continuous function $f(\mathbf{x})$ on Ω, we can define a linear mapping

$$\phi \mapsto \int_\Omega f(\mathbf{x})\phi(\mathbf{x}) \, d\mathbf{x} \tag{5.4}$$

from a suitable class of functions (which will be called test functions) into $\mathbb{R}$. We shall see that this mapping has certain continuity properties. A generalized function is then defined to be a linear mapping on the test functions with these same continuity properties.

Since we intend to use generalized functions to study differential equations, a key question is: how do we define derivatives of such functions? The answer is: by using integration by parts. Test functions will be required to vanish near $\partial\Omega$, so the derivative $\partial f/\partial x_j$ can be defined as the mapping

$$\phi \mapsto -\int_\Omega f(\mathbf{x})\frac{\partial\phi}{\partial x_j}(\mathbf{x})\,d\mathbf{x}. \tag{5.5}$$

Clearly, this definition requires no differentiability of f in the usual sense; the only differentiability requirement is on ϕ. We shall therefore choose the test functions to be functions with very nice smoothness properties.

5.1.2 Test Functions

Let Ω be a nonempty open set in $\mathbb{R}^m$. We make the following definition.

Definition 5.1. A function f defined on Ω is called a **test function** if $f \in C^\infty(\Omega)$ and there is a compact set $K \subset \Omega$ such that the support of f lies in K. The set of all test functions on Ω is denoted by $\mathcal{D}(\Omega) = C_0^\infty(\Omega)$.

Obviously, $\mathcal{D}(\Omega)$ is a linear space. To do analysis, we need a notion of convergence. It is possible to define open sets in $\mathcal{D}(\Omega)$ and use the notions of general topology. However, for most purposes in PDEs this is not necessary; only a definition for the convergence of sequences is required. This definition is as follows.

Definition 5.2. Let ϕ_n, $n \in \mathbb{N}$ and ϕ be elements of $\mathcal{D}(\Omega)$. We say that ϕ_n **converges** to ϕ in $\mathcal{D}(\Omega)$, if there is a compact subset K of Ω such that the supports of all the ϕ_n (and of ϕ) lie in K and, moreover, ϕ_n and derivatives of ϕ_n of arbitrary order converge uniformly to those of ϕ.

Remark 5.3. Note that the notion of convergence defined above does not come from a metric or norm.

It is often important to know that test functions with certain properties exist; for example one often needs a function that is positive in a small neighborhood of a given point $\mathbf{y}$ and zero outside that neighborhood. Such a function can be given explicitly:

$$\phi_{\mathbf{y},\epsilon}(\mathbf{x}) = \begin{cases} \exp\left(-\frac{\epsilon^2}{\epsilon^2-|\mathbf{x}-\mathbf{y}|^2}\right), & |\mathbf{x}-\mathbf{y}| < \epsilon \\ 0, & \text{otherwise.} \end{cases} \tag{5.6}$$

Indeed, this example can be used generate other examples of test functions. The following theorem states that any continuous function of compact support can be approximated uniformly by test functions.

Theorem 5.4. Let K be a compact subset of Ω and let $f \in C(\Omega)$ have support contained in K. For $\epsilon > 0$, let

$$f_\epsilon(\mathbf{x}) = \frac{1}{C(\epsilon)}\int_K \phi_{\mathbf{y},\epsilon}(\mathbf{x})f(\mathbf{y})\,d\mathbf{y}, \tag{5.7}$$

where

$$C(\epsilon) = \int_{\mathbb{R}^m} \phi_{\mathbf{y},\epsilon}(\mathbf{x}) \, d\mathbf{y}. \qquad (5.8)$$

If $\epsilon < \mathrm{dist}(K, \partial\Omega)$, then $f_\epsilon \in \mathcal{D}(\Omega)$; moreover, $f_\epsilon \to f$ uniformly as $\epsilon \to 0$.

The proof is left as an exercise.

In a similar fashion, we can construct test functions which are equal to 1 on a given set and equal to 0 on another.

Theorem 5.5. *Let K be a compact subset of Ω and let $U \subset \Omega$ be an open set containing K. Then there is a test function which is equal to 1 on K, is equal to 0 outside U and assumes values in $[0, 1]$ on $U \backslash K$.*

Proof. Let $\epsilon > 0$ be such that the ϵ-neighborhood of K is contained in U. Let K_1 be the closure of the $\epsilon/3$-neighborhood of K and define

$$f(\mathbf{x}) = 1 - \min\left\{1, \frac{3}{\epsilon}\mathrm{dist}(\mathbf{x}, K_1)\right\}. \qquad (5.9)$$

The function f is continuous, equal to 1 on K_1 and equal to zero outside of the $2\epsilon/3$-neighborhood of K. A function with the properties desired by the theorem is given by $f_{\epsilon/3}$ as defined by (5.7). $\qquad \square$

Many proofs in PDEs involve a reduction to local considerations in a small neighborhood of a point. (See, for example, Chapter 9.) The device by which this is achieved is known as a partition of unity.

Definition 5.6. *Let U_i, $i \in \mathbb{N}$ be a family of bounded open subsets of Ω such that*

1. *the closure of each U_i is contained in Ω,*

2. *every compact subset of Ω intersects only a finite number of the U_i (this property is called local finiteness), and*

3. *$\bigcup_{i \in \mathbb{N}} U_i = \Omega$.*

*A **partition of unity** subordinate to the covering $\{U_i\}$ is a set of test functions ϕ_i such that*

1. *$0 \le \phi_i \le 1$,*

2. *$\mathrm{supp}\, \phi_i \subset U_i$,*

3. *$\sum_{i \in \mathbb{N}} \phi_i(\mathbf{x}) = 1$ for every $\mathbf{x} \in \Omega$.*

The following theorem says that such partitions of unity exist.

Theorem 5.7. *Let U_i, $i \in \mathbb{N}$ be a collection of sets with the properties stated in Definition 5.6. Then there is a partition of unity subordinate to the covering $\{U_i\}$.*

Proof. We first construct a new covering $\{V_i\}$, where the V_i have all the properties of Definition 5.6 and the closure of V_i is contained in U_i. The V_i are constructed inductively. Suppose $V_1, V_2, \ldots, V_{k-1}$ have already been found such that U_j contains $\overline{V}_j$ and

$$\Omega = \bigcup_{j=1}^{k-1} V_j \cup \bigcup_{j=k}^{\infty} U_j. \tag{5.10}$$

Let F_k be the complement of the set

$$\bigcup_{j=1}^{k-1} V_j \cup \bigcup_{j=k+1}^{\infty} U_j. \tag{5.11}$$

Then F_k is a closed set contained in U_k. We choose V_k to be any open set containing F_k such that $\overline{V}_k \subset U_k$. Each point $\mathbf{x} \in \Omega$ is contained in only finitely many of the U_i; hence there is $N \in \mathbb{N}$ with $\mathbf{x} \notin \bigcup_{j=N+1}^{\infty} U_j$. But this implies that $\mathbf{x} \in \bigcup_{j=1}^{N} V_j$. Hence the V_i have property 3 of Definition 5.6; the other two properties follow trivially from the fact that $V_i \subset U_i$.

Let W_k be an open set such that $\overline{V}_k \subset W_k$, $\overline{W}_k \subset U_k$. According to Theorem 5.5, there is now a test function ψ_k, which is equal to 1 on $\overline{V}_k$, is equal to zero outside W_k and takes values between 0 and 1 otherwise. Let

$$\psi(\mathbf{x}) = \sum_{k \in \mathbb{N}} \psi_k(\mathbf{x}). \tag{5.12}$$

Because of property 2 in Definition 5.6, the right-hand side of (5.12) has only finitely many nonzero terms in the neighborhood of any given $\mathbf{x}$, and there is no issue of convergence. The functions $\phi_k := \psi_k/\psi$ yield the desired partition of unity. $\qquad\square$

5.1.3 *Distributions*

We now define the space of distributions. As we indicated in the introduction, the definition of a distribution is constructed very cleverly to achieve two seemingly contradictory goals. We wish to have a generalized notion of a "function" that includes objects that are highly singular or "rough." At the same time we wish to be able to define "derivatives" of arbitrary order of these objects.

Definition 5.8. A **distribution** or **generalized function** is a linear mapping $\phi \mapsto (f, \phi)$ from $\mathcal{D}(\Omega)$ to $\mathbb{R}$, which is continuous in the following sense: If $\phi_n \to \phi$ in $\mathcal{D}(\Omega)$, then $(f, \phi_n) \to (f, \phi)$. The set of all distributions is called $\mathcal{D}'(\Omega)$.

Example 5.9. Any continuous function f on Ω can be identified with a generalized function by setting

$$(f, \phi) = \int_\Omega f(\mathbf{x})\phi(\mathbf{x})\, d\mathbf{x}. \tag{5.13}$$

The continuity of the mapping follows from the familiar theorem concerning the limit of the intergral of a uniformly convergent sequence of functions. Indeed, the Lebesgue dominated convergence theorem allows us to make the same claim if f is merely locally integrable.

Example 5.10. Of course, there are many generalized functions which do not correspond to "functions" in the ordinary sense. The most important example is known as the Dirac delta function. We assume that Ω contains the origin, and we define

$$(\delta, \phi) = \phi(\mathbf{0}). \tag{5.14}$$

The continuity of the functional follow from the fact that convergence of a sequence of test functions implies pointwise convergence.

It is easy to show that there is no continuous function satisfying (5.14), (cf. Problem 5.5).

Remark 5.11. Generalized functions like the delta function do not take "values" like ordinary functions. Nevertheless, it is customary to use the language of ordinary functions and speak of "the generalized function $\delta(\mathbf{x})$,"[1] even though it does not make sense to plug in a specific $\mathbf{x}$. We shall also write $\int_\Omega \delta(\mathbf{x})\phi(\mathbf{x})\, d\mathbf{x}$ for (δ, ϕ).

Example 5.12. For any multiindex α, the mapping

$$\phi \mapsto D^\alpha \phi(\mathbf{0})$$

is a generalized function.

Example 5.13. Other singular distributions include such examples from physics as surface charge. If $\mathcal{S}$ is a smooth two-dimensional surface in $\mathbb{R}^3$ and $q : \mathcal{S} \to \mathbb{R}$ is integrable, then for $\phi \in \mathcal{D}(\mathbb{R}^3)$ we define

$$(q, \phi) = \int_{\mathcal{S}} q(\mathbf{x})\phi(\mathbf{x})\, da(\mathbf{x})$$

where $da(\mathbf{x})$ indicates integration with respect to surface area on $\mathcal{S}$.

Example 5.14. A current flowing along a curve $\mathcal{C} \subset \mathbb{R}^3$ is an example of a vector-valued distribution. If $\mathbf{j} : \mathcal{C} \to \mathbb{R}^3$ is integrable, then for $\phi \in \mathcal{D}(\mathbb{R}^3)^3$ we define

$$(\mathbf{j}, \phi) = \int_{\mathcal{C}} \mathbf{j}(\mathbf{x}) \cdot \phi(\mathbf{x})\, d\sigma(\mathbf{x})$$

---

[1] We apologize to those among our friends to whom such language is an abomination — even for ordinary functions!

where $d\sigma(\mathbf{x})$ indicates integration with respect to arclength on $\mathcal{C}$.

Remark 5.15. Of course, complex-valued distributions can be defined in the same fashion as real-valued distributions; in that case, however, it is customary to make the convention

$$(f, \phi) = \int_\Omega \overline{f(\mathbf{x})} \phi(\mathbf{x}) \, d\mathbf{x} \qquad (5.15)$$

in place of (5.13); the pairing of generalized functions and test functions thus takes the same form as the inner product in the Hilbert space $L^2(\Omega)$.[2]

An important property of distributions is that they are locally of "finite order."

Lemma 5.16. *Let $f \in \mathcal{D}'(\Omega)$ and let K be a compact subset of Ω. Then there exists $n \in \mathbb{N}$ and a constant C such that*

$$|(f, \phi)| \le C \sum_{|\alpha| \le n} \max_{\mathbf{x} \in K} |D^\alpha \phi(\mathbf{x})| \qquad (5.16)$$

for every $\phi \in \mathcal{D}(\Omega)$ with support contained in K.

Proof. Suppose not. Then for every n there exists ψ_n such that

$$|(f, \psi_n)| > n \sum_{|\alpha| \le n} \max_{\mathbf{x} \in K} |D^\alpha \psi_n(\mathbf{x})|.$$

Let $\phi_n := \psi_n / |(f, \psi_n)|$. Then $\phi_n \to 0$ in $\mathcal{D}(\Omega)$, but $(f, \phi_n) \equiv 1$. This is a contradiction, and the proof is complete. $\qquad\qquad\square$

We conclude this subsection with some straightforward definitions.

Definition 5.17. For distributions f and g and real number $\alpha \in \mathbb{R}$, we set

$$(f + g, \phi) = (f, \phi) + (g, \phi), \qquad (5.17)$$

$$(\alpha f, \phi) = (f, \alpha \phi). \qquad (5.18)$$

(If α is allowed to be complex, then the right-hand side of (5.18) is changed to $(f, \bar{\alpha}\phi)$.)

Remark 5.18. It is in general not possible to define the product of two generalized functions (cf. Problems 5.11, 5.12). However, we can define the product of a distribution and a smooth function.

---

[2]One of the oldest problems in Hilbert space theory is whether to put the complex conjugate on the first or on the second factor in the inner product. The convention made here is widely followed by physicists. Pure mathematicians tend to make the opposite convention.

Definition 5.19. For any function $a \in C^\infty(\Omega)$, we define

$$(af, \phi) = (f, a\phi). \tag{5.19}$$

If the graph of a function $f(x)$ is shifted by h, one obtains the graph of the function $f(x - h)$, i.e., x is shifted by $-h$. This can be generalized to distributions on $\mathbb{R}^m$.

Definition 5.20. Let $\mathbf{U}(\mathbf{x}) = \mathbf{A}\mathbf{x} + \mathbf{b}$ be a nonsingular linear transformation in $\mathbb{R}^m$, and let $\mathbf{U}^{-1}(\mathbf{y}) = \mathbf{A}^{-1}(\mathbf{y} - \mathbf{b})$ be the inverse transformation. Then we set

$$(\mathbf{U}f, \phi) = |\det A| \, (f(\mathbf{x}), \phi(\mathbf{U}(\mathbf{x}))). \tag{5.20}$$

This definition is motivated by the following formal calculation:

$$
\begin{aligned}
(\mathbf{U}f, \phi) &= (f(\mathbf{U}^{-1}(\mathbf{x})), \phi(\mathbf{x})) \\
&= \int_{\mathbb{R}^m} f(\mathbf{U}^{-1}(\mathbf{x}))\phi(\mathbf{x}) \, d\mathbf{x} \\
&= |\det A| \int_{\mathbb{R}^m} f(\mathbf{y})\phi(\mathbf{U}(\mathbf{y})) \, d\mathbf{y}.
\end{aligned}
$$

(We have substituted $\mathbf{x} = \mathbf{U}(\mathbf{y})$.)

Example 5.21. The translation $\delta(\mathbf{x} - \mathbf{x}_0)$ is defined as

$$(\delta(\mathbf{x} - \mathbf{x}_0), \phi(\mathbf{x})) = (\delta(\mathbf{x}), \phi(\mathbf{x} + \mathbf{x}_0)) = \phi(\mathbf{x}_0). \tag{5.21}$$

Remark 5.22. With this definition, we can define the symmetry of a generalized function; for example, f is even if $f(-\mathbf{x}) = f(\mathbf{x})$, i.e., $(f(\mathbf{x}), \phi(\mathbf{x})) = (f(\mathbf{x}), \phi(-\mathbf{x}))$.

5.1.4 Localization and Regularization

Although generalized functions cannot be evaluated at points, they can be restricted to open sets. This is quite straightforward. If G is an open subset of Ω, then $\mathcal{D}(G)$ is naturally embedded in $\mathcal{D}(\Omega)$, and hence every generalized function on Ω defines a generalized function on G by restriction. Consequently, we shall shall define the following.

Definition 5.23. We say that $f \in \mathcal{D}'(\Omega)$ **vanishes** on and open set $G \subset \Omega$ if $(f, \phi) = 0$ for every $\phi \in \mathcal{D}(G)$. Two distributions are **equal** on G if their difference vanishes on G.

It can be shown (cf. Problem 5.7) that if f vanishes locally near every point of G, i.e., if every point of G has a neighborhood on which f vanishes, then f vanishes on G. An immediate consequence is that if f vanishes on each of a family of open sets, it also vanishes on their union. Hence there is a largest open set N_f on which f vanishes.

Definition 5.24. The complement of N_f in Ω is called the **support** of f.

Example 5.25. The support of the delta function is the set $\{0\}$. Although the delta function cannot be evaluated at points, it makes sense to say that it vanishes except at the origin.

Remark 5.26. Functions with nonintegrable singularities are not defined as generalized functions by equation (5.13). However, it is often possible to define a generalized function which agrees with a singular function on any open set that does not contain the singularity. Such a generalized function is called a regularization. For example, a regularization of the function $1/x$ on $\mathbb{R}$ is given by the principal value integral

$$(f, \phi) = \int_{-\infty}^{-\epsilon} \frac{\phi(x)}{x} \, dx + \int_{-\epsilon}^{\epsilon} \frac{\phi(x) - \phi(0)}{x} \, dx + \int_{\epsilon}^{\infty} \frac{\phi(x)}{x} \, dx \qquad (5.22)$$

(cf. Problem 5.9).

5.1.5 Convergence of Distributions

Just as sequences of classical functions are central to PDEs, so are sequences of generalized functions.

Definition 5.27. A sequence f_n in $\mathcal{D}'(\Omega)$ **converges** to $f \in \mathcal{D}'(\Omega)$ if

$$(f_n, \phi) \to (f, \phi)$$

for every $\phi \in \mathcal{D}(\Omega)$.

Example 5.28. A uniformly convergent sequence of continuous functions (which define distributions as in Example 5.9) also converges in $\mathcal{D}'$.

Example 5.29. Consider the sequence

$$f_n(x) = \begin{cases} n, & 0 < x < 1/n \\ 0, & \text{otherwise.} \end{cases} \qquad (5.23)$$

We have

$$\int_{-\infty}^{\infty} f_n(x)\phi(x) \, dx = n \int_0^{1/n} \phi(x) \, dx, \qquad (5.24)$$

which converges to $\phi(0)$ as $n \to \infty$. Hence $f_n(x) \to \delta(x)$ in $\mathcal{D}'(\mathbb{R})$.

Remark 5.30. Problem 5.10 asks the reader to prove that every distribution is the limit of distributions with compact support. Later we shall actually see that every distribution is a limit of test functions; in other words, test functions are dense in $\mathcal{D}'(\Omega)$.

Another important result is the (sequential) completeness of $\mathcal{D}'(\Omega)$.

Theorem 5.31. *Let f_n be a sequence in $\mathcal{D}'(\Omega)$ such that (f_n, ϕ) converges for every $\phi \in \mathcal{D}(\Omega)$. Then there exists $f \in \mathcal{D}'(\Omega)$ such that $f_n \to f$.*

Proof. We define

$$(f, \phi) = \lim_{n \to \infty} (f_n, \phi). \tag{5.25}$$

Obviously, f is a linear mapping from $\mathcal{D}(\Omega)$ to $\mathbb{R}$. To verify that $f \in \mathcal{D}'(\Omega)$, we have to establish its continuity, i.e., we must show that if $\phi_n \to 0$ in $\mathcal{D}(\Omega)$, then $(f, \phi_n) \to 0$. Assume the contrary. Then, after choosing a subsequence which we again label ϕ_n,[3] we may assume $\phi_n \to 0$, but $|(f, \phi_n)| \geq c > 0$.

Now recall that convergence to 0 in $\mathcal{D}(\Omega)$ means that the supports of all the ϕ_n lie in a fixed compact subset of Ω and that all derivatives of the ϕ_n converge to zero uniformly. After again choosing a subsequence, we may assume that $|D^\alpha \phi_n(\mathbf{x})| \leq 4^{-n}$ for $|\alpha| \leq n$. Let now $\psi_n = 2^n \phi_n$. Then the ψ_n still converge to 0 in $\mathcal{D}(\Omega)$, but $|(f, \psi_n)| \to \infty$.

We shall now recursively construct a subsequence $\{f'_n\}$ of $\{f_n\}$ and a subsequence $\{\psi'_n\}$ of $\{\psi_n\}$. First we choose ψ'_1 such that $|(f, \psi'_1)| > 1$. Since $(f_n, \psi'_1) \to (f, \psi'_1)$, we may choose f'_1 such that $|(f'_1, \psi'_1)| > 1$. Now suppose we have chosen f'_j and ψ'_j for $j < n$. We then choose ψ'_n from the sequence $\{\psi_n\}$ such that

$$|(f'_j, \psi'_n)| < \frac{1}{2^{n-j}}, \quad j = 1, 2, \ldots, n-1, \tag{5.26}$$

$$|(f, \psi'_n)| > \sum_{j=1}^{n-1} |(f, \psi'_j)| + n. \tag{5.27}$$

This is possible because, on the one hand, $\psi_n \to 0$, and, on the other hand, $|(f, \psi_n)| \to \infty$. Since, moreover, $(f_n, \psi) \to (f, \psi)$, we can choose f'_n such that

$$|(f'_n, \psi'_n)| > \sum_{j=1}^{n-1} |(f'_n, \psi'_j)| + n. \tag{5.28}$$

Next we set

$$\psi = \sum_{n=1}^{\infty} \psi'_n. \tag{5.29}$$

---

[3]The use of the same symbol for both a sequence and any of its subsequences is a typical practice in PDEs. Its primary purpose is clarity of notation (since we often have to consider subsequences several levels deep), but it has the pleasant side effect of driving many classical analysts crazy. Of course, there are cases where it is important to distinguish between a sequence and its subsequence (as we do later in this proof) and we do so with appropriate notation.

It follows from the construction of the ψ_n' that the series on the right converges in $\mathcal{D}(\Omega)$. Hence

$$(f_n', \psi) = \sum_{j=1}^{n-1} (f_n', \psi_j') + (f_n', \psi_n') + \sum_{j=n+1}^{\infty} (f_n', \psi_j'). \tag{5.30}$$

From (5.26) we find that

$$\left| \sum_{j=n+1}^{\infty} (f_n', \psi_j') \right| < \sum_{j=n+1}^{\infty} 2^{n-j} = 1, \tag{5.31}$$

and this in conjunction with (5.30) and (5.28) implies that $|(f_n', \psi)| > n-1$. Hence the limit of (f_n', ψ) as $n \to \infty$ does not exist, a contradiction. $\qquad\square$

A similar contradiction argument can be used to prove the following lemma; the details of the proof are left as an exercise (cf. Problem 5.18).

Lemma 5.32. *Assume that $f_n \to 0$ in $\mathcal{D}'(\Omega)$ and $\phi_n \to 0$ in $\mathcal{D}(\Omega)$. Then $(f_n, \phi_n) \to 0$.*

We also have the following corollary.

Corollary 5.33. *If $f_n \to f$ and $\phi_n \to \phi$, then $(f_n, \phi_n) \to (f, \phi)$.*

Hence the pairing between distributions and test functions is continuous. (Of course separate continuity in each factor is obvious from the definitions, but joint continuity requires a proof.)

Proof. The corollary follows immediately from the identity

$$(f_n - f, \phi_n - \phi) = (f_n, \phi_n) - (f, \phi_n) - (f_n, \phi) + (f, \phi). \tag{5.32}$$

$\qquad\square$

5.1.6 Tempered Distributions

It is possible to define different spaces of test functions and, correspondingly, of distributions. In particular, for $\Omega = \mathbb{R}^m$, it is natural to replace the requirement of compact support by one of rapid decay at infinity. This leads to the following definition.

Definition 5.34. Let $\mathcal{S}(\mathbb{R}^m)$ be the space of all complex-valued functions on $\mathbb{R}^m$ which are of class C^∞ and such that $|\mathbf{x}|^k |D^\alpha \phi(\mathbf{x})|$ is bounded for every $k \in \mathbb{N}$ and every multi-index α. We say that a sequence ϕ_n in $\mathcal{S}(\mathbb{R}^m)$ converges to ϕ if the derivatives of all orders of the ϕ_n converge uniformly to those of ϕ and the constants $C_{k\alpha}$ in the bounds $|\mathbf{x}|^k |D^\alpha \phi_n(\mathbf{x})| \le C_{k\alpha}$ can be chosen independently of n.

Obviously, $\mathcal{D}(\mathbb{R}^m)$ is a subspace of $\mathcal{S}(\mathbb{R}^m)$. Moreover, $\mathcal{D}(\mathbb{R}^m)$ is dense in $\mathcal{S}(\mathbb{R}^m)$. To see this, let $e(\mathbf{x})$ be a C^∞-function which is equal to 1 in the unit ball and vanishes outside the ball of radius 2. Let $e_n(\mathbf{x}) = e(\mathbf{x}/n)$. Then, for any $f \in \mathcal{S}(\mathbb{R}^m)$, we have $f = \lim_{n \to \infty} f e_n$.

We now define the tempered distributions to be continuous linear functionals on $\mathcal{S}$.

Definition 5.35. A tempered distribution on $\mathbb{R}^m$ is a linear mapping $\phi \mapsto (f, \phi)$ from $\mathcal{S}(\mathbb{R}^m)$ to $\mathbb{C}$ with the continuity property that $(f, \phi_n) \to (f, \phi)$ if $\phi_n \to \phi$ in $\mathcal{S}(\mathbb{R}^m)$. The set of all tempered distributions is denoted by $\mathcal{S}'(\mathbb{R}^m)$. We say that $f_n \to f$ in $\mathcal{S}'(\mathbb{R}^m)$ if $(f_n, \phi) \to (f, \phi)$ for every $\phi \in \mathcal{S}(\mathbb{R}^m)$.

Clearly, every tempered distribution defines a distribution by restriction. Moreover, if two tempered distributions agree as elements of $\mathcal{D}'(\mathbb{R}^m)$, they also agree as elements of $\mathcal{S}'(\mathbb{R}^m)$; this follows from the fact that $\mathcal{D}(\mathbb{R}^m)$ is dense in $\mathcal{S}(\mathbb{R}^m)$. Hence $\mathcal{S}'(\mathbb{R}^m)$ is a linear subspace of $\mathcal{D}'(\mathbb{R}^m)$. Moreover, convergence in $\mathcal{S}'(\mathbb{R}^m)$ obviously implies convergence in $\mathcal{D}'(\mathbb{R}^m)$.

Problems

5.1. Show that $\phi_{\mathbf{y}, \epsilon} \in \mathcal{D}(\mathbb{R}^m)$.

5.2. Show that the sequence $\phi_n(\mathbf{x}) = n^{-1}\phi_{\mathbf{0}, \epsilon}(\mathbf{x})$ converges to zero in $\mathcal{D}(\mathbb{R}^m)$. Show that the sequence $\psi_n(\mathbf{x}) = n^{-1}\phi_{\mathbf{0}, \epsilon}(\mathbf{x}/n)$ converges to zero uniformly and so do all derivatives. Why does ψ_n nevertheless not converge to zero in $\mathcal{D}(\mathbb{R}^m)$? Does it converge to zero in $\mathcal{S}(\mathbb{R}^m)$?

5.3. Prove Theorem 5.4.

5.4. Let f and g be two different functions in $C(\Omega)$. Show that they also differ as generalized functions.

5.5. Show that the Dirac delta function cannot be identified with any continuous function.

5.6. Explain what it means for a generalized function to be periodic or radially symmetric.

5.7. Let f be a generalized function on Ω and let G be an open subset of Ω. Assume that every point in G has a neighborhood on which f vanishes. Prove that f vanishes on G. Hint: Use a partition of unity argument.

5.8. Prove that if ϕ vanishes in a neighborhood of the support of f, then $(f, \phi) = 0$. Would it suffice if ϕ vanishes on the support of f?

5.9. Show that (5.22) does indeed define a generalized function and that the definition does not depend on ϵ. How can one define a regularization of $1/x^2$?

5.10. Prove that every distribution is the limit of a sequence of distributions with compact support. Hint: Let $f_n = f\psi_n$, where ψ_n is a C^∞ cutoff function.

5.11. Show that

$$\lim_{n \to \infty} \sin(nx) = 0$$

in $\mathcal{D}'(\mathbb{R})$, but that

$$\lim_{n \to \infty} \sin^2(nx) \neq 0.$$

Hence multiplication of distributions is not a continuous operation even where it is defined.

5.12. Let f_n be the sequence defined by (5.23). Show that

$$\lim_{n \to \infty} f_n^2$$

does not exist in the sense of distributions. Show, however, that

$$\lim_{n \to \infty} f_n^2 - n\delta$$

exists.

5.13. Find

$$\lim_{n \to \infty} \sqrt{n} \exp(-nx^2)$$

in the sense of distributions.

5.14. Exhibit a sequence in $\mathcal{S}'(\mathbb{R})$ which converges to zero in $\mathcal{D}'(\mathbb{R})$, but not in $\mathcal{S}'(\mathbb{R})$.

5.15. Show that the sequence ϕ_n converges in $\mathcal{S}(\mathbb{R}^m)$ if and only if $|\mathbf{x}|^k D^\alpha \phi_n(\mathbf{x})$ converges uniformly for every $k \in \mathbb{N} \cup \{0\}$ and every α.

5.16. Show that $\mathcal{S}'(\mathbb{R}^m)$ is sequentially complete.

5.17. Let U_i, $i \in \mathbb{N}$ be open sets such that $\Omega = \bigcup_{i \in \mathbb{N}} U_i$. Let $f_i \in \mathcal{D}'(U_i)$ be given such that f_i and f_j agree on $U_i \cap U_j$. Show that there exists $f \in \mathcal{D}'(\Omega)$ such that $f = f_i$ on U_i.

5.18. Prove Lemma 5.32.

5.19. (a) Let Ω be any open subset of $\mathbb{R}^m$. Show that a family of subsets with the properties of Definition 5.6 exists.

(b) Let $\{U_i\}$ be any countable covering of Ω by open sets. A refinement of $\{U_i\}$ is a covering by open sets V_k, where each V_k is a subset of one of the U_i. Given any covering of Ω by open sets, show that there is a refinement satisfying the properties of Definition 5.6.

5.2 Derivatives and Integrals

5.2.1 Basic Definitions

In this section, we discuss differentiation of distributions and various applications. We shall confine our discussion to distributions in $\mathcal{D}'(\Omega)$; completely analogous considerations apply in $\mathcal{S}'(\mathbb{R}^m)$. We define the derivative of a distribution as follows.

Definition 5.36. Let $f \in \mathcal{D}'(\Omega)$. Then the **derivative** of f with respect to x_j is defined as

$$\left(\frac{\partial f}{\partial x_j}, \phi\right) = -\left(f, \frac{\partial \phi}{\partial x_j}\right). \tag{5.33}$$

Remark 5.37. If f is in $C^1(\Omega)$, this definition agrees with the classical derivative, as can be seen by an integration by parts. It is easy to see that $\partial f / \partial x_j$ is again in $\mathcal{D}'(\Omega)$.

Note that differentiation is a continuous operation in $\mathcal{D}'(\Omega)$, i.e., the reader can show the following.

Theorem 5.38. If $f_n \to f$ in $\mathcal{D}'(\Omega)$ then $\partial f_n / \partial x_j \to \partial f / \partial x_j$.

Thus, for distributions exchanging derivatives and limits is no problem, quite a contrast to the situation in classical calculus.

Remark 5.39. Higher derivatives are defined recursively; the equality of mixed partial derivatives is obvious from the definition and the equality of the mixed partial derivatives of test functions. In general, we have

$$(D^\alpha f, \phi) = (-1)^{|\alpha|}(f, D^\alpha \phi). \tag{5.34}$$

Remark 5.40. The classical derivative is defined as a limit of difference quotients. In a sense, distributional derivatives are still limits of difference quotients. In the previous section, we defined the translation of a distribution by

$$(f(\mathbf{x} + h\mathbf{e}_j), \phi(\mathbf{x})) = (f(\mathbf{x}), \phi(\mathbf{x} - h\mathbf{e}_j)). \tag{5.35}$$

This does not necessarily make sense, because $\mathbf{x} - h\mathbf{e}_j$ need not lie in Ω. For fixed $\phi \in \mathcal{D}(\Omega)$, however, $\phi(\mathbf{x} - h\mathbf{e}_j)$ is in $\mathcal{D}(\Omega)$ provided h is sufficiently small. Hence (5.35) is meaningful for small h, although how small h has to be depends on ϕ. We now find

$$\lim_{h \to 0} \frac{1}{h} [(f(\mathbf{x} + h\mathbf{e}_j), \phi(\mathbf{x})) - (f(\mathbf{x}), \phi(\mathbf{x}))]$$

$$= \lim_{h \to 0} \left(f(\mathbf{x}), \frac{1}{h}(\phi(\mathbf{x} - h\mathbf{e}_j) - \phi(\mathbf{x})) \right)$$

$$= \left(f(\mathbf{x}), \lim_{h \to 0} \frac{1}{h}(\phi(\mathbf{x} - h\mathbf{e}_j) - \phi(\mathbf{x})) \right)$$

$$= - \left(f, \frac{\partial \phi}{\partial x_j} \right)$$

$$= \left(\frac{\partial f}{\partial x_j}, \phi \right).$$

5.2.2 Examples

Example 5.41. Consider the function

$$H(x) = \begin{cases} 0, & x \le 0 \\ 1, & x > 0. \end{cases} \tag{5.36}$$

We compute

$$(H', \phi) = -(H, \phi') = - \int_0^\infty \phi'(x) \, dx = \phi(0) = (\delta, \phi), \tag{5.37}$$

i.e., the derivative of H is the delta function. The function H is called the Heaviside function.

Example 5.42. The kth derivative of the delta function is the functional $\phi \mapsto (-1)^k \phi^{(k)}(0)$.

Example 5.43. Let

$$x_+^\lambda = \begin{cases} 0, & x \le 0 \\ x^\lambda, & x > 0 \end{cases} \tag{5.38}$$

and $-1 < \lambda < 0$. Naively, one may expect that the derivative is $\lambda x_+^{\lambda-1}$, but this function has a nonintegrable singularity and hence it is not a distribution. The proper answer is an appropriate regularization. We compute

$$((x_+^\lambda)', \phi) = -(x_+^\lambda, \phi')$$

$$= -\int_0^\infty x^\lambda \phi'(x) \, dx$$

$$= -\lim_{\epsilon \to 0} \int_\epsilon^\infty x^\lambda \phi'(x) \, dx$$

$$= \lim_{\epsilon \to 0} \int_\epsilon^\infty \lambda x^{\lambda-1}(\phi(x) - \phi(\epsilon)) \, dx$$

$$= \int_0^\infty \lambda x^{\lambda-1}(\phi(x) - \phi(0)) \, dx.$$

Example 5.44. Let Ω be a domain with smooth boundary Γ. Let f be in $C^1(\overline{\Omega})$ and let $f = 0$ in the exterior of Ω. We regard f as a distribution on $\mathbb{R}^m$. We find

$$\left(\frac{\partial f}{\partial x_j}, \phi\right) = -\left(f, \frac{\partial \phi}{\partial x_j}\right)$$

$$= -\int_\Omega f(\mathbf{x}) \frac{\partial \phi}{\partial x_j}(\mathbf{x}) \, d\mathbf{x}$$

$$= \int_\Omega \frac{\partial f}{\partial x_j}(\mathbf{x})\phi(\mathbf{x}) \, d\mathbf{x} - \int_\Gamma f(\mathbf{x})\phi(\mathbf{x})n_j \, dS.$$

Here n_j is the jth component of the unit outward normal to Γ and dS is differential $m-1$ dimensional surface area. Thus, the distributional derivative of f involves one term corresponding to the ordinary derivative in Ω and another term involving a distribution supported on Γ. This latter term results from the jump of f across Γ.

Example 5.45. The previous example has some important applications in electromagnetism. Let $\Omega \subset \mathbb{R}^3$ be a domain with smooth boundary Γ. Suppose we have a **polarization** vector field $\mathbf{p} : \overline{\Omega} \to \mathbb{R}^3$ which is in $C^1(\overline{\Omega})$. By setting $\mathbf{p} = \mathbf{0}$ in the exterior of Ω, we can regard $\mathbf{p}$ as a distribution on $\mathbb{R}^3$. We then define the **polarization charge** to be the divergence of $\mathbf{p}$ *in the sense of distributions*. We calculate this as follows.

$$(\nabla \cdot \mathbf{p}, \phi) = -\sum_{i=1}^3 \left(p_i, \frac{\partial \phi}{\partial x_i}\right)$$

$$= -\int_\Omega \mathbf{p}(\mathbf{x}) \cdot \nabla \phi(\mathbf{x}) \, d\mathbf{x}$$

$$= \int_\Omega \nabla \cdot \mathbf{p}(\mathbf{x})\phi(\mathbf{x}) \, d\mathbf{x} - \int_\Gamma \mathbf{p}(\mathbf{x}) \cdot \mathbf{n}(\mathbf{x})\phi(\mathbf{x}) \, dA.$$

Here **n** is the unit outward normal to Γ and dA is differential surface area on Γ. Thus, the polarization charge involves one term corresponding to the ordinary divergence of **p** in Ω and surface charge given by the normal component of **p** on Γ. This latter term results from the jump of **p** across Γ. If **p** was piecewise smooth with surfaces of jump discontinuity in the interior of Ω, the normal components of the jumps along these surfaces would contribute polarization charge as well.

Example 5.46. Let $f(\mathbf{x}) = 1/|\mathbf{x}| = 1/r$ on $\mathbb{R}^3$. It is easy to check that $\Delta(1/r) = 0$ for $r \neq 0$. We shall evaluate the Laplacian of $1/r$ in the distributional sense. We compute

$$\left(\Delta\frac{1}{r},\phi\right) = \left(\frac{1}{r},\Delta\phi\right)$$

$$= \int_{\mathbb{R}^3}\frac{\Delta\phi(\mathbf{x})}{r}\,d\mathbf{x}$$

$$= \lim_{\epsilon\to 0}\int_{r\geq\epsilon}\frac{\Delta\phi}{r}\,d\mathbf{x}.$$

Integration by parts yields

$$\int_{r\geq\epsilon}\frac{\Delta\phi}{r}\,d\mathbf{x} = \int_{r\geq\epsilon}\Delta\left(\frac{1}{r}\right)\phi\,d\mathbf{x} - \int_{r=\epsilon}\frac{\partial\phi}{\partial r}\frac{1}{r}\,dS + \int_{r=\epsilon}\phi\frac{\partial}{\partial r}\frac{1}{r}\,dS. \quad (5.39)$$

On the right-hand side of (5.39), the first term vanishes, the second is of order ϵ as $\epsilon\to 0$ and the last term is equal to $-\epsilon^{-2}\int_{r=\epsilon}\phi\,dS$, i.e., to -4π times the average of ϕ on the sphere of radius ϵ. Letting $\epsilon\to 0$, we therefore obtain

$$\left(\Delta\frac{1}{r},\phi\right) = -4\pi\phi(\mathbf{0}), \quad (5.40)$$

i.e.,

$$\Delta\frac{1}{r} = -4\pi\delta. \quad (5.41)$$

Solutions of the equation $Lu = \delta$, where L is a partial differential operator with constant coefficients, are of considerable importance; we shall investigate more such solutions in the next two sections.

Example 5.47. In this and the following example, we exploit the fact that differentiation is a continuous operation. Let us consider the Fourier series

$$\cos x + \cos 2x + \cdots + \cos nx + \cdots. \quad (5.42)$$

Clearly, this series does not converge in the ordinary sense; for example, it diverges for $x = 0$. However, the series

$$\sin x + \frac{1}{2}\sin 2x + \frac{1}{3}\sin 3x + \cdots \quad (5.43)$$

converges to $(\pi - x)/2$ on $(0, 2\pi)$, uniformly on every compact subinterval, and the partial sums of the series (5.43) are uniformly bounded on $\mathbb{R}$. (We shall not prove these claims here; instead we refer to texts on Fourier series or advanced calculus or to the discussion of Fourier series in Chapter 6.) From this, it is clear that (5.43) converges in the sense of distributions to the 2π-periodic continuation of $(\pi - x)/2$; that is, a "sawtooth wave" with jumps at integer multiples of 2π. We can write this in terms of the Heaviside function.

$$\sin x + \frac{1}{2}\sin 2x + \frac{1}{3}\sin 3x + \cdots$$
$$= \frac{\pi - x}{2} + \pi \sum_{n=1}^{\infty} H(x - 2\pi n) - \pi \sum_{n=0}^{\infty} H(-2\pi n - x)$$

We now obtain (5.42) by differentiation and the symmetry of the dirac delta:

$$\cos x + \cos 2x + \cos 3x + \cdots = \frac{d}{dx}\left(\sin x + \frac{1}{2}\sin 2x + \frac{1}{3}\sin 3x + \cdots\right)$$
$$= -\frac{1}{2} + \pi \sum_{n \in \mathbb{Z}} \delta(x - 2\pi n) \tag{5.44}$$

(cf. Problem 5.20).

Example 5.48. To prove that a sequence of integrable functions $f_n : \mathbb{R} \to \mathbb{R}$ converges to the delta function, it suffices to show that the primitives converge to the Heaviside function. The following conditions are sufficient for this:

1. For any $\epsilon > 0$, we have

$$\lim_{n \to \infty} \int_{-\infty}^{-a} f_n(x)\, dx = 0, \quad \lim_{n \to \infty} \int_{a}^{\infty} f_n(x)\, dx = 0 \tag{5.45}$$

   uniformly for $a \in [\epsilon, \infty)$;

2.
$$\lim_{n \to \infty} \int_{-\infty}^{\infty} f_n(x)\, dx = 1; \tag{5.46}$$

3. $|\int_{-\infty}^{a} f_n(x)\, dx|$ is bounded by a constant independent of $a \in \mathbb{R}$ and $n \in \mathbb{N}$.

Examples of functions satisfying these conditions are

$$f_\epsilon(x) = \frac{\epsilon}{\pi(x^2 + \epsilon^2)}, \quad \epsilon \to 0, \tag{5.47}$$

$$f_t(x) = \frac{1}{2\sqrt{\pi t}}\exp\left(-\frac{x^2}{4t}\right), \quad t \to 0^+, \tag{5.48}$$

$$f_n(x) = \frac{\sin nx}{\pi x}, \quad n \to \infty. \tag{5.49}$$

5.2.3 Primitives and Ordinary Differential Equations

If the derivatives of a function vanish, the function is a constant. We shall now establish the analogous result for distributions.

Theorem 5.49. *Let Ω be connected, and let $u \in \mathcal{D}'(\Omega)$ be such that $\nabla u = 0$. Then u is a constant.*

Proof. We first consider the one-dimensional case. Let $\Omega = I$ be an interval. The condition that $u' = 0$ means that $(u, \phi') = 0$ for every $\phi \in \mathcal{D}(I)$. In other words, $(u, \psi) = 0$ for every test function ψ which is the derivative of a test function. It is easy to see that ψ is the derivative of a test function iff $\int_I \psi(x)\, dx = 0$. Let now ϕ_0 be any test function with unit integral. Then any $\phi \in \mathcal{D}(I)$ can be decomposed as

$$\phi(x) = \phi_0(x) \int_I \phi(s)\, ds + \psi(x), \tag{5.50}$$

where the integral of ψ is zero. Consequently,

$$(u, \phi) = (u, \phi_0) \int_I \phi(x)\, dx, \tag{5.51}$$

hence u is equal to the constant (u, ϕ_0).

We next consider the case where Ω is a product of intervals: $\Omega = (a_1, b_1) \times (a_2, b_2) \times \cdots \times (a_m, b_m)$. In this case, let $\phi_i \in \mathcal{D}(a_i, b_i)$ be a one-dimensional test function with unit integral. An arbitrary $\phi \in \mathcal{D}(\Omega)$ is now decomposed as follows:

$$\phi(x_1, \ldots, x_m) = \phi_1(x_1) \int_{a_1}^{b_1} \phi(s_1, x_2, \ldots, x_m)\, ds_1 + \psi_1(x_1, \ldots, x_m). \tag{5.52}$$

The function ψ_1 now has the property that

$$\int_{a_1}^{b_1} \psi_1(x_1, x_2, \ldots, x_m)\, dx_1 = 0 \tag{5.53}$$

for every $(x_2, \ldots, x_m)$; hence

$$\int_{a_1}^{x_1} \psi_1(s, x_2, \ldots, x_m)\, ds \tag{5.54}$$

is again a test function. Since $\partial u / \partial x_1 = 0$, it follows that $(u, \psi_1) = 0$. Next, we write

$$\phi_1(x_1) \int_{a_1}^{b_1} \phi(s_1, x_2, \dots, x_m) \, ds_1$$

$$= \phi_1(x_1) \phi_2(x_2) \int_{a_1}^{b_1} \int_{a_2}^{b_2} \phi(s_1, s_2, x_3, \dots, x_m) \, ds_2 \, ds_1$$

$$+ \phi_1(x_1) \psi_2(x_2, \dots, x_m), \tag{5.55}$$

where now

$$\int_{a_2}^{b_2} \psi_2(x_2, \dots, x_m) \, dx_2 = 0, \tag{5.56}$$

and hence $(u, \phi_1 \psi_2) = 0$. Proceeding thusly, we finally obtain

$$(u, \phi) = (u, \phi_1 \phi_2 \cdots \phi_m) \int_{\Omega} \phi(\mathbf{x}) \, d\mathbf{x}, \tag{5.57}$$

i.e., u is a constant.

For general Ω, it follows from the result just proved that every point has a neighborhood in which u is constant, and of course the constants must be the same if two neighborhoods overlap (Problem 5.4). The rest follows from Problem 5.7. $\square$

We next consider the existence of a primitive. Of course, we cannot define a definite integral of a generalized function. Nevertheless, primitives can be shown to exist.

Theorem 5.50. *Let $I = (a, b)$ be an open interval in $\mathbb{R}$ and let $f \in \mathcal{D}'(I)$. Then there exists $u \in \mathcal{D}'(I)$ such that $u' = f$. The primitive u is unique up to a constant.*

Proof. The uniqueness part is clear from the previous theorem. To construct a primitive, we use the decomposition (5.50)

$$\phi(x) = \phi_1(x) \int_I \phi(s) \, ds + \psi(x), \tag{5.58}$$

and we let

$$\chi(x) = \int_a^x \psi(y) \, dy. \tag{5.59}$$

We then define

$$(u, \phi) = C \int_I \phi(x) \, dx - (f, \chi), \tag{5.60}$$

where C is an arbitrary constant. If $\phi = \eta'$, then $\int_I \phi(x) \, dx = 0$ and $\psi = \phi$; hence $\chi = \eta$. We thus find

$$(u, \eta') = -(f, \eta); \tag{5.61}$$

hence $u' = f$.   □

The multidimensional result that any curl-free vectorfield on a simply connected domain is a gradient can also be extended to distributions; the proof is considerably more complicated than in the one-dimensional case and will not be given here.

The most elementary technique of solving an ODE is based on reducing it to the form $y' = f$; this is why solving an ODE is referred to as "integrating" it. Such procedures also work for distributional solutions. Consider, for example, the ODE

$$y' = a(x)y + f(x). \tag{5.62}$$

We assume that $a \in C^\infty(\mathbb{R})$ and $f \in \mathcal{D}'(\mathbb{R})$. We can now set

$$y(x) = z(x)\exp\left(\int_0^x a(s)\,ds\right); \tag{5.63}$$

note that multiplication of distributions by C^∞ functions is well defined and the product rule of differentiation is easily shown to hold. We thus obtain the new ODE

$$z'(x) = f(x)\exp\left(-\int_0^x a(s)\,ds\right). \tag{5.64}$$

From Theorem 5.50, we know that this ODE has a one-parameter family of solutions.

In particular, if f is a continuous function, then all distributional solutions of (5.62) are the classical ones. This is not necessarily true for singular ODEs; for example both the constant 1 and the Heaviside function are solutions of $xy' = 0$.

Problems

5.20. Let f be a piecewise continuous function with a piecewise continuous derivative. Describe the distributional derivative of f.

5.21. Find the distributional derivative of the function $\ln|x|$.

5.22. Let $u(x,t) = f(x+t)$, where f is any locally Riemann integrable function on $\mathbb{R}$. Show that $u_{tt} = u_{xx}$ in the sense of distributions.

5.23. Evaluate $\Delta(1/r^2)$ in $\mathbb{R}^3$.

5.24. Show that $e^x \cos e^x \in \mathcal{S}'(\mathbb{R})$.

5.25. Show that $\sum_{n \in \mathbb{N}} a_n \cos nx$ converges in the sense of distributions, provided $|a_n|$ grows at most polynomially as $n \to \infty$.

5.26. Fill in the details for Example 5.48.

5.27. Discuss how the substitution (5.64) is generalized to systems of ODEs.

5.28. Show that the general solution of $xy' = 0$ is $c_1 + c_2 H(x)$. Hint: Show first that if $\phi \in \mathcal{D}(\mathbb{R})$ vanishes at the origin, then $\phi(x)/x$ is a test function.

5.29. Let $f \in \mathcal{D}'(\mathbb{R})$ be such that $f(x + h) = f(x)$ for every positive h. Show that f is constant.

5.30. Let f_n be a convergent sequence in $\mathcal{D}'(\mathbb{R})$ and assume that $F_n' = f_n$. Assume, in addition, that there is a test function ϕ_0 with a nonzero integral such that the sequence (F_n, ϕ_0) is bounded. Show that F_n has a convergent subsequence.

5.31. Show that an even distribution on $\mathbb{R}$ has an odd primitive.

5.32. Assume that the support of the distribution f is the set $\{0\}$. Show that f is a linear combination of derivatives of the delta function. Hint: Let n be as given by Lemma 5.16 and assume that $D^\alpha \phi(0)$ vanishes for $|\alpha| \leq n$. Let e be a test function which equals 1 for $|\mathbf{x}| \leq 1$ and 0 for $|\mathbf{x}| \geq 2$. Now consider the sequence $\phi_k(\mathbf{x}) = \phi(\mathbf{x})e(k\mathbf{x})$. Show that $(f, \phi_k) \to 0$ and hence $(f, \phi) = 0$.

5.3   Convolutions and Fundamental Solutions

The classical definition of the convolution of two functions defined on $\mathbb{R}^m$ is

$$f * g(\mathbf{x}) = \int_{\mathbb{R}^m} f(\mathbf{x} - \mathbf{y})g(\mathbf{y})\,d\mathbf{y}. \tag{5.65}$$

In this section, we shall consider a generalization of the definition to generalized functions and we shall give applications to the solution of partial differential equations with constant coefficients.

5.3.1   The Direct Product of Distributions

In general, one cannot define the product of two generalized functions $f(\mathbf{x})$ and $g(\mathbf{x})$. However, it is always possible to multiply two generalized functions depending on different variables. That is, if $f \in \mathcal{D}'(\mathbb{R}^p)$ and $g \in \mathcal{D}'(\mathbb{R}^q)$, then $f(\mathbf{x})g(\mathbf{y})$ can be defined as a distribution on $\mathbb{R}^{p+q}$.

Definition 5.51. Let $f \in \mathcal{D}'(\mathbb{R}^p)$, $g \in \mathcal{D}'(\mathbb{R}^q)$. Then the **direct product** $f(\mathbf{x})g(\mathbf{y})$ is the distribution on $\mathbb{R}^{p+q}$ given by

$$(f(\mathbf{x})g(\mathbf{y}), \phi(\mathbf{x}, \mathbf{y})) = (f(\mathbf{x}), (g(\mathbf{y}), \phi(\mathbf{x}, \mathbf{y}))). \tag{5.66}$$

That is, we first regard $\phi(\mathbf{x}, \mathbf{y})$ as a function only of $\mathbf{y}$, which depends on $\mathbf{x}$ as a parameter. To this function we apply the functional g. The result is then a real-valued function $\psi(\mathbf{x})$, which obviously has compact support. It is easy to show that ψ is of class C^∞ (Problem 5.33). Hence

ψ is a test function and (f, ψ) is well defined. To justify the definition, it remains to be shown that $(f(\mathbf{x}), (g(\mathbf{y}), \phi_n(\mathbf{x}, \mathbf{y})))$ converges to zero if ϕ_n converges to zero in $\mathcal{D}(\mathbb{R}^{p+q})$. Since f is a distribution, it suffices to show that $\psi_n := (g(\mathbf{y}), \phi_n(\mathbf{x}, \mathbf{y}))$ converges to zero in $\mathcal{D}(\mathbb{R}^p)$. If $S_p \times S_q$ is a compact set containing the supports of all the ϕ_n, then S_p contains the supports of all the ψ_n. It remains to be shown that ψ_n and all its derivatives converge uniformly to zero. Let α be a p-dimensional multi-index and let $\beta = (\alpha, 0, \ldots, 0)$. Assume that $D^\alpha \psi_n$ does not converge uniformly to zero. After choosing a subsequence, we may assume that there is a sequence of points $\mathbf{x}_n$ such that

$$|D^\alpha \psi_n(\mathbf{x}_n)| = |(g(\mathbf{y}), D^\beta \phi_n(\mathbf{x}_n, \mathbf{y}))| \geq \epsilon. \tag{5.67}$$

But since the ϕ_n converge to zero with all their derivatives, the same is true for the sequence $\chi_n(\mathbf{y}) := D^\beta \phi_n(\mathbf{x}_n, \mathbf{y})$. Hence χ_n converges to zero in $\mathcal{D}(\mathbb{R}^q)$ and therefore (g, χ_n) converges to zero, a contradiction.

Example 5.52. As a simple example of a direct product, we note that

$$\delta(\mathbf{x})\delta(\mathbf{y}) = \delta(\mathbf{x}, \mathbf{y}).$$

If $\phi(\mathbf{x}, \mathbf{y})$ has the special form $\phi_1(\mathbf{x})\phi_2(\mathbf{y})$, we obtain

$$(f(\mathbf{x})g(\mathbf{y}), \phi(\mathbf{x}, \mathbf{y})) = (f, \phi_1)(g, \phi_2). \tag{5.68}$$

Linear combinations of the form $\phi_1(\mathbf{x})\phi_2(\mathbf{y})$ are actually dense in $\mathcal{D}(\mathbb{R}^{p+q})$. To see this, let ϕ have support in the set $Q := \{|\mathbf{x}| \leq a, \ |\mathbf{y}| \leq a\}$. By the Weierstraß approximation theorem (see Section 2.3.3), there is a sequence of polynomials which converges to ϕ uniformly on the set $Q' := \{|\mathbf{x}| \leq 2a, \ |\mathbf{y}| \leq 2a\}$. Moreover, the argument used in the proof of the theorem also shows that the derivatives of the polynomials converge uniformly to those of ϕ on Q'. We can thus choose polynomials p_n in such a way that on Q' we have

$$|D^\alpha p_n - D^\alpha \phi| \leq \frac{1}{n}, \quad \forall |\alpha| \leq n. \tag{5.69}$$

Let now $b_1(\mathbf{x})$, $b_2(\mathbf{y})$ be fixed test functions which are equal to 1 for $|\mathbf{x}| \leq a$ ($|\mathbf{y}| \leq a$) and equal to 0 for $|\mathbf{x}| \geq 2a$ ($|\mathbf{y}| \geq 2a$). Then the sequence

$$\phi_n(\mathbf{x}, \mathbf{y}) := b_1(\mathbf{x})b_2(\mathbf{x})p_n(\mathbf{x}, \mathbf{y}) \tag{5.70}$$

converges to ϕ in $\mathcal{D}(\mathbb{R}^{p+q})$.

This fact and continuity can be used to show properties of the direct product by verifying them only on the restricted set of test functions of the form $\phi_1(\mathbf{x})\phi_2(\mathbf{y})$. One immediate conclusion is that in the definition we can evaluate f and g in the opposite order, i.e., we also have

$$(f(\mathbf{x})g(\mathbf{y}), \phi(\mathbf{x}, \mathbf{y})) = (g(\mathbf{y}), (f(\mathbf{x}), \phi(\mathbf{x}, \mathbf{y}))); \tag{5.71}$$

we express this fact by the suggestive notation

$$f(\mathbf{x})g(\mathbf{y}) = g(\mathbf{y})f(\mathbf{x}). \tag{5.72}$$

Another obvious property is the associative law

$$f(\mathbf{x})(g(\mathbf{y})h(\mathbf{z})) = (f(\mathbf{x})g(\mathbf{y}))h(\mathbf{z}). \tag{5.73}$$

5.3.2 Convolution of Distributions

Let f and g be continuous functions on $\mathbb{R}^m$ which decay rapidly at infinity. We then have the following identity:

$$
\begin{aligned}
(f * g, \phi) &= \int_{\mathbb{R}^m} (f * g)(\mathbf{x})\phi(\mathbf{x}) \, d\mathbf{x} \\
&= \int_{\mathbb{R}^m} \int_{\mathbb{R}^m} f(\mathbf{x} - \mathbf{y})g(\mathbf{y})\phi(\mathbf{x}) \, d\mathbf{x} \, d\mathbf{y} \\
&= \int_{\mathbb{R}^m} \int_{\mathbb{R}^m} f(\mathbf{x})g(\mathbf{y})\phi(\mathbf{x} + \mathbf{y}) \, d\mathbf{x} \, d\mathbf{y}.
\end{aligned}
\tag{5.74}
$$

This identity is used as the definition of the convolution of two distributions.

"Definition" 5.53. Let $f, g \in \mathcal{D}'(\mathbb{R}^m)$. Then the convolution of f and g is defined by

$$(f * g, \phi) = (f(\mathbf{x})g(\mathbf{y}), \phi(\mathbf{x} + \mathbf{y})). \tag{5.75}$$

The quotes are meant to draw attention to the fact that this does not make any sense. We defined the direct product $f(\mathbf{x})g(\mathbf{y})$ as an element of $\mathcal{D}'(\mathbb{R}^{2m})$, but $\phi(\mathbf{x}+\mathbf{y})$ is not in $\mathcal{D}(\mathbb{R}^{2m})$; it does not have compact support. Indeed, the convolution of arbitrary distributions cannot be defined in a rational manner. There are, however, special cases where a meaning can be given to (5.75). In the simplest such case, the support of $\phi(\mathbf{x} + \mathbf{y})$ has a compact intersection with the support of $f(\mathbf{x})g(\mathbf{y})$. If this is the case, we may replace $\phi(\mathbf{x}+\mathbf{y})$ by any test function which agrees with $\phi(\mathbf{x}+\mathbf{y})$ in a neighborhood of $\mathrm{supp}(f(\mathbf{x})g(\mathbf{y}))$. In particular, (5.75) is meaningful under either of the following conditions:

1. Either f or g has compact support.

2. In one dimension, the supports of f and g are bounded from the same side (e.g., $f = 0$ for $x < a$ and $g = 0$ for $x < b$).

From the corresponding properties of the direct product, it follows that convolution is commutative and associative where it is defined.

Let us consider some special cases:

1. We have

$$
\begin{aligned}
(\delta * f, \phi) &= (\delta(\mathbf{x})f(\mathbf{y}), \phi(\mathbf{x} + \mathbf{y})) \\
&= (f(\mathbf{y}), (\delta(\mathbf{x}), \phi(\mathbf{x} + \mathbf{y}))) \\
&= (f(\mathbf{y}), \phi(\mathbf{y})) \\
&= (f, \phi),
\end{aligned}
\tag{5.76}
$$

i.e., $\delta * f = f$.

2. Let us consider $f * \psi$, where $\psi \in \mathcal{D}(\mathbb{R}^m)$. We have

$$(f * \psi, \phi) = (f(\mathbf{x})\psi(\mathbf{y}), \phi(\mathbf{x} + \mathbf{y}))$$

$$= \left(f(\mathbf{x}), \int_{\mathbb{R}^m} \psi(\mathbf{y})\phi(\mathbf{x} + \mathbf{y}) \, d\mathbf{y} \right)$$

$$= \left(f(\mathbf{x}), \int_{\mathbb{R}^m} \psi(\mathbf{y} - \mathbf{x})\phi(\mathbf{y}) \, d\mathbf{y} \right) \tag{5.77}$$

$$= \int_{\mathbb{R}^m} (f(\mathbf{x}), \psi(\mathbf{y} - \mathbf{x}))\phi(\mathbf{y}) \, d\mathbf{y}.$$

In the last step, we have used the continuity of the functional f to take it under the integral; see Problem 5.36. Hence $f * \psi(\mathbf{y})$ is equal to the function $(f(\mathbf{x}), \psi(\mathbf{y} - \mathbf{x}))$. This function is of class C^∞, and if f has compact support, it is a test function.

We next consider differentiation of a convolution. By definition, we have

$$(D^\alpha(f * g), \phi) = (-1)^{|\alpha|}(f * g, D^\alpha \phi)$$

$$= (-1)^{|\alpha|}(g(\mathbf{y}), (f(\mathbf{x}), D^\alpha \phi(\mathbf{x} + \mathbf{y})))$$

$$= (g(\mathbf{y}), (D^\alpha f(\mathbf{x}), \phi(\mathbf{x} + \mathbf{y}))) \tag{5.78}$$

$$= (D^\alpha f * g, \phi).$$

Thus $D^\alpha(f * g) = D^\alpha f * g$, and using commutativity, we also find $D^\alpha(f * g) = f * D^\alpha g$. A convolution is differentiated by differentiating either one of the factors.

The following lemma expresses continuity of the operation of convolution.

Lemma 5.54. *Assume that $f_n \to f$ in $\mathcal{D}'(\mathbb{R}^m)$ and that one of the following holds:*

1. *The supports of all the f_n are contained in a common compact set;*

2. *g has compact support;*

3. *$m = 1$ and the supports of the f_n and of g are bounded on the same side, independently of n.*

Then $f_n * g \to f * g$ in $\mathcal{D}'(\mathbb{R}^m)$.

The proof is left as an exercise (cf. Problem 5.37). A consequence is the following theorem.

Theorem 5.55. $\mathcal{D}(\mathbb{R}^m)$ *is dense in* $\mathcal{D}'(\mathbb{R}^m)$.

Proof. We first show that distributions of compact support are dense. To see this, simply let e_n be a test function which equals 1 on the set $\{|\mathbf{x}| \leq n\}$. Then $e_n f \to f$ for every $f \in \mathcal{D}'(\mathbb{R}^m)$, and the support of $e_n f$ is contained in the support of e_n, hence compact.

It therefore suffices to show that distributions of compact support are limits of test functions. Let f be a distribution of compact support, and let ϕ_n be a delta-convergent sequence of test functions; we may for example choose the sequence $\phi_n = C(1/n)^{-1}\phi_{0,1/n}$, where $\phi_{0,\epsilon}$ and $C(\epsilon)$ are defined by (5.6) and (5.8). Then $\phi_n * f$ is a test function, and by the previous lemma $\phi_n * f$ converges to $\delta * f = f$.   □

5.3.3 Fundamental Solutions

Definition 5.56. Let $L(D)$ be a differential operator with constant coefficients. Then a **fundamental solution** for L is a distribution $G \in \mathcal{D}'(\mathbb{R}^m)$ satisfying the equation $L(D)G = \delta$.

Of course, fundamental solutions are unique only up to a solution of the homogeneous equation $L(D)u = 0$; in choosing a specific fundamental solution one often selects the one with the "nicest" behavior at infinity. The significance of the fundamental solution lies in the fact that

$$L(D)(G * f) = (L(D)G) * f = \delta * f = f, \qquad (5.79)$$

provided that the convolution $G * f$ is defined. If, for example, f has compact support, then $G * f$ is a solution of the equation $L(D)u = f$.

The construction of fundamental solutions for general operators with constant coefficients is quite complicated, and we shall limit our discussion to some important examples.

Example 5.57. Ordinary differential equations. We seek a solution to the ODE

$$a_n G^{(n)}(x) + \cdots + a_0 G(x) = \delta(x). \qquad (5.80)$$

For both positive and negative x, G must agree with a solution of the homogeneous ODE. That is, if $u_1(x), \dots, u_n(x)$ are a complete set of solutions for the homogeneous ODE, then we must have

$$G(x) = \begin{cases} \alpha_1 u_1(x) + \cdots + \alpha_n u_n(x) & x > 0 \\ \beta_1 u_1(x) + \cdots + \beta_n u_n(x) & x < 0. \end{cases} \qquad (5.81)$$

We can now satisfy (5.80) by requiring that all derivatives of G up to the $(n-2)$nd are continuous at 0, but the $(n-1)$st derivative has a jump of magnitude $1/a_n$. With $\gamma_i = \alpha_i - \beta_i$, this yields the system

$$\gamma_1 u_1(0) + \cdots + \gamma_n u_n(0) = 0,$$
$$\gamma_1 u_1'(0) + \cdots + \gamma_n u_n'(0) = 0,$$
$$\vdots \qquad (5.82)$$
$$\gamma_1 u_1^{(n-2)}(0) + \cdots + \gamma_n u_n^{(n-2)}(0) = 0,$$
$$\gamma_1 u_1^{(n-1)}(0) + \cdots + \gamma_n u_n^{(n-1)}(0) = \frac{1}{a_n}.$$

The determinant of this system is the Wronskian of the complete set of solutions u_i and is hence nonzero.

Example 5.58. Laplace's equation. We now seek a solution of the equation

$$\Delta G(\mathbf{x}) = \delta(\mathbf{x}) \tag{5.83}$$

on $\mathbb{R}^m$. Of course this makes G a solution of the homogeneous Laplace equation except at the origin. Moreover, since δ is radially symmetric, it is natural to seek a radially symmetric G. For radially symmetric functions, Laplace's equation reduces to

$$G''(r) + \frac{m-1}{r}G'(r) = 0, \quad r > 0, \tag{5.84}$$

and we obtain the solution

$$G(r) = \begin{cases} c_1 + c_2 r^{2-m} & m \geq 3 \\ c_1 + c_2 \ln r & m = 2. \end{cases} \tag{5.85}$$

We can now satisfy (5.83) by an appropriate choice of c_2. For $m = 3$, we did this calculation in Example 5.46 of the previous section; the result was $c_2 = -1/4\pi$. The general result is obtained in an analogous fashion; one finds the fundamental solutions

$$G(\mathbf{x}) = \begin{cases} -r^{2-m}/(m-2)\Omega_m & m \geq 3 \\ \ln r/2\pi & m = 2. \end{cases} \tag{5.86}$$

Here $\Omega_m = 2\pi^{m/2}/\Gamma(m/2)$ is the surface area of the unit sphere (cf. Problem 4.16).

Example 5.59. The heat equation. For equations which are naturally posed as initial-value problems, a different definition of fundamental solution is used. Consider the equation

$$u_t(\mathbf{x}, t) = L(D)u(\mathbf{x}, t), \quad \mathbf{x} \in \mathbb{R}^m, \ t > 0, \tag{5.87}$$

where L is a constant coefficient differential operator on $\mathbb{R}^m$. Instead of regarding u as a distribution on $\mathbb{R}^m \times (0, \infty)$, we shall in the following regard u as a distribution on $\mathbb{R}^m$, depending on t as a parameter. We say that u depends continuously on t if

$$\int_{\mathbb{R}^m} u(\mathbf{x}, t)\phi(\mathbf{x}) \, d\mathbf{x} \tag{5.88}$$

is continuous in t for every $\phi \in \mathcal{D}(\mathbb{R}^m)$ and we say that u is differentiable with respect to t if (5.88) is differentiable for every ϕ. If u is differentiable with respect to t, the derivative is also a distribution on $\mathbb{R}^m$; this follows from the representation of the derivative as a limit of difference quotients and the completeness of the space of distributions.

If $u(\mathbf{x}, t)$ is a distribution on $\mathbb{R}^m$ depending continuously on $t > 0$, we can also regard u as a distribution on $\mathbb{R}^m \times (0, \infty)$. This is because every test function $\phi(\mathbf{x}, t) \in \mathcal{D}(\mathbb{R}^m \times (0, \infty))$ can also be thought of as a test function on $\mathbb{R}^m$ which depends continuously on the parameter t. Because of Lemma 5.32, this makes

$$\int_{\mathbb{R}^m} u(\mathbf{x}, t)\phi(\mathbf{x}, t) \, d\mathbf{x} \tag{5.89}$$

a continuous function of t; hence

$$\int_0^\infty \int_{\mathbb{R}^m} u(\mathbf{x}, t)\phi(\mathbf{x}, t) \, d\mathbf{x} \, dt \tag{5.90}$$

exists. Hence u defines a linear functional on $\mathcal{D}(\mathbb{R}^m \times (0, \infty))$; the continuity of this functional can be deduced, for example, by representing the outer integral in (5.90) as a limit of Riemann sums and using the completeness of the space of distributions.

We are now ready to define a fundamental solution.

Definition 5.60. We call $G : [0, \infty) \to \mathcal{D}'(\mathbb{R}^m)$ a **fundamental solution** of (5.87) if G is continuously differentiable on $[0, \infty)$, and moreover, G satisfies (5.87) with the initial condition $G(\mathbf{x}, 0) = \delta(\mathbf{x})$.

In this definition, we think of u_t in (5.87) as differentiation with respect to the parameter t. Nevertheless, it is easy to show that G also satisfies (5.87) in the sense of distributions on $\mathbb{R}^m \times (0, \infty)$.

A solution of the inhomogeneous problem

$$u_t = L(D)u + f(\mathbf{x}, t), \quad u(\mathbf{x}, 0) = u_0(\mathbf{x}), \tag{5.91}$$

where f and u_0 have compact support, and f is continuous from $[0, \infty)$ to $\mathcal{D}'(\mathbb{R}^m)$, can now be represented as follows:

$$u(\mathbf{x}, t) = \int_{\mathbb{R}^m} G(\mathbf{x} - \mathbf{y}, t)u_0(\mathbf{y}) \, d\mathbf{y} + \int_0^t \int_{\mathbb{R}^m} G(\mathbf{x} - \mathbf{y}, t - s)f(\mathbf{y}, s) \, d\mathbf{y} \, ds.$$
$$\tag{5.92}$$

The reader should verify that this is indeed a solution (cf. Problem 5.41).

We now consider the heat equation in one space dimension. Problem 1.24 asks for the solution of the problem

$$\begin{aligned} u_t &= u_{xx}, \quad x \in \mathbb{R}, \ t > 0, \\ u(x, 0) &= H(x), \quad x \in \mathbb{R}. \end{aligned} \tag{5.93}$$

The solution can be found by the ansatz $u(x, t) = \phi(x/\sqrt{t})$, which reduces the problem to an ODE. The result is

$$u(x, t) = \frac{1}{2\sqrt{\pi}} \int_{-\infty}^{x/\sqrt{t}} \exp\left(-\frac{v^2}{4}\right) dv. \tag{5.94}$$

To obtain the fundamental solution, we simply need to differentiate with respect to x. We thus obtain

$$G(x,t) = \frac{1}{2\sqrt{\pi t}} \exp\left(-\frac{x^2}{4t}\right). \tag{5.95}$$

The fundamental solution for the heat equation in several dimensions can be obtained as a direct product:

$$G(\mathbf{x},t) = \left(\frac{1}{2\sqrt{\pi t}}\right)^m \exp\left(-\frac{|\mathbf{x}|^2}{4t}\right). \tag{5.96}$$

Example 5.61. The wave equation. For second-order equations

$$u_{tt} = L(D)u \tag{5.97}$$

we define the fundamental solution G as a twice continuously differentiable function $[0,\infty) \to \mathcal{D}'(\mathbb{R}^m)$ such that G satisfies (5.97) with the initial conditions $G(\mathbf{x},0) = 0$, $G_t(\mathbf{x},0) = \delta(\mathbf{x})$. The solution of the inhomogeneous problem

$$u_{tt} = L(D)u + f(\mathbf{x},t),\ u(\mathbf{x},0) = u_0(\mathbf{x}),\ u_t(\mathbf{x},0) = u_1(\mathbf{x}) \tag{5.98}$$

is then represented by

$$u(\mathbf{x},t) = \int_{\mathbb{R}^m} G_t(\mathbf{x}-\mathbf{y},t)u_0(\mathbf{y})\,d\mathbf{y} + \int_{\mathbb{R}^m} G(\mathbf{x}-\mathbf{y},t)u_1(\mathbf{y})\,d\mathbf{y}$$
$$+ \int_0^t \int_{\mathbb{R}^m} G(\mathbf{x}-\mathbf{y},t-s)f(\mathbf{y},s)\,d\mathbf{y}\,ds. \tag{5.99}$$

For the wave equation in one space dimension,

$$G(x,t) = \begin{cases} 1/2 & |x| < t \\ 0 & |x| \geq t \end{cases} \tag{5.100}$$

is a fundamental solution. Indeed, it is obvious that $G(x,0) = 0$ and from the representation $G(x,t) = (H(x+t)-H(x-t))/2$ it follows that G satisfies the wave equation and that $G_t(x,t) = (\delta(x+t)+\delta(x-t))/2$, which equals $\delta(x)$ for $t = 0$. The fundamental solution for the wave equation in several dimensions will be discussed in the next section. We draw attention to the fact that the fundamental solutions for the Laplace and heat equations are (apart from the singularity at the origin) C^∞ functions, but that of the wave equation is not. This has important implications for the regularity of solutions.

Problems

5.33. Let $\phi(\mathbf{x},\mathbf{y}) \in \mathcal{D}(\mathbb{R}^{p+q})$, $g \in \mathcal{D}'(\mathbb{R}^q)$. Show that

$$\psi(\mathbf{x}) := (g(\mathbf{y}), \phi(\mathbf{x},\mathbf{y}))$$

is in $\mathcal{D}(\mathbb{R}^p)$.

5.34. Show that the direct product can also be defined in $\mathcal{S}'$.

5.35. Let F and G be the supports of f and g. Show that the support of the direct product is $F \times G$.

5.36. Let $\phi, \psi \in \mathcal{D}(\mathbb{R}^m)$. Show that $\int_{\mathbb{R}^m} \psi(\mathbf{y}-\mathbf{x})\phi(\mathbf{y}) \, d\mathbf{y}$ defines an element of $\mathcal{D}(\mathbb{R}^m)$. Moreover, show that, in the sense of convergence in $\mathcal{D}(\mathbb{R}^m)$, the integral is a limit of Riemann sums.

5.37. Prove Lemma 5.54.

5.38. Discuss how the proof of Theorem 5.55 needs to be modified to show that $\mathcal{D}(\Omega)$ is dense in $\mathcal{D}'(\Omega)$. Also show that $\mathcal{D}(\mathbb{R}^m)$ is dense in $\mathcal{S}'(\mathbb{R}^m)$.

5.39. Show that the direct product is jointly continuous in its factors.

5.40. Find a fundamental solution for the biharmonic equation $\Delta\Delta G = \delta(\mathbf{x})$ on $\mathbb{R}^m$.

5.41. Prove that (5.92) is indeed a solution of (5.91).

5.42. Let G be the fundamental solution corresponding to the initial-value problem of $u_t = L(D)u$. Show that the functional

$$F : \phi \to \int_0^\infty \int_{\mathbb{R}^m} G(\mathbf{x}, t)\phi(\mathbf{x}, t) \, d\mathbf{x} \, dt \qquad (5.101)$$

defines a distribution on $\mathbb{R}^{m+1}$ and that this distribution satisfies the equation $F_t - L(D)F = \delta(\mathbf{x}, t)$.

5.43. Specialize (5.99) to the one-dimensional wave equation.

5.44. Let f be a distribution with compact support and let P be a polynomial. Show that $P * f$ is a polynomial.

5.4 The Fourier Transform

5.4.1 Fourier Transforms of Test Functions

Definition 5.62. The Fourier transform of a continuous, absolutely integrable function $f : \mathbb{R}^m \to \mathbb{C}$ is defined by[4]

$$\hat{f}(\boldsymbol{\xi}) = \mathcal{F}[f](\boldsymbol{\xi}) := (2\pi)^{-m/2} \int_{\mathbb{R}^m} e^{-i\boldsymbol{\xi}\cdot\mathbf{x}} f(\mathbf{x}) \, d\mathbf{x}. \qquad (5.102)$$

In particular, this defines the Fourier transform of every $f \in \mathcal{S}(\mathbb{R}^m)$. In fact, we have the following result.

---

[4]Definitions in the literature differ as to whether or not the minus sign is included in the exponent and whether the factor $(2\pi)^{-m/2}$ is included.

Theorem 5.63. *If $f \in S(\mathbb{R}^m)$, then $\hat{f} \in S(\mathbb{R}^m)$. Moreover, the mapping $\mathcal{F}$ is continuous from $S(\mathbb{R}^m)$ into itself.*

Proof. If $f \in S(\mathbb{R}^m)$, then clearly $\hat{f}$ is a continuous, bounded function; moreover, if $f_n \to 0$ in $S(\mathbb{R}^m)$, then $\hat{f}_n \to 0$ uniformly. The rest follows from the identities

$$D^\alpha \hat{f}(\boldsymbol{\xi}) = \mathcal{F}[(-i\mathbf{x})^\alpha f](\boldsymbol{\xi}), \tag{5.103}$$

$$(i\boldsymbol{\xi})^\alpha \hat{f}(\boldsymbol{\xi}) = \mathcal{F}[D^\alpha f](\boldsymbol{\xi}). \tag{5.104}$$

The first of these identities is obtained by differentiating under the integral, the second by integration by parts.   □

Equation (5.104) is the principal reason why Fourier transforms are important; if $L(D)$ is a differential operator with constant coefficients, then

$$\mathcal{F}[L(D)u] = L(i\boldsymbol{\xi})\mathcal{F}[u], \tag{5.105}$$

where $L(i\boldsymbol{\xi})$ is the symbol of L defined in Section 2.1. Partial differential equations with constant coefficients can therefore be transformed into algebraic equations by Fourier transform. Of course, knowing the Fourier transform of a solution is of little use, unless we know how to invert the transform. This is addressed by the next theorem.

Theorem 5.64. *Let $g \in S(\mathbb{R}^m)$. Then there is a unique $f \in S(\mathbb{R}^m)$ such that $g = \mathcal{F}[f]$. Furthermore, the inverse Fourier transform of g is given by the formula*

$$f(\mathbf{x}) = (2\pi)^{-m/2} \int_{\mathbb{R}^m} e^{i\boldsymbol{\xi}\cdot\mathbf{x}} g(\boldsymbol{\xi}) \, d\boldsymbol{\xi}. \tag{5.106}$$

Except for the minus sign in the exponent, the formula for the inverse Fourier transform agrees with that for the Fourier transform itself. To evaluate the integrals arising in Fourier transforms, complex contour deformations are often useful; for examples, see Problem 5.46.

Proof. Let $Q_M = [-M, M]^m$, and let f be given by (5.106). Then we find

$$\begin{aligned}
\hat{f}(\boldsymbol{\xi}) &= (2\pi)^{-m/2} \int_{\mathbb{R}^m} e^{-i\boldsymbol{\xi}\cdot\mathbf{x}} f(\mathbf{x}) \, d\mathbf{x} \\
&= (2\pi)^{-m} \int_{\mathbb{R}^m} e^{-i\boldsymbol{\xi}\cdot\mathbf{x}} \int_{\mathbb{R}^m} e^{i\boldsymbol{\eta}\cdot\mathbf{x}} g(\boldsymbol{\eta}) \, d\boldsymbol{\eta} \, d\mathbf{x} \\
&= (2\pi)^{-m} \lim_{M\to\infty} \int_{Q_M} \int_{\mathbb{R}^m} e^{i(\boldsymbol{\eta}-\boldsymbol{\xi})\cdot\mathbf{x}} g(\boldsymbol{\eta}) \, d\boldsymbol{\eta} \, d\mathbf{x} \\
&= (2\pi)^{-m} \lim_{M\to\infty} \int_{\mathbb{R}^m} \int_{Q_M} e^{i(\boldsymbol{\eta}-\boldsymbol{\xi})\cdot\mathbf{x}} g(\boldsymbol{\eta}) \, d\mathbf{x} \, d\boldsymbol{\eta} \\
&= \pi^{-m} \lim_{M\to\infty} \int_{\mathbb{R}^m} \prod_{i=1}^{m} \frac{\sin M(\eta_i - \xi_i)}{\eta_i - \xi_i} g(\boldsymbol{\eta}) \, d\boldsymbol{\eta}.
\end{aligned} \tag{5.107}$$

As we have seen in Example 5.48, the limit of $\sin M(\eta_i - \xi_i)/(\eta_i - \xi_i)$ as $M \to \infty$ is $\pi\delta(\eta_i - \xi_i)$ in the sense of distributions, and also in the sense of tempered distributions. Using this fact and the continuity of the direct product, we find $\hat{f}(\boldsymbol{\xi}) = g(\boldsymbol{\xi})$, i.e., the Fourier transform of f is indeed g.

An analogous calculation shows that if $g = \hat{h}$ for some function $h \in \mathcal{S}(\mathbb{R}^m)$, then $h = f$ as given by (5.106).   □

An important property of the Fourier transform is that it preserves the inner product.

Theorem 5.65. Let $f, \phi \in \mathcal{S}(\mathbb{R}^m)$. Then $(\hat{f}, \hat{\phi}) = (f, \phi)$.

Proof. We have

$$
\begin{aligned}
(f, \phi) &= \int_{\mathbb{R}^m} \overline{f(\mathbf{x})}\phi(\mathbf{x}) \, d\mathbf{x} \\
&= (2\pi)^{-m/2} \int_{\mathbb{R}^m} \overline{f(\mathbf{x})} \int_{\mathbb{R}^m} \hat{\phi}(\boldsymbol{\xi}) e^{i\boldsymbol{\xi}\cdot\mathbf{x}} \, d\boldsymbol{\xi} \, d\mathbf{x} \\
&= (2\pi)^{-m/2} \int_{\mathbb{R}^m} \hat{\phi}(\boldsymbol{\xi}) \int_{\mathbb{R}^m} \overline{f(\mathbf{x}) e^{-i\boldsymbol{\xi}\cdot\mathbf{x}}} \, d\mathbf{x} \, d\boldsymbol{\xi} \\
&= \int_{\mathbb{R}^m} \overline{\hat{f}(\boldsymbol{\xi})}\hat{\phi}(\boldsymbol{\xi}) \, d\boldsymbol{\xi} = (\hat{f}, \hat{\phi}).
\end{aligned}
\tag{5.108}
$$

This completes the proof.   □

5.4.2 Fourier Transforms of Tempered Distributions

The previous theorem motivates the definition of the Fourier transform of a tempered distribution.

Definition 5.66. Let $f \in \mathcal{S}'(\mathbb{R}^m)$. Then the **Fourier transform** of f is defined by the functional

$$
(\mathcal{F}[f], \phi) = (f, \mathcal{F}^{-1}[\phi]), \quad \phi \in \mathcal{S}(\mathbb{R}^m).
\tag{5.109}
$$

It is clear from the definition that $\mathcal{F}$ is a continuous mapping from $\mathcal{S}'(\mathbb{R}^m)$ into itself. It is also easy to check that the formulas (5.103) and (5.104) still hold in $\mathcal{S}'(\mathbb{R}^m)$; the same is true for the inversion formula (5.106), which can be restated as

$$
\mathcal{F}\mathcal{F}[f(\mathbf{x})] = f(-\mathbf{x});
\tag{5.110}
$$

this form has meaning for generalized functions.

We shall now consider a number of examples.

Example 5.67. We have

$$
\begin{aligned}
(\mathcal{F}[\delta], \phi) &= (\delta, \mathcal{F}^{-1}[\phi]) = \mathcal{F}^{-1}[\phi](\mathbf{0}) \\
&= (2\pi)^{-m/2} \int_{\mathbb{R}^m} \phi(\mathbf{x}) \, d\mathbf{x},
\end{aligned}
\tag{5.111}
$$

i.e., the Fourier transform of δ is the constant $(2\pi)^{-m/2}$.

Example 5.68. We have

$$
\begin{aligned}
(\mathcal{F}[1], \phi) = (1, \mathcal{F}^{-1}[\phi]) &= \int_{\mathbb{R}^m} \mathcal{F}^{-1}[\phi](\mathbf{x}) \, d\mathbf{x} \\
&= (2\pi)^{m/2} \mathcal{F} \mathcal{F}^{-1}[\phi](\mathbf{0}) = (2\pi)^{m/2} \phi(\mathbf{0}),
\end{aligned}
\tag{5.112}
$$

i.e., the Fourier transform of 1 is $(2\pi)^{m/2}\delta$. From (5.103), (5.104), it is now clear that the Fourier transforms of polynomials are linear combinations of derivatives of the delta function and vice versa.

Example 5.69. A calculation similar to that in Example 5.68 shows that the Fourier transform of $\exp(i\boldsymbol{\eta} \cdot \mathbf{x})$ (viewed as a function of $\mathbf{x}$) is $(2\pi)^{m/2}\delta(\boldsymbol{\xi} - \boldsymbol{\eta})$. If f is a periodic distribution given by a Fourier series

$$
f(x) = \sum_{n=-\infty}^{\infty} c_n e^{inx},
\tag{5.113}
$$

we find that

$$
\mathcal{F}[f](\boldsymbol{\xi}) = \sqrt{2\pi} \sum_{n=-\infty}^{\infty} c_n \delta(\xi - n).
\tag{5.114}
$$

Example 5.70. Let f be a distribution with compact support. Then we can define (f, ϕ) for any $\phi \in C^\infty(\mathbb{R}^m)$; we set $(f, \phi) = (f, \phi_0)$, where ϕ_0 is any element of $\mathcal{D}(\mathbb{R}^m)$ which agrees with ϕ in a neighborhood of the support of f. It follows from the definition of the support that this definition does not depend on the choice of ϕ_0. In particular, this defines f as an element of $\mathcal{S}'(\mathbb{R}^m)$. We claim now that $\mathcal{F}[f]$ is the function

$$
\mathcal{F}[f](\boldsymbol{\xi}) = (2\pi)^{-m/2}(\overline{f(\mathbf{x})}, e^{-i\boldsymbol{\xi}\cdot\mathbf{x}}).
\tag{5.115}
$$

Here $(\bar{f}, \phi)$ is defined as the complex conjugate of $(f, \bar{\phi})$. To verify the claim, we must show that, for any $\phi \in \mathcal{S}(\mathbb{R}^m)$, we have

$$
\begin{aligned}
(2\pi)^{-\frac{m}{2}} \int_{\mathbb{R}^m} (f(\mathbf{x}), e^{i\boldsymbol{\xi}\cdot\mathbf{x}}) \phi(\boldsymbol{\xi}) \, d\boldsymbol{\xi} &= (f, \mathcal{F}^{-1}[\phi]) \\
&= (2\pi)^{-\frac{m}{2}} \left(f, \int_{\mathbb{R}^m} e^{i\boldsymbol{\xi}\cdot\mathbf{x}} \phi(\boldsymbol{\xi}) \, d\boldsymbol{\xi}\right).
\end{aligned}
\tag{5.116}
$$

That is, we have to justify taking f under the integral, which is accomplished in the usual way by approximating the integral by finite sums. We note that (5.115) defines an entire function of $\boldsymbol{\xi} \in \mathbb{C}^m$. The fact that a distribution of compact support has finite order (Lemma 5.16) implies that for real arguments this function has polynomial growth.

Example 5.71. Fourier transforms which cannot be defined classically as an integral can often be determined as limits of regularizations. As an example, we consider the Heaviside function $H(x)$. Clearly, we cannot define the Fourier transform as

$$\frac{1}{\sqrt{2\pi}} \int_0^\infty e^{-i\xi x}\, dx. \tag{5.117}$$

Observe, however, that

$$H(x) = \lim_{\epsilon \to 0+} H(x) e^{-\epsilon x} \tag{5.118}$$

in the sense of tempered distributions, and consequently

$$\mathcal{F}[H](\xi) = \frac{1}{\sqrt{2\pi}} \lim_{\epsilon \to 0+} \int_0^\infty e^{-\epsilon x - i\xi x}\, dx = \lim_{\epsilon \to 0+} \frac{1}{\sqrt{2\pi}} \frac{1}{\epsilon + i\xi}. \tag{5.119}$$

We can evaluate this limit a little more explicitly by applying it to a test function. For any $\delta > 0$, we have

$$\begin{aligned}
\lim_{\epsilon \to 0+} \int_{-\infty}^\infty \frac{\phi(\xi)}{\epsilon + i\xi}\, d\xi &= \int_{|\xi|>\delta} \frac{\phi(\xi)}{i\xi}\, d\xi + \int_{-\delta}^\delta \frac{\phi(\xi) - \phi(0)}{i\xi}\, d\xi \\
&\quad + \lim_{\epsilon \to 0+} \int_{-\delta}^\delta \frac{\phi(0)}{\epsilon + i\xi}\, d\xi \\
&= \int_{|\xi|>\delta} \frac{\phi(\xi)}{i\xi}\, d\xi + \int_{-\delta}^\delta \frac{\phi(\xi) - \phi(0)}{i\xi}\, d\xi + \pi\phi(0).
\end{aligned} \tag{5.120}$$

We conclude that

$$\mathcal{F}[H] = \frac{1}{\sqrt{2\pi}} \left(\frac{1}{i\xi} + \pi\delta \right), \tag{5.121}$$

where $1/(i\xi)$ is interpreted as a principal value.

Example 5.72. Let f be any continuous function which has polynomial growth at infinity. Then, in the sense of tempered distributions, f is the limit as $M \to \infty$ of

$$f_M(\mathbf{x}) = \begin{cases} f(\mathbf{x}), & |\mathbf{x}| \le M \\ 0, & |\mathbf{x}| > M. \end{cases} \tag{5.122}$$

As a consequence, we find that, in the sense of tempered distributions,

$$\hat{f}(\boldsymbol{\xi}) = (2\pi)^{-m/2} \lim_{M \to \infty} \int_{|\mathbf{x}| \le M} f(\mathbf{x}) e^{-i\boldsymbol{\xi} \cdot \mathbf{x}}\, d\mathbf{x}. \tag{5.123}$$

In particular, if f is integrable at infinity, the Fourier transform of f as a distribution agrees with the ordinary Fourier transform. Another way to evaluate the Fourier transform of functions with polynomial growth is therefore to approximate them by integrable functions, such as $f(\mathbf{x}) \exp(-\epsilon|\mathbf{x}|^2)$. See Problem 5.48 for examples.

Example 5.73. Let $\delta(r-a)$ represent a uniform mass distribution on the sphere of radius a, i.e.,

$$(\delta(r-a),\phi) = \int_{|\mathbf{x}|=a} \phi(\mathbf{x})\, dS. \tag{5.124}$$

(Of course, this is not consistent with our previous use of "δ" as a distribution on $\mathbb{R}^m$, but it is a standard abuse of notation with which the reader should become accustomed.) Then the Fourier transform of $\delta(r-a)$ is given by (5.115)

$$\mathcal{F}[\delta(r-a)](\boldsymbol{\xi}) = (2\pi)^{-m/2} \int_{|\mathbf{x}|=a} e^{-i\boldsymbol{\xi}\cdot\mathbf{x}}\, dS. \tag{5.125}$$

We want to evaluate this expression for $m = 3$. We use polar coordinates with the axis aligned with the direction of $\boldsymbol{\xi}$ so that $\boldsymbol{\xi}\cdot\mathbf{x} = a|\boldsymbol{\xi}|\cos\theta$; we shall use ρ to denote $|\boldsymbol{\xi}|$. We thus find

$$\mathcal{F}[\delta(r-a)](\boldsymbol{\xi}) = (2\pi)^{-3/2} a^2 \int_0^\pi \int_0^{2\pi} e^{-ia\rho\cos\theta} \sin\theta\, d\phi\, d\theta = \sqrt{\frac{2}{\pi}}\, a\, \frac{\sin a\rho}{\rho}. \tag{5.126}$$

Example 5.74. The Fourier transform of a direct product is the direct product of the Fourier transforms. To show this, it suffices to prove agreement for a dense set of test functions. We have

$$(\hat{f}(\boldsymbol{\xi})\hat{g}(\boldsymbol{\eta}), \hat{\phi}(\boldsymbol{\xi})\hat{\psi}(\boldsymbol{\eta})) = (\hat{f},\hat{\phi})(\hat{g},\hat{\psi}) = (f,\phi)(g,\psi) = (f(\mathbf{x})g(\mathbf{y}), \phi(\mathbf{x})\psi(\mathbf{y})). \tag{5.127}$$

5.4.3 The Fundamental Solution for the Wave Equation

The Fourier transform is obviously useful in obtaining fundamental solutions. If $L(D)$ is a constant coefficient operator, then the equation $L(D)u = \delta$ is transformed to $L(i\boldsymbol{\xi})\hat{u} = (2\pi)^{-m/2}$, i.e., to a purely algebraic equation. We immediately obtain

$$\hat{u}(\boldsymbol{\xi}) = \frac{1}{(2\pi)^{m/2} L(i\boldsymbol{\xi})}; \tag{5.128}$$

the only problem is that $L(i\boldsymbol{\xi})$ may have zeros. If (5.128) has nonintegrable singularities, we have to consider appropriate regularizations. Finally, one has to compute the inverse Fourier transform of $\hat{u}(\boldsymbol{\xi})$; this step is not necessarily easy.

Similarly, the Fourier transform can be used to find fundamental solutions for initial-value problems; we shall now do so for the wave equation in $\mathbb{R}^3$. The problem

$$G_{tt} = \Delta G, \quad G(\mathbf{x},0) = 0, \quad G_t(\mathbf{x},0) = \delta(\mathbf{x}) \tag{5.129}$$

is Fourier transformed in the spatial variables only; i.e., we define

$$\hat{G}(\boldsymbol{\xi}, t) = (2\pi)^{-m/2} \int_{\mathbb{R}^m} e^{-i\boldsymbol{\xi} \cdot \mathbf{x}} G(\mathbf{x}, t) \, d\mathbf{x}, \tag{5.130}$$

and apply the same type of transform to (5.129). The result is an ODE in the variable t,

$$\hat{G}_{tt}(\boldsymbol{\xi}, t) = -|\boldsymbol{\xi}|^2 \hat{G}(\boldsymbol{\xi}, t), \quad \hat{G}(\boldsymbol{\xi}, 0) = 0, \quad \hat{G}_t(\boldsymbol{\xi}, 0) = (2\pi)^{-3/2}. \tag{5.131}$$

With $|\boldsymbol{\xi}| = \rho$, the solution is easily obtained as

$$\hat{G}(\boldsymbol{\xi}, t) = (2\pi)^{-3/2} \frac{\sin \rho t}{\rho}. \tag{5.132}$$

Using Example 5.73 above, we find

$$G(\mathbf{x}, t) = \frac{\delta(r - t)}{4\pi t}. \tag{5.133}$$

It can be shown that, in any odd space dimension greater than 1, the fundamental solution of the wave equation can be expressed in terms of derivatives of $\delta(r - t)$; since there is little applied interest in solving the wave equation in more than three dimensions, we shall not prove this here. It is, however, of interest to solve the wave equation in two dimensions. In even space dimensions, it is not easy to evaluate the inverse Fourier transform of $\sin \rho t / \rho$ directly; instead, one uses a trick known as the method of descent. This trick is based on the simple observation that any solution of the wave equation in two dimensions can be regarded as a solution in three dimensions, simply by taking the direct product with the constant function 1. The fundamental solution in two dimensions can therefore be obtained by convolution of (5.133) with $\delta(x)\delta(y)1(z)$. Using the definition of convolution (5.75), we compute

$$\left(\delta(x)\delta(y)1(z) \frac{\delta(r' - t)}{4\pi t}, \phi(\mathbf{x} + \mathbf{x}') \right) = \frac{1}{4\pi t} \int_{-\infty}^{\infty} \int_{r'=t} \phi(x', y', z' + z) \, dS' dz. \tag{5.134}$$

With $\psi(x, y)$ denoting $\int_{-\infty}^{\infty} \phi(x, y, z) \, dz$, (5.134) simplifies to

$$\frac{1}{4\pi t} \int_{r'=t} \psi(x', y') \, dS', \tag{5.135}$$

and evaluation of this integral yields

$$\frac{1}{2\pi} \int_{x^2 + y^2 \leq t^2} \frac{\psi(x, y)}{\sqrt{t^2 - x^2 - y^2}} \, dx \, dy. \tag{5.136}$$

We have thus obtained the following fundamental solution in two space dimensions:

$$G(\mathbf{x}, t) = \begin{cases} 1/2\pi\sqrt{t^2 - r^2}, & r < t \\ 0, & r \geq t. \end{cases} \tag{5.137}$$

We note that the qualitative nature of the fundamental solution for the heat equation does not really change with the space dimension, but the fundamental solution of the wave equation changes dramatically. In any number of dimensions, the support of the fundamental solution for the wave equation is contained in $|\mathbf{x}| \le t$, but otherwise the fundamental solutions look quite different. Whereas the fundamental solution in three dimensions is supported only on the sphere $|\mathbf{x}| = t$, the support of (5.137) fills out the full circle. Television sets in Abbott's *Flatland* [Ab] would have to be designed quite differently from ours; in this context, see also [Mo].

5.4.4 Fourier Transform of Convolutions

Another useful property of the Fourier transform is that it turns convolutions into products and vice versa. We shall first consider test functions. It is easy to see that the product and convolution of functions in $\mathcal{S}(\mathbb{R}^m)$ are again in $\mathcal{S}(\mathbb{R}^m)$. Their behavior under Fourier transform is given by the next lemma.

Lemma 5.75. *Let* $\phi, \psi \in \mathcal{S}(\mathbb{R}^m)$. *Then*

$$\mathcal{F}[\phi * \psi] = (2\pi)^{m/2}\mathcal{F}[\phi]\mathcal{F}[\psi], \tag{5.138}$$

$$\mathcal{F}[\phi\psi] = (2\pi)^{-m/2}\mathcal{F}[\phi] * \mathcal{F}[\psi]. \tag{5.139}$$

Proof. We have

$$
\begin{aligned}
\mathcal{F}[\phi * \psi](\boldsymbol{\xi}) &= (2\pi)^{-m/2} \int_{\mathbb{R}^m} e^{-i\boldsymbol{\xi}\cdot\mathbf{x}} \int_{\mathbb{R}^m} \phi(\mathbf{x}-\mathbf{y})\psi(\mathbf{y})\, d\mathbf{y}\, d\mathbf{x} \\
&= (2\pi)^{-m/2} \int_{\mathbb{R}^m} \psi(\mathbf{y}) \int_{\mathbb{R}^m} e^{-i\boldsymbol{\xi}\cdot\mathbf{x}}\phi(\mathbf{x}-\mathbf{y})\, d\mathbf{x}\, d\mathbf{y} \\
&= (2\pi)^{-m/2} \int_{\mathbb{R}^m} \psi(\mathbf{y}) \int_{\mathbb{R}^m} e^{-i\boldsymbol{\xi}\cdot(\mathbf{z}+\mathbf{y})}\phi(\mathbf{z})\, d\mathbf{z}\, d\mathbf{y} \\
&= (2\pi)^{m/2}\mathcal{F}[\phi](\boldsymbol{\xi})\mathcal{F}[\psi](\boldsymbol{\xi}).
\end{aligned}
\tag{5.140}
$$

This yields (5.138). Applying the inverse Fourier transform, we can restate (5.138) as

$$\mathcal{F}^{-1}[\hat{\phi}\hat{\psi}] = (2\pi)^{-m/2}\mathcal{F}^{-1}[\hat{\phi}] * \mathcal{F}^{-1}[\hat{\psi}]. \tag{5.141}$$

This and the trivial identity

$$\mathcal{F}^{-1}[\hat{\phi}](\mathbf{x}) = \mathcal{F}[\hat{\phi}](-\mathbf{x}) \tag{5.142}$$

lead to (5.139). □

We shall now extend this result to distributions.

Theorem 5.76. *Let* $f, g \in \mathcal{S}'(\mathbb{R}^m)$ *and let* g *have compact support. Then* $f * g \in \mathcal{S}'(\mathbb{R}^m)$ *and*

$$\mathcal{F}[f * g] = (2\pi)^{m/2}\mathcal{F}[f]\mathcal{F}[g]. \tag{5.143}$$

We note that $\mathcal{F}[g]$ is a C^∞ function with polynomial growth (see Example 5.70 above) and therefore the right-hand side of (5.143) is well defined. A similar result can be established for tempered distributions on $\mathbb{R}$ whose supports are bounded from the same side; see Problem 5.52.

Proof. By definition, we have

$$(f * g, \phi) = (f(\mathbf{x}), (g(\mathbf{y}), \phi(\mathbf{x} + \mathbf{y}))). \tag{5.144}$$

Since g has compact support, it is of finite order, i.e., with Q denoting any compact set containing the support of g in its interior, there is $n \in \mathbb{N}$ and $C > 0$ such that

$$|(g(\mathbf{y}), \phi(\mathbf{x} + \mathbf{y}))| \le C \max_{\mathbf{y} \in Q} \sum_{|\alpha| \le n} |D^\alpha \phi(\mathbf{x} + \mathbf{y})|. \tag{5.145}$$

It is easy to see that, for every k and α, $|\mathbf{x}|^k |D^\alpha \phi(\mathbf{x} + \mathbf{y})|$ is bounded uniformly for $\mathbf{x} \in \mathbb{R}^m$ and $\mathbf{y} \in Q$; hence (5.145) leads to a uniform bound for $|\mathbf{x}|^k |(g(\mathbf{y}), \phi(\mathbf{x} + \mathbf{y}))|$. Also, we can replace ϕ in (5.145) by any of its derivatives. We conclude that the mapping $\phi \to (g(\mathbf{y}), \phi(\mathbf{x} + \mathbf{y}))$ is continuous from $\mathcal{S}(\mathbb{R}^m)$ into itself. Hence (5.144) is well defined and represents a continuous linear functional on $\mathcal{S}(\mathbb{R}^m)$.

From the definition of the Fourier transform, we now find

$$(\mathcal{F}[f * g](\boldsymbol{\xi}), \phi(\boldsymbol{\xi})) = (f * g(\mathbf{x}), \mathcal{F}^{-1}[\phi](\mathbf{x}))$$

$$= (2\pi)^{-m/2} \left(f(\mathbf{x}), \left(g(\mathbf{y}), \int_{\mathbb{R}^m} e^{i\boldsymbol{\xi} \cdot (\mathbf{x} + \mathbf{y})} \phi(\boldsymbol{\xi}) \, d\boldsymbol{\xi} \right) \right)$$

$$= \left(f(\mathbf{x}), \int_{\mathbb{R}^m} e^{i\boldsymbol{\xi} \cdot \mathbf{x}} \phi(\boldsymbol{\xi}) \overline{\mathcal{F}[g](\boldsymbol{\xi})} \, d\boldsymbol{\xi} \right)$$

$$= (2\pi)^{m/2} (f, \mathcal{F}^{-1}[\phi \overline{\mathcal{F}[g]}]) = (2\pi)^{m/2} (\mathcal{F}[f], \phi \overline{\mathcal{F}[g]})$$

$$= (2\pi)^{m/2} (\mathcal{F}[f] \mathcal{F}[g], \phi). \tag{5.146}$$

This gives us (5.143) and completes the proof. $\qquad\square$

5.4.5 Laplace Transforms

Let $f \in \mathcal{S}'(\mathbb{R})$ have support contained in $\{x \ge 0\}$. Then obviously $e^{-\mu x} f(x)$ is also in $\mathcal{S}'(\mathbb{R})$ for every $\mu > 0$. Formally, we have

$$\mathcal{F}[e^{-\mu x} f](\xi) = \frac{1}{\sqrt{2\pi}} \int_0^\infty f(x) e^{-i\xi x} e^{-\mu x} \, dx = \mathcal{F}[f](\xi - i\mu). \tag{5.147}$$

Hence it is sensible to define $\mathcal{F}[f](\xi - i\mu)$ as $\mathcal{F}[f \exp(-\mu x)](\xi)$. This defines $\mathcal{F}[f]$ in the lower half of the complex ξ-plane — as a generalized function of Re ξ depending on Im ξ as a parameter. Actually, however, $\mathcal{F}[f]$ is an analytic function of ξ in the open half-plane $\{\text{Im } \xi < 0\}$; this is shown by an argument similar to Example 5.70 (see Problem 5.52).

The Laplace transform is defined as

$$\mathcal{L}[f](s) := \sqrt{2\pi}\mathcal{F}[f](-is); \qquad (5.148)$$

for $f \in \mathcal{S}'(\mathbb{R})$ with support in $\{x \geq 0\}$ it is defined in the right half-plane. Formally, we have

$$\mathcal{L}[f](s) = \int_0^\infty e^{-sx} f(x) \, dx. \qquad (5.149)$$

If f is not in $\mathcal{S}'(\mathbb{R})$, but $\exp(-\mu x)f$ is in $\mathcal{S}'(\mathbb{R})$ for some positive μ, then we can define $\mathcal{L}[f]$ in the half-plane $\{\operatorname{Re} s \geq \mu\}$. We note that by inverting the Fourier transform in (5.148) we obtain

$$e^{-\mu x} f(x) = \frac{1}{\sqrt{2\pi}} \mathcal{F}^{-1}[\mathcal{L}[f](\mu + i\xi)] \qquad (5.150)$$

or, equivalently,

$$f(x) = \frac{1}{2\pi i} \int_{\mu-i\infty}^{\mu+i\infty} e^{sx} \mathcal{L}[f](s) \, ds. \qquad (5.151)$$

In using (5.151), care must be taken that the resulting expression really vanishes for $x < 0$, since this was our basic assumption. Typically, one shows this by closing the contour of integration by a half-circle to the right; e^{sx} decays rapidly in the right half-plane. For this argument to work, it is necessary to choose μ to the right of any singularities of $\mathcal{L}[f]$.

We now give a few examples of applications of Laplace transforms.

Example 5.77. Consider the initial-value problem

$$y'(x) = ay(x) + f(x), \quad y(0) = \alpha. \qquad (5.152)$$

We are interested in a solution for positive x. For negative x, we extend y and f by zero. The extended function does not satisfy (5.152); since it jumps from 0 to α at the origin, its derivative contains a contribution $\alpha\delta(x)$. Thus the proper equation for the extended functions is

$$y'(x) = ay(x) + f(x) + \alpha\delta(x). \qquad (5.153)$$

We now take Laplace transforms. We obtain

$$s\mathcal{L}[y](s) = a\mathcal{L}[y](s) + \mathcal{L}[f](s) + \alpha, \qquad (5.154)$$

and hence

$$\mathcal{L}[y](s) = \frac{\mathcal{L}[f](s) + \alpha}{s - a}. \qquad (5.155)$$

To obtain $y(x)$, we must now invert the Laplace transform; for instance, the inverse Laplace transform of $1/(s-a)$ is found from (5.151) as

$$\frac{1}{2\pi i} \int_{\mu-i\infty}^{\mu+i\infty} \frac{e^{sx}}{s - a} \, ds. \qquad (5.156)$$

This integral is easily evaluated by the method of residues; for $\mu > a$, we obtain e^{ax} for $x > 0$ and 0 for $x < 0$. (Note that if we choose $\mu < a$, we still get a solution of (5.153), but one that vanishes for $x > 0$ rather than $x < 0$; thus we do not get a solution of the original problem (5.152).) If we exploit the fact that the transform of a product is a convolution, we can now write the solution as

$$y(x) = \alpha e^{ax} + \int_0^x e^{a(x-t)} f(t)\, dt, \quad x > 0; \qquad (5.157)$$

of course we could have found this without using transforms.

Example 5.78. Abel's integral equation is

$$\int_0^x \frac{y(t)}{\sqrt{x-t}}\, dt = \sqrt{\pi} f(x), \qquad (5.158)$$

again we seek a solution for $x > 0$ and we think of y and f as being extended by zero for negative x. In order to have a solution, we must obviously have $f(0) = 0$. The left-hand side is the convolution of y and $x_+^{-1/2}$, and the Laplace transform of a convolution is the product of the Laplace transforms. To find the transform of $x_+^{-1/2}$, we compute

$$\int_0^\infty e^{-sx} x^{-1/2}\, dx = \frac{1}{\sqrt{s}} \int_0^\infty e^{-t} t^{-1/2}\, dt = \sqrt{\frac{\pi}{s}} \qquad (5.159)$$

for any real positive s and because of the uniqueness of analytic continuation this also holds for complex s. Hence the transformed equation reads

$$\frac{\mathcal{L}[y](s)}{\sqrt{s}} = \mathcal{L}[f](s), \qquad (5.160)$$

which we write as

$$\mathcal{L}[y](s) = \frac{s\mathcal{L}[f](s)}{\sqrt{s}}. \qquad (5.161)$$

Transforming back, we find

$$y(x) = \frac{1}{\sqrt{\pi}} \int_0^x \frac{f'(t)}{\sqrt{x-t}}\, dt. \qquad (5.162)$$

Example 5.79. The Laplace transform is also applicable to initial-value problems for PDEs. We first remark that Definition 5.66 is easily generalized to define the Fourier transform of a generalized function with respect to only a subset of the variables. For example, when dealing with an initial-value problem, we can take the Laplace transform with respect to time. Of course, to make sense of boundary conditions, one needs to know more about the solution than that it is a generalized function. For example, in the following problem, we may think of u as a generalized function of t depending on x as a parameter.

We consider the initial/boundary-value problem

$$u_t = u_{xx}, \quad x \in (0,1), \ t > 0,$$
$$u(x,0) = 0, \quad x \in (0,1), \qquad\qquad (5.163)$$
$$u(0,t) = u(1,t) = 1, \quad t > 0.$$

As usual, we extend u by zero for negative t. Laplace transform in time leads to the problem

$$s\mathcal{L}[u](x,s) = \mathcal{L}[u]_{xx}(x,s),$$
$$\mathcal{L}[u](0,s) = \mathcal{L}[u](1,s) = \frac{1}{s}. \qquad\qquad (5.164)$$

This equation has the solution

$$\mathcal{L}[u](x,s) = \frac{\cosh(\sqrt{s}(x-1/2))}{s\cosh(\sqrt{s}/2)}. \qquad\qquad (5.165)$$

The formula for the inverse transform yields

$$u(x,t) = \frac{1}{2\pi i}\int_{\mu-i\infty}^{\mu+i\infty} e^{st}\frac{\cosh(\sqrt{s}(x-1/2))}{s\cosh(\sqrt{s}/2)}\,ds. \qquad\qquad (5.166)$$

Here we can take μ to be any positive number. The integral cannot be evaluated in closed form, but of course it can be evaluated numerically; it can also be used to deduce qualitative properties of the solution such as its regularity or its asymptotic behavior as $t \to \infty$.

Problems

5.45. Let $f \in \mathcal{D}(\mathbb{R}^m)$. Under what conditions is $\mathcal{F}[f]$ also in $\mathcal{D}(\mathbb{R}^m)$? Hint: Consider $\mathcal{F}[f]$ as a function of a complex argument.

5.46. Find the Fourier transforms of the following functions: $\exp(-x^2)$, $1/(1+x^2)$, $\sin x/(1+x^2)$.

5.47. Check that (5.103), (5.104) and (5.110) hold for tempered distributions.

5.48. Find the Fourier transforms of $|x|$, $\sin(x^2)$, $x_+^{1/2}$. Hint: Try modifying the functions using factors like $e^{-\epsilon x^2}$ and passing to the limit.

5.49. Let $\mathbf{A}$ be a nonsingular matrix. How is the Fourier transform of $f(\mathbf{A}x)$ related to that of $f(\mathbf{x})$? Use the result to show that the Fourier transform of a radially symmetric tempered distribution is radially symmetric.

5.50. Use the Fourier transform to find the fundamental solution for the heat equation.

5.51. Use the Fourier transform to find the fundamental solution for Laplace's equation in $\mathbb{R}^3$.

5.52. (a) Let $f \in \mathcal{S}'(\mathbb{R})$ and assume that the support of f is contained in $\{x \geq 0\}$. Show that the Fourier transform of f is an analytic function in the half-plane Im $\xi < 0$.
(b) Let $f, g \in \mathcal{S}'(\mathbb{R})$ have support in $\{x \geq 0\}$. Show that $f * g$ also has support in $\{x \geq 0\}$ and that (5.138) holds (in the pointwise sense) in the lower half-plane.

5.53. Use the Laplace transform to find the fundamental solution of the heat equation in one space dimension.

5.54. Show that, for any $t > 0$, (5.166) represents a C^∞ function of x for $x \in [0, 1]$. Hint: First deform the contour into the left half-plane. Then differentiate under the integral.

5.55. In (5.163), replace the heat equation by the backwards heat equation $u_t = -u_{xx}$. Explain what goes wrong when you try to solve the problem by Laplace transform.

5.5   Green's Functions

In the previous two sections, we have considered fundamental solutions for PDEs with constant coefficients. Such fundamental solutions allow the solution of problems of the form $L(D)u = f$, posed on all of space. In practical applications, however, one does not usually want to solve problems posed on all of space; rather one wants to solve PDEs on some domain, subject to certain conditions on the boundary. Green's functions are the analogue of fundamental solutions for this situation. They can be found explicitly only in very special cases. Nevertheless, the concept of Green's functions is useful for theoretical investigations. At present, we do not have the methods available to discuss the existence, uniqueness and regularity of Green's functions for PDEs, and the discussions in this section will to a large extent be formal.

5.5.1   Boundary-Value Problems and their Adjoints

Definition 5.80. Let

$$L(\mathbf{x}, D) = \sum_{|\alpha| \leq k} a_\alpha(\mathbf{x}) D^\alpha \tag{5.167}$$

be a differential operator defined on Ω. Then the **formal adjoint** of L is the operator given by

$$L^*(\mathbf{x}, D)u = \sum_{|\alpha| \leq k} (-1)^{|\alpha|} D^\alpha (a_\alpha(\mathbf{x})u(\mathbf{x})). \tag{5.168}$$

The importance of this definition lies in the fact that

$$(\phi, L(\mathbf{x}, D)\psi) = (L^*(\mathbf{x}, D)\phi, \psi) \tag{5.169}$$

for every $\phi, \psi \in \mathcal{D}(\Omega)$. If the assumption of compact support is removed, then (5.169) no longer holds; instead the integration by parts yields additional terms involving integrals over the boundary $\partial\Omega$. However, these boundary terms vanish if ϕ and ψ satisfy certain restrictions on the boundary. We are interested in the case where the order of L is even, $k = 2p$ and p linear homogeneous boundary conditions on ψ are given. It is then natural to seek p boundary conditions to be imposed on ϕ which would make (5.169) hold. This leads to the notion of an adjoint boundary-value problem.

To make this idea concrete, let us first consider the case of ordinary differential equations. Let

$$L(x, D)u = \sum_{i=0}^{2p} a_i(x)\frac{d^i u}{dx^i}(x), \quad x \in (a, b). \tag{5.170}$$

We assume that $a_i \in C^i([a, b])$; this guarantees that the coefficients of L^* are continuous. Moreover, we assume that $a_{2p}(x) \neq 0$ for $x \in [a, b]$. For any functions $\phi, \psi \in C^{2p}[a, b]$, we compute

$$(\phi, L(x, D)\psi) - (L^*(x, D)\phi, \psi)$$

$$= \sum_{i=1}^{2p}\sum_{k=0}^{i-1}(-1)^k D^{i-k-1}\psi(b)D^k(a_i\phi)(b) \tag{5.171}$$

$$- \sum_{i=1}^{2p}\sum_{k=0}^{i-1}(-1)^k D^{i-k-1}\psi(a)D^k(a_i\phi)(a).$$

The boundary terms can be recast in the form

$$\sum_{k,l=1}^{2p} A_{kl}(b)D^{k-1}\phi(b)D^{l-1}\psi(b) - \sum_{k,l=1}^{2p} A_{kl}(a)D^{k-1}\phi(a)D^{l-1}\psi(a). \tag{5.172}$$

Here A_{kl} vanishes for $k + l > 2p + 1$, and $A_{k(2p+1-k)} = (-1)^{k-1}a_{2p}$. Since a_{2p} was assumed nonzero, this implies that the matrix $\mathbf{A}$ is nonsingular. Now assume that at the point b we have p linearly independent boundary conditions

$$\sum_{j=1}^{2p} b_{ij}D^{j-1}\psi(b) = 0, \quad i = 1, \ldots, p. \tag{5.173}$$

Let $\mathbf{u}$ denote the $2p$ vector with components $u_i = D^{i-1}\psi(b)$ and let $\mathbf{v}$ denote the $2p$ vector with components $v_i = D^{i-1}\phi(b)$. Then (5.173) constrains $\mathbf{u}$ to a p-dimensional subspace X of $\mathbb{R}^{2p}$. The image of X under $\mathbf{A}(b)$ is a p-dimensional subspace Y of $\mathbb{R}^{2p}$. In order to make the first term in (5.172) vanish for every ψ that satisfies (5.173), it is necessary and sufficient to have $\mathbf{v}$ in the orthogonal complement of Y. This yields a set of p boundary conditions for ϕ, which we call the adjoint boundary conditions. An analogous consideration applies at the point a.

As an example, let $L\psi = \psi''''$ with boundary conditions $\psi + \psi' = 2\psi + \psi'' = 0$ at each endpoint. In this case, $L^* = L$, and (5.171) specializes to

$$
\int_a^b \phi(x)\psi''''(x)\,dx = \int_a^b \phi''''(x)\psi(x)\,dx
$$
$$
+\phi(b)\psi'''(b) - \phi'(b)\psi''(b) + \phi''(b)\psi'(b) - \phi'''(b)\psi(b)
$$
$$
-\phi(a)\psi'''(a) + \phi'(a)\psi''(a) - \phi''(a)\psi'(a) + \phi'''(a)\psi(a).
$$
$$(5.174)$$

The matrix $\mathbf{A}$ in (5.172) is

$$
\mathbf{A} = \begin{pmatrix} 0 & 0 & 0 & 1 \\ 0 & 0 & -1 & 0 \\ 0 & 1 & 0 & 0 \\ -1 & 0 & 0 & 0 \end{pmatrix}, \tag{5.175}
$$

and the vector $\mathbf{u}$ is subject to the conditions $u_1 + u_2 = 2u_1 + u_3 = 0$. A basis for the space X of vectors satisfying these conditions is given by the vectors $(1, -1, -2, 0)$ and $(0, 0, 0, 1)$. The images of these vectors under $\mathbf{A}$ are $(0, 2, -1, -1)$ and $(1, 0, 0, 0)$. Thus the vector $\mathbf{v}$ has to satisfy the conditions $2v_2 - v_3 - v_4 = 0$, $v_1 = 0$, i.e., the adjoint boundary conditions are $\phi = 2\phi' - \phi'' - \phi''' = 0$.

Let now Ω be a bounded domain in $\mathbb{R}^m$ with a smooth boundary.[5] Let $L(\mathbf{x}, D)$ be a differential operator of order $2p$ with smooth coefficients defined on $\overline{\Omega}$. Moreover, let $B_j(\mathbf{x}, D)$, $j = 1, \ldots, p$, be differential operators of orders less than $2p$ which are defined for $\mathbf{x} \in \partial\Omega$. In the following, we are concerned with the boundary-value problem

$$
L(\mathbf{x}, D)u = f(\mathbf{x}), \quad \mathbf{x} \in \Omega,
$$
$$
B_j(\mathbf{x}, D)u = 0, \quad \mathbf{x} \in \partial\Omega, \; j = 1, \ldots, p. \tag{5.176}
$$

We assume that there are additional differential operators

$$
S_j(\mathbf{x}, D), \quad T_j(\mathbf{x}, D), \quad C_j(\mathbf{x}, D), \quad j = 1, \ldots, p,
$$

defined for $\mathbf{x} \in \partial\Omega$, with the following properties:

1. S_j, T_j and C_j have smooth coefficients and orders less than $2p$.

2. Given any set of smooth functions ϕ_j, $j = 1, \ldots, 2p$, defined on $\partial\Omega$, there exist functions $u, v \in C^{2p}(\overline{\Omega})$ such that on $\partial\Omega$ we have $B_j u =$

---

[5]Since this section focuses on introducing basic concepts without any precise statement of results, we shall be vague about smoothness assumptions. "Smooth" should therefore be interpreted to mean "as smooth as may be needed."

ϕ_j, $S_j u = \phi_{p+j}$, $j = 1, \ldots, p$, and, respectively, $C_j v = \phi_j$, $T_j v = \phi_{p+j}$, $j = 1, \ldots, p$.

3. For any $u, v \in C^{2p}(\overline{\Omega})$, we have

$$\int_\Omega vL(\mathbf{x}, D)u - uL^*(\mathbf{x}, D)v \; d\mathbf{x} = \int_{\partial\Omega} \sum_{j=1}^p S_j(\mathbf{x}, D)uC_j(\mathbf{x}, D)v$$
$$- B_j(\mathbf{x}, D)uT_j(\mathbf{x}, D)v \; dS. \tag{5.177}$$

Of course, the question of what assumptions a boundary-value problem must satisfy for such operators to exist is of crucial importance; we shall return to this issue later when we discuss elliptic boundary-value problems. For the moment, we simply take the existence of the S_j, T_j and C_j for granted.

Definition 5.81. Let the preceding assumptions hold. Then the boundary-value problem

$$L^*(\mathbf{x}, D)v = g(\mathbf{x}), \quad \mathbf{x} \in \Omega,$$
$$C_j(\mathbf{x}, D)v = 0, \quad \mathbf{x} \in \partial\Omega, \; j = 1, \ldots, p. \tag{5.178}$$

is called **adjoint** to (5.176).

We note that if u and v satisfy (5.176) and (5.178), respectively, then, according to (5.177),

$$\int_\Omega fv - gu \; d\mathbf{x} = 0. \tag{5.179}$$

We have made no claim that the operators C_j are unique, and indeed, even for ordinary differential equations, the adjoint boundary conditions are determined only up to linear recombination. We can, however, give an intrinsic characterization of the set of functions characterized by the conditions $C_j v = 0$.

Lemma 5.82. Let $v \in C^{2p}(\overline{\Omega})$ and let X_B denote the set of all $u \in C^{2p}(\overline{\Omega})$ such that $B_j u = 0$ on $\partial\Omega$ for $j = 1, \ldots, p$. Then $(v, Lu) = (L^*v, u)$ for every $u \in X_B$ iff $C_j v = 0$ for $j = 1, \ldots, p$.

Proof. One direction is obvious from (5.177). To see the converse, we note that by assumption we can construct $u \in C^{2p}(\overline{\Omega})$ such that $B_j u = 0$ and $S_j u = \phi_j$, where ϕ_j are given smooth functions. If $(v, Lu) = (L^*v, u)$, then (5.177) assumes the form

$$\int_{\partial\Omega} \sum_{j=1}^p \phi_j(\mathbf{x})C_j(\mathbf{x}, D)v \; dS = 0. \tag{5.180}$$

If this holds for arbitrary ϕ_j, then clearly $C_j v$ must be zero.   □

Thus, although there may be different sets of adjoint boundary conditions, they must be equivalent to each other. As a caution, we note that equivalent sets of boundary conditions need not be linear combinations of each other. For example, let $\partial\Omega$ be a closed curve in $\mathbb{R}^2$ and let s denote arclength. Then the conditions $v = 0$ and $dv/ds + v = 0$ are equivalent, although they are not multiples of each other.

We conclude this subsection with two examples:

Example 5.83. We have

$$\int_\Omega u\Delta v - v\Delta u \; dx = \int_{\partial\Omega} u\frac{\partial v}{\partial n} - v\frac{\partial u}{\partial n} \; dS. \tag{5.181}$$

Hence the Dirichlet and Neumann boundary-value problems for Laplace's equation are their own adjoints.

Example 5.84. Let Ω be a bounded domain in $\mathbb{R}^2$ bounded by a closed smooth curve. Let s denote arclength along the curve. Consider the boundary-value problem

$$\Delta u = f(\mathbf{x}), \; \mathbf{x} \in \Omega, \quad \frac{\partial u}{\partial n} + \frac{\partial u}{\partial s} = 0, \; \mathbf{x} \in \partial\Omega. \tag{5.182}$$

We find

$$
\begin{aligned}
\int_\Omega u\Delta v - v\Delta u \; dx &= \int_{\partial\Omega} u\frac{\partial v}{\partial n} - v\frac{\partial u}{\partial n} \; ds \\
&= \int_{\partial\Omega} u\left(\frac{\partial v}{\partial n} - \frac{\partial v}{\partial s}\right) - v\left(\frac{\partial u}{\partial n} + \frac{\partial u}{\partial s}\right) \; ds.
\end{aligned}
\tag{5.183}
$$

Hence the adjoint boundary-value problem is

$$\Delta v = g(\mathbf{x}), \; \mathbf{x} \in \Omega, \quad \frac{\partial v}{\partial n} - \frac{\partial v}{\partial s} = 0, \; \mathbf{x} \in \partial\Omega. \tag{5.184}$$

5.5.2   Green's Functions for Boundary-Value Problems

We shall consider the boundary-value problem (5.176), and we make the assumptions of the last section.

Definition 5.85. A **Green's function** $G(\mathbf{x}, \mathbf{y})$ for (5.176) is a solution of the problem

$$
\begin{aligned}
L(\mathbf{x}, D_{\mathbf{x}})G(\mathbf{x}, \mathbf{y}) &= \delta(\mathbf{x} - \mathbf{y}), \quad \mathbf{x}, \mathbf{y} \in \Omega, \\
B_j(\mathbf{x}, D_{\mathbf{x}})G(\mathbf{x}, \mathbf{y}) &= 0, \quad \mathbf{x} \in \partial\Omega, \; \mathbf{y} \in \Omega, \; j = 1, \ldots, p.
\end{aligned}
\tag{5.185}
$$

The first equation in (5.185) is to be interpreted in the sense of distributions. Of course, giving a meaning to the boundary conditions requires more smoothness of G than that it be a distribution. For elliptic boundary-value problems, however, it turns out that G is smooth as long as $\mathbf{x} \neq \mathbf{y}$, and hence the interpretation of the boundary conditions poses no problems. Clearly, the concept of a Green's function generalizes that of a fundamental solution. If L has constant coefficients, it is in fact often advantageous to think of the Green's function as a perturbation of the fundamental solution. Namely, if $\mathcal{G}(\mathbf{x} - \mathbf{y})$ is the fundamental solution, we set $G(\mathbf{x}, \mathbf{y}) = \mathcal{G}(\mathbf{x} - \mathbf{y}) + g(\mathbf{x}, \mathbf{y})$ where g satisfies

$$L(\mathbf{x}, D_{\mathbf{x}})g(\mathbf{x}, \mathbf{y}) = 0, \quad \mathbf{x}, \mathbf{y} \in \Omega, \tag{5.186}$$

and for $j = 1, \ldots, p$

$$B_j(\mathbf{x}, D_{\mathbf{x}})g(\mathbf{x}, \mathbf{y}) = -B_j(\mathbf{x}, D_{\mathbf{x}})\mathcal{G}(\mathbf{x} - \mathbf{y}), \quad \mathbf{x} \in \partial\Omega, \ \mathbf{y} \in \Omega. \tag{5.187}$$

If $\mathcal{G}$ is smooth for $\mathbf{x} \neq \mathbf{y}$, then the right-hand side of (5.187) is smooth for every $\mathbf{y} \in \Omega$. For elliptic problems, we shall see in Chapter 9 that this implies that g is also smooth. In the interior of Ω, the fundamental solution in a sense "dominates" the Green's function by contributing the most singular part.

It is of course of fundamental importance to identify classes of boundary-value problems for which Green's functions exist and are unique. At present, we do not have the techniques available which are required to address this question, but we shall address the issue of existence for elliptic equations in Chapter 9.

If a Green's function exists, then a formal solution of (5.176) is

$$u(\mathbf{x}) = \int_\Omega G(\mathbf{x}, \mathbf{y}) f(\mathbf{y}) \, d\mathbf{y}; \tag{5.188}$$

in fact, if $f \in \mathcal{D}(\Omega)$, then (5.188) gives a solution of (5.176) under fairly minimal assumptions on G. It suffices, for example, if $G(\cdot, \mathbf{y})$ as an element of $\mathcal{D}'(\Omega)$ depends continuously on $\mathbf{y}$ and G is smooth for $\mathbf{x} \neq \mathbf{y}$. In particular, if the boundary-value problem (5.176) is uniquely solvable, (5.188) leads to the identity

$$u(\mathbf{x}) = \int_\Omega G(\mathbf{x}, \mathbf{y}) L(\mathbf{y}, D_{\mathbf{y}}) u(\mathbf{y}) \, d\mathbf{y} \tag{5.189}$$

for all $u \in \mathcal{D}(\Omega)$. We shall now assume that G has sufficient regularity to establish (5.188) not only for $u \in \mathcal{D}(\Omega)$, but for every $u \in X_B$. Using (5.177), we conclude that

$$u(\mathbf{x}) = \int_\Omega u(\mathbf{y}) L^*(\mathbf{y}, D_{\mathbf{y}}) G(\mathbf{x}, \mathbf{y}) \, d\mathbf{y}$$
$$+ \int_{\partial\Omega} \sum_{j=1}^p S_j(\mathbf{y}, D_{\mathbf{y}}) u(\mathbf{y}) C_j(\mathbf{y}, D_{\mathbf{y}}) G(\mathbf{x}, \mathbf{y}) \, dS_{\mathbf{y}}. \tag{5.190}$$

If this holds for arbitrary $u \in X_B$, we find that, for every $\mathbf{x} \in \Omega$, we must have

$$L^*(\mathbf{y}, D_{\mathbf{y}}) G(\mathbf{x}, \mathbf{y}) = \delta(\mathbf{x} - \mathbf{y}), \quad \mathbf{x}, \mathbf{y} \in \Omega,$$
$$C_j(\mathbf{y}, D_{\mathbf{y}}) G(\mathbf{x}, \mathbf{y}) = 0, \quad \mathbf{y} \in \partial\Omega, \ \mathbf{x} \in \Omega, \ j = 1, \ldots, p. \tag{5.191}$$

That is, G, regarded as a function of $\mathbf{y}$ for fixed $\mathbf{x}$, satisfies the adjoint boundary-value problem.

Using (5.191) and setting $v(\mathbf{y}) = G(\mathbf{x}, \mathbf{y})$ in (5.177), we find

$$
\begin{aligned}
u(\mathbf{x}) = & \int_{\Omega} G(\mathbf{x}, \mathbf{y}) L(\mathbf{y}, D_{\mathbf{y}}) u(\mathbf{y}) \, d\mathbf{y} \\
& + \int_{\partial\Omega} \sum_{j=1}^{p} T_j(\mathbf{y}, D_{\mathbf{y}}) G(\mathbf{x}, \mathbf{y}) B_j(\mathbf{y}, D_{\mathbf{y}}) u(\mathbf{y}) \, dS_{\mathbf{y}}.
\end{aligned}
\tag{5.192}
$$

Thus, if the inhomogeneous boundary-value problem

$$
\begin{aligned}
L(\mathbf{x}, D)u &= f(\mathbf{x}), \quad \mathbf{x} \in \Omega, \\
B_j(\mathbf{x}, D)u &= \phi_j(\mathbf{x}), \quad \mathbf{x} \in \partial\Omega, \ j = 1, \ldots, p
\end{aligned}
\tag{5.193}
$$

has a solution, then the solution is represented by

$$
\begin{aligned}
u(\mathbf{x}) = & \int_{\Omega} G(\mathbf{x}, \mathbf{y}) f(\mathbf{y}) \, d\mathbf{y} \\
& + \int_{\partial\Omega} \sum_{j=1}^{p} T_j(\mathbf{y}, D_{\mathbf{y}}) G(\mathbf{x}, \mathbf{y}) \phi_j(\mathbf{y}) \, dS_{\mathbf{y}}.
\end{aligned}
\tag{5.194}
$$

As a caution, we note that in justifying the integration by parts which leads to (5.192), it is important that $\mathbf{x} \in \Omega$ so that $G(\mathbf{x}, \mathbf{y})$ is smooth for $\mathbf{y} \in \partial\Omega$. In general, (5.192) does not represent $u(\mathbf{x})$ for $\mathbf{x} \in \partial\Omega$.

For some simple equations in simple domains, Green's functions can be given explicitly. As an example, we consider Laplace's equation on the ball B_R of radius R with Dirichlet boundary conditions. In this case, the Green's function can be constructed by what is known as the method of images. The fundamental solution $\mathcal{G}(|\mathbf{x} - \mathbf{y}|)$ can be thought of as the potential of a point charge located at the point $\mathbf{y}$. The idea is now to put a second point charge at the reflected point $\bar{\mathbf{y}} = R^2\mathbf{y}/|\mathbf{y}|^2$ in such a way that the potentials of the two charges cancel each other on the sphere $|\mathbf{x}| = R$. This leads to the Green's function

$$
\begin{aligned}
G(\mathbf{x}, \mathbf{y}) &= \mathcal{G}(|\mathbf{x} - \mathbf{y}|) - \mathcal{G}\left(\frac{|\mathbf{y}|}{R}|\mathbf{x} - \bar{\mathbf{y}}|\right) \\
&= \mathcal{G}(\sqrt{|\mathbf{x}|^2 + |\mathbf{y}|^2 - 2\mathbf{x} \cdot \mathbf{y}}) - \mathcal{G}(\sqrt{(|\mathbf{x}||\mathbf{y}|/R)^2 + R^2 - 2\mathbf{x} \cdot \mathbf{y}}).
\end{aligned}
\tag{5.195}
$$

If $\mathbf{y} = \mathbf{0}$, we set $G(\mathbf{x}, \mathbf{y}) = \mathcal{G}(|\mathbf{x}|) - \mathcal{G}(R)$. If $|\mathbf{x}| = R$, then $G(\mathbf{x}, \mathbf{y}) = 0$; moreover, we compute

$$
\Delta_{\mathbf{x}} G(\mathbf{x}, \mathbf{y}) = \delta(\mathbf{x} - \mathbf{y}) - \frac{|\mathbf{y}|^2}{R^2} \delta(\mathbf{x} - \bar{\mathbf{y}}),
\tag{5.196}
$$

which agrees with $\delta(\mathbf{x} - \mathbf{y})$ if $\mathbf{x}$ and $\mathbf{y}$ are restricted to B_R. Hence G is indeed a Green's function for the Dirichlet problem. We see that G is symmetric in its arguments, reflecting the self-adjointness of the Dirichlet problem.

The solution of the Dirichlet problem

$$\Delta u = f(\mathbf{x}), \quad \mathbf{x} \in B_R, \quad u = \phi(\mathbf{x}), \quad \mathbf{x} \in \partial B_R \qquad (5.197)$$

is represented by (5.194):

$$u(\mathbf{x}) = \int_{B_R} G(\mathbf{x}, \mathbf{y}) f(\mathbf{y}) \, d\mathbf{y} + \int_{\partial B_R} \frac{\partial}{\partial n_\mathbf{y}} G(\mathbf{x}, \mathbf{y}) \phi(\mathbf{y}) \, dS_\mathbf{y}. \qquad (5.198)$$

Moreover, a direct calculation shows that

$$\frac{\partial}{\partial n_\mathbf{y}} G(\mathbf{x}, \mathbf{y}) = \frac{R^2 - |\mathbf{x}|^2}{\Omega_m R |\mathbf{x} - \mathbf{y}|^m} \qquad (5.199)$$

(cf. Problem 5.60). This leads to Poisson's formula, which we have already encountered in Section 4.2.

5.5.3 Boundary Integral Methods

Let Ω be a bounded domain with a smooth boundary. We want to solve the problem

$$\Delta u = 0, \ \mathbf{x} \in \Omega, \quad u = \phi(\mathbf{x}), \ \mathbf{x} \in \partial\Omega. \qquad (5.200)$$

If we knew the Green's function, we would have the representation

$$u(\mathbf{x}) = \int_{\partial\Omega} \frac{\partial}{\partial n_\mathbf{y}} G(\mathbf{x}, \mathbf{y}) \phi(\mathbf{y}) \, dS_\mathbf{y}. \qquad (5.201)$$

We make an ansatz analogous to (5.201), with the Green's function replaced by the fundamental solution of the Laplace equation

$$u(\mathbf{x}) = \int_{\partial\Omega} \frac{\partial}{\partial n_\mathbf{y}} \mathcal{G}(|\mathbf{x} - \mathbf{y}|) g(\mathbf{y}) \, dS_\mathbf{y}. \qquad (5.202)$$

Here the function g is unknown, and we are seeking an equation relating g to ϕ.

We note that for any $g \in C(\partial\Omega)$, the function u given by (5.202) is harmonic in Ω; we can simply take the Laplacian with respect to $\mathbf{x}$ under the integral. To satisfy the boundary condition, we must have

$$\phi(\mathbf{x}) = \lim_{\mathbf{z} \to \mathbf{x}, \mathbf{z} \in \Omega} \int_{\partial\Omega} \frac{\partial}{\partial n_\mathbf{y}} \mathcal{G}(|\mathbf{z} - \mathbf{y}|) g(\mathbf{y}) \, dS_\mathbf{y} \qquad (5.203)$$

for $\mathbf{x} \in \partial\Omega$; this is the desired equation relating g to ϕ. One cannot pass to the limit in (5.203) by simply substituting $\mathbf{x}$ for $\mathbf{z}$; although the integral exists for $\mathbf{z} \in \partial\Omega$, it is discontinuous there. Indeed, we shall show below that actually

$$\lim_{\mathbf{z} \to \mathbf{x}, \mathbf{z} \in \Omega} \int_{\partial\Omega} \frac{\partial}{\partial n_\mathbf{y}} \mathcal{G}(|\mathbf{z} - \mathbf{y}|) g(\mathbf{y}) \, dS_\mathbf{y}$$
$$= \int_{\partial\Omega} \frac{\partial}{\partial n_\mathbf{y}} \mathcal{G}(|\mathbf{x} - \mathbf{y}|) g(\mathbf{y}) \, dS_\mathbf{y} + \frac{1}{2} g(\mathbf{x}). \qquad (5.204)$$

Recall that a similar situation applies to Cauchy's formula; if C is a smooth closed curve in the plane and f is analytic, then

$$\frac{1}{2\pi i} \int_C \frac{f(\zeta)}{\zeta - z}\, d\zeta \qquad (5.205)$$

equals $f(z)$ for z inside C, 0 for z outside C and (in the sense of principal value) $f(z)/2$ on C. Inserting (5.204) in (5.203), we obtain

$$\phi(\mathbf{x}) - \frac{1}{2}g(\mathbf{x}) = \int_{\partial\Omega} \frac{\partial}{\partial n_{\mathbf{y}}} \mathcal{G}(|\mathbf{x} - \mathbf{y}|)g(\mathbf{y})\, dS_{\mathbf{y}}. \qquad (5.206)$$

We have thus replaced the partial differential equation (5.200) by the equivalent integral equation (5.206). This has two advantages. As we shall see in Chapter 9, it is fairly easy to develop an existence theory for integral equations such as (5.206). Moreover, a numerical approach based on (5.206) rather than (5.200) has the advantage of working with a problem in a lower space dimension, which translates into fewer gridpoints. Indeed, there is an extensive literature on "boundary-element methods" for Laplace's equation as well as for the Stokes equation.

It remains to verify (5.204). Let $\mathbf{z}$ be close to $\partial\Omega$, and let $\mathbf{x}$ be the point on $\partial\Omega$ nearest to $\mathbf{z}$; without loss of generality, we may choose the coordinate system in such a way that $\mathbf{x}$ is the origin and the normal to $\partial\Omega$ is in the mth coordinate direction. Let N be a neighborhood of the origin; we can then split up the right-hand side of (5.203) as follows:

$$\int_{\partial\Omega} \frac{\partial}{\partial n_{\mathbf{y}}} \mathcal{G}(|\mathbf{z} - \mathbf{y}|)g(\mathbf{y})\, dS_{\mathbf{y}} = \int_{\partial\Omega \cap N} \frac{\partial}{\partial n_{\mathbf{y}}} \mathcal{G}(|\mathbf{z} - \mathbf{y}|)g(\mathbf{y})\, dS_{\mathbf{y}}$$
$$+ \int_{\partial\Omega \setminus N} \frac{\partial}{\partial n_{\mathbf{y}}} \mathcal{G}(|\mathbf{z} - \mathbf{y}|)g(\mathbf{y})\, dS_{\mathbf{y}}. \qquad (5.207)$$

The second term is continuous at $\mathbf{z} = \mathbf{0}$. For the first term, we choose N small enough so that $\partial\Omega \cap N$ can be represented in the form $y_m = \phi(y_1, \ldots, y_{m-1})$; we set $\mathbf{u} = (y_1, \ldots, y_{m-1})$. This leads to

$$\mathbf{n_y} = (-\nabla\phi, 1)/\sqrt{1 + |\nabla\phi|^2}, \quad dS_{\mathbf{y}} = \sqrt{1 + |\nabla\phi|^2}\, d\mathbf{u}. \qquad (5.208)$$

We may choose N in such a way that $\partial\Omega \cap N = \{(\mathbf{u}, \phi(\mathbf{u})) \mid |\mathbf{u}| < \epsilon\}$. The first term on the right-hand side of (5.207) now assumes the form

$$\int_{\{|\mathbf{u}|<\epsilon\}} \nabla_{\mathbf{y}} \mathcal{G}(|\mathbf{z} - (\mathbf{u}, \phi(\mathbf{u}))|) \cdot (-\nabla\phi(\mathbf{u}), 1)g(\mathbf{u}, \phi(\mathbf{u}))\, d\mathbf{u}$$

$$\qquad (5.209)$$

$$= \int_{\{|\mathbf{u}|<\epsilon\}} \frac{-\mathbf{u} \cdot \nabla\phi(\mathbf{u}) + \phi(\mathbf{u}) - z_m}{\Omega_m(\sqrt{|\mathbf{u}|^2 + |\phi(\mathbf{u}) - z_m|^2})^m} g(\mathbf{u}, \phi(\mathbf{u}))\, d\mathbf{u}.$$

We note that if we set $z_m = 0$ in (5.209), then $-\mathbf{u} \cdot \nabla\phi(\mathbf{u}) + \phi(\mathbf{u})$ is of order $|\mathbf{u}|^2$ as $\mathbf{u} \to \mathbf{0}$, hence the integrand is of order $|\mathbf{u}|^{-(m-2)}$, i.e., it is integrable. Although the integral exists for $\mathbf{z} = \mathbf{0}$, we cannot take the limit $z_m \to 0$ under the integral. We shall now consider this limit with the constraint that $z_m < 0$. The term which needs to be investigated is

$$-z_m \int_{\{|\mathbf{u}|<\epsilon\}} \frac{1}{\Omega_m(\sqrt{|\mathbf{u}|^2 + |\phi(\mathbf{u}) - z_m|^2})^m} g(\mathbf{u}, \phi(\mathbf{u})) \, d\mathbf{u}. \qquad (5.210)$$

For small $|\mathbf{u}|$ and $|z_m|$, one has

$$\frac{1}{\sqrt{|\mathbf{u}|^2 + (\phi(\mathbf{u}) - z_m)^2}^m} = \frac{1}{\sqrt{|\mathbf{u}|^2 + z_m^2}^m} (1 + O(|z_m|) + O(|\mathbf{u}|^2)), \qquad (5.211)$$

and it is easily checked that only the leading contribution leads to a discontinuity in (5.210) as $z_m \to 0$. It thus remains to consider the integral

$$-z_m \int_{\{|\mathbf{u}|<\epsilon\}} \frac{1}{\Omega_m(\sqrt{|\mathbf{u}|^2 + z_m^2})^m} g(\mathbf{u}, \phi(\mathbf{u})) \, d\mathbf{u}. \qquad (5.212)$$

We define

$$I_r(g) = \frac{1}{r^{m-2}\Omega_{m-1}} \int_{\{|\mathbf{u}|=r\}} g(\mathbf{u}, \phi(\mathbf{u})) \, dS. \qquad (5.213)$$

We substitute $\mathbf{u} = -z_m \mathbf{v}$ in (5.212). This leads to the expression

$$\int_{\{|\mathbf{v}|\le -\epsilon/z_m\}} \frac{1}{\Omega_m(\sqrt{|\mathbf{v}|^2 + 1})^m} g(-z_m \mathbf{v}, \phi(-z_m \mathbf{v})) \, d\mathbf{v}$$

$$\qquad (5.214)$$

$$= \frac{\Omega_{m-1}}{\Omega_m} \int_0^{-\epsilon/z_m} \frac{r^{m-2}}{(r^2 + 1)^{m/2}} I_{-z_m r}(g) \, dr.$$

In the limit $z_m \to 0-$, we obtain

$$\frac{\Omega_{m-1}}{\Omega_m} g(\mathbf{0}) \int_0^\infty \frac{r^{m-2}}{(r^2 + 1)^{m/2}} \, dr = \frac{1}{2} g(\mathbf{0}). \qquad (5.215)$$

Here we have used that

$$\int_0^\infty \frac{r^{m-2}}{(r^2 + 1)^{m/2}} \, dr = \frac{\Gamma(\frac{m-1}{2})\sqrt{\pi}}{2\Gamma(\frac{m}{2})} \qquad (5.216)$$

(see [GR], p. 292) and the expression for Ω_m obtained in Problem 4.16.

Problems

5.56. On the interval $[0, 1]$, let $Lu = u'''' + u'$ with boundary conditions $u'' + u = u - u''' = 0$ at the endpoints. Find the adjoint operator and the adjoint boundary conditions.

5.57. Find the Green's function for the fourth derivative operator on $(0, 1)$ with boundary conditions $u(0) = u'(0) = u(1) = u'(1) = 0$.

5.58. Let Ω be a domain in $\mathbb{R}^2$ bounded by a smooth curve. Consider the equation $\Delta\Delta u = f$ with boundary conditions $\Delta u = \frac{\partial u}{\partial n} + \frac{\partial u}{\partial s} = 0$. Determine the adjoint boundary-value problem.

5.59. Define a Green's function for an initial/boundary-value problem of the form $u_t = L(\mathbf{x}, D)u + f$ with boundary conditions $B_j(\mathbf{x}, D)u = 0$. Give an analogue of the discussion in Section 5.5.2.

5.60. Verify (5.199).

5.61. Reformulate the Neumann problem as an integral equation on the boundary.

6
Function Spaces

Both the data and the solutions for problems in PDEs are functions defined on certain domains or manifolds. In order to formulate precise theorems of existence, uniqueness, continuous dependence, etc., it is essential to specify the spaces in which these functions lie and to give a precise meaning to convergence in those spaces. This issue has led to the development of what is now considered one of the main ("core") fields of pure mathematics, namely, functional analysis. In this and the subsequent chapter, we shall give a brief introduction to this field, with particular emphasis on those issues and concepts that are important in differential equations. This introduction is limited to what is needed in the rest of the book and is not meant as a substitute for a proper course in functional analysis.

6.1   Banach Spaces and Hilbert Spaces

6.1.1   Banach Spaces

The fundamental concept used to define a notion of distance between functions is that of a norm.

Definition 6.1. Let X be a real or complex vector space. A **norm** on X is a function $\| \cdot \| : X \to [0, \infty)$ such that

1. $\|x\| = 0$ if and only if $x = 0$,

2. $\|\lambda x\| = |\lambda| \|x\|$ for every $x \in X$, $\lambda \in \mathbb{R}$ (or $\mathbb{C}$),

3. $\|x + y\| \leq \|x\| + \|y\|$ for every $x, y \in X$.

It is easy to see (cf. Problem 6.1) that the function $d(x, y) = \|x - y\|$ defines a metric on X. From this, we also obtain a notion of convergence, i.e., we say that $x_n \to x$ if $\|x_n - x\| \to 0$. Terminology familiar from the finite-dimensional case will be adopted in a natural way, e.g., a (closed) ball centered at x with radius r is the set of all y with $\|y - x\| \leq r$, a set M is bounded, if $\{\|x\| \mid x \in M\}$ is bounded etc.

Definition 6.2. Two normed vector spaces X and Y are called **isometric**, if there is a linear bijection $L : X \to Y$ such that $\|L(x)\| = \|x\|$ for every $x \in X$.

Definition 6.3. Two norms $\|\cdot\|_1$ and $\|\cdot\|_2$ on a vector space X are called **equivalent** if there are constants $c, C > 0$ such that $c\|x\|_1 \leq \|x\|_2 \leq C\|x\|_1$ for every $x \in X$.

Procedures to solve partial differential equations are usually based on iteration methods, discretization or other means of approximation. In all these cases, one generates a sequence of approximate solutions and one wants to show that this sequence converges to a limit. Since it is usually not a priori clear that a solution to the PDE exists, one needs a convergence criterion that does not involve the limit. This leads to the notion of a Cauchy sequence. It is hence an important issue whether a normed vector space is complete.

Definition 6.4. A normed vector space X is called a **Banach space** if it is complete, i.e., if every sequence x_n such that $\lim_{m,n\to\infty} \|x_n - x_m\| = 0$ has a limit $x \in X$.

If a normed space is not complete, we can define its completion; the procedure for this is the usual one for completing a metric space, i.e., by considering equivalence classes of Cauchy sequences.

Theorem 6.5. *Let X be a normed vector space. Then there exists a normed vector space Y such that Y is complete and X is a dense subspace of Y. Up to isometry, the space Y is unique.*

Definition 6.6. The space Y given by Theorem 6.5 is called the **completion** of X.

Proof of Theorem 6.5: We only sketch the main ideas. We assume that our readers have seen the construction of the real numbers (a Banach space) from the rational numbers, and readers should keep this example in mind. We will use subscripts to identify the norms in the various spaces (i.e., $\|\cdot\|_X$ and $\|\cdot\|_Y$).

- We call two Cauchy sequences $\{x_n\}_{n=1}^{\infty}$ and $\{y_n\}_{n=1}^{\infty}$ in X *equivalent* if $\lim_{n\to\infty} \|x_n - y_n\|_X = 0$. We denote the set of equivalence classes of

Cauchy sequences by Y. The reader should verify that Y is a vector space.

- It follows from the triangle inequality (condition 3 of Definition 6.1) that

$$\big| \|x\|_X - \|y\|_X \big| \le \|x - y\|_X, \qquad (6.1)$$

and hence if $\{x_n\}_{n=1}^\infty$ is a Cauchy sequence in X, then $\{\|x_n\|\}_{n=1}^\infty$ is a Cauchy sequence in $\mathbb{R}$. We now define

$$\|\{x_n\}_{n=1}^\infty\|_Y = \lim_{n\to\infty} \|x_n\|_X \qquad (6.2)$$

for every Cauchy sequence $\{x_n\}_{n=1}^\infty$. Using (6.1) again, we see that equivalent Cauchy sequences have equal norms. Hence (6.2) defines a norm on the space Y. We leave it to the reader to check the properties 1-3 of Definition 6.1 and thus verify that Y is a normed space.

- The original space X becomes a subspace of Y by identification with (equivalence classes of) constant Cauchy sequences. It is clear that X is dense in Y.

- We need to show that Y is complete. Let $\{y_i\}_{i=1}^\infty = \{\{w_{i,n}\}_{n=1}^\infty\}_{i=1}^\infty$ be a Cauchy sequence (of equivalence classes of Cauchy sequences) in Y. Recall that X (thought of as a set of constant sequences) is dense in Y; hence for every i there is an element $x_i \in X$ such that $\lim_{n\to\infty} \|x_i - w_{i,n}\|_X < 1/i$. The reader should show that the sequence $\{x_i\}_{i=1}^\infty$ is Cauchy in X and hence can be identified with an element $y \in Y$. It can then be shown that this element y is the limit of the Cauchy sequence $\{y_i\}_{i=1}^\infty$ in Y.

- It remains to prove that Y is unique up to isometry. Let $\tilde{Y}$ be a Banach space containing X as a dense subspace. Then every $y \in \tilde{Y}$ is the limit of a Cauchy sequence $\{x_n\}_{n=1}^\infty$ in X, and we must have

$$\|y\| = \lim_{n\to\infty} \|x_n\|_X. \qquad (6.3)$$

On the other hand, every Cauchy sequence in X can be identified with an element of Y. The reader should use these ideas to define the bijection from $\tilde{Y}$ to Y and verify the isometry.

This completes the proof.

Remark 6.7. One of the greatest difficulties with the proof above is the rather cumbersome notation required. In practice, we dispense with the notation and refer to an element of Y as the *limit* of a Cauchy sequence in X; i.e., for any $y \in Y$, we will say that a sequence $\{y_n\} \subset X$ satisfies $y_n \to y$ rather than say that $\{y_n\}$ is a member of the equivalence class y. We work with the space Y directly and revive X only when absolutely necessary. Of course, you already do this unconsciously in the case of real numbers: $\sqrt{2}$ and π are thought of as single points on the number line or at worst as decimal expansions rather than as equivalence classes of rational sequences.

6.1.2 Examples of Banach Spaces

Example 6.8. Let Ω be an open set in $\mathbb{R}^m$ and let $C_b(\overline{\Omega})$ be the set of all bounded continuous functions on $\overline{\Omega}$. We define

$$\|u\| = \sup_{\mathbf{x} \in \overline{\Omega}} |u(\mathbf{x})|. \tag{6.4}$$

Then $C_b(\overline{\Omega})$ is a Banach space. Indeed, the properties of a norm are easy to check, and the completeness is a restatement of the well-known fact that uniform limits of continuous functions are continuous (note that convergence in $C_b(\overline{\Omega})$ means uniform convergence).

Example 6.9. Let Ω be an open set in $\mathbb{C}$ and let $A(\Omega)$ be the set of all bounded analytic functions on Ω. We define

$$\|u\| = \sup_{z \in \Omega} |u(z)|. \tag{6.5}$$

Then $A(\Omega)$ is a Banach space. Indeed, the fact that uniform limits of analytic functions are analytic is an easy consequence of Morera's theorem.

Example 6.10. Let Ω be an open, locally Jordan measurable (cf. [Bu]) (this means that the intersection with any ball is Jordan measurable) set in $\mathbb{R}^m$ and let $1 \le p < \infty$. Let $\tilde{L}^p(\Omega)$ be the set of all continuous functions $u : \overline{\Omega} \to \mathbb{R}$ for which

$$\|u\|_p := \left(\int_\Omega |u(\mathbf{x})|^p \, d\mathbf{x} \right)^{1/p} \tag{6.6}$$

is finite. We want to show that $\tilde{L}^p(\Omega)$ is a normed vector space. The first two properties of a norm are trivial, and the triangle inequality is trivial for $p = 1$. The triangle inequality for $p > 1$ takes some preparation.

Lemma 6.11. *Let* $a, b \ge 0$, $p, q > 1$ *and* $p^{-1} + q^{-1} = 1$. *Then*

$$ab \le \frac{a^p}{p} + \frac{b^q}{q}. \tag{6.7}$$

The proof is obtained by seeking the minimum of the function

$$b \to a^p/p + b^q/q - ab.$$

See Problem 6.4. As a consequence of Lemma 6.11, we shall now establish Hölder's inequality.

Theorem 6.12. *Let $f \in \tilde{L}^p(\Omega)$, $g \in \tilde{L}^q(\Omega)$, where $p, q \in (1, \infty)$ with $p^{-1} + q^{-1} = 1$. Then $fg \in L^1(\Omega)$, and*

$$\|fg\|_1 \le \|f\|_p \|g\|_q. \tag{6.8}$$

Proof. We only rule out trivial cases by assuming that $\|f\|_p \ne 0$, $\|g\|_q \ne 0$. By the previous lemma, we find

$$\frac{|f(x)|}{\|f\|_p} \frac{|g(x)|}{\|g\|_q} \le \frac{1}{p} \frac{|f(x)|^p}{\|f\|_p^p} + \frac{1}{q} \frac{|g(x)|^q}{\|g\|_q^q}. \tag{6.9}$$

The theorem follows by integrating over Ω.   □

Theorem 6.13. *Let $f, g \in \tilde{L}^p(\Omega)$. Then $f + g \in \tilde{L}^p(\Omega)$ and $\|f + g\|_p \le \|f\|_p + \|g\|_p$.*

Proof. We note that

$$|f + g|^p \le (|f| + |g|)^p \le 2^p(|f|^p + |g|^p); \tag{6.10}$$

hence $|f + g| \in \tilde{L}^p(\Omega)$. Moreover, we have

$$|f + g|^p \le |f + g|^{p-1}|f| + |f + g|^{p-1}|g|, \tag{6.11}$$

and with $p^{-1} + q^{-1} = 1$, we have $(p - 1)q = p$; hence $|f + g|^{p-1} \in \tilde{L}^q(\Omega)$. We can therefore apply Hölder's inequality and obtain

$$\int_\Omega |f + g|^p \, d\mathbf{x} \le \left(\int_\Omega |f + g|^p \, d\mathbf{x} \right)^{1/q} (\|f\|_p + \|g\|_p). \tag{6.12}$$

From this the triangle inequality follows immediately.   □

It is easy to see that the space $\tilde{L}^p(\Omega)$ is not complete (see Problem 6.6). We therefore make the following definition.

Definition 6.14. *Let $1 \le p < \infty$. Then $L^p(\Omega)$ is the completion of $\tilde{L}^p(\Omega)$.*

Remark 6.15. If a sequence f_n is Cauchy in $\tilde{L}^p(\Omega)$, then it follows from Hölder's inequality that $\int_\Omega f_n \phi \, d\mathbf{x}$ converges for every test function ϕ. Hence f_n converges in the sense of distributions, and by Theorem 5.31 every element of $L^p(\Omega)$ defines a distribution.

We need to show that different elements of $L^p(\Omega)$ correspond to different distributions.

Lemma 6.16. *Let f_n be a Cauchy sequence in $\tilde{L}^p(\Omega)$ $(1 \le p < \infty)$ and assume that $\int_\Omega f_n \phi \, d\mathbf{x} \to 0$ for every $\phi \in \mathcal{D}(\Omega)$. Then $\|f_n\|_p \to 0$.*

Proof. Assume the contrary, i.e., let $\lim_{n \to \infty} \|f_n\|_p = M > 0$. Choose N large enough so that $3M/2 > \|f_n\|_p > M/2$ and $\|f_n - f_m\|_p < \epsilon$ for $n, m \ge N$. Here ϵ is chosen small. By Problem 6.9, $\mathcal{D}(\Omega)$ is dense in $L^p(\Omega)$. Hence we can choose $\psi \in \mathcal{D}(\Omega)$ with $\|f_N - \psi\|_p < \epsilon$; more specifically, we can choose ψ to be the product of a polynomial and a cutoff function. Now let

$$g = \begin{cases} |\psi|^{p-1} \operatorname{sgn} \psi & p > 1 \\ \operatorname{sgn} \psi & p = 1. \end{cases} \tag{6.13}$$

Thus g is a piecewise continuous function of compact support.

If $p > 1$, we have $g \in L^q(\Omega)$, and we can choose $\phi \in \mathcal{D}(\Omega)$ with $\|g - \phi\|_q < \epsilon$; here $q = p/(p-1)$. We now estimate, for $n \geq N$,

$$\left| \int_\Omega f_n \phi \, d\mathbf{x} \right|$$

$$\geq \int_\Omega \psi g \, d\mathbf{x}$$

$$\qquad - \left| \int_\Omega \psi(\phi - g) + (f_n - f_N)\phi + (f_N - \psi)\phi \, d\mathbf{x} \right|$$

$$\geq \int_\Omega |\psi|^p \, d\mathbf{x} - \|\psi\|_p \|\phi - g\|_q \tag{6.14}$$

$$\qquad + \|f_n - f_N\|_p \|\phi\|_q + \|f_N - \psi\|_p \|\phi\|_q$$

$$\geq \|\psi\|_p^p - O(\epsilon)$$

$$\geq (M/2 - \epsilon)^p - O(\epsilon).$$

Since ϵ can be chosen arbitrarily small for large N, this contradicts the fact that the left-hand side of (6.14) should tend to zero as $n \to \infty$.

If $p = 1$, we choose ϕ such that $\max |\phi| \leq 1$ and such that ϕ is uniformly close to g except in a small neighborhood of those surfaces where g is discontinuous. We can then make $\int \psi(\phi - g) \, d\mathbf{x}$ as small as we wish. The rest goes as above, with the elementary inequality $|\int fg \, d\mathbf{x}| \leq \max |g| \|f\|_1$ in place of Hölder's inequality. $\square$

Remark 6.17. Consider the sequence $f_n(x) = \exp(-nx^2)$. It is easy to see that $f_n \to 0$ in $\tilde{L}^p(\mathbb{R})$ for $1 \leq p < \infty$, and thus the sequence can be identified with the constant function $0 \in L^p(\mathbb{R})$. However, note that $f_n(0) = 1$ for every n. This example should convince the reader that we cannot regard elements of L^p as functions with point values. Fortunately, using the results of measure theory, it can be shown that one can "almost" do so. In measure theory, elements of L^p are defined as equivalence classes of functions which differ only on "sets of measure zero" (for instance, at a finite number of points). More precisely, $L^p(\Omega)$ is the set of all (equivalence classes of) Lebesgue measurable functions f such that $|f|^p$ is integrable. This definition is equivalent to ours. To show this equivalence, it suffices to prove that continuous bounded functions are dense in the space L^p (as defined in integration theory). Since we do not assume familiarity with Lebesgue integration, we shall not pursue this point further.

On the basis of our definition of $L^p(\Omega)$, we can define "integrals" as limits. For example, if $f \in L^1(\Omega)$, then there is a sequence $f_n \in \tilde{L}^1(\Omega)$ such that $f_n \to f$, and we can define

$$\int_\Omega f \, d\mathbf{x} = \lim_{n \to \infty} \int_\Omega f_n \, d\mathbf{x}. \qquad (6.15)$$

Also, if $f \in L^p(\Omega)$ and $g \in L^q(\Omega)$, where $p, q \in (1, \infty)$ and $p^{-1} + q^{-1} = 1$, we can choose sequences $f_n \in \tilde{L}^p(\Omega)$ and $g_n \in \tilde{L}^q(\Omega)$ such that $f_n \to f$ and $g_n \to g$. Hölder's inequality implies that $f_n g_n$ converges in $L^1(\Omega)$, and it is natural to call the limit fg. With this interpretation Hölder's inequality holds for all $f \in L^p(\Omega)$ and $g \in L^q(\Omega)$. These examples are typical of the distributional approach. Properties of functions in L^p and similar spaces are proved by first considering continuous (or even smoother) functions and then considering the limit of an appropriate sequence.

Example 6.18. Let X_1 and X_2 be Banach spaces. Then $X_1 \times X_2$ is a Banach space with the norm $\|(x_1, x_2)\| = \|x_1\| + \|x_2\|$. In this situation, we shall sometimes identify X_1 with $X_1 \times \{0\}$ and X_2 with $\{0\} \times X_2$. We then write $x_1 + x_2$ for (x_1, x_2).

6.1.3 Hilbert Spaces

Banach spaces have geometric structure generated by the norm (which corresponds to a notion of "size") and the induced metric (which corresponds to a notion of "distance"). We now introduce more geometric structure by defining an inner product which generalizes the notion of "angle."

Definition 6.19. Let H be a vector space over $\mathbb{R}$ (or, respectively, $\mathbb{C}$). An **inner product** (x, y) is a mapping from $H \times H$ to $\mathbb{R}$ ($\mathbb{C}$) with the following properties:

1. For every $x \in H$, the mapping $y \to (x, y)$ is linear.

2. $(y, x) = \overline{(x, y)}$ for every $x, y \in H$.

3. $(x, x) \geq 0$ for every $x \in H$ with equality holding if and only if $x = 0$.

Inner product spaces are special cases of normed vector spaces. This is expressed by the following lemma.

Lemma 6.20. *Let H be a vector space with inner product $(\cdot, \cdot)$. Then for every $x, y \in H$ we have*

$$|(x, y)|^2 \leq (x, x)(y, y). \qquad (6.16)$$

Moreover, the expression

$$\|x\| = \sqrt{(x, x)} \qquad (6.17)$$

defines a norm on H.

The inequality (6.16) is known as the Cauchy-Schwarz inequality.

Proof. If either x or y is 0, then (6.16) is trivial; let us assume that $y \neq 0$. We note that

$$(x + \lambda y, x + \lambda y) = (x, x) + \bar{\lambda}(y, x) + \lambda(x, y) + |\lambda|^2 (y, y) \geq 0. \quad (6.18)$$

The Cauchy-Schwarz inequality follows by setting $\lambda = -(y, x)/(y, y)$. Using the Cauchy-Schwarz inequality, we find

$$\begin{aligned}
\|x + y\|^2 &= (x + y, x + y) \\
&= (x, x) + (x, y) + (y, x) + (y, y) \\
&\leq \|x\|^2 + 2\|x\|\|y\| + \|y\|^2 \\
&= (\|x\| + \|y\|)^2.
\end{aligned} \quad (6.19)$$

Hence the triangle inequality holds. The other properties of a norm are trivial. □

Definition 6.21. A Hilbert space is an inner product space which (as a normed vector space) is complete.

Example 6.22. Let ℓ^2 be the set of all complex-valued sequences x_n such that

$$\|x\|^2 := \sum_{n=1}^{\infty} |x_n|^2 < \infty. \quad (6.20)$$

The inner product is defined by

$$(x, y) = \sum_{n=1}^{\infty} \bar{x}_n y_n. \quad (6.21)$$

It is easy to show that ℓ^2 is an inner product space. We shall show it is complete. Let

$$u^{(n)} = (u_1^{(n)}, u_2^{(n)}, \ldots) \quad (6.22)$$

be a Cauchy sequence. Then for any $\epsilon > 0$, there is an $N(\epsilon)$ such that

$$\|u^{(n)} - u^{(m)}\| = \sqrt{\sum_{j=1}^{\infty} |u_j^{(n)} - u_j^{(m)}|^2} < \epsilon \quad (6.23)$$

for $m, n > N(\epsilon)$. This implies in particular that $u_j^{(n)}$ is a Cauchy sequence for every fixed j. Let

$$u_j = \lim_{n \to \infty} u_j^{(n)}. \quad (6.24)$$

From (6.23), it follows that

$$\sum_{j=1}^{k} |u_j^{(n)} - u_j^{(m)}|^2 \leq \epsilon^2 \quad (6.25)$$

for every $n, m > N(\epsilon)$ and every $k \in \mathbb{N}$. We let $m \to \infty$ and obtain

$$\sum_{j=1}^{k} |u_j^{(n)} - u_j|^2 < \epsilon^2 \qquad (6.26)$$

for $n > N(\epsilon)$, $k \in \mathbb{N}$. We now let $k \to \infty$ and conclude that $u^{(n)} \to u$ in H.

Example 6.23. The space $L^2(\Omega)$ defined in Example 6.10, with the inner product

$$(f, g) = \int_{\Omega} \overline{f(\mathbf{x})} g(\mathbf{x}) \, d\mathbf{x} \qquad (6.27)$$

is a Hilbert space. Here the integral in (6.27) is defined in the sense of Remark 6.17.

Definition 6.24. A Hilbert space (or, more generally, a Banach space) is called **separable** if it contains a countable, dense subset.

Most spaces arising in applications are separable. Separability is important for the practical solution of problems, say, by discretization, because only countably many (well, in the real world, only finitely many) elements of the space can be represented in such a fashion. It is easy to see that ℓ^2 is separable, because terminating sequences are dense. The space $L^2(\Omega)$ is also separable; see Problem 6.12.

Definition 6.25. Let H be a Hilbert space. We say that two elements of H, x and y are **orthogonal** if $(x, y) = 0$. For any subspace M of H, we define the **orthogonal complement** by

$$M^{\perp} = \{x \in H \mid (x, y) = 0 \; \forall y \in M\}. \qquad (6.28)$$

It is clear that $M^{\perp}$ is a closed subspace. If M is also closed, then H is the direct sum of M and $M^{\perp}$: $H = M \oplus M^{\perp}$.

Theorem 6.26 (Projection theorem). *Let H be a Hilbert space and let M be a closed subspace of H. Then every $u \in H$ has a unique decomposition $u = v + w$, where $v \in M$ and $w \in M^{\perp}$.*

Proof. From elementary geometry, we expect v to be the point in M that is closest to u. Let us assume $u \notin M$ and let

$$d := \inf_{v' \in M} \|u - v'\|^2. \qquad (6.29)$$

Then there is a sequence $v_n \in M$ such that $d_n := \|u - v_n\|^2$ converges to d. We shall prove that v_n is a Cauchy sequence and take v to be its limit.

Let y be an arbitrary element of M and let λ be a scalar. Then $v_n + \lambda y \in M$, and hence

$$d \le \|u - (v_n + \lambda y)\|^2 = \|u - v_n\|^2 - \bar{\lambda}(y, u - v_n)$$
$$- \lambda(u - v_n, y) + |\lambda|^2 \|y\|^2. \tag{6.30}$$

Setting $\lambda = (y, u - v_n)/\|y\|^2$, we conclude

$$d \le \|u - v_n\|^2 - \frac{|(u - v_n, y)|^2}{\|y\|^2}, \tag{6.31}$$

which can be rewritten as

$$|(u - v_n, y)| \le \|y\|\sqrt{d_n - d}. \tag{6.32}$$

Next, we observe that

$$|(v_n - v_m, y)| \le |(u - v_n, y)| + |(u - v_m, y)| \le \|y\|(\sqrt{d_n - d} + \sqrt{d_m - d}). \tag{6.33}$$

By setting $y = v_n - v_m$, we conclude that

$$\|v_n - v_m\| \le \sqrt{d_n - d} + \sqrt{d_m - d}, \tag{6.34}$$

i.e., v_n is a Cauchy sequence. Let v be its limit. By taking limits in (6.32), we find $(u - v, y) = 0$ for every $y \in M$, i.e., $u - v \in M^\perp$. This proves the existence of the desired decomposition.

Suppose we had two decompositions $u = v + w = v' + w'$, then $v - v' = w' - w$, and hence

$$\|v - v'\|^2 = (v - v', v - v') = (v - v', w' - w) = 0, \tag{6.35}$$

since $v - v' \in M$ and $w' - w \in M^\perp$. $\qquad\square$

Corollary 6.27. *Let H be a Hilbert space. A subspace M of H is dense iff $M^\perp = \{0\}$.*

Proof. Suppose $v \in M^\perp$. Then $(v, u) = 0$ for every $u \in M$. If M is dense, we conclude that $(v, u) = 0$ for every $u \in H$ by taking limits. Setting $u = v$ yields $v = 0$.

On the other hand, the closure of M has an orthogonal complement according to the preceding theorem. If M is not dense, then this orthogonal complement must contain nonzero vectors. $\qquad\square$

Problems

6.1. Let $(X, \|\cdot\|)$ be a normed vector space. Show that the function $d(x, y) = \|x - y\|$ defines a metric.

6.2. Provide the details for the proof of Theorem 6.5.

6.3. Let Ω be an open set in $\mathbb{R}^m$ and let $C_b^k(\overline{\Omega})$ be the set of all functions on $\overline{\Omega}$ which have continuous bounded derivatives up to order k. Define

$$\|u\| = \sum_{|\alpha| \leq k} \sup_{\mathbf{x} \in \overline{\Omega}} |D^\alpha u(\mathbf{x})|. \tag{6.36}$$

Show that $C_b^k(\overline{\Omega})$ is a Banach space.

6.4. Prove Lemma 6.11.

6.5. Let $f \in \tilde{L}^p(\Omega)$, $g \in \tilde{L}^q(\Omega)$, where $p, q \in (1, \infty)$ and $r^{-1} = p^{-1} + q^{-1} < 1$. Show that $fg \in \tilde{L}^r(\Omega)$.

6.6. Show that the space $\tilde{L}^p(\Omega)$ is not complete.

6.7. Let Ω be bounded and $1 \leq p < q < \infty$. Show that $L^q(\Omega) \subset L^p(\Omega)$, and that there exists a constant C depending only on Ω such that $\|u\|_p \leq C\|u\|_q$ for all $u \in L^q(\Omega)$.

6.8. Show that for every $u, v \in L^2(\Omega)$ and every $\epsilon > 0$ we have

$$\left| \int_\Omega uv \, d\mathbf{x} \right| \leq \epsilon \|u\|_2^2 + C(\epsilon) \|v\|_2^2,$$

where $C(\epsilon) := (4\epsilon)^{-1}$.

6.9. Prove that $\mathcal{D}(\Omega)$ is dense in $L^p(\Omega)$, $1 \leq p < \infty$.

6.10. Let H be an inner product space. Prove that the inner product is continuous on $H \times H$.

6.11. State the specific form of the Cauchy-Schwarz and triangle inequalities for ℓ^2 and $L^2(\Omega)$.

6.12. Prove that $L^2(\Omega)$ is separable. Hint: Use Problem 6.9.

6.13. Prove that $(M^\perp)^\perp = M$ iff M is closed.

6.14. Prove that all norms on $\mathbb{R}^n$ are equivalent.

6.2   Bases in Hilbert Spaces

6.2.1   The Existence of a Basis

From linear algebra, we know that every Euclidean vector space has a Cartesian basis. In this subsection, we shall extend this result to Hilbert spaces. We shall need the following definition.

Definition 6.28. Let H be a Hilbert space and I a (possibly uncountable) index set. Let $\{x_i\}_{i \in I}$ be a family of elements of H. We say that $\sum_{i \in I} x_i = x$ if at most countably many of the x_i are nonzero, and if for any enumeration of these nonzero elements we have $x = \sum_{j \in \mathbb{N}} x_{i(j)}$.

Remark 6.29. Note that while it is convenient for us to allow for the possibility of an uncountable index set, at most countably many elements can be nonzero if this notion of convergence is to make sense. To see this, note that for any series of real numbers to be absolutely convergent, it can have at most a finite number of terms with norm greater than, say, $1/n$ for and natural number n. Hence, it can have at most countably many nonzero terms. The above definition of convergence is a generalization of absolute convergence of a series of real number, and the following conditions are easily shown to be equivalent to that definition. We leave the proof to the reader (Problem 6.21).

Lemma 6.30. *The sum $x = \sum_{i \in I} x_i$ exists if and only if either of the following hold:*

1. *For every $\epsilon > 0$, there is a finite subset J_ϵ of I such that for any finite J with $J_\epsilon \subset J \subset I$ we have*

$$\left\| x - \sum_{i \in J} x_i \right\| < \epsilon. \qquad (6.37)$$

2. *For every $\epsilon > 0$ there is a finite subset J_ϵ of I such that*

$$\left\| \sum_{i \in J} x_i \right\| < \epsilon \qquad (6.38)$$

for any finite subset J of I with $J \cap J_\epsilon = \emptyset$.

In the following, we are interested in sums of orthogonal elements. We have the following lemma.

Lemma 6.31. *Let $\{x_i\}_{i \in I}$ be a family of mutually orthogonal elements of a Hilbert space H. Then $\sum_{i \in I} x_i$ exists if and only if $\sum_{i \in I} \|x_i\|^2 < \infty$. In this case we have, moreover,*

$$\left\| \sum_{i \in I} x_i \right\|^2 = \sum_{i \in I} \|x_i\|^2. \qquad (6.39)$$

Proof. For any finite subset J of I we use the fact that elements of $\{x_i\}_{i \in I}$ are mutually orthogonal to get

$$\left\| \sum_{i \in J} x_i \right\|^2 = \left(\sum_{i \in J} x_i, \sum_{l \in J} x_l \right) = \sum_{i \in J} (x_i, x_i) = \sum_{i \in J} \|x_i\|^2. \qquad (6.40)$$

The rest follows from Lemma 6.30. $\qquad \square$

Definition 6.32. A family $\{x_i\}_{i \in I}$ of mutually orthogonal elements of H is called **orthonormal** if $\|x_i\| = 1$ for every $i \in I$.

Theorem 6.33. *Let $\{x_i\}_{i \in I}$ be an orthonormal set in a Hilbert space H. Then*

1. $\sum_{i \in I} |(x_i, x)|^2 \leq \|x\|^2$ for every $x \in X$.

2. Equality in 1 holds if and only if $x = \sum_{i \in I} (x_i, x) x_i$.

The inequality in 1 is referred to as Bessel's inequality, or, in the case where equality holds, as Parseval's equality.

Proof. For finite subsets J of I we can use the fact that $\{x_i\}_{i \in I}$ is an orthonormal set to get the following:

$$
\begin{aligned}
0 &\leq \left\| x - \sum_{i \in J} (x_i, x) x_i \right\|^2 \\
&= \left(x - \sum_{i \in J} (x_i, x) x_i, \, x - \sum_{l \in J} (x_l, x) x_l \right) \\
&= \|x\|^2 - \sum_{i \in J} (x_i, x)(x, x_i) - \sum_{l \in J} (x, x_l)(x_l, x) \\
&\quad + \left(\sum_{i \in J} (x_i, x) x_i, \, \sum_{l \in J} (x_l, x) x_l \right) \\
&= \|x\|^2 - 2 \sum_{i \in J} |(x_i, x)|^2 + \sum_{i \in J} \sum_{l \in J} \overline{(x_i, x)} (x_l, x)(x_i, x_l) \\
&= \|x\|^2 - \sum_{i \in J} |(x_i, x)|^2.
\end{aligned}
$$

Hence $\sum_{i \in I} |(x_i, x)|^2$ exists, and Bessel's inequality holds. By Lemma 6.31, this implies that $\sum_{i \in I} (x_i, x) x_i$ also exists. Moreover, using the argument above,

$$
\left\| x - \sum_{i \in I} (x_i, x) x_i \right\|^2 = \|x\|^2 - \sum_{i \in I} |(x_i, x)|^2, \tag{6.41}
$$

and the second claim of the theorem is immediate. $\square$

Definition 6.34. An orthonormal set $\{x_i\}_{i \in I}$ in a Hilbert space H is called a **basis** if $x = \sum_{i \in I} (x_i, x) x_i$ for every $x \in H$.

In contrast to the usual definition of a vector space basis, we are allowing infinite series in the representation of x as a linear combination of the x_i. If there is danger of confusion, then a basis in the sense of Definition 6.34 is called a Hilbert basis, whereas a vector space basis in the sense of finite linear combinations is called a Hamel basis. In the following, a basis is always a Hilbert basis.

Theorem 6.35. Let $\{x_i\}_{i \in I}$ be an orthonormal set in a Hilbert space H. Then the following are equivalent:

(i) $\{x_i\}_{i \in I}$ is a basis.

(ii) For every $x, y \in H$, we have

$$(x, y) = \sum_{i \in I} (x, x_i)(x_i, y). \tag{6.42}$$

(iii) For every $x \in X$, we have

$$\|x\|^2 = \sum_{i \in I} |(x_i, x)|^2. \tag{6.43}$$

(iv) The set $\{x_i\}_{i \in I}$ is maximal, i.e., there is no orthonormal set containing it as a proper subset. In other words, if x is orthogonal to each x_i, then $x = 0$.

Proof. (i)$\Rightarrow$(ii): We have

$$
\begin{aligned}
(x, y) &= \left(\sum_{i \in I} (x_i, x)x_i, \sum_{j \in I} (x_j, y)x_j \right) \\
&= \sum_{i,j \in I} (x, x_i)(x_j, y)\delta_{ij} \\
&= \sum_{i \in I} (x, x_i)(x_i, y).
\end{aligned}
\tag{6.44}
$$

The exchange of summation and inner product is justified in the usual way by considering finite sums and then passing to the limit.

(ii)$\Rightarrow$(iii): Set $y = x$.

(iii)$\Rightarrow$(iv): If x is orthogonal to each x_i, then (6.43) implies $x = 0$.

(iv)$\Rightarrow$(i): Let

$$Y := \left\{ x \in H \mid x = \sum_{i \in I} (x_i, x)x_i \right\}. \tag{6.45}$$

Let $x^{(n)}$ be a Cauchy sequence in Y. Then there are at most countably many $i \in I$ for which $(x_i, x^{(n)}) \neq 0$ for any n. Let $\tilde{I}$ be this at most countable set and let

$$\tilde{Y} := \left\{ x \in H \mid x = \sum_{i \in \tilde{I}} (x_i, x)x_i \right\}. \tag{6.46}$$

Parseval's equality shows that $\tilde{Y}$ is either finite-dimensional or isometric to the sequence space ℓ^2 and hence complete; see Example 6.22. Therefore, the Cauchy sequence $x^{(n)}$ has a limit in $\tilde{Y} \subseteq Y$, i.e., Y is a closed subspace of H. On the other hand, (iv) says that $Y^\perp = \{0\}$, and by Theorem 6.26 we conclude that $Y = H$. $\qquad\square$

Corollary 6.36. *Every Hilbert space has a basis.*

Proof. A standard application of Zorn's lemma shows that there is a maximal orthonormal set. $\qquad\square$

For separable Hilbert spaces, a basis can be found in a more constructive way using the Schmidt orthogonalization procedure. Let $\{x_n\}_{n\in\mathbb{N}}$ be a countable dense set. We then drop from this sequence each element which can be represented as a linear combination of the preceding ones. We thus end up with a new sequence $\{y_n\}$ of linearly independent elements such that the linear span of the y_n is still dense in H. We now construct a sequence z_n as follows:

$$z_1 = y_1/\|y_1\|,$$
$$u_n = y_n - \sum_{i=1}^{n-1}(z_i, y_n)z_i, \tag{6.47}$$
$$z_n = u_n/\|u_n\|.$$

It is easy to see that the z_n are orthonormal, and their linear span is the same as that of the y_n, hence dense in H. Hence (iv) of Theorem 6.35 applies and the z_n form a basis.

6.2.2 Fourier Series

The most important example of expansions with respect to an orthonormal basis is the Fourier expansion.

Theorem 6.37. Let $\phi_0(x) = 1$, $\phi_n(x) = \sqrt{2}\cos(n\pi x)$, $n \in \mathbb{N}$. Then the functions ϕ_n, $n = 0, 1, 2, \ldots$, form a basis of $L^2(0,1)$.

Proof. An easy calculation shows that the ϕ_n are an orthonormal system. By Theorem 6.35 it therefore suffices to show that the linear span of the ϕ_n is dense in $L^2(0,1)$. Since $C([0,1])$ is dense in $L^2(0,1)$, we only need to show that every continuous function can be approximated by a linear combination of the ϕ_n. We make the substitution $\cos \pi x = u$, which bijectively maps $[0,1]$ to $[-1,1]$. By the Weierstraß approximation theorem, every continuous function on $[-1,1]$ can be approximated uniformly by polynomials; hence every continuous function on $[0,1]$ can be approximated uniformly by polynomials in $\cos \pi x$. Elementary trigonometric identities show that any expression $\sum_{n=0}^{N} a_n(\cos \pi x)^n$ can be rewritten in the form $\sum_{n=0}^{N} b_n \cos(n\pi x)$. $\square$

Functions in $L^2(0,1)$ can also be expanded in terms of a sine series instead of a cosine series.

Theorem 6.38. Let $\psi_n(x) = \sqrt{2}\sin(n\pi x)$, $n \in \mathbb{N}$. Then the functions ψ_n, $n \in \mathbb{N}$ form a basis of $L^2(0,1)$.

Proof. We use the fact that $\mathcal{D}(0,1)$ is dense in $L^2(0,1)$. If $f \in \mathcal{D}(0,1)$, then $f(x)/\sin(\pi x)$ is continuous on $[0,1]$ and from the proof of the last theorem we conclude that it can be uniformly approximated by expressions of the form $\sum_{n=0}^{N} a_n \cos(n\pi x)$. Hence $f(x)$ can be uniformly approximated by expressions of the form

$$\sum_{n=0}^{N} a_n \cos(n\pi x) \sin(\pi x) = \frac{1}{2} \sum_{n=0}^{N} a_n \Big(\sin((n+1)\pi x) - \sin((n-1)\pi x) \Big).$$

(6.48)

This completes the proof.   □

Theorems 6.37 and 6.38 yield the following simple consequence.

Corollary 6.39. *The functions* $(1/\sqrt{2})e^{in\pi x}$, $n \in \mathbb{Z}$, *form a basis of* $L^2(-1,1)$.

Proof. Any function in $L^2(-1,1)$ can be decomposed into an even and an odd part. Using the preceding two theorems, we can expand the even part in a cosine series and the odd part in a sine series.   □

In applications, it typically depends on boundary conditions whether expansion in a cosine or sine series is desirable; see the examples in Chapter 1 and also the comments below on pointwise convergence of Fourier series. The expansion in terms of sines and cosines provided by Corollary 6.39 is typically used for periodic functions.

It is nice to know that Fourier series converge in L^2, but this leaves a number of issues. For example:

1. Under what conditions does the Fourier series represent a function in a pointwise sense?

2. Can Fourier series be differentiated term by term? Of course they can be in the sense of distributions, but it is also of interest to know whether the differentiated series converges in L^2.

It is known from measure theory that a sequence converging in L^2 has a subsequence which converges almost everywhere. For Fourier series it is actually not necessary to take a subsequence; this is a hard theorem which was not proved until 1966. A much more elementary observation is that $\sum_{n=0}^{\infty} a_n \cos(n\pi x)$ converges uniformly on $[0,1]$ if $\sum_{n=0}^{\infty} |a_n|$ converges, and using the Cauchy-Schwarz inequality in ℓ^2, we can see that this is the case if $\sum_{n=1}^{\infty} |a_n|^2 n^\alpha$ converges for any $\alpha > 1$ (set $|a_n| = (|a_n| n^{\alpha/2})(n^{-\alpha/2})$; if $\alpha > 1$, then the sequence $n^{-\alpha/2}$ is in ℓ^2).

Now, let $f \in L^2(0,1)$ be such that the derivative of f (in the sense of distributions) is also in $L^2(0,1)$ (we shall study such functions extensively in the section on Sobolev spaces later). Then f' can be expanded in either a sine or a cosine series:

$$f'(x) = \sum_{n=1}^{\infty} a_n \sin(n\pi x),$$

$$f'(x) = \sum_{n=0}^{\infty} b_n \cos(n\pi x).$$

(6.49)

By integration, we find

$$f(x) = \sum_{n=1}^{\infty} -\frac{a_n}{n\pi} \cos(n\pi x) + \alpha,$$

$$f(x) = b_0 x + \sum_{n=1}^{\infty} \frac{b_n}{n\pi} \sin(n\pi x) + \beta.$$

(6.50)

The first of these expressions represents a cosine series for f, and since

$$\sum_{n=1}^{\infty} |a_n^2| < \infty$$

we have

$$\sum_{n=1}^{\infty} |a_n|/n\pi < \infty,$$

i.e., the first series in (6.50) converges uniformly. Hence any $f \in L^2(0,1)$ which has a derivative in $L^2(0,1)$ has a uniformly convergent cosine series. (In particular, this implies that any such f is continuous. This is a special case of the Sobolev embedding theorem.) Moreover, in the sense of L^2-convergence, the series can be differentiated term by term.

The second series in (6.50), on the other hand, is a sine series only if $\beta = b_0 = 0$. It is easy to see that $\beta = f(0)$ and $b_0 = \int_0^1 f'(x)\, dx = f(1)$. Hence any function $f \in L^2(0,1)$ such that $f' \in L^2(0,1)$ and in addition $f(0) = f(1) = 0$ has an absolutely convergent sine series. This shows that the convergence behavior of a Fourier series is influenced not only by the smoothness of the function but also by its behavior at the boundary.

6.2.3 Orthogonal Polynomials

According to the Weierstraß approximation theorem, polynomials are dense in $L^2(-1,1)$. It is therefore natural to apply the Schmidt orthogonalization procedure to the sequence $1, x, x^2, \ldots$ and obtain a basis consisting of polynomials. We claim that up to factors these orthogonal polynomials are given by

$$P_n(x) = \frac{1}{2^n n!} \frac{d^n (x^2 - 1)^n}{dx^n}.$$

(6.51)

First of all, it is obvious that P_n is a polynomial of degree n. Moreover, integration by parts shows that

$$\int_{-1}^{1} P_n(x) x^m \, dx = 0 \tag{6.52}$$

for every $m < n$; hence we also have

$$\int_{-1}^{1} P_n(x) P_m(x) \, dx = 0 \tag{6.53}$$

for $m < n$, i.e., the P_n are orthogonal in $L^2(-1,1)$. The P_n are called Legendre polynomials. The first few of them are

$$P_0(x) = 1, \qquad P_1(x) = x, \qquad P_2(x) = \frac{3}{2}x^2 - \frac{1}{2},$$

$$\tag{6.54}$$

$$P_3(x) = \frac{5}{2}x^3 - \frac{3}{2}x, \qquad P_4(x) = \frac{35}{8}x^4 - \frac{15}{4}x^2 + \frac{3}{8}.$$

The Legendre polynomials are not normalized. Using repeated integration by parts, one finds

$$\int_{-1}^{1} P_n^2(x) \, dx = \frac{(2n)!}{2^{2n}(n!)^2} \int_{-1}^{1} (1-x^2)^n \, dx. \tag{6.55}$$

The integral of $(1-x^2)^n$ can be evaluated by observing that $(1-x^2)^n = (1-x)^n(1+x)^n$ and using repeated integration by parts. The final result is

$$\int_{-1}^{1} P_n^2(x) \, dx = \frac{2}{2n+1}. \tag{6.56}$$

A variety of other orthogonal polynomials are also important for applications. These polynomials are orthogonal in weighted L^2-spaces.

Definition 6.40. Let Ω be an open set in $\mathbb{R}^m$, and let w be a continuous function from Ω to $\mathbb{R}^+$. Then we define

$$\tilde{L}_w^2(\Omega) := \left\{ u \in C(\bar{\Omega}) \mid \int_{\Omega} w(\mathbf{x}) |u(\mathbf{x})|^2 \, d\mathbf{x} < \infty \right\}. \tag{6.57}$$

The inner product is defined by

$$(u, v) = \int_{\Omega} w(\mathbf{x}) \overline{u(\mathbf{x})} v(\mathbf{x}) \, d\mathbf{x}. \tag{6.58}$$

The space $L_w^2(\Omega)$ is the completion of $\tilde{L}_w^2(\Omega)$.

For any weight function w and any interval (a, b), we can now define orthogonal polynomials by orthogonalizing the sequence $1, x, x^2, \ldots$ (provided of course, that w is such that polynomials are in $L_w^2(\Omega)$). The following cases are particularly important:

1. $a = -1$, $b = 1$, $w(x) = 1/\sqrt{1 - x^2}$. This leads to the Chebyshev polynomials

$$T_n(x) = \frac{1}{2^{n-1}} \cos(n \arccos x). \tag{6.59}$$

2. $a = -\infty$, $b = \infty$, $w(x) = \exp(-x^2)$. This leads to the Hermite polynomials

$$H_n(x) = (-1)^n \exp(x^2) \frac{d^n}{dx^n} \exp(-x^2). \tag{6.60}$$

3. $a = 0$, $b = \infty$, $w(x) = \exp(-x)$. This leads to the Laguerre polynomials

$$L_n(x) = e^x \frac{d^n}{dx^n} (x^n e^{-x}). \tag{6.61}$$

There are various other orthogonal polynomials with specific names, e.g., Jacobi and Gegenbauer polynomials. We leave it to the reader to verify the orthogonality of the Chebyshev, Hermite and Laguerre polynomials; see Problem 6.18. There are numerous facts known about orthogonal polynomials, e.g., formulas for their coefficients, "generating functions," differential equations which orthogonal polynomials satisfy, recursion relations and relationships to various special functions. We shall not discuss these issues here and instead refer to the literature.

We have yet to address the completeness of the polynomials introduced above. For the Chebyshev polynomials, this is clear from the Weierstraß approximation theorem, since uniform convergence implies convergence in $L_w^2(-1, 1)$. For the Hermite and Laguerre polynomials, however, we are dealing with infinite intervals, and we need a somewhat different argument.

We first consider the Laguerre polynomials. We have the identity

$$\sum_{n=0}^{\infty} \frac{L_n(x)}{n!} t^n = g(x, t) := \frac{1}{1 - t} \exp\left(-\frac{xt}{1 - t}\right); \tag{6.62}$$

see Problem 6.19. Since the Laguerre polynomials are "generated" by Taylor expansion of the right-hand side of (6.62), this expression is referred to as a generating function. We note that the convergence radius of the Taylor series is 1. An explicit calculation shows that

$$\int_0^{\infty} e^{-x} \left(g(x, t) - \sum_{n=0}^{N} \frac{L_n(x)}{n!} t^n\right)^2 dx = \frac{1}{1 - t^2} - \sum_{n=0}^{N} t^{2n} \tag{6.63}$$

(see Problem 6.20); hence the series in (6.62) converges also in $L_w^2(0, \infty)$ if $|t| < 1$. It follows that any function $e^{-\alpha x}$, $\alpha > -1/2$, can be approximated in $L_w^2(0, \infty)$ by Laguerre polynomials. However, if $f \in L_w^2$ is orthogonal to $e^{-\alpha x}$ for every α, then the Laplace transform of f is zero, and therefore f is zero. Hence linear combinations of exponentials $e^{-\alpha x}$ are dense in $L_w^2(0, \infty)$.

For the Hermite polynomials, we have the identity

$$e^{-t^2+2tx} = \sum_{n=0}^{\infty} \frac{H_n(x)}{n!} t^n \tag{6.64}$$

corresponding to (6.62) and an analogous argument applies. In this case, the convergence radius of the Taylor series is infinite, and one needs to show that linear combinations of the functions e^{2tx}, $t \in \mathbb{C}$, are dense in $L_w^2(-\infty, \infty)$. First observe that $\mathcal{D}(\mathbb{R})$ is dense. Every function ϕ in $\mathcal{D}(\mathbb{R})$ can be represented by a convergent Fourier integral:

$$\phi(x) = \frac{1}{\sqrt{2\pi}} \int_{-\infty}^{\infty} e^{i\xi x} \hat{\phi}(\xi) \, d\xi, \tag{6.65}$$

and the integral can be approximated by Riemann sums

$$\phi(x) \sim \frac{1}{\sqrt{2\pi}} \sum_{k=1}^{n} e^{i\xi_k x} \hat{\phi}(\xi_k)(\xi_k - \xi_{k-1}). \tag{6.66}$$

It is easy to see that the discrete sums converge to the integral in the sense of convergence in $L_w^2(-\infty, \infty)$.

Problems

6.15. Let $f \in L^2(-1, 1)$. Show that the Fourier series of f given by Corollary 6.39 converges uniformly if $f' \in L^2(-1, 1)$ and in addition $f(-1) = f(1)$.

6.16. Find the Fourier sine series of the function $f(x) = x$ on the interval $[0,1]$. Show that the series converges uniformly on $[0, 1 - \delta]$ for any $\delta > 0$. Hint: Consider also the Fourier sine series for the function $g(x) = x(1 + \cos \pi x)$.

6.17. Let $f \in C^1[0, 1]$. Show that the Fourier sine series for f converges uniformly except near the endpoints of the interval. Hint: Write f as the sum of a function which vanishes at the endpoints and a function whose Fourier series you can compute explicitly.

6.18. Verify the orthogonality of the Chebyshev, Hermite and Laguerre polynomials and find the factors necessary to normalize them.

6.19. Verify (6.62).

6.20. Fill in the details for showing the completeness of the Laguerre and Hermite polynomials.

6.21. Prove Lemma 6.30.

6.22. Is the span of x, x^2, x^3, etc., dense in $L^2(0, 1)$?

6.23. Prove that all separable, infinite-dimensional Hilbert spaces are isometric.

6.3 Duality and Weak Convergence

We have already encountered many of the ideas of duality (studying a function by studying how linear functionals act upon it) in the theory of distributions. These ideas are very powerful in the study of Banach spaces and Hilbert spaces as well.

6.3.1 Bounded Linear Mappings

Definition 6.41. Let X, Y be normed vector spaces. A linear mapping $L : X \to Y$ is called **bounded** if there is a constant C such that $\|Lx\| \leq C\|x\|$ for every $x \in X$.

For linear mappings, the ideas of continuity and boundedness are essentially the same.

Definition 6.42. We say that an operator $L : X \to Y$ is **continuous** at a point $x \in X$ if, whenever $x_n \in X$ is a sequence such that $x_n \to x$, we have $L(x_n) \to L(x)$.

Remark 6.43. Definition 6.42 is often called the definition of *sequential* continuity to distinguish it from the topological version of the definition.

Lemma 6.44. *Let $L : X \to Y$ be a linear mapping.*

1. *If L is continuous at the origin, it is continuous at every $x \in X$.*

2. *L is continuous if and only if it is bounded.*

Proof. To prove part 1 we assume that L is continuous at the origin. Note that the linearity of the operator implies $L(0) = 0$. Then for any $x \in X$ and any sequence $x_n \in X$ with $\lim_{n \to \infty} x_n = x$ we have

$$\lim_{n \to \infty} L(x_n) - L(x) = \lim_{n \to \infty} L(x_n - x) = L(0) = 0, \qquad (6.67)$$

so L is continuous at x.

To prove part 2 we note that the definition of boundedness implies continuity at zero. Thus, by part 1, boundedness implies continuity. On the other hand, suppose L is not bounded. Then for every n there exists y_n such that $\|Ly_n\| > n^2 \|y_n\|$. Note that $x_n := y_n(n\|y_n\|)^{-1} \to 0$ but $\|Lx_n\| > n$. If follows that L is not continuous at the origin. Thus, continuity implies boundedness. $\square$

It is natural to consider the set of all bounded linear mappings. It follows immediately from the definition that this set forms a vector space. Also, if we take the smallest possible constant in Definition 6.41, then this quantity gives us a measure of the "size" of a linear mapping. This motivates the following definition.

Definition 6.45. By $\mathcal{L}(X,Y)$ we denote the set of all bounded linear mappings from X to Y. If $X = Y$, we also write $\mathcal{L}(X)$ for $\mathcal{L}(X,X)$. Moreover, if $L \in \mathcal{L}(X,Y)$, we set

$$\|L\| := \sup_{\|x\|=1} \frac{\|Lx\|}{\|x\|} = \sup_{\|x\|\neq 0} \frac{\|Lx\|}{\|x\|}. \tag{6.68}$$

Theorem 6.46. *Let X, Y be Banach spaces. Then $\mathcal{L}(X,Y)$, with the norm defined by (6.68), is also a Banach space.*

The proof is straightforward and is left as an exercise (Problem 6.24). Linear mappings, also called linear operators, will be studied more extensively in the next chapter. In this section, we are interested in the special case of linear mappings from a Banach space to its scalar field.

Definition 6.47. Let X be a real (complex) Banach space. Then a **linear functional** on X is a bounded linear mapping from X to $\mathbb{R}$ ($\mathbb{C}$). The space of all linear functionals on X is called the **dual space** of X and is denoted by X^*.

6.3.2 Examples of Dual Spaces

Example 6.48. Let $1 < p < \infty$ and $p^{-1}+q^{-1} = 1$. It follows from Hölder's inequality that every $f \in L^p(\Omega)$ can be identified with a linear functional l_f on $L^q(\Omega)$ by the correspondence

$$l_f(g) := \int_\Omega \overline{f(\mathbf{x})}g(\mathbf{x})\, d\mathbf{x}. \tag{6.69}$$

The complex conjugate is included to make the definition analogous to the inner product in Hilbert space. It is clear that $\|l_f\| \leq \|f\|_p$. Assume now that $f \neq 0$ and let f_n be a Cauchy sequence in $\tilde{L}^p(\Omega)$ which converges to f in L^p. Let $g_n = f_n|f_n|^{p-2}$. Then $g_n \in \tilde{L}^q(\Omega)$ and $\|g_n\|_q^q = \|f_n\|_p^p$. Further, we find

$$|l_f(g_n) - l_{f_n}(g_n)| \leq \|f - f_n\|_p\|g_n\|_q = \|f - f_n\|_p\|f_n\|_p^{p/q} \tag{6.70}$$

and

$$l_{f_n}(g_n) = \|f_n\|_p^p. \tag{6.71}$$

It follows that $l_f(g_n)$ converges to $\|f\|_p^p$ and since $\|g_n\|_q$ converges to $\|f\|_p^{p/q}$, we conclude that actually $\|l_f\| = \|f\|_p$. Hence we have an isometric embedding from $L^p(\Omega)$ into the dual space of $L^q(\Omega)$. It can be shown that actually all functionals on $L^q(\Omega)$ can be represented in this way. We state this theorem without proof.

Theorem 6.49. *Let Ω be an open, locally Jordan measurable set in $\mathbb{R}^m$, $1 < p < \infty$ and $p^{-1} + q^{-1} = 1$. Then the dual space of $L^p(\Omega)$ agrees with $L^q(\Omega)$ by means of the correspondence (6.69).*

Example 6.50. Let f be a bounded continuous function on Ω. Then we have

$$\left| \int_\Omega \overline{f(\mathbf{x})} g(\mathbf{x}) \, d\mathbf{x} \right| \leq \|g\|_1 \sup_{\mathbf{x} \in \Omega} |f(\mathbf{x})| \qquad (6.72)$$

for every $g \in L^1(\Omega)$. Moreover, by taking g to be non-negative with support near a point where $|f|$ is close to its supremum, we can see that

$$\|f\|_\infty := \sup_{\mathbf{x} \in \Omega} |f(\mathbf{x})| \qquad (6.73)$$

represents the norm of f as a linear functional on $L^1(\Omega)$. However, not all linear functionals on $L^1(\Omega)$ can be represented in this way; for example we could take f's which are discontinuous across some surface. We make the following definition.

Definition 6.51. Let Ω be an open, locally Jordan measurable set in $\mathbb{R}^m$. The dual space of $L^1(\Omega)$ is denoted by $L^\infty(\Omega)$.

Since test functions are dense in $L^1(\Omega)$, and convergence in the sense of test functions implies L^1-convergence, $L^\infty(\Omega)$ is a space of distributions. Moreover, since a uniform limit of continuous functions is also continuous, continuous functions cannot be dense in $L^\infty(\Omega)$; in fact, $L^\infty(\Omega)$ is not separable. In measure theory, L^∞ is characterized as the space of bounded measurable functions.

The dual space of a Hilbert space is very simple: it is isometric to the space itself. This result is known as the Riesz Representation Theorem and is one of the foundations of the theory of Hilbert spaces.

Theorem 6.52 (Riesz representation). *The dual space of a Hilbert space is isometric to the Hilbert space itself. In particular, for every $x \in H$ the linear functional on H defined by*

$$l_x(y) := (x, y) \qquad (6.74)$$

is bounded with norm $\|l_x\|_{H^*} = \|x\|_H$. Moreover, for every $l \in H^*$, there exists a unique $x \in H$ such that

$$l(y) = (x, y) \quad \text{for every } y \in H, \qquad (6.75)$$

and, furthermore, $\|x\|_H = \|l\|_{H^*}$.

Proof. As the statement of the theorem indicates, for every $x \in H$ we can define a linear functional l_x using (6.74). It follows from the Cauchy-Schwarz inequality that $l_x \in H^*$ and that $\|l_x\| \leq \|x\|$. Furthermore, by setting $y = x$, we find $\|l_x\| = \|x\|$.

On the other hand, let l be a nonzero linear functional on H. We define $M = \{y \in H \mid l(y) = 0\}$. Then M is a closed subspace of H. Furthermore, $M^\perp$ is one-dimensional since for any u and v in $M^\perp$ we have $l(v)u - l(u)v \in M$. (This implies $l(v)u - l(u)v = 0$.)

Now, let $\tilde{x}$ be a unit vector in $M^\perp$. For every $y \in H$, we have the decomposition $y = (\tilde{x}, y)\tilde{x} + z$, where $z \in M$, and hence $l(y) = (\tilde{x}, y)l(\tilde{x})$. Thus, if we let

$$x = \overline{l(\tilde{x})}\tilde{x}, \tag{6.76}$$

we see that $l = l_x := (x, \cdot)$. The equality of the norms has already been established. $\qquad\square$

Remark 6.53. In the previous proof we constructed isomorphic mappings $A_H : H^* \to H$ and $A_H^{-1} : H \to H^*$ between a Hilbert space and the corresponding element of its dual; these are called the *Riesz mappings*.

Example 6.54. Let X be a Banach space. Then every $x \in X$ defines a linear functional on X^* by the correspondence $l_x y = y(x)$. It is clear that $\|l_x\| \leq \|x\|$, and it follows from the Hahn-Banach theorem, stated in the next subsection, that there is a $y \in X^*$ with $\|y\| = 1$ and $y(x) = \|x\|$. This implies that $\|l_x\| = \|x\|$, and hence we have an isometry from X to a subspace of X^{**}. It is important to characterize those Banach spaces for which this isometry is onto.

Definition 6.55. A Banach space is called **reflexive** if the isometry $x \mapsto l_x$ from $X \to X^{**}$ defined by $l_x y = y(x)$ is surjective.

It is clear from Theorem 6.52 that all Hilbert spaces are reflexive. Also, the spaces $L^p(\Omega)$, $1 < p < \infty$ are reflexive. By contrast, it can be shown that $L^1(\Omega)$ and $L^\infty(\Omega)$ are not reflexive. Reflexive spaces have many useful properties that make them advantageous for analysis. In the remainder of the book, we shall mostly work in Hilbert spaces, and we shall not pursue reflexive Banach spaces further. The reader is referred to the literature, e.g., [DS].

6.3.3   The Hahn-Banach Theorem

In the last example of the preceding section, we used the following result.

Theorem 6.56. *Let X be a Banach space and let M be a linear subspace of X. Let l_M be a bounded linear mapping from M to $\mathbb{R}$ ($\mathbb{C}$). Then there is a linear functional $l \in X^*$ such that $l|_M = l_M$ and $\|l\| = \|l_M\|$.*

In other words, linear functionals defined on any linear subspace can be extended without increasing their norm. This result and various more general ones from which it follows are known as the Hahn-Banach theorem. Some of the more general versions are relevant in convex analysis and optimization; since we shall not pursue these topics in the book, we do not state those results. The Hahn-Banach theorem guarantees the existence of many linear functionals, because we can always define linear functionals on any finite-dimensional subspace and then extend them. In particular, given any nonzero $x \in X$, there exists $f \in X^*$ with $f(x) \neq 0$.

For the proof of Theorem 6.56, we refer the reader to texts on functional analysis. For the case of a Hilbert space, however, the result is almost trivial. We first extend l to the closure of M by continuity, set $l = 0$ on $M^\perp$ and then define l on the whole space by linearity. We leave it to the reader to verify that the l defined in this way has the properties claimed in the theorem; see Problem 6.25.

6.3.4   The Uniform Boundedness Theorem

It is easy to construct a sequence f_n of functions on $[0,1]$ such that the sequence $f_n(x)$ is bounded for each x, but the f_n are not uniformly bounded. For instance,

$$f_n(x) := \begin{cases} 0, & x \in [0, 1/n) \\ 1/x, & x \in [1/n, 1]. \end{cases} \tag{6.77}$$

The uniform boundedness theorem says that, in contrast, for linear mappings between Banach spaces pointwise bounds imply uniform bounds.

Theorem 6.57. Let X, Y be Banach spaces. Let $T_n \in \mathcal{L}(X, Y)$, $n \in \mathbb{N}$, be such that $\|T_n x\|$ is bounded uniformly in n for each fixed $x \in X$. Then $\|T_n\|$ is bounded uniformly in n.

Corollary 6.58. Let Z be a Banach space and let x_n be a sequence in Z such that $f(x_n)$ is bounded uniformly in n for every $f \in Z^*$. Then $\|x_n\|$ is bounded uniformly in n.

The proof of the corollary is obtained by setting $X = Z^*$ and $Y = \mathbb{R}$ ($\mathbb{C}$) in Theorem 6.57.

Corollary 6.59. Let X, Y be Banach spaces and let $T_n \in \mathcal{L}(X, Y)$ be such that $|f(T_n x)|$ is bounded uniformly in n for each $f \in Y^*$ and $x \in X$. Then $\|T_n\|$ is bounded uniformly in n.

For the proof, apply first Corollary 6.58 to find that $\|T_n x\|$ is bounded uniformly in n and then use Theorem 6.57.

The proof of Theorem 6.57 is based on the Baire category theorem. We shall not carry out this proof here, but refer to the literature instead. In the following, we give a much simpler proof in a special case, namely, the case of Corollary 6.58, where Z is a Hilbert space. Assume that the conclusion is false, i.e., $\|x_n\|$ is not uniformly bounded. Let $M = \{x_n \mid n \in \mathbb{N}\}$, and let

$$\alpha(f) = \sup_{n \in \mathbb{N}} |(f, x_n)| \tag{6.78}$$

for $f \in Z$. Since M is not uniformly bounded, there exist $y_1 \in M$ and a unit vector e_1 with $|(e_1, y_1)| \geq 1$. Let Q be the vector space spanned by e_1 and y_1. Every x_n can be decomposed as $x_n = x_n^1 + x_n^2$, where $x_n^1 \in Q$

and $x_n^2 \in Q^{\perp}$. Since $|(e_1, x_n)|$ and $|(y_1, x_n)|$ are uniformly bounded, the x_n^1 are uniformly bounded and hence the x_n^2 are not. We can therefore find $y_2 \in M$ and a unit vector $e_2 \in Q^{\perp}$ such that

$$|(e_2, y_2)| \geq 2(\alpha(e_1) + 2). \tag{6.79}$$

Inductively, we can find $y_{n+1} \in M$ and a unit vector e_{n+1}, orthogonal to $e_1, e_2, \ldots, e_n; y_1, y_2, \ldots, y_n$, such that

$$|(e_{n+1}, y_{n+1})| \geq (n+1)\Big(\sum_{i=1}^{n} \frac{1}{i}\alpha(e_i) + n + 1\Big). \tag{6.80}$$

We now put

$$f = \sum_{i=1}^{\infty} \frac{1}{i} e_i. \tag{6.81}$$

This series converges because the e_i are orthonormal and $\sum_{i=1}^{\infty} i^{-2} < \infty$. We obtain further

$$
\begin{aligned}
|(f, y_{n+1})| &= \Big|\sum_{i=1}^{n} \frac{1}{i}(e_i, y_{n+1}) + \frac{1}{n+1}(e_{n+1}, y_{n+1})\Big| \\
&\geq -\sum_{i=1}^{n} \frac{1}{i}\alpha(e_i) + \frac{1}{n+1}(n+1)\Big(\sum_{i=1}^{n} \frac{1}{i}\alpha(e_i) + n + 1\Big) \\
&= n+1.
\end{aligned}
\tag{6.82}
$$

Hence the sequence $|(f, x_n)|$ cannot be bounded, a contradiction.

6.3.5   Weak Convergence

Procedures to solve differential equations usually involve a sequence of solutions to approximate problems. Very often it is possible to obtain uniform bounds for the approximates in the norm of some Banach space, but it is not so easy to show that they actually converge in that Banach space. In these situations, a weaker notion of convergence, known as weak convergence, is extremely useful. Weak convergence of a sequence of functions means that certain linear functionals, when applied to that sequence, yield convergent sequences of numbers. We have already encountered such a notion when we defined a sequence of distributions f_n to be convergent if (f_n, ϕ) converges for every test function ϕ. In the context of Banach spaces, we make the following definition.

Definition 6.60. Let X be a Banach space. A sequence x_n in X **converges weakly** to x if $f(x_n)$ converges to $f(x)$ for every $f \in X^*$. A sequence f_n in X^* **converges weakly-∗** to f if $f_n(x)$ converges to $f(x)$ for every $x \in X$.

To distinguish notations, one writes $x_n \to x$ for convergence in norm, $x_n \rightharpoonup x$ for weak convergence, and $x_n \overset{*}{\rightharpoonup} x$ for weak-$*$ convergence.

Remark 6.61. We list here a number of important consequences of the definition of weak convergence. We give only very brief indications of the (usually very short) proofs of these assertions.

- It follows directly form the definition that a weakly convergent sequence in X^* also converges weakly-$*$. The converse is false in general, but it is true if X is reflexive.

- The uniqueness of weak limits follows from the Hahn-Banach theorem; the uniqueness of weak-$*$ limits is obvious from the definition.

- It follows from the uniform boundedness principle that weakly or weakly-$*$ convergent sequences are bounded.

- If $f(x_n)$ is a Cauchy sequence for every $f \in X^*$, then

$$l(f) := \lim_{n \to \infty} f(x_n)$$

defines a linear functional on X^*, but we do not necessarily know that this linear functional can be represented in the form $l(f) = f(x)$, where $x \in X$. Thus a Cauchy criterium does not necessarily hold for weak convergence. The spaces for which it does hold are called weakly (sequentially) complete. Obviously, reflexive spaces are weakly complete. On the other hand, a Cauchy criterium always holds for weak-$*$ convergence.

Example 6.62. Let H be a separable, infinite-dimensional Hilbert space and let e_n be an orthonormal basis of H. Then e_n converges weakly to zero (Problem 6.28).

Example 6.63. A particular case of the previous example is the sequence $f_n(x) = \sin nx$ in $L^2(0,1)$. This sequence of increasingly oscillatoy functions converges weakly to zero. Of course, it does not converge to zero in any pointwise sense, though one could say it converges to zero "on average."

The usefulness of the concept of weak-$*$ convergence is to a large extent based on the following result.

Theorem 6.64 (Weak compactness, Alaoglu). *Let X be a separable Banach space and let f_n be a bounded sequence in X^*. Then f_n has a weakly-$*$ convergent subsequence.*

Proof. Let x_k, $k \in \mathbb{N}$ be a sequence that is dense in X. We first note that if $f_n(x_k)$ converges for every $k \in \mathbb{N}$, then $f_n(x)$ converges for every $x \in X$. To see this, let M be an upper bound for $\|f_n\|$. For given $\epsilon > 0$ and $x \in X$, we can choose k such that $\|x - x_k\| \leq \epsilon/(3M)$. We then have

$$
\begin{aligned}
|f_n(x) - f_m(x)| &\leq |f_n(x) - f_n(x_k)| + |f_n(x_k) - f_m(x_k)| \\
&\quad + |f_m(x_k) - f_m(x)| \quad\quad\quad\quad (6.83) \\
&\leq \frac{2}{3}\epsilon + |f_n(x_k) - f_m(x_k)|,
\end{aligned}
$$

and the last term on the right-hand side is less than $\epsilon/3$ if m and n are large enough. Hence $f_n(x)$ converges for every $x \in X$. The rest of the proof is a standard diagonal argument. Since $f_n(x_1)$ is bounded, we can extract a subsequence f_{n1} such that $f_{n1}(x_1)$ converges. From this subsequence, we extract another subsequence f_{n2} such that $f_{n2}(x_2)$ converges. We proceed like this inductively. The diagonal sequence f_{nn} has the property that $f_{nn}(x_k)$ converges for every k. $\qquad\square$

Remark 6.65. In the typical applications of this theorem, the f_n are approximate solutions to a PDE. While it may be difficult to show that the sequence converges, one can often obtain bounds on the sequence which imply a weakly convergent subsequence. The limit of the subsequence is then shown to be the solution of the original PDE.

Problems

6.24. Prove Theorem 6.46.

6.25. Verify the remarks at the end of Section 6.3.3.

6.26. Let M be a linear manifold in a Banach space X. Show that M is not dense in X if and only if there exists a nonzero $l \in X^*$ such that

$$l(x) = 0 \text{ for every } x \in M.$$

6.27. Let H be an infinite-dimensional Hilbert space. Prove that $\mathcal{L}(H)$ is not separable.

6.28. Verify the claim of Example 6.62.

6.29. Let X, Y be Banach spaces and let $T \in \mathcal{L}(X,Y)$ and let x_n be a weakly convergent sequence in X. Show that Tx_n converges weakly in Y.

6.30. Let M be a closed subspace of the Banach space X and let $x_n \in M$ converge weakly to x. Show that $x \in M$. Hint: Show first that if $f(x) = 0$ for every $f \in X^*$ with $f|_M = 0$, then $x \in M$.

6.31. Let H be a Hilbert space. Let x_n converge weakly to x and assume in addition that $\|x_n\| \to \|x\|$. Show that $x_n \to x$ in norm (strongly).

6.32. Let f_n be a bounded sequence in $L^2(\Omega)$, and assume that f_n converges in the sense of distributions. Show that f_n converges weakly.

6.33. Let $p \in (1, \infty)$, $q = p/(p-1)$. Suppose

$$f_n \rightharpoonup \bar{f} \text{ (weakly) in } L^p(\Omega),$$

$$g_n \to \bar{g} \text{ (strongly) in } L^q(\Omega).$$

Show that

$$f_n g_n \to \bar{f}\bar{g} \text{ in } \mathcal{D}'(\Omega).$$

6.34. (a) Show that $\delta(x) \notin L^1(\mathbb{R})$. Hint: For given $g \in \tilde{L}^1(\mathbb{R})$, find a continuous function f with

$$\left| \int_{-\infty}^{\infty} f(x)g(x) \, dx - f(0) \right| \geq \frac{1}{2} \max_{x \in \mathbb{R}} |f(x)|. \qquad (6.84)$$

(b) Show that $L^1(\mathbb{R})$ is not reflexive.

7
Sobolev Spaces

The analysis of partial differential equations naturally involves function spaces which are defined not only in terms of properties of the functions themselves, but also of their derivatives. In Problem 6.3, for example, we introduced the Banach space $C_b^k(\overline{\Omega})$. Unfortunately, these spaces turn out to be rather unsuitable for analyzing PDEs. For example, if $\Delta u = f$ with f continuous, it is in general not true that u is C^2. Sobolev spaces turn out to be much more useful. They are defined in an analogous fashion as C_b^k, but with L^p taking over the role of the continuous bounded functions. Throughout this section, Ω will be an open, locally Jordan measurable set in $\mathbb{R}^m$.

This chapter is intended to give the reader a brief overview of Sobolev spaces. In order to arrange the chapter to make it easier for the reader to get the main ideas before juming into the proofs we have arranged the material as follows.

- We begin with the basic definitions of the "positive" Sobolev spaces, examples of typical functions contained in (and excluded from) these spaces, and basic properties of the spaces.

- We then have have a section on various important properties of Sobolev spaces. We address the following questions.

  1. If Ω is all of $\mathbb{R}^m$ we can define the Fourier transform of a function in a Sobolev space. What do we know about the transform of a function in a particular space?

2. What imbedding relations exist between Sobolev spaces with different values of p? How large must k be so that $W^{k,p}(\Omega)$ consists of continuous functions?

3. Compactness theorems: Recall the Arzela-Ascoli theorem which implies that a sequence of uniformly bounded continuous functions with uniformly bounded derivatives has a convergent subsequence. Can we replace uniform bounds on derivatives by L^p-bounds?

4. Trace theorems: Since $\mathcal{D}(\Omega)$ is dense in L^p, it is not meaningful to talk about boundary values of arbitrary L^p-functions. However, $\mathcal{D}(\Omega)$ is generally not dense in $W^{k,p}(\Omega)$. What can one say about boundary values of functions in Sobolev spaces? How can $W_0^{k,p}(\Omega)$ be characterized in terms of the behavior at the boundary?

- We then have a section on the dual spaces of Sobolev spaces, the so-called "negative" spaces.

- Finally, we conclude with a section of technical results. These results are interesting in their own right, but are included here primarily to make it possible to prove some of the main results of the previous sections. The technical questions addressed are as follows.

1. Density theorems: Are functions in $C^k(\overline{\Omega})$ dense in $W^{k,p}(\Omega)$? If Ω is unbounded, are functions of bounded support dense?

2. Coordinate transformations: How do functions in Sobolev spaces change under transformations to different domains?

3. Extension theorems: If $f \in W^{k,p}(\Omega)$, does there exist $F \in W^{k,p}(\mathbb{R}^m)$ such that $F|_\Omega = f$?

7.1 Basic Definitions

Definition 7.1. Let k be a non-negative integer and let $1 \le p \le \infty$. Then we define $W^{k,p}(\Omega)$ to be the set of all distributions $u \in L^p(\Omega)$ such that $D^\alpha u \in L^p(\Omega)$ for $|\alpha| \le k$. In $W^{k,p}(\Omega)$, we define a norm by

$$\|u\|_{k,p}^p := \sum_{|\alpha|\le k} \|D^\alpha u\|_p^p, \quad p < \infty,$$

$$\|u\|_{k,\infty} := \max_{|\alpha|\le k} \|D^\alpha u\|_\infty, \tag{7.1}$$

and for $p = 2$, we define an inner product by

$$(u,v)_k := \sum_{|\alpha|\le k} \int_\Omega D^\alpha \overline{u(\mathbf{x})} D^\alpha v(\mathbf{x}) \, d\mathbf{x}. \tag{7.2}$$

For $W^{k,2}(\Omega)$ we also use the notation $H^k(\Omega)$.

Example 7.2. For $1 \le p < \infty$ we have

$$\|u\|_{1,p} = \left\{ \int_\Omega \left(|u|^p + \sum_{i=1}^m \left| \frac{\partial u}{\partial x_i} \right|^p \right) dx \right\}^{1/p},$$

$$\|u\|_{2,p} = \left\{ \int_\Omega \left(|u|^p + \sum_{i=1}^m \left| \frac{\partial u}{\partial x_i} \right|^p + \sum_{i,j=1}^m \left| \frac{\partial^2 u}{\partial x_i \partial x_j} \right|^p \right) d\mathbf{x} \right\}^{1/p}.$$

Example 7.3. The Heaviside function $H(x)$ is in $L^p(-1,1)$ for any $p \in [1, \infty]$. However, it is *not* in $W^{1,p}(-1,1)$ since its distributional derivative, the Dirac delta, cannot be represented by an $L^p(-1,1)$ function . Similarly, the absolute value function $f(x) = |x|$ is in $W^{1,p}(-1,1)$ since its distributional derivative can be represented by the $L^p(-1,1)$ function

$$f'(x) = \begin{cases} -1 & x < 0 \\ 1 & x > 0 \end{cases}$$

However, f is not in $W^{2,p}(-1,1)$ since its second distributional derivative (2δ) cannot be represented by a function in L^p.

Example 7.4. Consider the function

$$f(\mathbf{x}) := |\mathbf{x}|^\alpha = \left(\sum_{i=1}^m x_i^2 \right)^{\alpha/2}$$

on $B = B_1(0) \subset \mathbb{R}^m$, the ball of radius one in $\mathbb{R}^m$. One can easily check using spherical coordinates that if $\alpha + m > 0$ this function is integrable. Thus, if $p\alpha > -m$ we have $f \in L^p(B)$.

Formally taking partial derivatives we get

$$\frac{\partial f}{\partial x_i}(\mathbf{x}) = \alpha x_i \left(\sum_{i=1}^m x_i^2 \right)^{\alpha/2-1} = \alpha |\mathbf{x}|^{(\alpha-2)/2} x_i$$

$$\frac{\partial^2 f}{\partial x_i \partial x_j}(\mathbf{x}) = \alpha(\alpha-2)x_i x_j \left(\sum_{i=1}^m x_i^2 \right)^{\alpha/2-2} + \alpha \left(\sum_{i=1}^m x_i^2 \right)^{\alpha/2-1} \delta_{ij}$$

where δ_{ij} is the Kronecker delta.

We can use standard techniques of improper integrals to show that if the formal derivative has an integrable singularity, then that function is a representative of the distributional derivative. Using this and calculating the norms we get (after some work)

- $f \in W^{1,p}(B)$ if and only if $p(\alpha - 1) > -m$ (i.e. $\alpha > 1 - m/p$).
- $f \in W^{2,p}(B)$ if and only if $p(\alpha - 2) > -m$ (i.e. $\alpha > 2 - m/p$).

Similar calculations show that $|\mathbf{x}|^\alpha \in W^{k,p}(B)$ if and only if $\alpha > k - m/p$.

Theorem 7.5. *The space $W^{k,p}(\Omega)$ is a Banach space. If $p < \infty$, it is separable.*

Proof. For $u \in W^{k,p}(\Omega)$, let $u_\alpha = D^\alpha u$. Then the mapping $u \to (u_\alpha)_{|\alpha| \le k}$ is an isometry from $W^{k,p}(\Omega)$ onto a subspace M of the direct product $\prod_{|\alpha| \le k} L^p(\Omega)$. The completeness of $W^{k,p}(\Omega)$ follows if we show that M is closed. For this, one has to show that if $u_n \in W^{k,p}$ and $D^\alpha u_n \to v^\alpha$ in $L^p(\Omega)$, then $v^\alpha = D^\alpha v^0$. But this is clear, since L^p-convergence implies distributional convergence.

If $p < \infty$, then $\prod_{|\alpha| \le k} L^p(\Omega)$ is separable. Hence we only need to show that a subspace of a separable Banach space is separable. This is done in the following lemma to complete the proof of this theorem.   □

Lemma 7.6. *Let X be a separable Banach space and let M be a linear subspace of X. Then M is separable.*

Proof. For any given $\epsilon > 0$, there is a countable set $\{x_n\}_{n \in \mathbb{N}}$ in X such that for any $x \in X$ there is $n \in \mathbb{N}$ with $\|x - x_n\| \le \epsilon$. From the sequence x_n, discard all elements such that the ball of radius ϵ centered at x_n does not intersect M. The remaining points form a new sequence y_n. Now let u_n be any point in M which lies in the ϵ-neighborhood of y_n. Then every point in M lies within ϵ of one of the y_n and hence within 2ϵ of one of the u_n. By letting ϵ run through any sequence that converges to zero, we can generate a countable dense set in M.   □

If $p < \infty$, then $\mathcal{D}(\Omega)$ is dense in $L^p(\Omega)$. This is generally not the case for the spaces $W^{k,p}(\Omega)$. Indeed we have the following result.

Lemma 7.7. *Let Ω be bounded, let $k \ge 1$ and $1 \le p \le \infty$. Then $\mathcal{D}(\Omega)$ is not dense in $W^{k,p}(\Omega)$.*

Proof. If u is any function in $C^\infty(\overline{\Omega})$, then the mapping

$$L : \phi \mapsto \int_\Omega \sum_{|\alpha| \le k} D^\alpha \bar{u}(\mathbf{x}) D^\alpha \phi(\mathbf{x}) \, d\mathbf{x} \tag{7.3}$$

defines a bounded linear functional on $W^{k,p}(\Omega)$. Moreover, if

$$\sum_{|\alpha| \le k} (-1)^{|\alpha|} D^{2\alpha} \bar{u} = 0, \tag{7.4}$$

then $L(\phi)$ vanishes for every $\phi \in \mathcal{D}(\Omega)$. A nonzero solution u of (7.4) in any parallelepiped containing Ω can easily be found by separation of variables; see Chapter 1. Moreover, $L(u) > 0$ and $u \in W^{k,p}(\Omega)$; hence $\mathcal{D}(\Omega)$ cannot be dense.   □

The last lemma motivates the following definition.

Definition 7.8. By $W_0^{k,p}(\Omega)$ we denote the closure of $\mathcal{D}(\Omega)$ in $W^{k,p}(\Omega)$.

7.2 Characterizations of Sobolev Spaces

The basic definition of a Sobolev space describes it as a subspace of $L^p(\Omega)$. Of course, there is much more to be said, and in this section we describe some of the most important ways that functions in a Sobolev space can be characterized.

In most of this section, we shall confine our discussions to the case $p = 2$. Many of the results we discuss have analogues for general p, for which we refer to the literature.

7.2.1 Some Comments on the Domain Ω

The answers to a number of questions about Sobolev spaces depend on assumptions on the regularity of the boundary of Ω.[1] Most of the time, we shall assume a smooth boundary. Specifically, we make the following definition.

Definition 7.9. We say that Ω is of class C^k, $k \geq 1$, if every point on $\partial\Omega$ has a neighborhood N so that $\partial\Omega \cap N$ is a C^k-surface and, moreover, $\Omega \cap N$ is "on one side" of $\partial\Omega \cap N$.

If Ω is a bounded domain, i.e., connected, the last assumption is redundant; cf. Remark 4.8.

There are two classes of problems in applications, where nonsmooth domains are relevant:

1. domains with corners, and

2. free boundary problems where Ω is a priori unknown.

It turns out that in fact many results on Sobolev spaces do not require a smooth boundary. Instead, various geometric conditions such as the "segment property" and "cone property" (cf. [Fr]) need to be assumed. We shall not discuss these conditions here, but we shall state some results for Lipschitz domains.

Definition 7.10. We say that Ω is **Lipschitz** if every point on $\partial\Omega$ has a neighborhood N such that, after an affine change of coordinates (translation and rotation), $\partial\Omega \cap N$ is described by the equation $x_m =$

---

[1]In this rather short treatment of Sobolev spaces, we have chosen to avoid most questions of boundary smoothness. For a more complete study of the subject we recommend the paper of Fraenkel [Fr].

$\phi(x_1, \ldots, x_{m-1})$, where ϕ is **uniformly Lipschitz continuous**. Moreover, $\Omega \cap N$ is on one side of $\partial\Omega \cap N$, e.g., $\Omega \cap N = \{\mathbf{x} \in N \mid x_m < \phi(x_1, \ldots, x_{m-1})\}$.

If Ω is unbounded, then, in addition to smoothness conditions on $\partial\Omega$, one needs to impose conditions which say that Ω is well behaved at infinity. We shall not give a general discussion of such conditions, and many results will be stated only for the case when $\partial\Omega$ is bounded.

Finally, we define a characterization of the domain Ω that will be very useful as a concise technical hypothesis.

Definition 7.11. We say that Ω has the k-**extension property** if there is a bounded linear mapping $E : H^k(\Omega) \to H^k(\mathbb{R}^m)$ such that $Eu|_\Omega = u$ for every $u \in H^k(\Omega)$.

It is of course trivial that, conversely, the restriction of every function in $H^k(\mathbb{R}^m)$ is in $H^k(\Omega)$. The extension property will be investigated in a later subsection; it turns out that bounded Lipschitz domains have the extension property for every k.

7.2.2   Sobolev Spaces and Fourier Transform

We now consider Sobolev spaces of all of $\mathbb{R}^m$. Clearly, it follows from Theorem 5.65 that the Fourier transform maps $L^2(\mathbb{R}^m)$ to itself; indeed, it is an isometry in $L^2(\mathbb{R}^m)$. Moreover, the Fourier transform of $D^\alpha u$ is $(i\xi)^\alpha \hat{u}$. Hence we immediately obtain the following result.

Theorem 7.12. *The Fourier transform $\mathcal{F}$ is a homeomorphism from $H^k(\mathbb{R}^m)$ onto the weighted space $L_w^2(\mathbb{R}^m)$ (cf. Definition 6.40), where $w(\xi) = 1 + |\xi|^{2k}$.*

We shall use the notation L_k^2 to denote this weighted L^2-space. It is easy to see that $\mathcal{S}(\mathbb{R}^m)$ is dense in $L_k^2(\mathbb{R}^m)$. Theorem 7.12 then implies that $\mathcal{S}(\mathbb{R}^m)$, and hence also $\mathcal{D}(\mathbb{R}^m)$, is dense in $H^k(\mathbb{R}^m)$.

Corollary 7.13. *$\mathcal{D}(\mathbb{R}^m)$ is dense in $H^k(\mathbb{R}^m)$.*

Another application of the theorem is the definition of fractional order Sobolev spaces.

Definition 7.14. We say that $u \in H^s(\mathbb{R}^m)$, $s \in \mathbb{R}^+$, if $\mathcal{F}[u]$ is in the weighted L^2-space $L_w^2(\mathbb{R}^m) =: L_s^2(\mathbb{R}^m)$ with $w(\xi) = 1 + |\xi|^{2s}$.

There is an intrinsic characterization of the fractional Sobolev spaces, which is basically an L^2-analogue of Hölder continuity. It can be shown that an equivalent inner product on $H^s(\mathbb{R}^m)$ is given by

$$(u, v)_s = (u, v)_{[s]}$$

$$+ \sum_{|\alpha|=[s]} \int_{\mathbb{R}^m} \int_{\mathbb{R}^m} \frac{(D^\alpha \bar{u}(\mathbf{x}) - D^\alpha \bar{u}(\mathbf{y}))(D^\alpha v(\mathbf{x}) - D^\alpha v(\mathbf{y}))}{|\mathbf{x}-\mathbf{y}|^{m+2(s-[s])}} \, d\mathbf{x} \, d\mathbf{y}. \tag{7.5}$$

Here $[s]$ denotes the largest integer $\leq s$, and the integral in (7.5) is to be interpreted, as usual, as a limit for sequences of smooth functions which approximate u and v, respectively. Clearly, a definition in terms of (7.5) can be extended immediately to arbitrary Ω.

7.2.3   The Sobolev Imbedding Theorem

We begin with a definition of the term "imbedding."

Definition 7.15. Let X and Y be Banach spaces, we say X is **continuously imbedded** in Y and write

$$X \hookrightarrow Y$$

if $X \subset Y$ and there is a constant C such that

$$\|u\|_Y \leq C\|u\|_X \quad \forall u \in X. \tag{7.6}$$

Remark 7.16. If $X \subset Y$ then the identity operator $I : X \to Y$ is well defined. Condition (7.6) says that the output of the operator is bounded by a constant times the norm of the input. Thus, the identity is a bounded operator.

The Sobolev imbedding theorem says that if Ω is nice, then for sufficiently large k, $H^k(\Omega)$ is imbedded in a space of bounded, continuous functions. We first consider the case $\Omega = \mathbb{R}^m$. As a preparation, we state the following lemma.

Lemma 7.17. *If $\mathcal{F}[u] \in L^1(\mathbb{R}^m)$, then u is a continuous bounded function. Moreover, $\|u\|_\infty \leq (2\pi)^{-m/2}\|\mathcal{F}[u]\|_1$.*

Proof. If $\mathcal{F}[u] \in \mathcal{S}(\mathbb{R}^m)$, then $u \in \mathcal{S}(\mathbb{R}^m)$ and

$$\|u\|_\infty = \sup_{\mathbf{x}\in\mathbb{R}^m} (2\pi)^{-m/2}\left|\int_{\mathbb{R}^m} \exp(i\xi\cdot\mathbf{x})\mathcal{F}[u](\xi)\, d\xi\right| \leq (2\pi)^{-m/2}\|\mathcal{F}[u]\|_1. \tag{7.7}$$

Since $\mathcal{S}(\mathbb{R}^m)$ is dense in $L^1(\mathbb{R}^m)$, the lemma follows by taking limits. $\quad\square$

Theorem 7.18 (Sobolev imbedding theorem). *Let $s > m/2$. Then*

$$H^s(\mathbb{R}^m) \hookrightarrow C_b(\mathbb{R}^m).$$

That is

$$H^s(\mathbb{R}^m) \subset C_b(\mathbb{R}^m)$$

and there is a constant C such that $\|u\|_\infty \leq C\|u\|_{s,2}$ for every $u \in H^s(\mathbb{R}^m)$.

Proof. Let $u \in H^s(\mathbb{R}^m)$. Then $\mathcal{F}[u] \in L^2_s(\mathbb{R}^m)$. By the previous lemma, it therefore suffices to show that $L^2_s(\mathbb{R}^m)$ is continuously imbedded in $L^1(\mathbb{R}^m)$. Let $w(\xi) = 1 + |\xi|^{2s}$. Then Hölder's inequality yields

$$\|\mathcal{F}[u]\|_1 \leq \|w^{1/2}\mathcal{F}[u]\|_2\|w^{-1/2}\|_2. \tag{7.8}$$

Since we have assumed $s > m/2$, $w^{-1/2}$ is in $L^2(\mathbb{R}^m)$.   □

Corollary 7.19. *Assume that $k > m/2$ and that Ω has the k-extension property. Then*

$$H^k(\Omega) \hookrightarrow C_b(\overline{\Omega}).$$

If $k > (m/2) + j$, then

$$H^k(\Omega) \hookrightarrow C_b^j(\overline{\Omega}).$$

Remark 7.20. In Theorem 7.18, the bound $s > m/2$ is optimal; cf. Problem 7.4.

Example 7.21. In Example 7.3 we noted that the Heaviside function $H(x)$ was not in $H^1(-1, 1)$. The same argument could be used to show that no function with a jump discontinuity is in $H^1(-1, 1)$. The Sobolev imbedding theorem gives a much stronger result, stating that if $s > 1/2$ every function in $H^s(-1, 1)$ is bounded and continuous.

Example 7.22. Let B be the ball of radius one in $\mathbb{R}^m$. In example 7.4 we showed that $|\mathbf{x}|^\alpha \in W^{k,p}(B)$ if and only if $\alpha > k - m/p$. If we set $p = 2$ we note that the condition of the Sobolev imbedding theorem requiring $k > m/p$ is exactly the condition that ensures $\alpha > 0$ so that our radial function would be continuous.

Example 7.23. In $\mathbb{R}^3$ the critical "number of derivatives" is $s = 3/2$. Functions in $H^1(\mathbb{R}^3)$ are not necessarily continuous, but all functions in $H^2(\mathbb{R}^3)$ are bounded and continuous.

Remark 7.24. More general imbedding theorems for $W^{k,p}$ spaces can be established. In particular, it can be shown that

$$W^{k,p}(\mathbb{R}^m) \hookrightarrow L^{mp/(m-kp)}(\mathbb{R}^m) \quad \text{for } kp < m,$$

and

$$W^{k,p}(\mathbb{R}^m) \hookrightarrow C_b(\mathbb{R}^m) \quad \text{for } kp > m,$$

and that each of these imbeddings is continuous. Again this extends to more general domains Ω, e.g., those which have an extension property.

7.2.4 Compactness Properties

In this subsection, we show that certain imbeddings involving Sobolev spaces are compact. This is often useful in applications since it implies that certain sequences have convergent subsequences; the limit of such a subsequence is typically the solution that is being sought. It is also useful in defining certain equivalent norms for Sobolev spaces as we do at the end of this subsection.

We begin with the definition of a compact imbedding.

Definition 7.25. Let X, Y be Banach spaces such that X is continuously imbedded in Y. We say that X is **compactly imbedded** in Y and write

$$X \overset{c}{\hookrightarrow} Y$$

if the unit ball in X is precompact in Y or, equivalently, every bounded sequence in X has a subsequence that converges in Y.

Before identifying compact imbeddings of Sobolev spaces, we establish a lemma.

Lemma 7.26. *Let $s > m/2$ and $0 < \alpha < \min(s - m/2, 1)$. Then there is a constant C such that*

$$|u(\mathbf{x}) - u(\mathbf{y})| \leq C\|u\|_{s,2}|\mathbf{x} - \mathbf{y}|^{\alpha} \tag{7.9}$$

for every $\mathbf{x}, \mathbf{y} \in \mathbb{R}^m$ and $u \in H^s(\mathbb{R}^m)$.

Proof. Throughout the proof, let C denote a generic constant independent of $\mathbf{x}$, $\mathbf{y}$ and u. We first observe that, for any $\alpha \in (0,1)$,

$$|\exp(i\xi\mathbf{x}) - \exp(i\xi\mathbf{y})| \leq C|\mathbf{x} - \mathbf{y}|^{\alpha}|\xi|^{\alpha}. \tag{7.10}$$

With $\hat{u}$ denoting the Fourier transform of u, we have

$$\begin{aligned}
|u(\mathbf{x}) - u(\mathbf{y})| &\leq (2\pi)^{-m/2} \int_{\mathbb{R}^m} |\exp(i\xi\mathbf{x}) - \exp(i\xi\mathbf{y})||\hat{u}(\xi)| \, d\xi \\
&\leq C|\mathbf{x} - \mathbf{y}|^{\alpha} \int_{\mathbb{R}^m} |\hat{u}(\xi)||\xi|^{\alpha} \, d\xi \\
&\leq C|\mathbf{x} - \mathbf{y}|^{\alpha}\|(1 + |\xi|^s)\hat{u}\|_2 \||\xi|^{\alpha}/(1 + |\xi|^s)\|_2 \\
&\leq C|\mathbf{x} - \mathbf{y}|^{\alpha}\|u\|_{s,2}.
\end{aligned} \tag{7.11}$$

$\square$

We now prove one of the most basic compact imbedding theorems.

Theorem 7.27. *Let $k > m/2$ and let Ω be bounded and such that it has the k-extension property. Then*

$$H^k(\Omega) \overset{c}{\hookrightarrow} C_b(\overline{\Omega}),$$

i.e. $H^k(\Omega)$ is compactly imbedded in $C_b(\overline{\Omega})$.

Proof. The theorem follows immediately from the Sobolev imbedding theorem, Lemma 7.26, and the Arzela-Ascoli theorem. $\square$

Example 7.28. The assumption that Ω is bounded is essential. To see this let $\Omega = \mathbb{R}^m$ and consider the sequence of translated functions

$$u_n(\mathbf{x}) = u(\mathbf{x} + n\mathbf{e}),$$

where $\mathbf{e}$ is any fixed unit vector and u any nonzero element of $H^k(\mathbb{R}^m)$. This sequence is, of course, bounded in $H^k(\mathbb{R}^m)$. However, the sequence can have no uniformly convergent subsequence. (To prove this, one needs to show that no such sequence that does converge uniformly can be in $H^k(\mathbb{R}^m)$ because of its behavior at infinity.)

The next theorem concerns compactness of imbeddings between Sobolev spaces.

Theorem 7.29. *Let Ω be bounded and let k be a non-negative integer. Assume that Ω has the $(k+1)$-extension property. Then*

$$H^{k+1}(\Omega) \overset{c}{\hookrightarrow} H^k(\Omega),$$

i.e. $H^{k+1}(\Omega)$ is compactly imbedded in $H^k(\Omega)$.

Proof. Let E be an extension operator; we can always choose E in such a way that it maps to functions supported in some compact set S. If u_n is a bounded sequence in $H^{k+1}(\Omega)$, then Eu_n is a bounded sequence in $H^{k+1}(\mathbb{R}^m)$ and the support of Eu_n is contained in S. We shall show that Eu_n has a subsequence which converges in $H^k(\mathbb{R}^m)$. Let M_R denote the operator of multiplication by the characteristic function of the ball $\{|\xi| < R\}$. For $v \in H^{k+1}(\mathbb{R}^m)$, let $v^R = \mathcal{F}^{-1}[M_R\mathcal{F}[v]]$. Then it is easy to show that there is a constant C with

$$\|u - u^R\|_{k,2} \leq \frac{C}{R}\|u\|_{k+1,2}. \tag{7.12}$$

Moreover, for any fixed R and any positive integer l, we have $u^R \in C_b^l(\mathbb{R}^m)$ and there is a constant $C(R,l)$ such that

$$\|u^R\|_{l,\infty} \leq C(R,l)\|u\|_{k+1,2}. \tag{7.13}$$

Hence if Eu_n is bounded in $H^{k+1}(\mathbb{R}^m)$, and D is any bounded open set containing S, then $(Eu_n)^R$ has a subsequence which converges in $C_b^k(\overline{D})$ (Arzela-Ascoli theorem) and hence in $H^k(D)$. By choosing a sequence $R_n \to \infty$ and applying a diagonal argument, we find a subsequence n_j such that $(Eu_{n_j})^{R_n}$ converges in $H^k(D)$ for every n. In conjunction with (7.12), this implies that Eu_{n_j} converges in $H^k(D)$ and hence in $H^k(\mathbb{R}^m)$, since the support of each Eu_{n_j} is contained in D. □

As a consequence of Theorem 7.29, we shall establish Ehrling's lemma, which allows us to introduce an equivalent norm on $H^k(\Omega)$. We first give an abstract version.

Theorem 7.30 (Ehrling's lemma). *Let X, Y and Z be Banach spaces. Assume that X is compactly imbedded in Y and Y is continuously imbedded in Z*

$$X \overset{c}{\hookrightarrow} Y \hookrightarrow Z.$$

Then for every $\epsilon > 0$ there exists a constant $c(\epsilon)$ such that

$$\|x\|_Y \leq \epsilon\|x\|_X + c(\epsilon)\|x\|_Z \tag{7.14}$$

for every $x \in X$.

Proof. Assume the claim fails for some $\epsilon_0 > 0$. Then there is a sequence x_n in X such that $\|x_n\|_X = 1$ and

$$\|x_n\|_Y > \epsilon_0 + n\|x_n\|_Z. \tag{7.15}$$

Since the imbedding from X to Y is continuous, x_n is bounded in Y, and (7.15) implies that x_n must converge to 0 in Z. After passing to a subsequence, we may assume that x_n converges in Y, the limit must then be 0. But this contradicts (7.15). □

By setting $X = H^k(\Omega)$, $Y = H^{k-1}(\Omega)$ and $Z = L^1(\Omega)$, we can derive the following consequence.

Corollary 7.31. *Assume that Ω is bounded and*

$$H^k(\Omega) \overset{c}{\hookrightarrow} H^{k-1}(\Omega).$$

Then the following norms on $H^k(\Omega)$ are equivalent:

$$
\begin{aligned}
\|u\|_{k,2}^2 &= \sum_{|\alpha| \le k} \|D^\alpha u\|_2^2, \\
\|u\|_{k,2,*}^2 &= \sum_{|\alpha| = k} \|D^\alpha u\|_2^2 + \|u\|_1^2.
\end{aligned}
\tag{7.16}
$$

We leave the proof as an exercise (Problem 7.9).

In the following result we show that for the space H_0^k, we can leave out the term $\|u\|_1^2$ in the norms above. Moreover, we do not need to assume that Ω is bounded; it suffices that it be bounded in one direction. This result is known as Poincaré's inequality.

Theorem 7.32 (Poincaré's inequality). *Let Ω be contained in the strip $|x_1| \le d < \infty$. Then there is a constant c, depending only on k and d, such that*

$$\|u\|_{k,2}^2 \le c \sum_{|\alpha| = k} \|D^\alpha u\|_2^2 \tag{7.17}$$

for every $u \in H_0^k(\Omega)$.

Proof. We give the proof for $k = 1$, the general case follows by induction. By density, it suffices to consider $u \in \mathcal{D}(\Omega)$. An integration by parts yields

$$\|u\|_2^2 = \int_\Omega 1 \cdot |u(\mathbf{x})|^2 \, d\mathbf{x} = -\int_\Omega x_1 \frac{\partial}{\partial x_1} |u|^2 \, d\mathbf{x} \le 2d\|u\|_2 \left\| \frac{\partial u}{\partial x_1} \right\|_2. \tag{7.18}$$

□

Remark 7.33. For $p \in (1, \infty)$, an analogous result holds. In other words, if Ω satisfies the hypotheses of Theorem 7.32, then there is a constant C

depending only on d, k and p such that

$$\|u\|_{k,p}^p \leq C \sum_{|\alpha|=k} \|D^\alpha u\|_p^p \qquad (7.19)$$

for every $u \in W_0^{k,p}(\Omega)$.

7.2.5 The Trace Theorem

In this subsection, we address the question of whether functions in Sobolev spaces can be restricted to the boundary of the domain, or more generally, to other surfaces. We are also interested in the converse question, namely, how smooth boundary data have to be so that a function in $H^k(\Omega)$ can assume those boundary data. We first consider functions defined on $\mathbb{R}^m$ and their restriction to $\mathbb{R}^{m-1} \times \{0\}$.

Theorem 7.34. *Let $s > 1/2$ be real. Then there exists a continuous linear map $T : H^s(\mathbb{R}^m) \to H^{s-1/2}(\mathbb{R}^{m-1})$, called the trace operator, with the property that for any $\phi \in \mathcal{D}(\mathbb{R}^m)$, we have*

$$T\phi(x_1,\ldots,x_{m-1}) = \phi(x_1,x_2,\ldots,x_{m-1},0). \qquad (7.20)$$

It can be shown that for $s \leq 1/2$ the result is false; in fact, functions which vanish in a neighborhood of $x_m = 0$ are then dense in $H^s(\mathbb{R}^m)$.

Proof. Let $\phi \in \mathcal{D}(\mathbb{R}^m)$ and let $g(\mathbf{x}') = \phi(\mathbf{x}',0)$. Let $\tilde{\phi}$ denote the Fourier transform of ϕ with respect to the mth variable only, i.e.,

$$\tilde{\phi}(\mathbf{x}',\xi_m) = \frac{1}{\sqrt{2\pi}} \int_{-\infty}^{\infty} \phi(\mathbf{x}',x_m)e^{-ix_m\xi_m}\,dx_m, \qquad (7.21)$$

let $\hat{\phi}$ and $\hat{g}$ denote the Fourier transforms of ϕ and g in $\mathbb{R}^m$ and, respectively, $\mathbb{R}^{m-1}$. The Fourier inversion formula yields

$$g(\mathbf{x}') = \phi(\mathbf{x}',0) = \frac{1}{\sqrt{2\pi}} \int_{-\infty}^{\infty} \tilde{\phi}(\mathbf{x}',\xi_m)\,d\xi_m. \qquad (7.22)$$

Applying the Fourier transform, we find

$$\hat{g}(\xi') = \frac{1}{\sqrt{2\pi}} \int_{-\infty}^{\infty} \hat{\phi}(\xi)\,d\xi_m. \qquad (7.23)$$

We now estimate

$$
\begin{aligned}
\|g\|_{s-1/2,2}^2 &\leq C_1 \int_{\mathbb{R}^{m-1}} |\hat{g}(\xi')|^2 (1+|\xi'|^2)^{s-1/2} \, d\xi' \\
&= \frac{C_1}{2\pi} \int_{\mathbb{R}^{m-1}} \left| \int_{-\infty}^{\infty} \hat{\phi}(\xi) \, d\xi_m \right|^2 (1+|\xi'|^2)^{s-1/2} \, d\xi' \\
&\leq C_2 \int_{\mathbb{R}^{m-1}} (1+|\xi'|^2)^{s-1/2} \int_{-\infty}^{\infty} |\hat{\phi}(\xi)|^2 (1+|\xi|^2)^s \, d\xi_m \\
&\quad \times \int_{-\infty}^{\infty} (1+|\xi|^2)^{-s} \, d\xi_m \, d\xi'.
\end{aligned}
\tag{7.24}
$$

If $s > 1/2$, we have

$$
\begin{aligned}
\int_{-\infty}^{\infty} (1+|\xi|^2)^{-s} \, d\xi_m &= \int_{-\infty}^{\infty} (1+|\xi'|^2 + \xi_m^2)^{-s} \, d\xi_m \\
&= (1+|\xi'|^2)^{-s+1/2} \int_{-\infty}^{\infty} (1+y^2)^{-s} \, dy.
\end{aligned}
\tag{7.25}
$$

By inserting into (7.24), we find that $\|T\phi\|_{s-1/2,2} \leq C\|\phi\|_{s,2}$ for every $\phi \in \mathcal{D}(\mathbb{R}^m)$. Hence T can be extended by continuity to all of $H^s(\mathbb{R}^m)$. □

Lemma 7.35. *If $u \in H^s(\mathbb{R}^m) \cap C(\mathbb{R}^m)$, $s > 1/2$, then $Tu(\mathbf{x}') = u(\mathbf{x}', 0)$. Moreover* supp $Tu \subset$ supp $u \cap (\mathbb{R}^{m-1} \times \{0\})$ *for every $u \in H^s(\mathbb{R}^m)$.*

We leave the proof as an exercise.

A natural question is now which elements of $H^{s-1/2}(\mathbb{R}^{m-1})$ can be obtained as restrictions, or "traces" of functions in $H^s(\mathbb{R}^m)$. The answer is that all elements of $H^{s-1/2}(\mathbb{R}^{m-1})$ are obtained in this way.

Theorem 7.36. *Let $s > 1/2$. Then there exists a bounded linear mapping $Z : H^{s-1/2}(\mathbb{R}^{m-1}) \to H^s(\mathbb{R}^m)$ such that TZ is the identity.*

Proof. We shall construct Z explicitly in terms of Fourier transforms. By density, it suffices to define $Z\phi$ for $\phi \in \mathcal{D}(\mathbb{R}^{m-1})$; we can then extend by continuity. We put

$$
Z\phi := u(\mathbf{x}) := \frac{1}{(2\pi)^{(m-1)/2} K_s} \int_{\mathbb{R}^m} e^{i\xi \cdot \mathbf{x}} \frac{(1+|\xi'|^2)^{s-1/2}}{(1+|\xi|^2)^s} \hat{\phi}(\xi') \, d\xi,
\tag{7.26}
$$

where we have set

$$
K_s = \int_{-\infty}^{\infty} (1+y^2)^{-s} \, dy.
\tag{7.27}
$$

If $x_m = 0$, we can carry out the integration with respect to ξ_m and obtain

$$
u(\mathbf{x}', 0) = (2\pi)^{-(m-1)/2} \int_{\mathbb{R}^{m-1}} e^{i\xi' \cdot \mathbf{x}'} \hat{\phi}(\xi') \, d\xi' = \phi(\mathbf{x}').
\tag{7.28}
$$

This shows that $TZ\phi = \phi$. It remains to prove the continuity of Z. We have

$$\|u\|_{s,2}^2 \le C_1 \int_{\mathbb{R}^m} |\hat{u}(\xi)|^2 (1 + |\xi|^2)^s \, d\xi$$

$$= C_2 \int_{\mathbb{R}^{m-1}} (1 + |\xi'|^2)^{2s-1} |\hat{\phi}(\xi')|^2 \int_{-\infty}^{\infty} (1 + |\xi|^2)^{-s} \, d\xi_m \, d\xi' \quad (7.29)$$

$$\le C_3 \int_{\mathbb{R}^{m-1}} (1 + |\xi'|^2)^{s-1/2} |\hat{\phi}(\xi')|^2 \, d\xi'.$$

This completes the proof. □

If $s > 1/2 + k$, we can define traces of all derivatives up to order k. Hence there is a continuous trace operator

$$T_k : H^s(\mathbb{R}^m) \to \prod_{j=0}^{k} H^{s-j-1/2}(\mathbb{R}^{m-1}) \quad (7.30)$$

such that

$$T_k \phi(\mathbf{x}') = \left(\phi(\mathbf{x}', 0), \frac{\partial \phi}{\partial x_m}(\mathbf{x}', 0), \dots, \frac{\partial^k \phi}{\partial x_m^k}(\mathbf{x}', 0) \right) \quad (7.31)$$

for smooth functions ϕ. Again the inverse question of constructing a function with given trace is of interest. We have

Theorem 7.37. *The trace operator T_k has a bounded right inverse Z_k.*

Proof. We first define

$$\tilde{Z}_l \phi := \frac{x_m^l}{l!(2\pi)^{(m-1)/2} K_{s-l}} \int_{\mathbb{R}^m} e^{i\xi \cdot \mathbf{x}} \hat{\phi}(\xi') \frac{(1 + |\xi'|^2)^{s-l-1/2}}{(1 + |\xi|^2)^{s-l}} \, d\xi, \quad (7.32)$$

where K_{s-l} is as given by (7.27). An argument analogous to the proof of Theorem 7.36 shows that $\tilde{Z}_l$ is continuous from $H^{s-l-1/2}(\mathbb{R}^{m-1})$ to $H^s(\mathbb{R}^m)$ and that, for $\phi \in \mathcal{D}(\mathbb{R}^{m-1})$,

$$(\tilde{Z}_l \phi)(\mathbf{x}', 0) = 0, \quad \frac{\partial \tilde{Z}_l \phi}{\partial x_m}(\mathbf{x}', 0) = 0, \quad \dots, \quad \frac{\partial^l \tilde{Z}_l \phi}{\partial x_m^l}(\mathbf{x}', 0) = \phi(\mathbf{x}'). \quad (7.33)$$

We now construct $Z_k(\phi_0, \phi_1, \dots, \phi_k)$ recursively by the algorithm

$$v_0 = \tilde{Z}_0 \phi_0, \quad v_{j+1} = \tilde{Z}_{j+1}\left(\phi_{j+1} - \frac{\partial^{j+1} v_j}{\partial x_m^{j+1}} \right) + v_j, \quad Z_k(\phi_0, \phi_1, \dots, \phi_k) = v_k. \quad (7.34)$$

□

We note the following corollary of the trace theorem:

Corollary 7.38. *Let Φ be a k-diffeomorphism of $\mathbb{R}^m$. Then Φ^* is a bounded linear mapping from $H^{k-1/2}(\mathbb{R}^m)$ to itself.*

Proof. We simply extend $\boldsymbol{\Phi}$ to $\mathbb{R}^{m+1}$ by defining

$$\boldsymbol{\Psi}(\mathbf{x}', x_{m+1}) = (\boldsymbol{\Phi}(\mathbf{x}'), x_{m+1}).$$

Then $\boldsymbol{\Psi}$ is a k-diffeomorphism of $\mathbb{R}^{m+1}$ and $\boldsymbol{\Psi}^*$ is continuous from $H^k(\mathbb{R}^{m+1})$ into itself. The rest follows by taking traces. $\square$

We remark that since there is an extension operator from $H^k(\mathbb{R}_+^m)$ to $H^k(\mathbb{R}^m)$, we also have a trace operator which maps a function in $H^k(\mathbb{R}_+^m)$ to its boundary values in $H^{k-1/2}(\mathbb{R}^{m-1})$. By using a partition of unity argument, we can extend this result to domains with bounded boundary.

Theorem 7.39. *Let k be a positive integer. Assume that Ω is of class C^k and $\partial\Omega$ is bounded. Then there is a bounded trace operator $T : H^k(\Omega) \to H^{k-1/2}(\partial\Omega)$. Moreover, T has a bounded right inverse.*

If $l < k$, then the lth derivatives have traces in $H^{k-l-1/2}(\partial\Omega)$. It is customary to formulate trace theorems involving higher derivatives in terms of derivatives in the direction normal to $\partial\Omega$.

Theorem 7.40. *Let k, l be positive integers such that $k > l$. Let Ω be of class C^k and let $\partial\Omega$ be bounded. Then there exists a continuous trace operator*

$$T_l : H^k(\Omega) \to \prod_{j=0}^{l} H^{k-j-1/2}(\partial\Omega) \tag{7.35}$$

with the property that

$$T\phi = \left(\phi, \frac{\partial\phi}{\partial n}, \dots, \frac{\partial^l\phi}{\partial n^l}\right) \tag{7.36}$$

for every smooth ϕ. The operator T_l has a bounded right inverse.

We can now characterize $H_0^k(\Omega)$ in terms of boundary conditions.

Theorem 7.41. *Let Ω be of class C^k and let $\partial\Omega$ be bounded. Then $H_0^k(\Omega)$ is the set of all those functions in $u \in H^k(\Omega)$ for which*

$$u = \frac{\partial u}{\partial n} = \cdots = \frac{\partial^{k-1} u}{\partial n^{k-1}} = 0 \tag{7.37}$$

on $\partial\Omega$ in the sense of trace.

Proof. If $u \in \mathcal{D}(\Omega)$, it is clear that (7.37) holds. By continuity, (7.37) then holds for $u \in H_0^k(\Omega)$. We need to establish the converse. By using a partition of unity and local coordinate transformations, we are reduced to the case $\Omega = \mathbb{R}_+^m$. Let now $k = 1$ and let $u \in H^1(\mathbb{R}_+^m)$ be such that $u(\mathbf{x}', 0) = 0$ in the sense of trace. Let Eu be the extension of u by zero. To show that $Eu \in H^1(\mathbb{R}^m)$, it suffices to establish that $\partial(Eu)/\partial x_i = E(\partial u/\partial x_i)$. This is clear for $i < m$. For $i = m$, we have, for any $\phi \in \mathcal{D}(\mathbb{R}^m)$:

$$\int_{\mathbb{R}^m} \frac{\partial \phi}{\partial x_m} (Eu) \, d\mathbf{x} = \int_{\mathbb{R}^m_+} \frac{\partial \phi}{\partial x_m} u \, d\mathbf{x}$$

$$= \int_{\mathbb{R}^{m-1}} \phi(\mathbf{x}', 0) u(\mathbf{x}', 0) \, d\mathbf{x}' - \int_{\mathbb{R}^m_+} \phi \frac{\partial u}{\partial x_m} \, d\mathbf{x} \quad (7.38)$$

$$= - \int_{\mathbb{R}^m} \phi E \frac{\partial u}{\partial x_m} \, d\mathbf{x}.$$

An analogous argument applies to higher derivatives. Once we know that $Eu \in H^k(\mathbb{R}^m)$, the rest follows by considering the sequence $u_n = Eu(\mathbf{x} - \frac{1}{n}\mathbf{e}_m)$. Since the support of u_n is bounded away from $\partial \mathbb{R}^m_+$, it is easy to approximate u_n by test functions. □

7.3   Negative Sobolev Spaces and Duality

According to the Riesz representation theorem, Hilbert spaces are isometric to their dual spaces. Hence every linear functional on $H^k(\Omega)$ has a representation of the form $l(v) = (u, v)_k$. However, the inner product $(u, v)_k$ does not agree with the action of u as a distribution. In fact, since test functions are generally not dense in $H^k(\Omega)$, linear functionals are not necessarily distributions; there are nonzero linear functionals which vanish on all test functions. We make the following definition.

Definition 7.42. By $H^{-k}(\Omega)$, we denote the set of all linear functionals on $H^k_0(\Omega)$. Moreover, if M is $\mathbb{R}^m$ or a compact manifold of class C^k, $k > s$, then $H^{-s}(M)$ denotes the dual space of $H^s(M)$.

Since $\mathcal{D}(\Omega)$ is dense in $H^k_0(\Omega)$, $H^{-k}(\Omega)$ is a space of distributions. As we will see in the following examples, negative Sobolev spaces contain singular distributions.

Example 7.43. Suppose $k > m/2$ and $\Omega \subset \mathbb{R}^m$ has the k-extension property and contains the origin. Then the Dirac delta is in $H^{-k}(\Omega)$. To see this we note that the Sobolev imbedding themorem ensures that $H^k(\Omega)$ (and hence $H^k_0(\Omega)$) is continuously imbedded in $C_b(\overline{\Omega})$. This ensures that the delta distribution in well defined. It is also a bounded linear functional on $H^k_0(\Omega)$ since for every $u \in H^k_0(\Omega)$

$$|(\delta, u)| := |u(\mathbf{0})| \leq \|u\|_\infty \leq C\|u\|_{H^k(\Omega)}.$$

Example 7.44. Let S be a smooth, bounded surface in the interior of $\Omega \subset \mathbb{R}^3$ and let $g : S \to R$ be in $L^2(S)$. (We can think of g as a distribution of surface charge on S.) Then the distribution generated by g is in $H^{-1}(\Omega)$. To see this we use the trace theorem to note that for any smooth function

ϕ we have

$$
\begin{aligned}
|(g,\phi)| &= \left| \int_{S} g(\mathbf{x})\phi(\mathbf{x}) \, dA(\mathbf{x}) \right| \\
&\leq \|g\|_{L^2(S)} \|\phi\|_{L^2(S)} \\
&\leq \|g\|_{L^2(S)} \|\phi\|_{H^{1/2}(S)} \\
&\leq \|g\|_{L^2(S)} \|\phi\|_{H^1(\Omega)}
\end{aligned}
$$

Thus, the surface distribution g defines a bounded linear functional on functions in $H^1(\Omega)$.

We can also characterize functions in negative Sobolev spaces as derivatives of functions in positive Sobolev spaces. Let $f \in H^{-k}(\Omega)$. By the Riesz representation theorem, there is then a unique $u \in H_0^k(\Omega)$ with the property that

$$
(f,v) = (u,v)_k \tag{7.39}
$$

for every $v \in H_0^k(\Omega)$. How is u related to f? From (7.39) we find that, for any test function ϕ,

$$
(f,\phi) = \sum_{|\alpha| \leq k} (D^\alpha u, D^\alpha \phi) = \sum_{|\alpha| \leq k} (-1)^{|\alpha|} (D^{2\alpha} u, \phi), \tag{7.40}
$$

i.e.,

$$
f = \sum_{|\alpha| \leq k} (-1)^{|\alpha|} D^{2\alpha} u. \tag{7.41}
$$

For any given $f \in H^{-k}(\Omega)$, there is therefore a unique $u \in H_0^k(\Omega)$ satisfying the partial differential equation (7.41). Recall that the condition $u \in H_0^k(\Omega)$ can be interpreted as a boundary condition: $u = \partial u/\partial n = \cdots = \partial^{k-1} u/\partial n^{k-1} = 0$ on $\partial\Omega$ (Theorem 7.41). Considerations similar to the one just given form the starting point of the modern existence theory for elliptic boundary-value problems; we shall return to this in Chapter 9.

We conclude with a simple statement about differentiation of distributions in negative Sobolev spaces.

Lemma 7.45. *Let $u \in H^k(\Omega)$, $k \in \mathbb{Z}$. Then $\partial u/\partial x_i \in H^{k-1}(\Omega)$.*

The proof follows trivially from the definitions. Lemma 7.45 has a converse.

Lemma 7.46. *Let $f \in H^{-k}(\Omega)$, $k \in \mathbb{N}$. Then there exist functions $g_\alpha \in L^2(\Omega)$ such that $f = \sum_{|\alpha| \leq k} D^\alpha g_\alpha$.*

For the proof, we simply set $g_\alpha = (-1)^{|\alpha|} D^\alpha u$ in (7.41).

7.4   Technical Results

7.4.1   Density Theorems

In this subsection, we shall show that C^∞-functions with bounded support
are dense in $H^k(\Omega)$. No assumptions on boundary regularity are needed.
The same proofs work for $W^{k,p}(\Omega)$ if $p < \infty$. We first show that functions
with bounded support are dense; of course, this is only of interest if Ω is
unbounded.

Lemma 7.47. *Functions of bounded support are dense in $H^k(\Omega)$.*

Proof. Let $\phi \in \mathcal{D}(\mathbb{R}^m)$ be a function which equals 1 for $|\mathbf{x}| \leq 1$, 0 for
$|\mathbf{x}| \geq 2$, and which takes values in between otherwise. For any $u \in H^k(\Omega)$,
consider the sequence $u_n(\mathbf{x}) = u(\mathbf{x})\phi_n(\mathbf{x})$, where $\phi_n(\mathbf{x}) = \phi(\mathbf{x}/n)$. We first
show that $u_n \to u$ in $L^2(\Omega)$. Let $\epsilon > 0$ be given and let f be a continuous
function such that $\|u - f\|_2 \leq \epsilon$. We find

$$\|u\phi_n - u\|_2 \leq \|(u-f)\phi_n\|_2 + \|u-f\|_2 + \|f - f\phi_n\|_2 \leq 2\epsilon + \|f - f\phi_n\|_2. \quad (7.42)$$

It is easy to see that $f\phi_n$ converges to f as $n \to \infty$. Hence $u\phi_n$ converges
to u in L^2. The convergence of derivatives is obtained by using the product
rule of differentiation and a straightforward bootstrap argument.   □

Theorem 7.48. *$C^\infty(\Omega) \cap H^k(\Omega)$ is dense in $H^k(\Omega)$.*

Proof. Let $\{\phi_i\}_{i\in\mathbb{N}}$ be a partition of unity on Ω, subordinate to a locally
finite covering $\{U_i\}_{i\in\mathbb{N}}$. Then for any $u \in H^k(\Omega)$, we have $u = \sum_{i\in\mathbb{N}} u\phi_i$
(in the sense of distributions, *not* in the sense of convergence in $H^k(\Omega)$!).
Moreover, $u\phi_i$ is in $H^k(\Omega)$, and we claim it is actually in $H_0^k(\Omega)$. Indeed,
$u\phi_i$ can be extended by zero outside Ω; this yields a function in $H^k(\mathbb{R}^m)$.
By Corollary 7.13, there is a sequence $f_n \in \mathcal{D}(\mathbb{R}^m)$ such that $f_n \to u\phi_i$ in
$H^k(\mathbb{R}^m)$. Let now ψ be a test function which equals 1 on the support of ϕ_i
and 0 outside U_i; then we have $f_n\psi \to u\phi_i\psi = u\phi_i$, and $f_n\psi \in \mathcal{D}(\Omega)$.
   For every $i \in \mathbb{N}$, we can therefore choose $u_i^{(n)} \in \mathcal{D}(\Omega)$ such that

$$\|u_i^{(n)} - u\phi_i\|_{k,2} \leq \frac{1}{n2^i}. \quad (7.43)$$

It follows that

$$\sum_{i\in\mathbb{N}} u_i^{(n)} - u\phi_i = \sum_{i\in\mathbb{N}} u_i^{(n)} - u \quad (7.44)$$

is in $H^k(\Omega)$, with norm $\leq 1/n$. Hence $u^{(n)} := \sum_{i\in\mathbb{N}} u_i^{(n)}$ converges to u in
$H^k(\Omega)$, and because of local finiteness $u^{(n)}$ is of class C^∞.   □

Hence any function in $H^k(\Omega)$ can be approximated by functions whose
derivatives exist in the classical sense. This is often useful in proofs. One can
do everything for smooth functions first and then argue the general case by

taking limits. However, Theorem 7.48 is still unsatisfactory. In many cases one would like to have approximation by functions which are smooth up to the boundary, rather than just in the interior of Ω. This cannot be done without some assumptions on the boundary of Ω. The result we give now is not optimal, but will suffice for our purposes.

Lemma 7.49. *Assume that Ω has the k-extension property. Then functions in $C^\infty(\overline{\Omega})$ with bounded support are dense in $H^k(\Omega)$.*

The proof follows immediately from Corollary 7.13.

7.4.2 Coordinate Transformations and Sobolev Spaces on Manifolds

If $\partial\Omega$ is smooth, a standard trick is to make local changes of coordinates which make $\partial\Omega$ a coordinate surface. This requires us to consider how Sobolev spaces behave under coordinate transformations. Also, for boundary-value problems in PDEs we need to consider spaces of functions defined on $\partial\Omega$. Hence we want to define Sobolev spaces on manifolds. These will of course be defined in terms of local coordinate charts; again the behavior under coordinate transformations is a crucial issue.

Definition 7.50. Let Ω, Ω' be open sets in $\mathbb{R}^m$. A bijection $\mathbf{\Phi} : \Omega \to \Omega'$ is called a k-diffeomorphism $(k \geq 1)$ if

1. $\mathbf{\Phi}$ and $\mathbf{\Phi}^{-1}$ are continuous on $\overline{\Omega}$ and $\overline{\Omega'}$, respectively,

2. their derivatives of order 1 through k are bounded and continuous on $\overline{\Omega}$ and $\overline{\Omega'}$, respectively, and

3. there are positive constants c and C such that $c \leq |\det \nabla\mathbf{\Phi}| \leq C$ in Ω.

Definition 7.51. Let $\mathbf{\Phi} : \Omega \to \Omega'$ be a k-diffeomorphism. Then the pullback operators $\mathbf{\Phi}^*$ and $(\mathbf{\Phi}^{-1})^*$ are defined by

$$\mathbf{\Phi}^*u = u \circ \mathbf{\Phi}, \qquad (\mathbf{\Phi}^{-1})^*v = v \circ \mathbf{\Phi}^{-1} \tag{7.45}$$

for functions u, v defined, respectively, on Ω' and Ω.

Theorem 7.52. *Let $\mathbf{\Phi} : \Omega \to \Omega'$ be a k-diffeomorphism. Then $\mathbf{\Phi}^*$ and $(\mathbf{\Phi}^{-1})^*$ are bounded linear mappings from $H^k(\Omega')$ to $H^k(\Omega)$ and, respectively, from $H^k(\Omega)$ to $H^k(\Omega')$.*

Proof. Since the inverse of a k-diffeomorphism is also a k-diffeomorphism, it suffices to prove the claim for $\mathbf{\Phi}^*$. It is easy to see that $\mathbf{\Phi}^*$ is a bounded linear mapping from $\tilde{L}^2(\Omega')$ to $\tilde{L}^2(\Omega)$ and hence, by continuity, from $L^2(\Omega')$ to $L^2(\Omega)$. We note that $(u,v)_{\Omega'} = (\mathbf{\Phi}^*u, (\det \nabla\mathbf{\Phi})\mathbf{\Phi}^*v)_\Omega$.

It remains to consider derivatives. For this, we have to establish that the derivatives of $\mathbf{\Phi}^*u$ can be evaluated in the usual way by means of the chain and product rule. This is clear if $u \in C^\infty(\Omega)$, and we have already established that C^∞-functions are dense in $H^k(\Omega)$. For $u \in H^k(\Omega) \cap C^\infty(\Omega)$, any pth derivative of $\mathbf{\Phi}^* u$ ($p \leq k$) is a linear expression involving derivatives of u of orders $\leq p$, with coefficients depending on derivatives of $\mathbf{\Phi}$; in particular, all the coefficients are bounded, continuous functions. In order to extend the result to all of $H^k(\Omega)$, we need to be able to take limits. The following lemma establishes that we can do this.

Lemma 7.53. *Let $u \in L^2(\Omega) \cap C(\Omega)$ and $a \in C_b(\Omega)$. Then $au \in L^2(\Omega)$ and $\|au\|_2 \leq \|a\|_\infty \|u\|_2$.*

We omit the trivial proof of the lemma. Since $L^2(\Omega) \cap C(\Omega)$ is dense in $L^2(\Omega)$, the lemma allows us to define au for every $u \in L^2(\Omega)$. This completes the proof of Theorem 7.52. $\qquad\square$

One of the applications of the transformation theorem is that we can now define Sobolev spaces on manifolds. Let M be a compact p-dimensional surface in $\mathbb{R}^m$. We assume that M is smooth of class C^k, $k \geq 1$. Then every point in M has a neighborhood within which M can be represented in the form $\mathbf{x} = \mathbf{g}(\mathbf{y})$, where $\mathbf{y} \in \mathbb{R}^p$ and $\mathbf{g}$ is of class C^k. We can cover M with a finite number of such neighborhoods; let us call them U_i, $1 \leq i \leq N$, and let $\mathbf{g}_i : N_i \to U_i \cap M$ be the corresponding parametric representations of the surface. We make the following definition:

Definition 7.54. Let S be a closed subset of $\mathbb{R}^m$ and let $\{U_j\}$ be a locally finite covering of S with bounded open subsets of $\mathbb{R}^m$ (not of S). A **partition of unity** subordinate to the covering $\{U_j\}$ is a set of functions $\phi_j \in \mathcal{D}(\mathbb{R}^m)$ such that

1. $0 \leq \phi_j \leq 1$,

2. supp $\phi_j \subset U_j$, and

3. $\sum_j \phi_j = 1$ in a neighborhood of S.

The proof of the existence of a partition of unity is analogous to that of Theorem 5.7. Note that if S is compact, we can reduce every covering $\{U_j\}$ to a finite one.

Let $\{\phi_i\}$ be a partition of unity on M subordinate to the covering $\{U_i\}$.

Definition 7.55. Let M be as described above and let $l \leq k$ be a nonnegative integer. We say that $u \in H^l(M)$ if $\phi_i(u \circ \mathbf{g}_i) \in H^l(\mathbb{R}^p)$ for every $i = 1, \ldots, N$.

Here, it is of course understood that $\phi_i(u \circ \mathbf{g}_i)$ is set equal to zero outside N_i. The transformation theorem guarantees that this definition depends only on the surface and not on the parameterizations chosen. An inner product on $H^l(M)$ can be defined naturally by

$$(u, v)_{M,k} = \sum_{i=1}^{N} (\phi_i(u \circ \mathbf{g}_i), \phi_i(v \circ \mathbf{g}_i))_k. \qquad (7.46)$$

This definition does of course depend on the parameterizations, but all the inner products defined in this way are equivalent (Problem 7.23). We note that the definition of $H^l(M)$ is easily extended to fractional l, except that for this case we have not established a transformation theorem. We state such a theorem without proof.

Theorem 7.56. *Let $0 \leq l \leq k$ with $k \in \mathbb{N}$ and l real. Let $\mathbf{\Phi}$ be a k-diffeomorphism of $\mathbb{R}^m$. Then the pullback operator $\mathbf{\Phi}^*$ is a bounded linear mapping from $H^l(\mathbb{R}^m)$ onto itself.*

A proof can be based either on the theory of interpolation spaces or on (7.5). In applications, we are interested in Sobolev spaces on manifolds primarily for boundary data, and l is a half-integer. For l a half-integer, we shall obtain Theorem 7.56 later as a corollary of the trace theorem.

7.4.3   Extension Theorems

We now address the question whether a function in $H^k(\Omega)$ can be extended to a function in $H^k(\mathbb{R}^m)$. We first note a trivial extension result.

Lemma 7.57. *Let $u \in H_0^k(\Omega)$. Then u can be extended to a function in $H^k(\mathbb{R}^m)$ by defining it to be zero outside of $\overline{\Omega}$, i.e., we set $(u, \phi) = \int_{\Omega} u(\mathbf{x})\phi(\mathbf{x})\, d\mathbf{x}$ for every $\phi \in \mathcal{D}(\mathbb{R}^m)$.*

Proof. Take a sequence $u_n \in \mathcal{D}(\Omega)$ which converges to u in $H^k(\Omega)$. Then u_n also converges in $H^k(\mathbb{R}^m)$.   □

We next consider the case of a half-space. Let $\mathbb{R}^m_+ = \{x_m > 0\}$. We have the following result.

Theorem 7.58. *Let $k \geq 0$ be an integer. Then there exists a bounded linear mapping $E : H^k(\mathbb{R}^m_+) \to H^k(\mathbb{R}^m)$ with the property that $(Eu)|_{\mathbb{R}^m_+} = u$ for every $u \in H^k(\mathbb{R}^m_+)$. Moreover, for any given $K \in \mathbb{N}$, E can be chosen independently of k for $0 \leq k \leq K$.*

Proof. Let us first consider how we would extend smooth functions. A continuous function can be extended to a continuous function on the whole space by even continuation, i.e., $u(\mathbf{x}', x_m) = u(\mathbf{x}', -x_m)$ for $x_m < 0$. Here and in the following, $\mathbf{x}'$ stands for $(x_1, x_2, \ldots, x_{m-1})$. However, even continuation will generally lead to discontinuous derivatives. If we set $u(\mathbf{x}', x_m) = 4u(\mathbf{x}', -x_m/2) - 3u(\mathbf{x}', -x_m)$ for $x_m < 0$, then both u and $\partial u/\partial x_m$ turn out to be continuous across $x_m = 0$ if u is smooth. In general, we make the ansatz

$$u(\mathbf{x}', x_m) = \sum_{j=0}^{K} \alpha_j u\left(\mathbf{x}', -\frac{x_m}{j+1}\right) \qquad (7.47)$$

for $x_m < 0$, and we want to choose the α_j such that for smooth u the derivatives up to order K match across $x_m = 0$. This leads to the equations

$$\sum_{j=0}^{K} (-\frac{1}{j+1})^i \alpha_j = 1, \quad i = 0, 1, \ldots, K. \qquad (7.48)$$

The matrix of this system is known as a Vandermonde matrix and is well known to be nonsingular (Problem 7.24). Hence (7.48) is uniquely solvable.

It is easy to see that if $u \in C^k(\overline{\mathbb{R}^m_+})$ ($k \leq K$) with bounded support, then $Eu \in C^k(\mathbb{R}^m)$ with bounded support, and $\|Eu\|_{k,2} \leq C\|u\|_{k,2}$ for some constant C that is independent of u. If we can show that functions with bounded support in $C^k(\overline{\mathbb{R}^m_+})$ are dense in $H^k(\mathbb{R}^m_+)$, we are done.   □

Lemma 7.59. *Let* $u \in H^k(\mathbb{R}^m_+)$. *Then there is a sequence* $\phi_n \in \mathcal{D}(\mathbb{R}^m)$ *such that the restrictions of* ϕ_n *to* $\mathbb{R}^m_+$ *converge to* u.

Proof. Since we already know that $\mathcal{D}(\mathbb{R}^m)$ is dense in $H^k(\mathbb{R}^m)$, it suffices to show that every $u \in H^k(\mathbb{R}^m_+)$ can be approximated by restrictions of functions in $H^k(\mathbb{R}^m)$. Moreover, we know functions of bounded support are dense; hence let us assume u has bounded support. Let $u_n(\mathbf{x}', x_m) = u(\mathbf{x}', x_m + 1/n)$ for $\mathbf{x} \in \mathbb{R}^m_+$. Then $\|u_n\|_2 \leq \|u\|_2$, and $u_n \to u$ for $u \in \mathcal{D}(\mathbb{R}^m_+)$. By density, $u_n \to u$ in L^2 for every $u \in L^2(\mathbb{R}^m_+)$. Since, moreover, $D^\alpha u_n(\mathbf{x}', x_m) = D^\alpha u(\mathbf{x}', x_m + 1/n)$, the same argument shows that $u_n \to u$ in $H^k(\mathbb{R}^m_+)$. We claim that u_n can be extended. First we note that u_n is actually defined for $x_m > -1/n$. Let

$$(v_n, \phi) = \int_{x_m > -1/n} \overline{u_n(\mathbf{x})} \phi(\mathbf{x}) \, d\mathbf{x} \qquad (7.49)$$

be the extension of u_n by zero. Then clearly $D^\alpha v_n = 0$ for $x_m < -1/n$ and $D^\alpha v_n = D^\alpha u_n$ for $x_m > -1/n$. However, we cannot claim that $D^\alpha v_n$ is the extension of $D^\alpha u_n$ by zero. What we can claim is that

$$(D^\alpha v_n, \phi) = \int_{x_m > -1/n} \overline{D^\alpha u_n(\mathbf{x})} \phi(\mathbf{x}) \, d\mathbf{x} + (g_{n\alpha}, \phi), \qquad (7.50)$$

where $g_{n\alpha}$ is a distribution with bounded support contained in the plane $x_m = -1/n$. By Lemma 5.16, $g_{n\alpha}$ is of finite order. Let now $\psi_n(x_m)$ be a C^∞-function which vanishes for $x_m \leq -1/n$ and equals 1 for $x_m \geq 0$. Then $g_{n\alpha}\psi_n$ vanishes, and hence $w_n = v_n\psi_n$ is in $H^k(\mathbb{R}^m)$. It is clear that the restriction of w_n to $\mathbb{R}^m_+$ is u_n. This concludes the proof.   □

We now extend the result to more general domains.

Theorem 7.60. *Assume that Ω is of class C^K and $\partial\Omega$ is bounded. Then, for any integer k with $0 \leq k \leq K$, there exists a bounded operator*

$$E : H^k(\Omega) \to H^k(\mathbb{R}^m)$$

such that $Eu|_\Omega = u$. Moreover, E can be chosen independent of k.

Note that we only require $\partial\Omega$ to be bounded, not necessarily Ω. For example, the theorem includes the case of exterior domains.

Proof. We cover $\partial\Omega$ with small neighborhoods U_j such that within each U_j there is a k-diffeomorphism which maps $\partial\Omega \cap U_j$ to a subset of $\{x_m = 0\}$. Let $\{\phi_j\}$ be a partition of unity subordinate to the covering $\{U_j\}$ of $\partial\Omega$. Let $u \in H^k(\Omega)$. Then, for each j, $\phi_j u$ can be extended to all of $\mathbb{R}^m$ by using the transformation theorem and Theorem 7.58. More precisely, let Φ_j be the diffeomorphism on U_j which maps $\partial\Omega \cap U_j$ to a part of the plane $x_m = 0$. Then $(\Phi_j^{-1})^*(u\phi_j)$ (extended by zero outside $\Phi_j(U_j)$) is in $H^k(\mathbb{R}_+^m)$ and can be extended to a function in $H^k(\mathbb{R}^m)$; let us call the extended function v_j. We multiply v_j by a function $\psi_j \in \mathcal{D}(\mathbb{R}^m)$ which has support in $\Phi_j(U_j)$, but equals one in $\Phi_j(\text{supp }\phi_j)$. Then $\Phi_j^*(v_j\psi_j)$ is in $H^k(\mathbb{R}^m)$ with support contained in U_j. Hence each $\phi_j u$ can be extended. Finally $u - \sum_j \phi_j u$ has compact support in Ω, hence it can be extended by zero (see Problem 7.3). □

Remark 7.61. The same result (with a much harder proof) holds if $\partial\Omega$ is Lipschitz rather than of class C^K.

7.4.4 Problems

7.1. Let $u \in H^1(\mathbb{R})$. Show that $u'(x) = \lim_{h\to 0}(u(x+h) - u(x))/h$ in the sense of L^2-convergence.

7.2. Let $u, v \in H^1(\mathbb{R})$. Prove that

$$\int_{-\infty}^\infty u(x)v'(x)\,dx = -\int_{-\infty}^\infty u'(x)v(x)\,dx.$$

7.3. Assume $u \in H^k(\Omega)$ has compact support in Ω. Prove that $u \in H_0^k(\Omega)$.

7.4. Show that $H^1(\mathbb{R}^2)$ is not a subset of $C(\mathbb{R}^2)$. Hint: Consider powers of $|\ln|\mathbf{x}||$.

7.5. Assume $u, v \in H^1(\mathbb{R})$. Show that the product uv is also in $H^1(\mathbb{R})$.

7.6. Show that the Sobolev imbedding theorem fails in general if $\partial\Omega$ has a cusp. Hint: Consider functions with a singularity at the cusp.

7.7. Let $u \in H^s(\mathbb{R}^m)$ with $s > m/2$. Show that $\lim_{\mathbf{x}\to\infty} u(\mathbf{x}) = 0$.

7.8. Let Ω satisfy the assumptions of Theorem 7.29. Show that every weakly convergent sequence in $H^{k+1}(\Omega)$ converges strongly in $H^k(\Omega)$.

7.9. Prove Corollary 7.31.

7.10. Let Ω be bounded (no assumptions on $\partial\Omega$). Prove that $H_0^{k+1}(\Omega)$ is compactly imbedded in $H^k(\Omega)$.

7.11. Prove Lemma 7.35.

7.12. Show that there is no continuous trace operator mapping $H^k(\mathbb{R}^m)$ to $H^k(\mathbb{R}^{m-1})$.

7.13. Assume that Ω is bounded in one direction. Establish an existence result for a weak solution of the Dirichlet problem $\Delta u = f$, $u|_{\partial\Omega} = 0$.

7.14. Let D be open and bounded such that $\overline{D} \subset \Omega$ and let $f \in \mathcal{D}'(\Omega)$. Show that there exists an integer k such that $f|_D \in H^{-k}(D)$. Hint: Use Lemma 5.8 and the Sobolev imbedding theorem.

7.15. Assume that Ω is bounded, connected, and has the 1-extension property. Let

$$V = \left\{ u \in H^1(\Omega) \mid \int_\Omega u(\mathbf{x})\, d\mathbf{x} = 0 \right\}.$$

Show that there is a constant C such that

$$\int_\Omega |u(\mathbf{x})|^2\, d\mathbf{x} \leq C \int_\Omega |\nabla u(\mathbf{x})|^2 \quad \text{for all } u \in V.$$

For what other subsets V of $H^1(\Omega)$ will such an inequality hold?

7.16. Show that if Ω is bounded, then $L^2(\Omega)$ is compactly embedded in $H^{-1}(\Omega)$.

7.17. Show that there is no constant C such that

$$\int_\Omega |u(\mathbf{x})|^2\, d\mathbf{x} \leq C \int_\Omega |\nabla u(\mathbf{x})|^2 \quad \text{for all } u \in H^1(\mathbb{R}^m).$$

7.18. Let

$$\Omega = \{(x_1, x_2) \in \mathbb{R}^2 \mid 0 < x_1 < \infty,\ 0 < x_2 < e^{-x_1}\}.$$

Show that for each $p \in (2, \infty)$, there is a function u_p such that $u_p \in H^k(\Omega)$ for all $k \geq 0$, but $u_p \notin L^p(\Omega)$.

7.19. A classical theorem of Titchmarsh asserts that if $p \in [1, 2)$, then the Fourier transform maps $L^p(\mathbb{R}^m)$ into $L^q(\mathbb{R}^m)$ where $\frac{1}{p} + \frac{1}{q} = 1$. Use this result to show that $H^1(\mathbb{R}^3)$ is continuously embedded in $L^p(\mathbb{R}^3)$ for all $p \in [2, 6)$. (Note: $H^1(\mathbb{R}^3)$ is also embedded continuously in $L^6(\mathbb{R}^3)$.)

7.20. Define Sobolev spaces of periodic functions on $\mathbb{R}$ and characterize them in terms of Fourier series. How are Sobolev spaces of periodic functions related to Sobolev spaces on $[0, 2\pi]$? Hint: Recall Problem 6.15.

7.21. Give an example of an open set such that $H^1(\Omega) \cap C^\infty(\overline{\Omega})$ is not dense in $H^1(\Omega)$.

7.22. Discuss possible redundancies in the definition of a k-diffeomorphism.

7.23. Verify that all the inner products defined by (7.46) are equivalent.

7.24. Let $A_{ij} = a_j^i$, $i, j = 0, \ldots, K$, where the a_j are distinct real numbers. (Use the convention $0^0 = 1$.) Show that $\det \mathbf{A} \neq 0$.

8
Operator Theory

In this chapter we give a brief discussion of the theory of linear operators A from a Banach space X to a Banach space Y. Our primary concerns center on the equation

$$Ax = y, \tag{8.1}$$

where $y \in Y$ is given, and the main issues we address are existence, multiplicity, and computability of solutions $x \in X$. Of course, most readers have already addressed these issues in studying linear algebra. There, the spaces X and Y are the finite-dimensional vector spaces $\mathbb{R}^n$ and $\mathbb{R}^m$, respectively, and A is represented by an $m \times n$ matrix. We have already considered a more general type of operator in this text when we defined a bounded linear operator from one (possibly infinite-dimensional) Banach space to another in Definition 6.41. However, as we shall see below, many important operators in PDEs (and ODEs) are unbounded. The reader is strongly encouraged to compare the results of this section with the results of his or her old linear algebra text while keeping in mind the two main extensions of the theory: to spaces that are infinite-dimensional and to operators that are unbounded.

Note: Although we have defined operators to be maps between Banach spaces, most of the applications of operator theory that we address in this book will be to maps between separable Hilbert spaces. Thus, in many of the theorems below, we have given either statements or proofs only for the case of Hilbert spaces or separable Hilbert spaces. This practice greatly reduces the amount of machinery we need to develop, but it also limits the possible applications one can address using only material from this book. This is one of the prices you pay for learning functional analysis "in the street."

In the following, we will use the notations X and Y to refer to Banach spaces and H to refer to a Hilbert space unless we specify otherwise.

8.1 Basic Definitions and Examples

8.1.1 Operators

In order to accommodate unbounded operators we begin this section with the following extended definition.

Definition 8.1. Let X and Y be Banach spaces. A **linear operator** from X to Y is a pair $(\mathcal{D}(A), A)$ consisting of a subspace $\mathcal{D}(A) \subset X$ (called the **domain** of the operator) and a linear transformation $A : \mathcal{D}(A) \to Y$.

Many mathematics students have had to endure a calculus teacher who insisted that there was a profound difference between the function $f(x) = x$ with domain $[0, 1]$ and the same function defined on the whole real line. The students soon realize that in most cases the distinction can be ignored. In the course of this chapter, we shall see that including the domain in the definition of an operator is more than just pedantry. For unbounded operators, the specification of the domain can make a real difference. However, after having made such a big deal of the importance of the domain in the definition of a operator, we will often use sloppy language which ignores the point. That is, we will often refer to "the operator A" and leave the domain unspecified. This usage is standard and unambiguous in the study of bounded operators (whose domain, we see in Theorem 8.7 below, can be extended to all of X), and when there is no chance of confusion, we simply stick with the shorter nomenclature even for unbounded operators.

We will use both of the notations Ax and $A(x)$ to indicate the action of an operator on elements of its domain.

Definition 8.2. The **range** of $(\mathcal{D}(A), A)$ is a subspace $\mathcal{R}(A) \subset Y$ defined by

$$\mathcal{R}(A) := \{u \in Y \mid u = A(x), \quad \text{for some } x \in \mathcal{D}(A)\}. \qquad (8.2)$$

The **null space** of $(\mathcal{D}(A), A)$ is the subspace $\mathcal{N}(A) \subset X$ defined by

$$\mathcal{N}(A) := \{x \in X \mid A(x) = 0\}. \qquad (8.3)$$

With the range thus defined, we can use the following notation for the operator $(\mathcal{D}(A), A)$:

$$X \supset \mathcal{D}(A) \ni x \mapsto A(x) \in \mathcal{R}(A) \subset Y.$$

The sets X and Y are sometimes referred to as the *corange* and the *codomain* in order to distinguish them from their subspaces, the domain and range, respectively. Although we agree with the importance of the distinction, we shall not adopt these terms.

8.1.2 Inverse Operators

Recall that we say that a mapping $A : \mathcal{D}(A) \to \mathcal{R}(A)$ is one-to-one or injective if distinct points in $\mathcal{D}(A)$ get mapped to distinct points in $\mathcal{R}(A)$; i.e., if for any $x_1, x_2 \in \mathcal{D}(A)$ we have

$$x_1 \neq x_2 \quad \Rightarrow \quad Ax_1 \neq Ax_2. \tag{8.4}$$

For any such mapping we can define an **inverse** mapping $(\mathcal{R}(A), A^{-1})$ which maps any point $y \in \mathcal{R}(A)$ to the unique point $x \in \mathcal{D}(A)$ such that $Ax = y$. This definition implies

$$A^{-1}Ax = x \tag{8.5}$$

for every $x \in \mathcal{D}(A)$ and

$$AA^{-1}y = y \tag{8.6}$$

for every $y \in \mathcal{R}(A)$.

The following simple but important theorem is left to the reader (Problem 8.4).

Theorem 8.3. *Let X and Y be Banach spaces. Let $(\mathcal{D}(A), A)$ be a linear operator from X to Y with range $\mathcal{R}(A)$. Then the following hold.*

1. *The inverse operator $(\mathcal{R}(A), A^{-1})$ exists if and only if $\mathcal{N}(A) = \{0\}$.*

2. *If the inverse operator exists, it is linear.*

8.1.3 Bounded Operators, Extensions

We now modify our definition of a bounded operator and the norm of a bounded operator to fit our more general definition of operator.

Definition 8.4. A linear operator $(\mathcal{D}(A), A)$ from X to Y is said to be **bounded** if there exists a constant C such that

$$\|A(x)\|_Y \leq C\|x\|_X \quad \text{for every } x \in \mathcal{D}(A). \tag{8.7}$$

If no such C exists, the operator is said to be **unbounded**.

The **norm** of a bounded operator is the smallest C for which (8.7) holds; i.e., the non-negative number

$$\|A\| := \sup_{\substack{x \in \mathcal{D}(A) \\ \|x\|_X \neq 0}} \frac{\|A(x)\|_Y}{\|x\|_X} = \sup_{\substack{x \in \mathcal{D}(A) \\ \|x\|=1}} \|A(x)\|_Y. \tag{8.8}$$

Note that if $\mathcal{D}(A) = X$, this definition agrees with Definition 6.41. Definitions of properties such as continuity extend readily to the new version of the definition.

The following lemma tells us why we need to consider unbounded operators only when we study infinite-dimensional spaces.

Lemma 8.5. *All operators with finite-dimensional domain $\mathcal{D}(A)$ are bounded.*

Proof. Let $\{e_i\}_{i=1}^n$ be a basis for $D(A)$. By Lemma 6.44 it is enough to show that $A(x_j) \to 0$ whenever $x_j \to 0$. To see this note that we can write $x_j = \sum_{i=1}^n \alpha_{i,j} e_i$. Furthermore, whenever $\lim_{j\to\infty} x_j = 0$, we have

$$\lim_{j\to\infty} \alpha_{i,j} = 0. \tag{8.9}$$

Thus,

$$\lim_{j\to\infty} A(x_j) = \lim_{j\to\infty} \sum_{i=1}^n \alpha_{i,j} A(e_i) = 0. \tag{8.10}$$

$\square$

We now make explicit the idea of extending an operator.

Definition 8.6. An operator $(\mathcal{D}(\tilde{A}), \tilde{A})$ is said to be an **extension** of $(\mathcal{D}(A), A)$ if $\mathcal{D}(A) \subseteq \mathcal{D}(\tilde{A})$ and

$$A(x) = \tilde{A}(x) \quad \text{for every } x \in \mathcal{D}(A). \tag{8.11}$$

The following theorem states the we can extend a large class of bounded operators to all of X without changing their norms.

Theorem 8.7. *If either $\mathcal{D}(A)$ is dense in X or X is a Hilbert space, then a bounded operator $(\mathcal{D}(A), A)$ can be extended to all of X without changing its norm; i.e., there exists a bounded operator $(\mathcal{D}(\tilde{A}), \tilde{A})$ that extends A and such that $\mathcal{D}(\tilde{A}) = X$ and $\|\tilde{A}\| = \|A\|$.*

Proof. **Case 1.** *The domain $\mathcal{D}(A)$ is dense in X.* In this case we simply define the operator on all of X by continuity; i.e., for every $x \in X$ there exists a sequence $x_n \in \mathcal{D}(A)$ such that $x_n \to x$. Since our operator is bounded, the sequence Ax_n is Cauchy, and hence has a limit. Thus, we can define

$$Ax := \lim_{n\to\infty} Ax_n.$$

To see that this definition is unambiguous, note that for any other sequence $\tilde{x}_n \in \mathcal{D}(A)$ that converged to x, we would have $x_n - \tilde{x}_n \to 0$. Thus, by continuity

$$\lim_{n\to\infty} [Ax_n - A\tilde{x}_n] = \lim_{n\to\infty} [A(x_n - \tilde{x}_n)] = 0.$$

The extended operator has the same norm since

$$x \mapsto \frac{\|Ax\|}{\|x\|}$$

is continuous at $x \neq 0$.

Case 2. $X = H$, *a Hilbert space.* If $\mathcal{D}(A)$ is not closed, we extend the operator to $\overline{\mathcal{D}(A)}$ by continuity, as in case 1 above. Since H is a Hilbert space we can now use the projection theorem and define the operator $(\mathcal{D}(\tilde{A}), \tilde{A})$ with $\mathcal{D}(\tilde{A}) = H$ by

$$\tilde{A}(x) = \begin{cases} A(x), & x \in \overline{\mathcal{D}(A)} \\ 0, & x \in \overline{\mathcal{D}(A)}^{\perp} \end{cases} \tag{8.12}$$

and extend to the rest of H by linearity. The proof that $\|\tilde{A}\| = \|A\|$ is left to the reader. $\square$

Because Theorem 8.7 covers most situations encounted in applications, the domain of a bounded operator is almost always assumed to be the entire space X.

In Chapter 6, we saw that $\mathcal{L}(X, Y)$, equipped with the operator norm, is a Banach space. In the following, we shall sometimes also be interested in pointwise convergence rather than norm convergence. We make the following definition.

Definition 8.8. *We say that $A_n \in \mathcal{L}(X, Y)$ converges strongly to $A \in \mathcal{L}(X, Y)$ if $A_n x \to Ax$ for every $x \in X$.*

For example, let $X = Y = \ell^2$ and let A_n be the operator which truncates a sequence after n terms. Let A_n converges strongly to the identity, but not in the sense of the operator norm.

8.1.4 Examples of Operators

Example 8.9. For any Banach space X, any element of the dual space X^* is by definition in $\mathcal{L}(X, \mathbb{R})$.

Example 8.10. Of course, the **identity operator** I from X to X defined by $Ix = x$ is a trival example of a bounded operator. More interestingly, if $X \subset Y$, we can define an **identity mapping** from X to Y to be the map that takes each $x \in X$ to the same x (though this time thought of as an element of Y).

The following lemma is an immediate consequence of Definition 7.25.

Lemma 8.11. *Let X and Y be Banach spaces. Then X is continuously imbedded in Y if and only if the identity mapping from X to Y is well defined (i.e., $X \subset Y$) and is bounded.*

We use this in the following example.

Example 8.12. Corollary 7.19 implies that if $k > m/2$ and $\Omega \subset \mathbb{R}^m$ has the k-extension property, then the identity mapping from $H^k(\Omega)$ to $C_b(\overline{\Omega})$ is bounded.

Example 8.13. The Riesz representation theorem states that every Hilbert space H is isometric to its dual space H^*. In Remark 6.53 we constructed the map $A_H : H^* \to H$ that, for every linear functional $l \in H^*$, gives the unique element $x = A_H(l) \in H$ such that

$$l(y) = (x, y) \quad \text{for all } y \in H.$$

Note that the operator A_H is actually *conjugate linear*; i.e., $A_H(\alpha l) = \bar{\alpha} A_H(l)$. The operator is bounded; the Riesz representation theorem assures us that its norm is 1.

Example 8.14. For any $x = \{x_1, x_2, \ldots, x_n, \ldots\} \in \ell^2$ we define the **right shift** operator $S_r : \ell^2 \to \ell^2$ by

$$S_r(x) = \{0, x_1, x_2, \ldots, x_n, \ldots\}.$$

The range is the set of all elements in ℓ^2 with first component 0; i.e., $\{1, 0, 0, \ldots\}^\perp$. The nullspace is the singleton $\{0\}$. Since $\|S_r(x)\| = \|x\|$, we have $\|S_r\| = 1$.

We define the **left shift** operator $S_l : \ell^2 \to \ell^2$ by

$$S_l(x) = \{x_2, \ldots, x_n, \ldots\}.$$

The range of S_l is ℓ^2 and its nullspace is spanned by the element $\{1, 0, 0, 0, \ldots\}$. Since $\|S_l(x)\| \leq \|x\|$ and since $\|S_l(x)\| = \|x\|$ for any $x \in \{1, 0, 0, 0, \ldots\}^\perp$ we have $\|S_l\| = 1$.

Clearly, S_r is invertible and $S_r^{-1} = S_l$. If we take $\mathcal{D}(S_l) = \ell^2$, then S_l is not invertible. (Its nullspace is nontrivial.) However, if we take $\mathcal{D}(S_l) := \{1, 0, 0, 0, \ldots\}^\perp$, then S_l is invertible and $S_l^{-1} = S_r$.

Example 8.15. Let H be a Hilbert space and let $M \subset H$ be a closed subspace. Then we define the **orthogonal projection** operator $P_M : H \to M$ so that for any $x \in H$, $P_M x$ is the unique element of M such that $(x - P_M x) \in M^\perp$.

Example 8.16. Let $X = C_b([0, L])$ and define the **integration operator** $\mathcal{I} : X \to X$ by

$$\mathcal{I}(f)(x) := \int_0^x f(s)ds, \quad x \in [0, L]. \tag{8.13}$$

Then $\mathcal{N}(\mathcal{I}) = \{0\}$, and $\mathcal{R}(\mathcal{I}) = \{g \in C_b^1([0, L]) \mid g(0) = 0\}$. The operator is bounded; the reader is asked to calculate its norm (Problem 8.6).

Note that we could also think of this operator as mapping $L^2(0, L)$ to $L^2(0, L)$. (Hölder's inequality tells us that the $\mathcal{I}(f)(x)$ is well defined for every $f \in L^2(0, L)$ and every $x \in (0, L)$.) Once again using Hölder's inequality we see that the operator is bounded since

$$
\begin{aligned}
\|\mathcal{I}(f)\|_{L^2(0,L)}^2 &= \int_0^L |\mathcal{I}(f)(x)|^2 dx \\
&= \int_0^L \left(\int_0^x f(s) ds \right)^2 dx \\
&\le \int_0^L \left\{ \int_0^x ds \right\} \left\{ \int_0^x f(s)^2 \, ds \right\} dx \\
&\le \frac{L^2}{2} \int_0^L f(s)^2 \, ds \\
&= \frac{L^2}{2} \|f\|_{L^2(0,L)}^2
\end{aligned}
$$

Using Theorem 5.50 we see that $\mathcal{N}(\mathcal{I}) = \{0\}$.

Example 8.17. We now define a **differentiation** operator. Of course, we can simply define an operator from, say, $C_b^1([0, L])$ to $C_b([0, L])$ by

$$
Du(x) := \frac{du}{dx}(x). \tag{8.14}
$$

With these spaces as domain and range it is easy to show that the differentiation operator is bounded. However, such a proof depends on the fact that we have imposed a "stronger" norm on the domain than on the range. In many contexts it turns out that such a restriction is unsatisfactory. Thus, let us consider the differentiation operator as mapping $L^2(0, L)$ to $L^2(0, L)$ with the domain of the operator restricted to $\tilde{\mathcal{D}}(D) := C_b^1([0, L])$ and range $C_b([0, L])$; i.e., we consider

$$
L^2(0, L) \supset C_b^1([0, L]) \ni u \mapsto \frac{du}{dx} \in C_b([0, L]) \subset L^2(0, L).
$$

It is easy to see that this operator is unbounded. We simply take

$$
u_n(x) = e^{-nx} \tag{8.15}
$$

and calculate

$$
\frac{\left\| \frac{du_n}{dx} \right\|_{L^2(0,L)}}{\|u_n\|_{L^2(0,L)}} = n. \tag{8.16}
$$

Note that unboundedness is not a real problem in inverting the differentiation operator. (We are obviously going to use an integration operator to accomplish the task.) The most important difficulty is that the operator defined above has a nontrivial nullspace: the set of all constant functions. To solve this we simply restrict the domain using a boundary condition; for instance we can consider the domain $\mathcal{D}(D) := \{u \in C_b^1([0, L]) \mid u(0) = 0\}$. With this definition the differentiation operator is injective and the integration operator $\mathcal{I}$ defined above is its inverse.

Another area of concern is that the domain of the operator we have chosen is not "optimal." We have left some very obvious candidates out: for instance, piecewise C^1 functions. This issue is addressed in the section on closed operators below.

Example 8.18. We define an **integral operator** $K : L^2(\Omega) \to L^2(\Omega)$ by

$$Ku := \int_\Omega k(\mathbf{x}, \mathbf{y})u(\mathbf{y}) \, d\mathbf{y}. \qquad (8.17)$$

Here $k : \Omega \times \Omega \to \mathbb{C}$ is called the **kernel** of the operator.

We can make a number of hypotheses on the kernel to ensure that the integral operator is, for instance, bounded. We now give two lemmas based on hypotheses which will be important in applications.

Lemma 8.19. *Let $\Omega \subset \mathbb{R}^n$ be a bounded domain and suppose the kernel k satisfies*

$$k_1 := \sup_{\mathbf{x} \in \Omega} \int_\Omega |k(\mathbf{x}, \mathbf{y})| \, d\mathbf{y} < \infty, \qquad (8.18)$$

$$k_2 := \sup_{\mathbf{y} \in \Omega} \int_\Omega |k(\mathbf{x}, \mathbf{y})| \, d\mathbf{x} < \infty. \qquad (8.19)$$

Then the operator $K : L^2(\Omega) \to L^2(\Omega)$ defined by (8.17) is bounded.

Proof. We simply use Hölder's inequality to get

$$
\begin{aligned}
\|Ku\|_{L^2(\Omega)}^2 &= \int_\Omega \left(\int_\Omega k(\mathbf{x}, \mathbf{y})u(\mathbf{y}) \, d\mathbf{y} \right)^2 \, d\mathbf{x} \\
&\leq \int_\Omega \left(\int_\Omega \sqrt{|k(\mathbf{x}, \mathbf{y})|} \sqrt{|k(\mathbf{x}, \mathbf{y})|} u(\mathbf{y}) \, d\mathbf{y} \right)^2 \, d\mathbf{x} \\
&\leq \int_\Omega \left\{ \int_\Omega |k(\mathbf{x}, \mathbf{y})| \, d\mathbf{y} \right\} \left\{ \int_\Omega |k(\mathbf{x}, \mathbf{y})| u(\mathbf{y})^2 \, d\mathbf{y} \right\} \, d\mathbf{x} \\
&\leq k_1 \int_\Omega \int_\Omega |k(\mathbf{x}, \mathbf{y})| u(\mathbf{y})^2 \, d\mathbf{y} \, d\mathbf{x} \\
&= k_1 \int_\Omega u(\mathbf{y})^2 \int_\Omega |k(\mathbf{x}, \mathbf{y})| \, d\mathbf{x} \, d\mathbf{y} \\
&\leq k_1 k_2 \|u\|_{L^2(\Omega)}^2.
\end{aligned}
$$

This completes the proof.   □

Lemma 8.20. *Let $\Omega \subset \mathbb{R}^n$ be a domain. Let k be* **Hilbert-Schmidt;** *i.e., suppose*

$$\int_\Omega \int_\Omega |k(\mathbf{x}, \mathbf{y})|^2 \, d\mathbf{x} \, d\mathbf{y} = C < \infty. \tag{8.20}$$

Then the operator $K : L^2(\Omega) \to L^2(\Omega)$ defined by (8.17) is bounded.

Proof. Once again we use Hölder's to get

$$
\begin{aligned}
\|Ku\|_{L^2(\Omega)}^2 &= \int_\Omega \left(\int_\Omega k(\mathbf{x}, \mathbf{y}) u(\mathbf{y}) \, d\mathbf{y} \right)^2 d\mathbf{x} \\
&\leq \int_\Omega \left\{ \int_\Omega |k(\mathbf{x}, \mathbf{y})|^2 \, d\mathbf{y} \right\} \left\{ \int_\Omega u(\mathbf{y})^2 \, d\mathbf{y} \right\} d\mathbf{x} \\
&= \left[\int_\Omega \int_\Omega |k(\mathbf{x}, \mathbf{y})|^2 \, d\mathbf{y} \, d\mathbf{x} \right] \int_\Omega u(\mathbf{y})^2 \, d\mathbf{y} \\
&= C \|u\|_{L^2(\Omega)}^2.
\end{aligned}
$$

This completes the proof.   □

We have already encountered integral operators in Section 5.5 where the kernel was a Green's function. We shall study integral operators in more detail in Section 8.5.

Example 8.21. We now discuss the differential operators considered in Section 5.5.1 We let Ω be a domain with smooth boundary in $\mathbb{R}^m$, and let $p \in \mathbb{N}$ be given. We let $B_j(\mathbf{x}, D)$, $j = 1, \ldots, p$, be differential operators of order less than $2p$ which are well defined for $\mathbf{x} \in \partial\Omega$. We then define the domain

$$\mathcal{D}_B(L) := \{u \in C_b^{2p}(\Omega) | B(\mathbf{x}, D)u(\mathbf{x}) = 0, \mathbf{x} \in \partial\Omega, j = 1, \ldots, p\}. \tag{8.21}$$

We then define the operator $(\mathcal{D}_B(L), L)$ from $L^2(\Omega)$ to $L^2(\Omega)$ by

$$L(\mathbf{x}, D)u(\mathbf{x}) := \sum_{|\alpha| \leq 2p} a_\alpha(\mathbf{x}) D^\alpha u(\mathbf{x}). \tag{8.22}$$

Example 8.22. In particular, we consider the **Laplacian**

$$\Delta u := \frac{\partial^2 u}{\partial x_1^2} + \cdots + \frac{\partial^2 u}{\partial x_n^2}$$

again as an operator from $L^2(\Omega)$ to $L^2(\Omega)$. (For the moment, we let Ω be a bounded domain.) We consider two types of domains for the operator, the first corresponding to Dirichlet data:

$$\tilde{\mathcal{D}}_D(\Delta) := \{u \in C_b^2(\Omega) \mid u(\mathbf{x}) = 0 \text{ for } \mathbf{x} \in \partial\Omega\}; \tag{8.23}$$

and the second corresponding to Neumann data:

$$\tilde{\mathcal{D}}_N(\Delta) := \{u \in C_b^2(\Omega) \mid \frac{\partial u}{\partial \mathbf{n}}(\mathbf{x}) = 0 \text{ for } \mathbf{x} \in \partial\Omega\}. \tag{8.24}$$

For the moment we cannot say too much about the invertibility of the operator other than to observe that the operator $(\tilde{\mathcal{D}}_N(\Delta), \Delta)$ has a nontrivial nullspace: the constant functions. We will examine the invertibility of these operators in Chapter 9.

8.1.5 Closed Operators

The following concepts are very useful in studying unbounded operators.

Definition 8.23. The **graph** of a linear operator $(\mathcal{D}(A), A)$ is the set of ordered pairs

$$\Gamma(A) := \{(x, Ax) \mid x \in \mathcal{D}(A)\} \subset X \times Y. \qquad (8.25)$$

Note that the graph is a subspace of $X \times Y$.

The following lemma is a direct consequence of the definitions of extensions and graphs of operators.

Lemma 8.24. *The operator $(\mathcal{D}(\tilde{A}), \tilde{A})$ is an extension of $(\mathcal{D}(A), A)$ if and only if $\Gamma(\tilde{A}) \supset \Gamma(A)$.*

The proof is left to the reader.

Definition 8.25. We say that an operator $(\mathcal{D}(A), A)$ is **closed** if its graph is closed as a subset of $X \times Y$. We call $(\mathcal{D}(A), A)$ **closable** if it has a closed extension. Every closable operator has a smallest closed extension which we call its **closure** and denote by $(\mathcal{D}(\overline{A}), \overline{A})$.

It is useful to supplement this definition with a "sequential" notion of a closed operator. For instance, the following lemma is a direct consequence of the definitions of a closed operator and a closed set in a product space.

Lemma 8.26. *An operator $(\mathcal{D}(A), A)$ is closed if and only if it has the following property. Whenever there is sequence $x_n \in \mathcal{D}(A)$ such that*

1. *$x_n \to x$ and*

2. *$Ax_n \to f$,*

then

1. *$x \in \mathcal{D}(A)$ and*

2. *$Ax = f$.*

We have a similar characterization of a closable operator.

Lemma 8.27. *An operator $(\mathcal{D}(A), A)$ is closable if for every sequence $x_n \in \mathcal{D}(A)$ such that $x_n \to 0$, we have either*

1. *$Ax_n \to 0$, or*

2. *$\lim_{n \to \infty} Ax_n$ does not exist.*

Proof. To prove this we construct a closed extension $(\mathcal{D}(\tilde{A}), \tilde{A})$ of $(\mathcal{D}(A), A)$ as follows. If $x_n \in \mathcal{D}(A)$ is a sequence such that

1. $x_n \to x$ and

2. $Ax_n \to f$

for some $f \in Y$, then we let

1. $x \in \mathcal{D}(\tilde{A})$, and define

2. $\tilde{A}x = f$.

We need to assure ourselves that $\tilde{A}x$ is unambiguously defined. However, our hypothesis assures us that if there is another sequence $\hat{x}_n \in \mathcal{D}(A)$ and $\hat{x}_n \to x$, then either

1. $A\hat{x}_n \to f$, or

2. $\lim_{n \to \infty} A\hat{x}_n$ does not exist

(since $x_n - \hat{x}_n \to 0$). The operator $(\mathcal{D}(\tilde{A}), \tilde{A})$ is closed by Lemma 8.26, and it was constructed to be an extension of $(\mathcal{D}(A), A)$.   □

In fact, it is easy to see that $\Gamma(\tilde{A}) = \overline{\Gamma(A)}$. Thus, since $\overline{\Gamma(A)} \subset \Gamma(\overline{A})$ (this follows directly from Definition 8.25), and since the closure is defined to be the smallest possible extension, our construction actually yields the closure; i.e., $(\mathcal{D}(\tilde{A}), \tilde{A}) = (\mathcal{D}(\overline{A}), \overline{A})$. In fact, we have shown the following.

Corollary 8.28. *If $(\mathcal{D}(A), A)$ is closable, then $\Gamma(\overline{A}) = \overline{\Gamma(A)}$.*

Example 8.29. We now return to the differentiation operator $(\mathcal{D}(D), D)$ defined in Example 8.17. In our comments above, we alluded to the fact that the domain of the operator was not "optimal," in that it did not include such obvious functions as piecewise differentiable functions. We can now see that the problem is that the operator is not closed. We construct the closure as we did in the proof of Lemma 8.27 above. That is, we define the domain $\mathcal{D}(\overline{D})$ of the closure to be the functions $u \in L^2(0, L)$ such that there exists a sequence $u_n \in \mathcal{D}(D)$ and an element $v \in L^2(0, L)$ such that

1. $u_n \to u$ in $L^2(0, L)$, and

2. $\frac{du_n}{dx} \to v$ in $L^2(0, L)$.

However, this is simply the definition of the Sobolev space

$$\mathcal{D}(\overline{D}) := \{u \in H^1(0, L) \mid u(0) = 0\},$$

where the boundary condition is taken in the sense of trace.

Example 8.30. Closing the Laplacian operator is accomplished in much the same way. We consider the problem of Dirichlet conditions. Suppose there is a sequence u_n in $\tilde{\mathcal{D}}_D(\Delta)$ and functions u and v in $L^2(\Omega)$ such that

$$u_n \to u \text{ in } L^2(\Omega), \tag{8.26}$$

and

$$\Delta u_n \to v \text{ in } L^2(\Omega). \tag{8.27}$$

In Chapter 9 we shall show that the extended domain is given by $\mathcal{D}_D(\Delta) = H^2(\Omega) \cap H_0^1(\Omega)$. Although we cannot completely justify this assertion at this time, we can give parts of the proof. First, we note that if Ω is of class C^1, we have $\mathcal{D}_D(\Delta) \supset H^2(\Omega) \cap H_0^1(\Omega)$. To see this, note $\mathcal{D}_D(\Delta)$ is a subset of the completion of $\tilde{\mathcal{D}}_D(\Delta)$ in the $H^2(\Omega)$ norm. Also note that by Theorem 7.41, $H^2(\Omega) \cap H_0^1(\Omega)$ is the set of all $u \in H^2(\Omega)$ such that $u = 0$ on $\partial\Omega$ in the sense of trace. However, using the trace theorem again, we see that this set is also the completion of $\tilde{\mathcal{D}}_D(\Delta)$ in the $H^2(\Omega)$ norm. Second, we note that $\mathcal{D}_D(\Delta) \subset H^1(\Omega)$. To see this, note that we can use Green's identity to get

$$\int_\Omega |\nabla u_n|^2 = -\int_\Omega u_n \Delta u_n. \tag{8.28}$$

Thus, an application of Hölder's inequality shows that the sequence $u_n \in \mathcal{D}_D(\Delta)$ satisfying 1 and 2 is bounded in $H^1(\Omega)$. Thus, by Theorem 6.64 (and the fact that $H^1(\Omega)$ is reflexive), u_n has a weakly convergent subsequence $u_n \rightharpoonup \bar{u}$ in $H^1(\Omega)$. However, by the uniqueness of weak limits we must have $u = \bar{u}$.

Example 8.31. Note that all differential operators with smooth coeffcients are closable. To see this, suppose that $(\mathcal{D}(L), L)$ is a differential operator of order m with $\mathcal{D}(L) \subset L^2(\Omega)$. Then if $x_n \to 0$ in $L^2(\Omega)$, we have $L(x_n) \to 0$ in $H^{-m}(\Omega)$. Since $L^2(\Omega)$ is continuously imbedded into $H^{-m}(\Omega)$, either $\lim L(x_n)$ (thought of as a limit in $L^2(\Omega)$) is equal to 0 or does not exist. Thus, by Lemma 8.27, the operator is closable.

Although all differential operators are closable, the domain of the closure is not necessarily a Sobolev space as one might expect. For instance, consider the wave operator

$$L(u) := u_{tt} - u_{xx}, \tag{8.29}$$

with domain $H^2(\mathbb{R} \times (0, T))$, with $T < \infty$. The reader should verify that such obvious wave-like solutions as $u(x, t) = f(x - t)$, where $f \in L^2(\mathbb{R})$ is discontinuous, are in the closure of the operator. Of course, such functions are neither in $H^2(\mathbb{R} \times (0, T))$ nor $H^1(\mathbb{R} \times (0, T))$. (They do not have a well defined trace along the lines of discontinuity.)

Example 8.32. While most operators we encounter in practice are closable, there are examples of operators which are not. Let $X = L^2(\mathbb{R})$, $Y = \mathbb{R}$, and define $(\mathcal{D}(A), A)$ by letting $\mathcal{D}(A)$ be the set of bounded functions with compact support and $Au := \int_{-\infty}^{\infty} u(x)\, dx$. Then if we let

$$u_n(x) = \begin{cases} 1/n, & |x| \le n \\ 0, & |x| > n, \end{cases} \tag{8.30}$$

we see that $u_n \to 0$ in $L^2(\mathbb{R})$ but $Au_n \equiv 2$.

Problems

8.1. Show that every bounded operator is closable, but that the range of a bounded linear operator need not be closed.

8.2. Let $A \in \mathcal{L}(X, Y)$. We say that A is *bounded below* on a subspace $M \subset X$ if there exists a constant $k > 0$ such that

$$\|Ax\| \ge k\|x\|, \quad \text{for all } x \in M.$$

Let $(\mathcal{D}(A), A)$ be an operator from H to H. Show that if A is closed and bounded below on $\mathcal{N}(A)^\perp \cap \mathcal{D}(A)$, then $\mathcal{R}(A)$ is closed.

8.3. Let $A \in \mathcal{L}(X, Y)$ be surjective. Show that if A is bounded below on X, then A^{-1} exists and is bounded.

8.4. Prove Theorem 8.3.

8.5. Let X, Y and Z be Banach spaces, and let $A : X \to Y$ and $B : Y \to Z$ be bijective operators. Let $BA : X \to Z$ be the composition of A and B. Show that

$$(BA)^{-1} = A^{-1}B^{-1}.$$

8.6. Find the norm of the integration operator $\mathcal{I}$ defined in Example 8.16.

8.7. For any linear operator $(\mathcal{D}(A), A)$ from X to Y we define the **graph norm** on $\mathcal{D}(A) \subset X$ by

$$\|x\|_G := \|x\|_X + \|A(x)\|_Y \quad \text{for any } x \in \mathcal{D}(A).$$

(a) Show that this does indeed define a norm on $\mathcal{D}(A)$.
(b) Show that $(\mathcal{D}(A), A)$ is closed if and only if $\mathcal{D}(A)$ equipped with the graph norm is a Banach space.
(c) Every normed linear space has a completion (hence $\mathcal{D}(A)$ equipped with the graph norm does as well) yet not every operator is closable. Why is this not a contradiction?

8.8. Show that if the inverse of a closed operator exists, it is closed.

8.9. Show that the nullspace of a closed operator is closed.

8.10. Let X and Y be Banach spaces and let $(\mathcal{D}(A), A)$ be a closed operator from X to Y. Show that the image under A of a compact set in X is a closed set in Y. Show that the inverse image of a compact set in Y is a closed set in X.

8.11. Let A and B be operators from X to Y. We say that B is bounded relative to A if $\mathcal{D}(B) \supseteq \mathcal{D}(A)$ and there exists $a, b > 0$ such that

$$\|Bx\|_Y \leq a\|Ax\|_Y + b\|x\|_X$$

for every $x \in \mathcal{D}(A)$. Prove the following: If A is closed and B is bounded relative to A with $a < 1$, then $A + B$ (with domain $\mathcal{D}(A)$) is closed.

8.2 The Open Mapping Theorem

Let X and Y be Banach spaces. The following three theorems on operators from X to Y have important consequences in our attempts to invert operators.

Theorem 8.33 (Open mapping theorem). *Let $A \in \mathcal{L}(X, Y)$, and suppose A is surjective (onto). Then the image of every open set $S \subset X$ is open in Y.*

Theorem 8.34 (Bounded inverse theorem). *Let $A \in \mathcal{L}(X, Y)$, and suppose A is bijective (one-to-one and onto). Then the inverse map A^{-1} is bounded.*

Theorem 8.35 (Closed graph theorem). *Let $(\mathcal{D}(A), A)$ be a linear operator from X to Y. Then if $(\mathcal{D}(A), A)$ is a closed operator and its domain $\mathcal{D}(A)$ is closed in X, then the operator is bounded.*

In a typical course in functional analysis, the open mapping theorem is proved using the Baire category theorem, and the bounded inverse theorem and closed graph theorem are derived as consequences. In fact, as we show in Lemma 8.36 below, all three theorems are equivalent. Furthermore, after we have developed the machinery of adjoints in Section 8.4, we will be able to prove the bounded inverse theorem for Hilbert spaces. Since most of our applications will use Hilbert spaces, we will limit ourselves to the proof for this case and ask the reader to refer to the literature for proofs of the more general cases.

Lemma 8.36. *The open mapping theorem, the bounded inverse theorem, and the closed graph theorem are equivalent.*

The proof we give here is good only if X is a Hilbert space. However, the first three parts of the proof apply directly to Banach spaces, and the final part can be generalized by using equivalence classes. (This is a standard technique for getting around the lack of a projection theorem in Banach spaces, but one we won't go into in this text.)

Proof. **Open mapping theorem $\Rightarrow$ bounded inverse theorem.** It is immediately clear from the hypotheses of the bounded inverse theorem that a linear inverse operator A^{-1} with domain $\mathcal{R}(A)$ exists. The nontrivial assertion is that A^{-1} is bounded. However, this follows from the open mapping theorem, the equivalence of boundedness and continuity for linear operators, and the topological version of the definition of continuity: that an operator T is continuous if and only if the inverse image of open sets in $\mathcal{R}(T)$ is open in $\mathcal{D}(T)$. (The inverse image of an open set in $\mathcal{R}(A^{-1})$ ($=X = \mathcal{D}(A)$) under A^{-1} is the same as the image of the set under A.)

Bounded inverse theorem $\Rightarrow$ closed graph theorem. We first observe that the product space $X \times Y$ is a Banach space with norm

$$\|(x, y)\| = \|x\| + \|y\|. \tag{8.31}$$

Our hypothesis is that $\Gamma(A)$ is a closed subspace in $X \times Y$ and $\mathcal{D}(A)$ is a closed subspace in X. Thus, $\Gamma(A)$ and $\mathcal{D}(A)$ are Banach spaces. We now define a projection map

$$P : \Gamma(A) \to \mathcal{D}(A) \tag{8.32}$$

by

$$P(x, Ax) := x. \tag{8.33}$$

Note that P is linear and bijective. If fact, its inverse

$$P^{-1} : \mathcal{D}(A) \to \Gamma(A) \tag{8.34}$$

is defined by

$$P^{-1}x := (x, Ax). \tag{8.35}$$

The mapping P is also bounded since

$$\|P(x, Ax)\| = \|x\| \le \|x\| + \|Ax\| = \|(x, Ax)\|. \tag{8.36}$$

Thus, by the bounded inverse theorem (8.34) there is a constant C such that

$$\|(x, Ax)\| = \|P^{-1}x\| \le C\|x\|. \tag{8.37}$$

But this implies A is bounded since

$$\|Ax\| \le \|(x, Ax)\| \le C\|x\| \tag{8.38}$$

for every $x \in \mathcal{D}(A)$.

Closed graph theorem $\Rightarrow$ bounded inverse theorem. This part is left as an exercise. (Problem 8.12.)

Bounded inverse theorem $\Rightarrow$ open mapping theorem. We prove this only in the case where X is a Hilbert space. Since A is bounded, $\mathcal{N}(A)$ is closed (cf. Problem 8.9). Thus, we can use the projection theorem to decompose X into $X = \mathcal{N}(A) \oplus \mathcal{N}(A)^{\perp}$. We then let $P : X \to \mathcal{N}(A)^{\perp}$ be the orthogonal projection operator and define $\tilde{A}$ to be the restriction of A to the domain $\mathcal{N}(A)^\perp$. Observe that A can be written as the composition of these two operators; i.e.,

$$Ax = \tilde{A}(Px)$$

for every $x \in X$. The proof now hinges on two facts which we ask the reader to verify.

1. The projection map P maps open sets in X to open sets in $\mathcal{N}(A)^\perp$ (Problem 8.13).

2. The operator $\tilde{A}$ is a continuous bijection from $\mathcal{N}(A)^\perp$ to Y (Problem 8.14).

Now, an open set in X gets mapped by P to an open set in $\mathcal{N}(A)^\perp$, and by the bounded inverse theorem, this set gets mapped by $\tilde{A}$ to an open set in Y. (The image of a set under $\tilde{A}$ is the inverse image of a set under $\tilde{A}^{-1}$.) Hence, the map A, which is the composition of the two maps, takes open sets to opens sets. □

Problems

8.12. Show that the closed graph theorem implies the bounded inverse theorem.

8.13. Let M be a closed subspace of a Hilbert space H. Without using the open mapping theorem, show that the orthogonal projection operator $P : H \to M$ maps open sets in H to open sets in M.

8.14. Let $A : H \to Y$ be a bounded linear operator from a Hilbert space H *onto* a Banach space Y. Let $\tilde{A} : \mathcal{N}(A)^\perp \to Y$ be the restriction A to the domain $\mathcal{N}(A)^\perp$. Show that $\tilde{A}$ is a continuous bijection.

8.15. We call a mapping open if it maps every open set to an open set. Show that an open mapping need not map closed sets to closed sets.

8.16. Let X to be the space of sequences $x = \{x_1, x_2, x_3, \dots\}$ with only finitely many nonzero terms and norm

$$\|x\| := \sup_{i \in \mathbb{N}} |x_i|.$$

Let $T : X \to X$ be defined by

$$Tx := \left\{ x_1, \frac{x_2}{2}, \frac{x_3}{3}, \dots \right\}.$$

Show that T is linear and bounded but that T^{-1} is unbounded. Why does this not contradict the bounded inverse theorem?

8.3   Spectrum and Resolvent

In this section we generalize the eigenvalue problems of linear algebra to operators on Banach spaces. One of our main goals is to generalize the following theorem.

Theorem 8.37. *Let A be an $n \times n$ symmetric matrix. Then A has n eigenvalues $\lambda_1, \ldots, \lambda_n$ (counted with respect to algebraic multiplicity), and all of these eigenvalues are real. Furthermore, there is an orthonormal basis $\{\mathbf{e}_1, \ldots, \mathbf{e}_n\}$ for $\mathbb{R}^n$, such that $\mathbf{e}_i$ is an eigenvector corresponding to λ_i.*

The proof of this is given in any good elementary linear algebra text. The result will be a corollary to the theorems we prove below about self-adjoint compact operators.

One of our first tasks is to generalize the concept of eigenvalues and eigenvectors to accommodate the operators considered in this section (which may be defined on infinite-dimensional spaces and may be unbounded).

Definition 8.38. Let X be a complex Banach space. Let $(\mathcal{D}(A), A)$ be an operator from X to X. For any $\lambda \in \mathbb{C}$ we define the operator $(\mathcal{D}(A), A_\lambda)$ by

$$A_\lambda := A - \lambda I, \tag{8.39}$$

where I is the identity operator on X. If A_λ has an inverse (i.e., if it is one-to-one), we denote the inverse by $R_\lambda(A)$, and call it the **resolvent** of A.

Definition 8.39. Let $X \neq \{0\}$ be a complex Banach space and let $(\mathcal{D}(A), A)$ be a linear operator from X to X. Consider the following three conditions:

1. $R_\lambda(A)$ exists,

2. $R_\lambda(A)$ is bounded,

3. the domain of $R_\lambda(A)$ is dense in X.

We decompose the complex plane $\mathbb{C}$ into the following two sets.

- The **resolvent set** of the operator A is the set

$$\rho(A) := \{\lambda \in \mathbb{C} \mid (1), (2), \text{ and } (3) \text{ hold}\}. \tag{8.40}$$

  Elements $\lambda \in \rho(A)$ in the resolvent set are called **regular values** of the operator A.

- The **spectrum** of the operator A is the complement of the resolvent set

$$\sigma(A) := \mathbb{C} \backslash \rho(A). \tag{8.41}$$

The spectrum can be further decomposed into three disjoint sets.

- The **point spectrum** or **discrete spectrum** is the set

$$\sigma_p(A) := \{\lambda \in \sigma(A) \mid (1) \text{ does not hold}\}. \qquad (8.42)$$

That is, the point spectrum is the set of $\lambda \in \mathbb{C}$ for which $\mathcal{N}(A_\lambda)$ is nontrivial. Elements of the point spectrum are called **eigenvalues**. If $\lambda \in \sigma_p(A)$, elements $x \in \mathcal{N}(A_\lambda)$ are called **eigenvectors** or **eigenfunctions** of A. The dimension of $\mathcal{N}(A_\lambda)$ is called the (geometric) **multiplicity** of λ.

- The **continuous spectrum** is the set

$$\sigma_c(A) := \{\lambda \in \sigma(A) \mid (1) \text{ and } (3) \text{ hold but } (2) \text{ does not}\}. \qquad (8.43)$$

- The **residual spectrum** or **compression spectrum** is the set

$$\sigma_r(A) := \{\lambda \in \sigma(A) \mid (1) \text{ holds but } (3) \text{ does not}\}. \qquad (8.44)$$

Since $\overline{\mathcal{R}(A_\lambda)} \neq X$ we say that the range has been compressed.

Definition 8.40. If X is a Hilbert space, we refer to the dimension of $\mathcal{R}(A_\lambda)^\perp$ as the **deficiency** of $\lambda \in \mathbb{C}$.

Note that by our definition, $\lambda \in \sigma(A)$ can have nonzero deficiency and not be in the compression spectrum. Some authors define the compression spectrum to be all $\lambda \in \mathbb{C}$ such that the deficiency is nonzero, but in this case the point spectrum and compression spectrum are not necessarily disjoint.

Example 8.41. One of the fundamental results of linear algebra is that for a linear operator A on a finite-dimensional space the continuous spectrum and the compression spectrum of the operator are empty; i.e., the complex plane can be decomposed into regular values and eigenvalues of the operator.

Example 8.42. For a simple example of an operator with a spectral value that is not an eigenvalue, consider the right-shift operator $S_r : \ell^2 \to \ell^2$. The complex number $\lambda = 0$ is an element of the spectrum. To see this we recall that the resolvent operator $R_0(S_r)$ is simply the left-shift operator S_l operating on the domain $\{1, 0, 0, \ldots\}^\perp$, and while this operator is bounded, its domain is not dense in ℓ^2. Thus, $\lambda = 0$ is in the compression spectrum of S_r and has deficiency 1.

Spectral theory is a very broad and well studied subject. Our treatment of it here is of necessity very cursory; our aim is primarily to develop the tool of eigenfunction expansions. Thus, we begin with a basic theorem about eigenvectors.

Theorem 8.43. *If* λ_i, $i = 1, \ldots, n$, *are distinct eigenvalues of the operator* $(\mathcal{D}(A), A)$ *and* $x_i \in \mathcal{N}(A_{\lambda_i})$ *are corresponding eigenvectors, then the set*

$$\{x_1, x_2, \ldots, x_n\}$$

is linearly independent.

Proof. Suppose not. Then there is an integer $k \in [2, n]$ such that the set $\{x_1, \ldots, x_{k-1}\}$ is linearly independent, whereas x_k can be expanded in this set; i.e.,

$$x_k = \alpha_1 x_1 + \alpha_2 x_2 + \cdots + \alpha_{k-1} x_{k-1}, \tag{8.45}$$

where the coefficients α_i are not all zero. We now apply $(A - \lambda_k I)$ to both sides of the equation to get

$$
\begin{aligned}
0 &= (A - \lambda_k I) x_k \\
&= (A - \lambda_k I)[\alpha_1 x_1 + \alpha_2 x_2 + \cdots + \alpha_{k-1} x_{k-1}] \\
&= \alpha_1 (\lambda_1 - \lambda_k) x_1 + \alpha_2 (\lambda_2 - \lambda_k) x_2 + \cdots + \alpha_{k-1}(\lambda_{k-1} - \lambda_k) x_{k-1}.
\end{aligned}
$$

Since $\{x_1, \ldots, x_{k-1}\}$ is linearly independent we have

$$(\lambda_i - \lambda_k)\alpha_i = 0, \quad i = 1, \ldots, k - 1. \tag{8.46}$$

However, since $\lambda_i \neq \lambda_m$ this implies $\alpha_i = 0$, $i = 1, \ldots, k - 1$. This is a contradiction and completes our proof. $\qquad\square$

8.3.1 The Spectra of Bounded Operators

We now study the properties of the spectra of bounded operators. Many of our most important results about the spectrum (including the results for the results below for compact operators) are derived by using a power series expansion for the resolvent. We now prove a fundamental theorem that is the analogue of the elementary calculus result on the convergence of geometric series.

Theorem 8.44. *Let X be a Banach space and suppose $A \in \mathcal{L}(X)$ satisfies $\|A\| < 1$. Then $(I - A)^{-1}$ exists and is bounded, and the following power series expansion for $(I - A)^{-1}$ converges in the operator norm.*

$$(I - A)^{-1} = \sum_{k=0}^{\infty} A^k = I + A + A^2 + \cdots. \tag{8.47}$$

Proof. The main idea in this proof is that if a series in a Banach space *converges absolutely* (i.e., the sum of the norms of the terms converges), then the original series converges. (The proof of this fact is identical to the elementary calculus proof for series of real numbers.) In our case, the Banach space in question is $\mathcal{L}(X)$, and we have

$$\sum_{k=0}^{\infty} \|A^k\| \leq \sum_{k=0}^{\infty} \|A\|^k. \tag{8.48}$$

Since $\|A\| < 1$, the geometric series on the right converges. Hence, the series on the right of (8.47) is absolutely convergent and therefore convergent. We need only show that its limit is indeed $(I - A)^{-1}$. Once again the proof is essentially the same as the elementary calculus result for geometric series; i.e., we have

$$
\begin{aligned}
I - A^{k+1} &= (I - A)(I + A + A^2 + \cdots + A^k) \\
&= (I + A + A^2 + \cdots + A^k)(I - A).
\end{aligned}
$$

Now since $\|A\| < 1$ we have $\lim_{k \to \infty} A^{k+1} = 0$. Thus

$$
I = (I - A)\left(\sum_{k=0}^{\infty} A^k\right) = \left(\sum_{k=0}^{\infty} A^k\right)(I - A), \tag{8.49}
$$

and the theorem is proved.   □

This theorem immediately gives us the following result, which says that the spectrum $\sigma(A)$ of a bounded operator A lies in a bounded disk in the complex plane.

Corollary 8.45. *Let $A \in \mathcal{L}(X)$, and suppose $\lambda \in \sigma(A) \subset \mathbb{C}$. Then*

$$
|\lambda| \leq \|A\|. \tag{8.50}
$$

Proof. Suppose $|\lambda| > \|A\|$. Then we can show that $\lambda \in \rho(A)$ by using Theorem 8.44 to construct the resolvent as follows:

$$
R_\lambda(A) = (A - \lambda I)^{-1} = -\frac{1}{\lambda}\left(I - \frac{1}{\lambda}A\right)^{-1} = -\frac{1}{\lambda}\sum_{k=0}^{\infty}\left(\frac{1}{\lambda}A\right)^k. \tag{8.51}
$$

Here we have used the fact that $\|\frac{1}{\lambda}A\| < 1$. This completes the proof.   □

Since we have just shown that the spectrum of a bounded operator is contained in a disk, it is natural to ask whether this disk is optimal. Thus, we give the following definition.

Definition 8.46. The **spectral radius** of an operator from X to X is defined to be

$$
r_\sigma(A) := \sup_{\lambda \in \sigma(A)} |\lambda|. \tag{8.52}
$$

Thus, for $A \in \mathcal{L}(X)$, Corollary 8.45 simply says

$$
r_\sigma(A) \leq \|A\|. \tag{8.53}
$$

In general, equality does not hold in (8.53), but it does hold for a class of operators called *normal*. Problem 8.33 below establishes equality for self-adjoint operators.

In Corollary 8.45 we used the fact that we could expand $R_\lambda(A)$ in a power series if $|\lambda| > \|A\|$. In fact, we can do much better.

Theorem 8.47. *Let $A \in \mathcal{L}(X)$ and $\lambda_0 \in \rho(A)$. Suppose $\lambda \in \mathbb{C}$ lies in the disk*

$$
|\lambda - \lambda_0| < \frac{1}{\|R_{\lambda_0}\|}. \tag{8.54}
$$

Then $\lambda \in \rho(A)$ *and*

$$R_\lambda(A) = \sum_{k=0}^{\infty} (\lambda - \lambda_0)^k R_{\lambda_0}(A)^{k+1}. \tag{8.55}$$

Proof. Let $\lambda_0 \in \rho(A)$ and $\lambda \in \mathbb{C}$ satisfying (8.54) be given. We then write

$$
\begin{aligned}
A - \lambda I &= A - \lambda_0 I - (\lambda - \lambda_0)I \\
&= (A - \lambda_0 I)[I - (\lambda - \lambda_0)(A - \lambda_0 I)^{-1}] \\
&= (A - \lambda_0 I)[I - (\lambda - \lambda_0)R_{\lambda_0}(A)],
\end{aligned}
$$

or simply

$$A - \lambda I = (A - \lambda_0 I)B, \tag{8.56}$$

where

$$B := [I - (\lambda - \lambda_0)R_{\lambda_0}(A)]. \tag{8.57}$$

Now since $\|(\lambda - \lambda_0)R_{\lambda_0}(A)\| < 1$, we can use Theorem 8.44 to show that B has a bounded inverse and

$$B^{-1} = \sum_{k=0}^{\infty} (\lambda - \lambda_0)^k R_{\lambda_0}(A)^k. \tag{8.58}$$

Now, we use this and (8.56) to get

$$R_\lambda(A) = (A - \lambda I)^{-1} = B^{-1}(A - \lambda_0 I)^{-1} = \sum_{k=0}^{\infty} (\lambda - \lambda_0)^k R_{\lambda_0}(A)^{k+1}. \tag{8.59}$$

This completes the proof. $\qquad\square$

This immediately implies the following.

Corollary 8.48. *The resolvent set $\rho(A) \subset \mathbb{C}$ of a bounded operator A is open.*

Combining this with Theorem 8.45 and the Heine-Borel theorem gives us another important result.

Corollary 8.49. *The spectrum $\sigma(A) \subset \mathbb{C}$ of a bounded operator A is a compact set.*

We will be able to use the power series representation of Theorem 8.47 to employ some elementary techniques of complex variables, but first we need to give a definition of an analytic operator-valued function of a complex variable. The definition we give here holds for a mapping from the complex plane to any Banach space: A mapping to the Banach space of bounded operators $\mathcal{L}(X)$ is a special case.

Definition 8.50. Let $G \subset \mathbb{C}$ be a domain and let Y be a Banach space. Then a mapping

$$\mathbb{C} \supset G \ni \lambda \mapsto B(\lambda) \in Y \qquad (8.60)$$

is said to be **analytic** at a point $\lambda_0 \in \mathbb{C}$ if

$$\lim_{\lambda \to \lambda_0} \frac{B(\lambda) - B(\lambda_0)}{\lambda - \lambda_0} \qquad (8.61)$$

exists.

As we implied, our main result is the following.

Theorem 8.51. *Let $A \in \mathcal{L}(X)$. Then the resolvent operator $R_\lambda(A)$ (thought of as a function of λ) is analytic on the resolvent set $\rho(A)$.*

Proof. The existence of the limit of the difference quotient follows directly form the power series representation shown in Theorem 8.47. □

We now assert that the techniques and results developed for analytic functions in a standard complex variables course can be used with impunity on analytic functions with values in a Banach space. For a more thorough development of this idea; see e.g., [DS]. As an example of an application of old techniques in this new setting we now prove the following.

Theorem 8.52. *The spectrum of a bounded operator on a nonzero Banach space has at least one element.*

Proof. Let $A \in \mathcal{L}(X)$ and suppose $\sigma(A)$ is empty; i.e., the resolvent set is the entire complex plane. By Theorem 8.51, the resolvent operator $R_\lambda(A)$ (thought of as a function of λ) is *entire*; i.e., analytic on the entire complex plane. We now note that $\lambda \mapsto R_\lambda(A)$ is bounded on all of $\mathbb{C}$. To see this, note that by (8.51) we can get

$$\|R_\lambda(A)\| \leq \frac{1}{\|A\|} \quad \text{for } |\lambda| \geq 2\|A\|. \qquad (8.62)$$

In addition, $\lambda \mapsto R_\lambda(A)$ must be bounded on any bounded disk since it is analytic. Thus, we can use Liouville's theorem to deduce that $\lambda \mapsto R_\lambda(A)$ is a constant. This is a contradiction and completes the proof. □

Remark 8.53. Theorems 8.47 and 8.51 can be extended (with similar proofs) to closed operators (cf. Problem 8.23). However, it is possible for an unbounded operator to have an empty spectrum. For example, let $X = L^2(0,1)$ and let

$$\mathcal{D}(S) := \{ y \in H^1(0,1) \mid y(0) = 0 \} \qquad (8.63)$$

and

$$Sy = i\frac{dy}{dx}. \qquad (8.64)$$

The reader should verify that for any $\lambda \in \mathbb{C}$, the operator L_λ given by

$$L_\lambda(y)(x) := -i \int_0^x e^{-i\lambda(x-s)} y(s) ds \qquad (8.65)$$

with domain

$$\mathcal{D}(L_\lambda) := L^2(0, 1) \qquad (8.66)$$

is indeed the resolvent operator $R_\lambda(S)$.

Problems

8.17. Describe the spectrum $\sigma(P_M)$ of the projection operator described in Example 8.15.

8.18. (a) Define a multiplication operator $M : C_b([0, 1]) \to C_b([0, 1])$ by

$$M(u)(x) := xu(x),$$

for every $u \in C_b([0, 1])$. Describe $\sigma(M)$.

(b) Let $v \in C_b([0, 1])$ be given. Define an operator $M_v : C_b([0, 1]) \to C_b([0, 1])$ by

$$M_v(u)(x) := v(x)u(x),$$

for every $u \in C_b([0, 1])$. Describe $\sigma(M_v)$.

8.19. Suppose that $(\mathcal{D}(\tilde{A}), \tilde{A})$ is an extension of a bounded operator $(\mathcal{D}(A), A)$. Show the following:
 (a) $\sigma_p(\tilde{A}) \supset \sigma_p(A)$.
 (b) $\sigma_r(\tilde{A}) \subset \sigma_r(A)$.
 (c) $\sigma_c(A) \subset \sigma_c(\tilde{A}) \cup \sigma_p(\tilde{A})$.
 (d) $\rho(\tilde{A}) \subset \rho(A) \cup \sigma_r(A)$.

8.20. Let $A \in \mathcal{L}(X)$. Show that $\|R_\lambda(A)\| \to 0$ as $|\lambda| \to \infty$.

8.21. Let

$$\mathcal{D}(A) = \{u \in H^2(0, 1) \mid u(0) = u(1) = 0\}.$$

Define the operator $(\mathcal{D}(A), A)$ from $L^2(0, 1)$ to $L^2(0, 1)$ by

$$Au = u''$$

for $u \in \mathcal{D}(A)$. Show that $\sigma(A)$ is not compact. Does your answer contradict Corollary 8.49?

8.22. Let $G \subset \mathbb{C}$ be a domain and let X be a Banach space. Then a mapping

$$\mathbb{C} \supset G \ni \lambda \mapsto B(\lambda) \in X \qquad (8.67)$$

is said to be **weakly analytic** at $\lambda_0 \in \mathbb{C}$ if, for every $g \in X^*$, the complex-valued function defined by

$$f(\lambda) := g(B(\lambda)) \tag{8.68}$$

is analytic (in the usual sense) in a neighborhood of λ_0. The function $B(\lambda)$ is analytic on G if it is analytic at each point in G.

(a) Show that (strong) analyticity implies weak analyticity.

(b) Show that weak analyticity implies (strong) analyticity.

8.23. Extend Theorems 8.47 and 8.51 to unbounded closed operators.

8.4 Symmetry and Self-adjointness

8.4.1 The Adjoint Operator

We now define the *adjoint* of an operator.

Definition 8.54. Let $(\mathcal{D}(A), A)$ be an operator from a Banach space X to a Banach space Y such that $\mathcal{D}(A)$ is dense in X. We define $\mathcal{D}(A^\times)$ to be the set of all $v \in Y^*$ for which there exists $w \in X^*$ such that

$$v(Au) = w(u) \tag{8.69}$$

for all $u \in \mathcal{D}(A)$. Note that since $\mathcal{D}(A)$ is dense, w is uniquely determined by $v \in \mathcal{D}(A^\times)$ and (8.69). Thus, we can define an operator $(\mathcal{D}(A^\times), A^\times)$ from Y^* to X^* by

$$A^\times(v) := w \tag{8.70}$$

for every $v \in \mathcal{D}(A^\times)$. We call $(\mathcal{D}(A^\times), A^\times)$ the **adjoint** of $(\mathcal{D}(A), A)$.

It is clear that $\mathcal{D}(A^\times)$ is nonempty since $\{0\} \in \mathcal{D}(A^\times)$. Also, it follows directly from the definition that $A^\times$ is linear. Furthermore, for bounded operators we can show the following.

Theorem 8.55. *For any bounded operator $A \in \mathcal{L}(X, Y)$ we have $\mathcal{D}(A^\times) = Y^*$ and $A^\times : Y^* \to X^*$ is a bounded operator with $\|A^\times\| = \|A\|$.*

The proof depends on the following lemma, which is a direct consequence of the Hahn-Banach theorem.

Lemma 8.56. *Let X be a Banach space and let $\bar{x}$ be any nonzero element of X. Then there exists a linear functional $l \in X^*$ such that*

$$\|l\| = 1 \quad and \quad l(\bar{x}) = \|\bar{x}\|. \tag{8.71}$$

Proof. Let $M := \{\alpha \bar{x} \mid \alpha \in \mathbb{R}\}$ be the subspace spanned by $\bar{x}$. We define a linear functional $\tilde{l}$ on M by

$$\tilde{l}(\alpha \bar{x}) = \alpha \|\bar{x}\|. \tag{8.72}$$

It is easy to see that $\tilde{l}$ has norm 1. The Hahn-Banach theorem assures us that $\tilde{l}$ has an extension l to all of X with norm less than or equal to 1. Since $l(\bar{x}) = \tilde{l}(\bar{x}) = \|\bar{x}\|$ we see that in fact the norm is equal to 1, and the lemma is proved.   □

We now prove Theorem 8.55.

Proof. For any bounded linear functional $v \in Y^*$ we see that

$$u \mapsto v(A(u)) := w(u) \qquad (8.73)$$

is a linear map from X to $\mathbb{R}$. We further see that this map is bounded since

$$|w(u)| = |v(A(u))| \le \|v\| \|A(u)\| \le \|v\| \|A\| \|u\|. \qquad (8.74)$$

Thus, $v \in \mathcal{D}(A^\times)$ and $w = A^\times(v)$. We can also get from (8.74) that

$$\|A^\times(v)\| \le \|A\| \|v\|. \qquad (8.75)$$

Thus $\|A^\times\| \le \|A\|$. Now, by the previous lemma, for every $\bar{u} \in X$ there exists $\bar{v} \in Y^*$ such that $\|\bar{v}\| = 1$ and $\bar{v}(A(\bar{u})) = \|A(\bar{u})\|$. We now use the definition of the adjoint to get

$$
\begin{aligned}
\|A(\bar{u})\| &= \bar{v}(A(\bar{u})) \\
&= (A^\times(\bar{v}))(\bar{u}) \\
&\le \|A^\times(\bar{v})\| \, \|\bar{u}\| \\
&\le \|A^\times\| \, \|\bar{v}\| \, \|\bar{u}\| \\
&= \|A^\times\| \, \|\bar{u}\|.
\end{aligned}
$$

In the last equality we have used the fact that $\|\bar{v}\| = 1$. Since $\bar{u}$ was arbitrary we now have $\|A\| \le \|A^\times\|$, which completes the proof.   □

We now state a theorem on the relationship between the adjoint of an operator and its closure.

Theorem 8.57. *Let $(\mathcal{D}(A), A)$ be an operator from X to Y with dense domain. Then the adjoint operator $A^\times$ is closed. Furthermore, if X and Y are reflexive Banach spaces, then $(\mathcal{D}(A), A)$ is closable if and only if $\mathcal{D}(A^\times)$ is dense, in which case $\overline{A} = A^{\times\times}$.*

Proof. To show that $A^\times$ is closed we use Lemma 8.26. That is, suppose there is a sequence v_n in $\mathcal{D}(A^\times) \subset Y^*$ such that $v_n \to v$ for some $v \in Y^*$ and $A^\times v_n \to f$ for some $f \in X^*$. We need to show that $v(Au) = f(u)$ for every $u \in \mathcal{D}(A)$. But since convergence in Y^* (X^*) implies weak-∗ convergence in Y^* (X^*), we have

$$v(Au) = \lim_{n \to \infty} v_n(Au) = \lim_{n \to \infty} A^\times v_n(u) = f(u)$$

for any $u \in \mathcal{D}(A)$. Thus, $A^\times$ is closed.

The rest of the proof is left to the reader (Problem 8.27).   □

8.4.2   The Hilbert Adjoint Operator

In this section we consider only operators from one Hilbert space to another. In this case we define the following operator, which is closely related to the adjoint.

Definition 8.58. Let $(\mathcal{D}(A), A)$ be a densely defined operator from a Hilbert space H_1 to a Hilbert space H_2. We define $\mathcal{D}(A^*)$ to be the set of all $v \in H_2$ such that there exists $w \in H_1$ such that

$$(v, Au)_{H_2} = (w, u)_{H_1} \qquad (8.76)$$

for all $u \in \mathcal{D}(A)$. Note that since $\mathcal{D}(A)$ is dense, w is uniquely determined by $v \in \mathcal{D}(A^*)$ and (8.76). Thus, we can define an operator $(\mathcal{D}(A^*), A^*)$ from H_2 to H_1 by

$$A^* v := w \qquad (8.77)$$

for every $v \in \mathcal{D}(A^*)$. We call $(\mathcal{D}(A^*), A^*)$ the **Hilbert adjoint** of $(\mathcal{D}(A), A)$.

The relationship between the adjoint and the Hilbert adjoint can easily be obtained by using the Riesz maps A_{H_1} and A_{H_2} defined in Example 8.13:

$$A^* = A_{H_1} \circ A^\times \circ A_{H_2}^{-1}. \qquad (8.78)$$

(Here, $\circ$ denotes composition of the operators.) Since a Hilbert space and its dual are isometric we rarely use the dual space directly, and when studying operators on Hilbert space we almost always use the Hilbert adjoint rather than the adjoint. In fact, it is common practice to refer to Hilbert adjoint as the adjoint, ignoring the distinction entirely. We adopt this convention, though we will use distinct notation for the two types of operators.

Theorem 8.59. *For any densely defined operator $(\mathcal{D}(A), A)$ from H to H, the orthogonal complement of the range is the nullspace of the adjoint:*

$$\mathcal{R}(A)^\perp = \mathcal{N}(A^*). \qquad (8.79)$$

Furthermore, if $\mathcal{R}(A)$ is closed, then

$$\mathcal{R}(A) = \mathcal{N}(A^*)^\perp, \qquad (8.80)$$

i.e., the equation $Ax = f$ has a solution x if and only if $f \in \mathcal{N}(A^*)^\perp$.

Proof. We first show $\mathcal{R}(A)^\perp \subset \mathcal{N}(A^*)$. Let $z \in \mathcal{R}(A)^\perp$. Then

$$(z, Au) = 0 = (0, u) \qquad (8.81)$$

for all $u \in \mathcal{D}(A)$. Thus, $z \in \mathcal{D}(A^*)$ and $A^* z = 0$; i.e., $z \in \mathcal{N}(A^*)$. In a similar fashion, we see that if $z \in \mathcal{N}(A^*)$, then for every $u \in \mathcal{D}(A)$ we have

$$(z, Au) = (A^* z, u) = (0, u) = 0. \qquad (8.82)$$

Thus, $z \in \mathcal{R}(A)^\perp$ and $\mathcal{N}(A^*) \subset \mathcal{R}(A)^\perp$.

Finally, if $\mathcal{R}(A)$ is closed, we have

$$\mathcal{R}(A) = \mathcal{R}(A)^{\perp\perp} = \mathcal{N}(A^*)^\perp. \tag{8.83}$$

This completes the proof.   □

Definition 8.60. An operator $(\mathcal{D}(A), A)$ from H to H with dense domain $\mathcal{D}(A)$ is said to be **symmetric** if $(\mathcal{D}(A^*), A^*)$ is an extension of $(\mathcal{D}(A), A)$, or equivalently if

$$(Au, v) = (u, Av) \quad \text{for every } u, v \in \mathcal{D}(A). \tag{8.84}$$

Definition 8.61. An operator is said to be **self-adjoint** if

$$(\mathcal{D}(A), A) = (\mathcal{D}(A^*), A^*). \tag{8.85}$$

The following lemma is a direct consequence of the definitions.

Lemma 8.62. *An operator $(\mathcal{D}(A), A)$ is self-adjoint if and only if it is symmetric and $\mathcal{D}(A) = \mathcal{D}(A^*)$.*

For a bounded operator, symmetry and self-adjointness are the same thing.

Theorem 8.63. *An operator $A \in \mathcal{L}(H)$ is self-adjoint if and only if it is symmetric.*

Proof. By definition self-adjointness implies symmetry. Furthermore, if $\mathcal{D}(A) = H$ and A is symmetric, then, by (8.84), $\mathcal{D}(A^*) = \mathcal{D}(A) = H$.   □

As we shall see in Problem 8.29, there are examples of unbounded symmetric operators that are not self-adjoint.

Example 8.64. We now give an example of the computation of the adjoint of an unbounded operator. Let $H = L^2(0, 1)$ and let

$$\mathcal{D}(D) = \{u \in H^1(0, 1) \mid u(0) = 0\}.$$

Here the boundary condition is taken in the sense of trace. Define the differentiation operator by $(\mathcal{D}(D), D)$ from H to H by

$$D(u) := u'$$

for all $u \in \mathcal{D}(D)$. We begin computing the adjoint by doing a formal calculation which gives us a "guess" as to the identity of the adjoint. We then do a rigorous proof that the guess was right.

Both the formal calculation and the rigorous proof are based on the identity

$$(v, Du) = (D^*v, u) \quad \text{for all } u \in \mathcal{D}(D), \ v \in \mathcal{D}(D^*). \tag{8.86}$$

For the formal calculation we begin on the left of (8.86) and proceed to integrate by parts (even though we don't yet know anything about v other than that it is in $L^2(0,1)$). We get

$$
\begin{aligned}
(v, Du) &= \int_0^1 u'(x)v(x)\,dx \\
&= -\int_0^1 u(x)v'(x)\,dx + u(1)v(1) - u(0)v(0) \\
&= (u, -v') + u(1)v(1).
\end{aligned}
$$

In the last equality we used the fact that $u(0) = 0$. Examining (8.86) (and requiring that our formal calculations make sense) leads us to guess that the adjoint is given by $(\mathcal{D}(B), B)$ where

$$
\mathcal{D}(B) := \{v \in H^1(0,1) \mid v(1) = 0\}, \quad B(v) = -v'. \tag{8.87}
$$

This guess is correct, but all we have shown at this time is that the adjoint is an *extension* of the operator $(\mathcal{D}(B), B)$. To prove that $(\mathcal{D}(D^*), D^*) = (\mathcal{D}(B), B)$ we let $D^*v := f$ (making no assumptions on f other than that it lies in $L^2(0,1)$) and define

$$
F(x) = \int_x^1 f(s)\,ds. \tag{8.88}
$$

Of course, we note from (8.87) that $F \in \mathcal{D}(B)$ and that $B(F) = f$. We now work from the *right* of (8.86) and get that for every $v \in \mathcal{D}(D^*)$ and $u \in \mathcal{D}(D)$ we have

$$
\begin{aligned}
(D^*v, u) &= (f, u) \\
&= \int_0^1 u(x)f(x)\,dx \\
&= -\int_0^1 u(x)F'(x)\,dx \\
&= \int_0^1 u'(x)F(x)\,dx \\
&= (F, Du) \\
&= (v, Du).
\end{aligned}
$$

Since this is true for all $u \in \mathcal{D}(D)$ we must have $F - v \in \mathcal{R}(D)^\perp$. However, $\mathcal{R}(D) = L^2(0,1)$, so $F = v$. Hence, the assertion is proved.

Example 8.65. For a more general partial differential operator, such as those defined in Example 8.21, calculation of the adjoint is complicated by the fact that we cannot yet calculate the closure of the operator. However, without closing the operator, one can show that the Hilbert adjoint is an *extension* of the formal adjoint defined in Definition 5.54.

As we shall see below, self-adjoint operators have many properties that are useful in application. However, as the example above indicates, it is often difficult to determine whether an unbounded operator is self-adjoint. (We must first close the operator, and as we have seen, finding the domain of the closure is often nontrivial.) Fortunately, we can define a related property called *essential self-adjointness* for which there is a relatively easy test.

Definition 8.66. A symmetric operator $(\mathcal{D}(A), A)$ is said to be **essentially self-adjoint** if its closure is self-adjoint. If $(\mathcal{D}(A), A)$ is closed, a subset $D \subset \mathcal{D}(A)$ of the domain is called a **core** of the operator if $\overline{(D, A)} = (\mathcal{D}(A), A)$; i.e., the closure of the restriction of the operator to D is the original operator.

The easiest test for essential self-adjointness involves quantities called the deficiency indices.

Definition 8.67. Let $(\mathcal{D}(A), A)$ be a densely defined operator from H to H, and let γ^+ be the dimension of $\mathcal{R}(A - iI)^\perp$ and γ^- the dimension of $\mathcal{R}(A + iI)^\perp$. The numbers (γ^+, γ^-) are called the **deficiency indices** of the operator.

Theorem 8.68. *An operator $(\mathcal{D}(A), A)$ is essentially self-adjoint if and only if its deficiency indices are both 0.*

The proof is left to the reader (Problem 8.37).

8.4.3 Adjoint Operators and Spectral Theory

We can gain some understanding of the spectrum of an operator by studying its adjoint. Our first result relates the compression spectrum of an operator to the point spectrum of its Hilbert adjoint.

Theorem 8.69. *Let $(\mathcal{D}(A), A)$ be a densely defined operator from H to H. A complex number λ has deficiency m if and only if $\bar{\lambda}$ is an eigenvalue of A^* with multiplicity m.*

The proof of this is a simple application of Theorem 8.59 and is left to the reader (Problem 8.31).

We can say a great deal about the spectrum of symmetric operators.

Theorem 8.70. *Let $(\mathcal{D}(A), A)$ be a densely defined operator from H to H. If $(\mathcal{D}(A), A)$ is symmetric, then:*

1. *(Ax, x) is real for every $x \in \mathcal{D}(A)$.*

2. *All eigenvalues of A are real.*

3. *Eigenvectors of A corresponding to distinct eigenvalues are orthogonal.*

4. *The continuous spectrum of A is real.*

Proof. To prove 1, we note that for every $x \in \mathcal{D}(A)$ we have

$$(Ax, x) = (x, Ax) = \overline{(Ax, x)}. \tag{8.89}$$

To prove 2, we note that if λ is an eigenvalue and $x \in \mathcal{N}(A_\lambda)$ is a corresponding eigenfunction, then, using the fact that they are real, we have

$$\lambda = \frac{(x, Ax)}{\|x\|^2}. \tag{8.90}$$

Thus, by part 1, λ is real.

To prove 3, let λ_1 and λ_2 be eigenvalues of A and let $x_1 \in \mathcal{N}(A_{\lambda_1})$ and $x_2 \in \mathcal{N}(A_{\lambda_2})$ be corresponding eigenvectors. Then

$$\lambda_1(x_1, x_2) = (Ax_1, x_2) = (x_1, Ax_2) = \lambda_2(x_1, x_2). \tag{8.91}$$

Thus, either $\lambda_1 = \lambda_2$ or $(x_1, x_2) = 0$.

To prove 4, let $\lambda = \gamma + i\mu$, where γ and μ are real. Then, using the symmetry of A, one can show that

$$\|(A - \lambda I)x\|^2 = \|Ax - \gamma x\|^2 + \mu^2 \|x\|^2 \geq \mu^2 \|x\| \tag{8.92}$$

for $x \in \mathcal{D}(A)$. If $|\mu| > 0$, we have $A - \lambda I$ bounded below. Thus, by Problem 8.3, $R_\lambda(A)$ exists and is bounded. This completes the proof.   □

If an operator is self-adjoint we can say even more.

Theorem 8.71. *Let $(\mathcal{D}(A), A)$ be a densely defined operator from H to H. If $(\mathcal{D}(A), A)$ is self-adjoint, then every $\lambda \in \mathbb{C}$ with nonzero imaginary part is in the resolvent set of A. Furthermore, the compression spectrum is empty.*

Proof. We first note that Theorem 8.70 says that the continuous spectrum of A is real and that all eigenvalues of A are real. Next, Theorem 8.69 says that if λ has nonzero deficiency, then $\overline{\lambda}$ is an eigenvalue of $A(= A^*)$. Hence λ must be real and must lie in the point spectrum rather than the compression spectrum.   □

8.4.4   *Proof of the Bounded Inverse Theorem for Hilbert Spaces*

In this section we prove the result promised in Section 8.2.

Theorem 8.72. *If X and Y are Hilbert spaces and A is a continuous bijection from X to Y, then the inverse of A is bounded.*

Proof. Since $A = A^{**}$, Problem 8.36 implies that it is enough to show that A^* has a bounded inverse. Since the kernel of A is trivial, the range of A^*

is dense in X. Thus, it is enough to show that there exists $\delta > 0$ such that

$$\|A^*y\| \geq \delta\|y\| \tag{8.93}$$

for all $y \in Y$. Suppose not, then there exists a sequence y^n such that

$$\|A^*y^n\| = 1 \tag{8.94}$$

and

$$\|y^n\| \to \infty. \tag{8.95}$$

But now, for any $f \in Y$ we use the fact that A is onto and let x be the solution of $Ax = f$. Then

$$|(y^n, f)| = |(y^n, Ax)| = |(A^*y^n, x)| \leq \|x\|; \tag{8.96}$$

i.e., the sequence y^n is weakly bounded. By the uniform boundedness principle y^n must be bounded in norm, a contradiction.   □

Problems

8.24. Let A be an $m \times n$ complex matrix, and define an operator (also called A) from $\mathbb{C}^n \to \mathbb{C}^m$ by matrix multiplication. What is the relationship amoung the adjoint, the Hilbert adjoint of the operator A and the matrix A?

8.25. If A and B are in $\mathcal{L}(H)$ show that

$$(AB)^* = B^*A^*.$$

8.26. Show that if $(\mathcal{D}(B), B)$ is an extension of $(\mathcal{D}(A), A)$, then $(\mathcal{D}(A^\times), A^\times)$ is an extension of $(\mathcal{D}(B^\times), B^\times)$.

8.27. Complete the proof of Theorem 8.57.

8.28. Compute the Hilbert adjoint of the right shift operator S_r defined in Example 8.14

8.29. Let $H = L^2(0,1)$ and let

$$\mathcal{D}(A) = \{u \in H^2(0,1) \mid u(0) = u'(0) = u(1) = 0\}.$$

Here the boundary conditions are taken in the sense of trace. Define $A : \mathcal{D}(A) \to H$ by

$$A(u) := -\frac{d^2u}{dx^2}.$$

Find the Hilbert adjoint of $(\mathcal{D}(A), A)$. Is the operator symmetric, self-adjoint?

8.30. Show that $(\tilde{\mathcal{D}}_D(\Delta), \Delta)$ and $(\tilde{\mathcal{D}}_N(\Delta), \Delta)$ defined in Example 8.22 are symmetric.

8.31. Prove Theorem 8.69.

8.32. Let $A \in \mathcal{L}(H)$. Show that $\|A^*A\| = \|A\|^2$.

8.33. It can be shown that for an operator $A \in \mathcal{L}(X)$

$$r_\sigma(A) = \lim_{n \to \infty} \|A^n\|^{1/n}.$$

Use this fact and Problem 8.32 to show that if $A \in \mathcal{L}(H)$ is self-adjoint, then $r_\sigma(A) = \|A\|$.

8.34. Suppose $A, B \in \mathcal{L}(X)$ and that $AB = BA$. Show that

$$r_\sigma(AB) \leq r_\sigma(A) r_\sigma(B).$$

Show that the commutivity assumption in this result is essential.

8.35. Show that every symmetric operator is closable.

8.36. Let X and Y be Hilbert spaces and suppose $A \in \mathcal{L}(X,Y)$ is a bijection. Show that A has a bounded inverse if and only if A^* does.

8.37. Prove Theorem 8.68.

8.38. Describe the spectra of the right and left shift operators described in Example 8.14.

8.5   Compact Operators

Definition 8.73. Let X and Y be Banach spaces, and let $(\mathcal{D}(A), A)$ be a linear operator from X to Y. Then we say the operator A is **compact** if it maps bounded sets into precompact sets; i.e., if for every bounded set $\Omega \subset \mathcal{D}(A)$, we have $\overline{A(\Omega)} \subset Y$ compact.

It is often convenient to characterize compact operators in terms of sequences rather than in terms of sets.

Theorem 8.74. *An operator $(\mathcal{D}(A), A)$ from X to Y is compact if and only if it is **sequentially compact**; i.e., if and only if given any bounded sequence x_n in $\mathcal{D}(A)$, it follows that $A(x_n)$ has a convergent subsequence.*

Proof. The proof of this theorem follows directly from the topological result that a precompact set can be characterized by sequences; i.e., a set S in a normed linear space is precompact if and only if every sequence contained in S has a convergent subsequence. $\qquad \square$

As we shall see below, the most fundamental examples of compact operators are integral operators. However, we shall need to develop a bit of machinery in order to study them more fully. In the meantime, we have been provided with some very important examples of compact operators by our study of compact imbeddings in Section 7.2.4. In order to interpret them we need the following lemma.

Lemma 8.75. *Let X and Y be Banach spaces. Then X is compactly imbedded in Y if and only if the identity mapping from X to Y is well defined and compact.*

The proof follows immediately from Definition 7.25 and the definition of the identity mapping in Example 8.10.

Example 8.76. It follows from Theorem 7.27 that if $k > m/2$ and $\Omega \subset \mathbb{R}^m$ is bounded and has the k-extension property, then the identity mapping from $H^k(\Omega)$ to $C_b(\overline{\Omega})$ is compact. Thus by Theorem 8.74, every sequence of functions u_n that is bounded in the $H^k(\Omega)$ norm has a uniformly convergent subsequence.

Example 8.77. It follows from Theorem 7.29 that if k is a non-negative integer and $\Omega \subset \mathbb{R}^m$ is bounded and has the $k+1$-extension property, then the identity mapping from $H^{k+1}(\Omega)$ to $H^k(\Omega)$ is compact. Using Theorem 8.74 again, we see that every sequence of functions u_n that is bounded in the $H^{k+1}(\Omega)$ norm has a subsequence that converges strongly in the $H^k(\Omega)$ norm.

We now obtain the following elementary result.

Lemma 8.78. *Every compact operator is bounded.*

Proof. Suppose not, then there is a sequence $x_n \in \mathcal{D}(A)$ such that $\|x_n\| = 1$ and $\|A(x_n)\| \to \infty$. In fact, by eliminating superfluous elements of the sequence and relabeling, we can ensure that $\|A(x_{n+1})\| > \|A(x_n)\| + 1$. Thus, no subsequence of $A(x_n)$ could converge since no subsequence could be Cauchy.   □

Recall that by Theorem 8.7, every bounded operator can be extended to all of X without changing its norm. We leave it to the reader to show that when a compact operator is extended using the methods described in the proof of Theorem 8.7, the extended operator is also compact (Problem 8.43). Thus, we will usually assume that a compact operator is in $\mathcal{L}(X, Y)$.

Note that Lemma 8.78 and Lemma 6.44 tell us that every compact operator is continuous. However, the converse of this result is false. In particular, we have the following.

Lemma 8.79. *If X is any infinite-dimensional Banach space, then the identity operator is not compact.*

Proof. The proof follows immediately from the fact that in an infinite-dimensional space, the unit ball is not compact. We prove this only in the case of an infinite-dimensional Hilbert space and leave the general result to the reader (Problem 8.47). Recall that, by Corollary 6.36, in an infinite-dimensional Hilbert space there exists an infinite orthonormal set $\{x_i\}_{i=1}^{\infty}$. This set is contained in the closed unit ball, and if x_i and x_j are two distinct elements of the basis, we have $\|x_i - x_j\|^2 = 2$. Thus, no subsequence of x_i could converge since no subsequence could be Cauchy.   □

The fact that a compact operator is "more than" continuous motivated the use of the term *completely continuous* operator for a compact operator. This terminology was common years ago but is used less frequently today.

The connection between compact operators and the dimension of the domain and range of the operator is even closer than Lemma 8.79 suggests.

Theorem 8.80. *Let $(\mathcal{D}(A), A)$ be a linear operator from X to Y. Then we have the following:*

1. *If $(\mathcal{D}(A), A)$ is bounded and the range $\mathcal{R}(A)$ is finite-dimensional, then the operator $(\mathcal{D}(A), A)$ is compact.*

2. *If the domain $\mathcal{D}(A)$ is finite-dimensional, then the operator $(\mathcal{D}(A), A)$ is compact.*

Proof. For part 1, let $x_n \in \mathcal{D}(A)$ be a given bounded sequence. Since the operator $(\mathcal{D}(A), A)$ is bounded, the sequence $A(x_n) \in \mathcal{R}(A)$ is also bounded. Since $\mathcal{R}(A)$ is finite-dimensional, the Bolzano-Weierstraß theorem implies that $A(x_n)$ has a convergent subsequence. Thus, $(\mathcal{D}(A), A)$ is compact.

For part 2, we note that the dimension of the range of an operator is less than or equal to the dimension of the domain. (If $\{x_i\}_{i=1}^k$ is a basis for $\mathcal{D}(A)$, then $\{A(x_i)\}_{i=1}^k$ spans $\mathcal{R}(A)$.) Also, by Lemma 8.5, any operator with a finite-dimensional domain is bounded. Thus, we can use part 1 to complete the proof. □

Definition 8.81. If $A \in \mathcal{L}(X, Y)$ and $\mathcal{R}(A)$ is finite-dimensional, we say the operator A has **finite rank**.

One common way of proving an operator is compact is by approximating by other operators (such as operators of finite rank) which are known to be compact. In using such an approximation scheme one usually employs the following result.

Theorem 8.82. *Let $A_n \in \mathcal{L}(X, Y)$ be a sequence of compact operators. Suppose A_n converges in the operator norm to an operator A. Then A is compact.*

Proof. We employ a "diagonal sequence" argument. Let $\{x_n\}_{n=1}^\infty \subset X$ be a given bounded sequence. Then since A_1 is compact, the sequence $A_1(x_n)$ has a convergent subsequence. We label this subsequence $\{A_1(x_{1,n})\}_{n=1}^\infty$. Now, since $\{x_{1,n}\}_{n=1}^\infty$ is bounded and A_2 is compact, we see that $A_2(x_{1,n})$ has a convergent subsequence. We label this subsequence $\{A_2(x_{2,n})\}_{n=1}^\infty$. We now repeat the process, taking further subsequences of subsequences so that $\{x_{k,n}\}_{n=1}^\infty$ is a subsequence of $\{x_{j,n}\}_{n=1}^\infty$ if $j < k$ and so that $\{A_k(x_{k,n})\}_{n=1}^\infty$ converges. (Recall that since $\{A_k(x_{k,n})\}_{n=1}^\infty$ is convergent it is Cauchy.)

Now consider the diagonal sequence $\{x_{n,n}\}_{n=1}^\infty$. We denote $z_n := x_{n,n}$. Note that this is indeed a subsequence of the original sequence x_n. We claim that $A(z_n)$ is Cauchy and hence convergent. (This will complete the proof since x_n was an arbitrary bounded sequence.) Let $\epsilon > 0$ be given. We note that for any i, j and k, we have

$$\|A(z_i) - A(z_j)\| \le \|A(z_i) - A_k(z_i)\| + \|A_k(z_i) - A_k(z_j)\| + \|A(z_j) - A_k(z_j)\|. \tag{8.97}$$

Since z_n is a bounded sequence, and since $A_k \to A$ in the operator norm, we can pick k sufficiently large so that

$$\|A(z_n) - A_k(z_n)\| < \epsilon/3 \tag{8.98}$$

for every element of the sequence z_n. We now note that for fixed k, the sequence $A_k(z_n)$ is Cauchy. This is true since $\{z_n\}_{n=k}^{\infty}$ is a subsequence of $\{x_{k,m}\}_{m=1}^{\infty}$. Thus, we can pick i and j sufficiently large so that

$$\|A_k(z_i) - A_k(z_j)\| < \epsilon/3. \tag{8.99}$$

Combining (8.97) with (8.98) and (8.99) completes the proof.   □

In particular, we can use this theorem to get the following result.

Theorem 8.83. *Let the kernel $k : \Omega \times \Omega \to \mathbb{R}$ be Hilbert-Schmidt. Then the integral operator $K \in \mathcal{L}(L^2(\Omega))$ defined by*

$$K(u)(\mathbf{x}) := \int_{\Omega} k(\mathbf{x}, \mathbf{y}) u(\mathbf{y}) \, d\mathbf{y}$$

is compact.

Proof. Let $\{\phi_i(\mathbf{x})\}$ be an orthonormal basis for $L^2(\Omega)$. Then, using the methods of Section 5.3.1, one can show that $\{\phi_i(\mathbf{x})\phi_j(\mathbf{y})\}$ is a basis for $L^2(\Omega \times \Omega)$. Expanding k with respect to this basis gives us

$$k(\mathbf{x}, \mathbf{y}) = \sum_{i,j=1}^{\infty} k_{ij} \phi_i(\mathbf{x}) \phi_j(\mathbf{y}) \tag{8.100}$$

where the convergence of the sum is in the $L^2(\Omega \times \Omega)$ norm and

$$k_{ij} := \int_{\Omega} \int_{\Omega} k(\mathbf{x}, \mathbf{y}) \phi_i(\mathbf{x}) \phi_j(\mathbf{y}) \, d\mathbf{x} \, d\mathbf{y}. \tag{8.101}$$

Furthermore, by (6.43) we have

$$\int_{\Omega} \int_{\Omega} |k(\mathbf{x}, \mathbf{y})|^2 \, d\mathbf{x} \, d\mathbf{y} = \sum_{i,j=1}^{\infty} |k_{ij}|^2. \tag{8.102}$$

We now define the operator $K_n \in \mathcal{L}(L^2(\Omega))$ by

$$K_n(u)(\mathbf{x}) := \int_{\Omega} k_n(\mathbf{x}, \mathbf{y}) u(\mathbf{y}) \, d\mathbf{y}, \tag{8.103}$$

where

$$k_n(\mathbf{x}, \mathbf{y}) = \sum_{i,j=1}^{n} k_{ij} \phi_i(\mathbf{x}) \phi_j(\mathbf{y}). \tag{8.104}$$

We refer to k_n and K_n as **separable kernels** and **separable operators**, respectively. It is easy to see that a separable operator has finite rank and is thus compact.

We now use the techniques of Lemma 8.20 to get

$$\|K - K_n\|^2 \le \int_\Omega \int_\Omega |k(\mathbf{x}, \mathbf{y}) - k_n(\mathbf{x}, \mathbf{y})|^2 \, d\mathbf{x} \, d\mathbf{y}. \tag{8.105}$$

Now we use (8.102) to get

$$\lim_{n \to \infty} \int_\Omega \int_\Omega |k(\mathbf{x}, \mathbf{y}) - k_n(\mathbf{x}, \mathbf{y})|^2 \, d\mathbf{x} \, d\mathbf{y} = \lim_{n \to \infty} \sum_{i,j=n+1}^{\infty} |k_{ij}|^2 = 0. \tag{8.106}$$

Thus, K_n converges to K in the operator norm, so Theorem 8.82 implies that K is compact. □

Another useful property of compact operators is that they map weakly convergent sequences into strongly convergent sequences

Theorem 8.84. *Suppose $A \in \mathcal{L}(X, Y)$ is compact and that*

$$x_n \rightharpoonup x \text{ (weakly) in } X. \tag{8.107}$$

Then

$$A(x_n) \to A(x) \text{ (strongly) in } Y. \tag{8.108}$$

Proof. Our first step will be to show that

$$A(x_n) \rightharpoonup A(x) \text{ (weakly) in } Y. \tag{8.109}$$

Let $f \in Y^*$ be given. We must show that

$$\lim_{n \to \infty} f(A(x_n)) = f(A(x)). \tag{8.110}$$

To do this we define $g : X \to \mathbb{R}$ by

$$g(z) = f(A(z)), \quad z \in X. \tag{8.111}$$

Now g is linear since f and A are both linear, and g is bounded since

$$|g(z)| = |f(A(z))| \le \|f\| \|A(z)\| \le \|f\| \|A\| \|z\|. \tag{8.112}$$

Thus, $g \in X^*$, and since $x_n \rightharpoonup x$ in X we have

$$\begin{aligned}
\lim_{n \to \infty} f(A(x_n)) &= \lim_{n \to \infty} g(x_n) \\
&= g(x) \\
&= f(A(x)).
\end{aligned}$$

Since f was arbitrary, $A(x_n) \rightharpoonup A(x)$.

Now suppose that $A(x_n)$ does not converge strongly to $A(x)$ in Y. Then there exists an $\epsilon > 0$ and a subsequence $A(x_{n_k})$ such that

$$\|A(x_{n_k}) - A(x)\| \geq \epsilon. \tag{8.113}$$

Now, since x_n converges weakly to x so does x_{n_k}. Since x_{n_k} is weakly convergent it is bounded. Thus, since A is compact $A(x_{n_k})$ has a strongly convergent subsequence. However, since strong convergence implies weak convergence, and since weak limits are unique, this subsequence must converge to $A(x)$. However, this contradicts (8.113) and completes the proof.
$\square$

We can combine this result with Theorems 7.27 and 7.29 to get the following corollaries.

Corollary 8.85. *Suppose that $k > m/2$ and $\Omega \subset \mathbb{R}^m$ is bounded and has the k-extension property. Let*

$$u_n \rightharpoonup u \ (weakly) \ in \ H^k(\Omega). \tag{8.114}$$

Then

$$u_n \rightarrow u \ uniformly \ on \ \overline{\Omega}. \tag{8.115}$$

Corollary 8.86. *Suppose that k is a non-negative integer and $\Omega \subset \mathbb{R}^m$ is bounded and has the $k+1$-extension property. Let*

$$u_n \rightharpoonup u \ (weakly) \ in \ H^{k+1}(\Omega). \tag{8.116}$$

Then

$$u_n \rightarrow u \ (strongly) \ in \ H^k(\Omega). \tag{8.117}$$

We can also show that for compact operators on a Hilbert space the converse of Theorem 8.82 is true.

Theorem 8.87. *Let $A \in \mathcal{L}(X, H)$ be compact. Then there is a sequence of operators $A_n \in \mathcal{L}(X, H)$, each having finite rank, such that*

$$\lim_{n \to \infty} \|A_n - A\| = 0. \tag{8.118}$$

Proof. We assume that A does not have finite rank. Since A is compact, its range is a countable union of precompact sets and hence separable. Let $\{\phi_i\}_{i=1}^\infty$ be an orthonormal basis for $\overline{\mathcal{R}(A)}$. Let P_n be the orthogonal projection from $\overline{\mathcal{R}(A)}$ onto

$$M_n = \text{span}\{\phi_1, \dots, \phi_n\}, \tag{8.119}$$

and let $A_n = P_n A$. Obviously, A_n has finite rank. We claim that $A_n \rightarrow A$. If not, there is (after taking an appropriate subsequence) $u_n \in X$ with

$\|u_n\| = 1$ and $\|(A - A_n)u_n\| \geq \epsilon > 0$. After taking a subsequence, we may assume that Au_n converges to some limit v. We now find

$$(A - A_n)u_n = (I - P_n)Au_n = (I - P_n)v + (I - P_n)(Au_n - v). \quad (8.120)$$

Since the right-hand side of this equation converges to zero, we find that the left-hand side converges to zero, a contradiction.   □

Remark 8.88. Theorem 8.87 does not hold for general Banach spaces. On the other hand, we do not have to restrict the image space to be a Hilbert space. All we have actually used is the existence of finite-dimensional projections which converge strongly to the identity. Such projections actually exist in most of the Banach spaces which are important in applications.

The following result can be shown for general Banach spaces X and Y.

Theorem 8.89. *Let* $A \in \mathcal{L}(X, Y)$ *be compact. Then* $A^\times$ *is compact.*

We ask the reader to prove this in the special case where X and Y are Hilbert spaces (Problem 8.44).

8.5.1 The Spectrum of a Compact Operator

In this section we prove a number of results about the spectrum of a compact operator. Since compact operators are bounded, the spectrum of a compact operator has all of the properties described in Section 8.3.1. Of course, with the added hypothesis of compactness, we can say a good bit more. We restrict ourselves to the case of operators on Hilbert space, though many of the results we give can be generalized to operators on Banach spaces.

In Hilbert spaces we can make use of the projection theorem and its consequences. In order to make use of this, we begin with a description of the spectrum of an operator of finite rank.

Lemma 8.90. *Suppose* $A \in \mathcal{L}(H)$ *has finite rank. Then for every* $\lambda \in \mathbb{C} \backslash \{0\}$ *exactly one of the following holds: either*

1. $\lambda \in \rho(A)$, or

2. $\lambda \in \sigma_p(A)$. In this case λ is an eigenvalue of finite multiplicity.

The proof follows directly from the corresponding result of linear algebra and is left to the reader (Problem 8.40).

We now prove a slightly different version of the Fredholm alternative theorem for operators of finite rank. This version is really just a technical result which will be useful in proving the analytic Fredholm theorem below.

Lemma 8.91. *Let* $G \subset \mathbb{C}$ *be a domain, and suppose*

$$\mathbb{C} \supset G \ni \lambda \mapsto F(\lambda) \in \mathcal{L}(H) \quad (8.121)$$

is analytic in G. Further suppose that, for every $\lambda \in G$, $F(\lambda)$ is of finite rank and that

$$\mathcal{R}(F(\lambda)) \subseteq M, \tag{8.122}$$

where M is a finite-dimensional subspace of H, independent of λ. Then either

1. $(I - F(\lambda))^{-1}$ exists for no $\lambda \in G$, or

2. $(I - F(\lambda))^{-1}$ exists for every $\lambda \in G\backslash S$ where S is a discrete set in G (i.e., it has no limit point in G). In this case the function $\lambda \to (I - F(\lambda))^{-1}$ is analytic on $G\backslash S$, and if $\lambda \in S$, then $F(\lambda)\phi = \phi$ has a finite-dimensional family of solutions.

Proof. Let $\{\psi_i\}_{i=1}^N$ be a basis for M. Then there are analytic vector functions

$$G \ni \lambda \to \gamma_i(\lambda) \in H, \quad i = 1, \ldots, N, \tag{8.123}$$

such that

$$F(\lambda)\phi = \sum_{i=1}^n (\gamma_i(\lambda), \phi)\psi_i. \tag{8.124}$$

Let $\Lambda(\lambda)$ be the $N \times N$ matrix with components

$$\Lambda_{ij}(\lambda) = (\gamma_j(\lambda), \psi_i). \tag{8.125}$$

The reader should verify that $F(\lambda)\phi = \phi$ has a nontrivial solution if and only if

$$d(\lambda) := \det (I - \Lambda(\lambda)) = 0. \tag{8.126}$$

However, $d(\lambda)$ is analytic on G. Hence, by a standard result of complex variables, either d is identically zero in G, or the zeros of d form a discrete set.

Since the range of F is finite-dimensional, so is the solution space of $F(\lambda)\phi = \phi$. This completes the proof. $\qquad\square$

We now prove a result which is sometimes called the *analytic Fredholm theorem*. This is the basis for two important results: the Fredholm alternative theorem and the Hilbert-Schmidt theorem.

Theorem 8.92 (Analytic Fredholm theorem). *Let $G \subset \mathbb{C}$ be a domain. Suppose the mapping*

$$\mathbb{C} \supset G \ni \lambda \mapsto B(\lambda) \in \mathcal{L}(H) \tag{8.127}$$

is analytic on G and that $B(\lambda)$ is compact at each $\lambda \in G$. Then, either

1. $(I - B(\lambda))^{-1}$ exists for no $\lambda \in G$, or

2. $(I - B(\lambda))^{-1}$ *exists for every* $\lambda \in G \backslash S$ *where* S *is a discrete set in* G *(i.e., it has no limit point in* G*). In this case the function* $\lambda \to$ $(I - B(\lambda))^{-1}$ *is analytic on* $G \backslash S$, *and if* $\lambda \in S$, *then* $B(\lambda)\psi = \psi$ *has a finite-dimensional family of solutions.*

Proof. We give the proof only in a neighborhood of a point $\lambda_0 \in G$. Standard connectedness arguments can be used to extend the result to all of G.

Let $\lambda_0 \in G$ be given. Since $\lambda \mapsto B(\lambda)$ is continuous, we can choose $r > 0$ such that

$$\|B(\lambda) - B(\lambda_0)\| < \frac{1}{2} \tag{8.128}$$

for all λ in the disk $D_r = \{\lambda \in G \mid |\lambda - \lambda_0| < r\}$. Using the construction of Theorem 8.87, we see that there is an operator of finite rank B_N such that

$$\|B_N - B(\lambda_0)\| < 1/2. \tag{8.129}$$

Now, using the geometric series techniques of the proof of Theorem 8.51, the reader can verify that

$$(I - B(\lambda) + B_N)^{-1} \tag{8.130}$$

exists as a bounded operator and is analytic on D_r.

Now let

$$F(\lambda) := B_N \circ (I - B(\lambda) + B_N)^{-1}. \tag{8.131}$$

Note that

$$(I - B(\lambda)) = (I - F(\lambda))(I - B(\lambda) + B_N). \tag{8.132}$$

Thus $I - B(\lambda)$ is invertible if and only if $I - F(\lambda)$ is. However, F has finite rank, so, by Lemma 8.91, $I - F(\lambda)$ is either invertible at no $\lambda \in G$ or is invertible off of a discrete $S \subset G$.

The proof that the solution space of $B(\lambda)\psi = \psi$ is finite-dimensional follows from the compactness of $B(\lambda)$ and is left to the reader (Problem 8.41). This completes the proof. $\square$

We now use the analytic Fredholm theorem to derive the following characterization of the spectrum of a compact operator on a Hilbert space.

Theorem 8.93 (Fredholm alternative theorem). *Let* $A \in \mathcal{L}(H)$ *be compact. Then* $\sigma(A)$ *is a compact set having no limit point except perhaps* $\lambda = 0$. *Furthermore, given any* $\lambda \in \mathbb{C} \backslash \{0\}$, *either*

1. $\lambda \in \rho(A)$, *or*

2. $\lambda \in \sigma_p(A)$ *is an eigenvalue of finite multiplicity.*

Proof. Let $G = \mathbb{C} \backslash \{0\}$ and

$$B(\lambda) = \frac{1}{\lambda} A. \tag{8.133}$$

Then note that

$$(\lambda I - A)^{-1} = \frac{1}{\lambda}\left(I - \frac{1}{\lambda}A\right)^{-1} = \frac{1}{\lambda}(I - B(\lambda))^{-1}. \qquad (8.134)$$

The result follows directly from Theorem 8.91.   □

We can use these results to prove the following eigenfunction expansion theorem. This will prove very useful in solving elliptic boundary-value problems.

Theorem 8.94 (Hilbert-Schmidt theorem). *Let H be a Hilbert space and let $A \in \mathcal{L}(H)$ be compact, self-adjoint operator.*

Then there is a sequence of nonzero real eigenvalues $\{\lambda_i\}_{i=1}^N$ with N equal to the rank of the operator A, such that $|\lambda_i|$ is monotone nonincreasing, and if $N = \infty$,

$$\lim_{i \to \infty} \lambda_i = 0. \qquad (8.135)$$

Furthermore, if each eigenvalue of A is repeated in the sequence according to its multiplicity, then there exists an orthonormal set $\{\phi_i\}_{i=1}^N$ of corresponding eigenfunctions; i.e.,

$$A\phi_i = \lambda_i\phi_i. \qquad (8.136)$$

Moreover, $\{\phi_i\}_{i=1}^N$ is an orthonormal basis for $\mathcal{R}(A)$; and A can be represented by

$$Au = \sum_{i=1}^N \lambda_i(\phi_i, u)\phi_i. \qquad (8.137)$$

Proof. Note that by Theorem 8.70, the eigenvalues are real of A are real since A is self-adjoint. By the Fredholm alternative theorm, the nonzero eigenvalues are discrete, bounded, and have finite multiplicity. Thus, we can list them (repeating according to multiplicity) in a sequence $\{\lambda_i\}_{i=1}^N$ of decreasing absolute value, with N possibly infinite. Since the eigenvalues can have no accumulation point other than zero, (8.135) must hold if N is infinite.

We now choose an orthonormal basis for the eigenspace corresponding to each distinct nonzero eigenvalue, and use the collection of these bases (numbered according to the eigenvalue to which they correspond) to make up the sequence $\{\phi_i\}_{i=1}^N$. By Theorem 8.70, the entire set is orthonormal.

Let M be the closure of the span of $\{\phi_i\}_{i=1}^N$. We claim that $M \supseteq \mathcal{R}(A)$. Note that since A is self-adjoint, both M and $M^\perp$ are invariant under A. Let $\hat{A}$ be the restriction of A to $M^\perp$. The operator $\hat{A} \in \mathcal{L}(M^\perp)$ is self-adjoint and compact since A is. Thus, by Theorem 8.93, any nonzero spectral value of $\hat{A}$ is an eigenvalue. However, any eigenvalue of $\hat{A}$ is also an eigenvalue of A. Thus, the spectral radius of $\hat{A}$ is zero. By Problem 8.33, this implies that $\hat{A}$ is the zero operator. Thus, every element of $M^\perp$ is an eigenvector corresponding to the eigenvalue 0. Thus, $M^\perp = \mathcal{N}(A)$ and $\{\phi_i\}_{i=1}^{N}$ forms a basis for $\mathcal{R}(A)$.

Now, since $\{\phi_i\}_{i=1}^{N}$ forms a basis for $\mathcal{R}(A)$, we have

$$
\begin{aligned}
A(u) &= \sum_{i=1}^{N} (\phi_i, A(u))\phi_i \\
&= \sum_{i=1}^{N} (A(\phi_i), u) \\
&= \sum_{i=1}^{N} \lambda_i (\phi_i, u)\phi_i.
\end{aligned}
$$

This completes the proof. $\qquad\qquad\qquad\qquad\qquad\qquad\qquad$ □

The following important corollary gives us us a method for solving the nonhomogeneous problem.

Corollary 8.95. *Let $A \in \mathcal{L}(H)$ be a compact, self-adjoint operator, and let $\{\lambda_i\}_{i=1}^{N}$ be the nonzero eigenvalues and $\{\phi_i\}_{i=1}^{N}$ the corresponding eigenfunctions as describen in the previous theorem. For any $f \in H$ let*

$$
f_{\mathcal{N}(A)} := f - \sum_{i=1}^{N} (\phi_i, f)\phi_i \tag{8.138}
$$

be the projection of f onto the nullspace of A. Then the following alternative holds for the nonhomogeneous problem

$$
Au - \lambda u = f, \tag{8.139}
$$

for $\lambda \neq 0$. Either

1. *λ is not an eigenvalue of A, in which case (8.139) has the unique solution*

$$
u = \sum_{i=1}^{\infty} \frac{(\phi_i, f)}{\lambda_i - \lambda}\phi_i - \frac{1}{\lambda}f_{\mathcal{N}(A)}; \quad or \tag{8.140}
$$

2. *λ is an eigenvalue of A. In this case, we let J be the finite index set of natural numbers j such that $\lambda_j = \lambda$. Then (8.139) has a solution if and only if*

$$
(\phi_j, f) = 0 \quad for\ all\ j \in J. \tag{8.141}
$$

In this case (8.139) has a family of solutions

$$
u = \sum_{j \in J} c_j \phi_j + \sum_{i \in \mathbb{N}\backslash J} \frac{(\phi_i, f)}{\lambda_i - \lambda}\phi_i - \frac{1}{\lambda}f_{\mathcal{N}(A)}, \tag{8.142}
$$

where $\{c_j\}_{j \in J}$ are arbitrary constants.

Proof. The proof of this follows immediately from the Fredholm alternative and Hilbert-Schmidt theorems by writing

$$u = \sum_{i=1}^{N} (\phi_i, u)\phi_i + u_{N(A)}, \tag{8.143}$$

expanding (8.139), and equating coefficients. The details are left to the reader.   □

Problems

8.39. Let $A \in \mathcal{L}(X)$ be compact and let $B \in \mathcal{L}(X)$ be bounded. Show that AB and BA are compact.

8.40. Prove Lemma 8.90. Use appropriate results from linear algebra.

8.41. Let $A \in \mathcal{L}(X)$ be compact. Show that for $\lambda \neq 0$ the solution space of $A\phi = \lambda\phi$ is finite-dimensional.

8.42. Let $A \in \mathcal{L}(H)$ be compact. Show that there exist orthonormal sets $\{\psi_i\}_{i=1}^{N}$ and $\{\phi_i\}_{i=1}^{N}$ and positive real numbers $\{\lambda_i\}_{i=1}^{N}$ (here N may be finite or infinite) such that

$$A(u) = \sum_{i=1}^{N} \lambda_i (\psi_i, u)\phi_i. \tag{8.144}$$

Hint: A^*A is compact and self-adjoint.

8.43. Show that if a compact operator $(\mathcal{D}(A), A)$ from X to Y is extended using the methods defined in case 1 and case 2 of the proof of Theorem 8.7, the extension is also a compact operator.

8.44. Let H be a Hilbert space. Prove Theorem 8.89 in the case where $X = Y = H$. Hint: Use Theorem 8.87.

8.45. We say that B is compact relative to A if $\mathcal{D}(B) \supseteq \mathcal{D}(A)$ and if Bx_n has a convergent subsequence whenever $x_n \in \mathcal{D}(A)$ and $\|Ax_n\|_Y + \|x_n\|_X$ is bounded. Assume that A is closed, B is closable and that B is compact relative to A. Show that B is bounded relative to A and the constant a in Problem 8.23 can be made arbitrarily small. Hint: Try to imitate the proof of Ehrling's lemma 7.30.

8.46. Prove the following results due to F. Riesz. Let S_1 and S_2 be subspaces of a normed linear space. Suppose that S_1 is closed and that S_1 is a proper subset of S_2. Then for every $\theta \in (0, 1)$ there is an $x \in S_2$ such that $\|x\| = 1$ and

$$\|x - y\| \geq \theta \quad \text{for all } y \in S_1. \tag{8.145}$$

Hint: Let $S_2 \ni w \notin S_1$, and let $d = \text{dist}(w, S_1)$. Show that there exists $v \in S_1$ such that

$$d \leq \|w - v\| \leq \frac{d}{\theta}. \tag{8.146}$$

8.47. Show that the unit ball in a normed space X is compact if and only if X is finite-dimensional. Hint: Use Problem 8.46.

8.6 Sturm-Liouville Boundary-Value Problems

We now study a class of second-order ODE boundary-value problems which arise from separation of variables. A **Sturm-Liouville problem** (or S-L problem) involves the ordinary differential equation

$$-\frac{d}{dx}\left(p(x)\frac{du}{dx}(x)\right) + q(x)u(x) - \lambda w(x)u(x) = f(x) \tag{8.147}$$

on the interval (a, b) and appropriate boundary conditions which we describe below. We assume the following:

1. The functions p, p', q and w are real-valued and continuous on the open interval (a, b).

2. The functions p and w are positive on (a, b).

We say the S-L problem is **regular** if both a and b are finite and assumptions 1 and 2 hold on the *closed* interval $[a, b]$. If not, we say the problem is **singular**.

We formally define the differential operator

$$Lu := \frac{1}{w}[-(pu')' + qu], \tag{8.148}$$

and we note that (8.147) can now be written in the form

$$Lu - \lambda u = \frac{f}{w}. \tag{8.149}$$

We intend to use the theory just developed above to analyze this as an eigenvalue problem for an operator from the weighted space $L_w^2(a, b)$ to itself. However, since the analysis of singular problems emphasizes methods other than those we have described we will discuss only regular problems.

We use the weighted space $L_w^2(a, b)$, but in regular problems this is really nothing more than a notational convenience since

$$\min_{x \in [a,b]} w(x)\|u\|_{L^2(a,b)}^2 \leq \|u\|_{L_w^2(a,b)}^2 \leq \max_{x \in [a,b]} w(x)\|u\|_{L^2(a,b)}^2, \tag{8.150}$$

so that the L^2 and L_w^2 norms are equivalent. We will use this to define domains for the operator L. We will examine the most common type of boundary conditions for S-L problems encountered in applications, namely, those of **unmixed type.** We require

$$\cos \alpha u(a) - \sin \alpha u'(a) = 0, \qquad (8.151)$$
$$\cos \beta u(b) + \sin \beta u'(b) = 0. \qquad (8.152)$$

We now define the domain

$$\mathcal{D}(L) := \{u \in H^2(a,b) \mid (8.151) \text{ and } (8.152) \text{ are satisfied}\}. \qquad (8.153)$$

We now prove the following theorem.

Theorem 8.96. *Let $(\mathcal{D}(L), L)$ be defined by (8.148) and (8.153). The following hold:*

1. *The eigenvalues of $(\mathcal{D}(L), L)$ are real.*

2. *The eigenvalues of $(\mathcal{D}(L), L)$ are bounded below by a constant $\lambda_G \in \mathbb{R}$.*

3. *Eigenfunctions corresponding to distinct eigenvalues are mutually orthogonal in $L_w^2(a,b)$.*

4. *Each eigenvalue has multiplicity one.*

Proof. To begin, we integrate by parts to prove Lagrange's identity; i.e., that for every u and v is $H^2(a,b)$ we have

$$(Lu, v)_w - (u, Lv)_w = p(x)[u(x)v'(x) - u'(x)v(x)]|_a^b. \qquad (8.154)$$

Thus, if u and v are in $\mathcal{D}(L)$ we can use the boundary conditions (8.151) and (8.152) to get

$$(Lu, v)_w = (u, Lv)_w, \qquad (8.155)$$

proving that $(\mathcal{D}(L), L)$ is symmetric. Hence, Theorem 8.70 immediately gives us parts 1 and 3.

To prove part 2 we prove an energy estimate of the form

$$(Lu, u)_w \geq \lambda_G \|u\|_{L_w^2(a,b)}^2 \qquad (8.156)$$

for all $u \in \mathcal{D}(L)$. (This is an analogue of Gårding's inequality in elliptic PDEs (cf. Section 9.2.3), hence the notation λ_G.) We prove this only in the case $\tan \alpha, \tan \beta \in [0, \infty)$ and leave the proof of other cases to the reader

(Problem 8.50). For any $u \in \mathcal{D}(L)$ we have

$$
\begin{aligned}
(Lu, u) &= \int_a^b -(pu')'u + q|u|^2 \, dx \\
&= \int_a^b p|u'|^2 + q|u|^2 \, dx + p(a)u'(a)u(a) - p(b)u'(b)u(b) \\
&= \int_a^b p|u'|^2 + q|u|^2 \, dx + p(a)\tan\alpha|u'(a)|^2 + p(b)\tan\beta|u'(b)|^2 \\
&\geq \int_a^b q(x)|u(x)|^2 \, dx \\
&\geq \frac{\min_{x\in[a,b]} q(x)}{\max_{x\in[a,b]} w(x)} \|u\|_{L_w^2(a,b)}^2 .
\end{aligned}
$$

To get part 2 we simply observe that for any eigenvalue λ we have

$$(Lu, u)_w = (\lambda wu, u) = \lambda(u, u)_w. \tag{8.157}$$

Hence $\lambda \geq \lambda_G$.

Part 4 follows immediately from the uniqueness theorem for initial-value problems for ODEs, which implies that either of the boundary conditions (8.151) or (8.152) determines a solution of the homogeneous ODE $Lu = \lambda u$ up to a multiplicative constant. $\square$

We can prove the following result using Green's functions and the theory of compact operators.

Theorem 8.97. *Let $(\mathcal{D}(L), L)$ be defined by (8.148) and (8.153). The following hold:*

1. *The spectrum consists entirely of eigenvalues.*

2. *The eigenvalues are countable and can be listed in a sequence*

$$\lambda_1 < \lambda_2 < \cdots < \lambda_n < \cdots \tag{8.158}$$

   with

$$\lim_{n\to\infty} \lambda_n = \infty. \tag{8.159}$$

3. *The set of normalized eigenfunctions $\{\phi_i\}$ such that $\phi_i \in \mathcal{N}(L_{\lambda_i})$ (where $L_\lambda := L - \lambda I$) is an orthonormal basis for $L_w^2(a,b)$.*

4. *For the equation*

$$Lu - \lambda u = f, \tag{8.160}$$

   exactly one of the following alternatives hold:

   (a) *If λ is not an eigenvalue of $(\mathcal{D}(L), L)$, then (8.160) has a unique solution in $\mathcal{D}(L)$ for every $f \in L_w^2(a,b)$. This solution is given* by

$$u(x) = \sum_{i=1}^{\infty} \frac{(\phi_i, f)_w}{\lambda_i - \lambda} \phi_i(x). \qquad (8.161)$$

(b) If $\lambda = \lambda_j$ is an eigenvalue of $(\mathcal{D}(L), L)$, then (8.160) has a solution in $\mathcal{D}(L)$ provided

$$(\phi_j, f) = 0. \qquad (8.162)$$

In this case there is a one-parameter family of solutions given by

$$u(x) = C\phi_j(x) + \sum_{\substack{i=1 \\ i \neq j}}^{\infty} \frac{(\phi_i, f)_w}{\lambda_i - \lambda} \phi_i(x). \qquad (8.163)$$

Proof. We begin by constructing a Green's function for the ODE

$$-pu'' - p'u' + (q - \lambda w)u = 0 \qquad (8.164)$$

and the boundary conditions (8.151) and (8.152). Let $v_l(x; \lambda)$ satisfy the ODE (8.164) and the left-hand boundary condition (8.151) and $v_r(x; \lambda)$ satisfy (8.164) and the right-hand end condition. We assume that λ is not an eigenvalue. In this case, the ODE uniqueness theorem for initial-value problems implies that v_l and v_r are linearly independent.

The Green's functions will have the form

$$g(x, y; \lambda) := \begin{cases} a_l(y; \lambda)v_l(x; \lambda), & a < x < y \\ a_r(y; \lambda)v_r(x; \lambda), & y < x < b. \end{cases} \qquad (8.165)$$

Thus $g(\cdot, y; \lambda)$ satisfies (8.151) and (8.152). To get

$$-p\frac{\partial^2}{\partial x^2}g(x, y; \lambda) - p'\frac{\partial}{\partial x}g(x, y; \lambda) + (q - \lambda w)g(x, y; \lambda) = \delta(x - y) \quad (8.166)$$

we require

$$\begin{aligned} a_l(y; \lambda)v_l(y; \lambda) - a_r(y; \lambda)v_r(y; \lambda) &= 0, \\ a_l(y; \lambda)v_l'(y; \lambda) - a_r(y; \lambda)v_r'(y; \lambda) &= \frac{1}{p(y)}. \end{aligned}$$

Solving this gives us

$$\begin{aligned} a_l(y; \lambda) &= (W(v_l(y; \lambda), v_r(y; \lambda))p(y))^{-1}v_r(y; \lambda), & (8.167) \\ a_r(y; \lambda) &= (W(v_l(y; \lambda), v_r(y; \lambda))p(y))^{-1}v_l(y; \lambda), & (8.168) \end{aligned}$$

where

$$W(v_l(y; \lambda), v_r(y; \lambda)) := v_l(y; \lambda)v_r'(y; \lambda) - v_l'(y; \lambda)v_r(y; \lambda) \qquad (8.169)$$

is the Wronskian of v_l and v_r. A classical ODE result called *Abel's formula* (which we do not prove) gives us

$$W(v_l(y; \lambda), v_r(y; \lambda))p(y) = C^{-1} \qquad (8.170)$$

where C is a constant. Thus, we have

$$g(x,y;\lambda) := \begin{cases} Cv_r(y;\lambda)v_l(x;\lambda), & a < x < y \\ Cv_l(y;\lambda)v_r(x;\lambda), & y < x < b. \end{cases} \tag{8.171}$$

Note that g is bounded, and since $w(x)$ is bounded, the kernel $w(y)g(x,y;\lambda)$ is Hilbert-Schmidt. Thus, the operator

$$L^2_w(a,b) \in u \mapsto K_\lambda u := \int_a^b w(y)g(\cdot,y;\lambda)u(y)\,dy \in L^2_w(a,b) \tag{8.172}$$

is bounded. A direct computation shows that $K_\lambda = R_\lambda(L)$. Since K_λ has dense domain, and is well defined if and only if λ is not an eigenvalue of $(\mathcal{D}(L), L)$, we have established 1.

By part 2 of the previous theorem we can assume without loss of generality that $\lambda = 0$ is not an eigenvalue (otherwise make an appropriate change in q). Let $g_0(x,y) := g(x,y;0)$. Note that the integral equation

$$u(x) = \lambda \int_a^b w(y)g_0(x,y)u(y)dy \tag{8.173}$$

has a nontrivial solution u if and only if λ is an eigenvalue of $(\mathcal{D}(L), L)$ and u a corresponding eigenfunction. By letting

$$v(x) := \sqrt{w(x)}u(x),$$
$$k(x,y) := \sqrt{w(x)}g_0(x,y)\sqrt{w(y)},$$

we can write (8.173) in the form

$$Gv(x) := \int_a^b k(x,y)v(y)dy = \frac{1}{\lambda}v(x). \tag{8.174}$$

Since k is symmetric and bounded, G is a self-adjoint, compact operator from $L^2(a,b)$ to $L^2(a,b)$. Eigenvalues of $(\mathcal{D}(L), L)$ are clearly reciprocals of the eigenvalues of G, and normalized eigenfunctions of L can be obtained from those of G via the formula $\phi_i(x) = w(x)^{-1/2}v_i(x)$. Furthermore, since for any $f \in L^2_w(a,b)$ we have $\sqrt{w}f \in L^2(a,b)$, so we can write

$$\sqrt{w(x)}f(x) = \sum_{i=1}^\infty (v_i, \sqrt{w}f)v_i(x) = \sqrt{w(x)} \sum_{i=1}^\infty (\phi_i, f)_w \phi_i(x). \tag{8.175}$$

Hence,

$$f(x) = \sum_{i=1}^\infty (\phi_i, f)_w \phi_i(x). \tag{8.176}$$

This proves 2 and 3.

Existence and multiplicity of solutions for the nonhomogeneous equation follows from the Fredholm alternative theorem and the equivalent integral formulations. To derive the expansion formulas, assume first that λ is not an eigenvalue. Let

$$u(x) = \sum_{i=1}^{\infty} (\phi_i, u)_w \phi_i(x) \tag{8.177}$$

and

$$f(x) = \sum_{i=1}^{\infty} (\phi_i, f)_w \phi_i(x). \tag{8.178}$$

Then

$$(L - \lambda I)u = \sum_{i=1}^{\infty} (\phi_i, u)_w (\lambda_i - \lambda)\phi_i(x) = \sum_{i=1}^{\infty} (\phi_i, f)_w \phi_i(x). \tag{8.179}$$

Equating coefficients give us (8.161). A similar computation yields (8.163). This completes the proof.   □

Example 8.98. Consider the Sturm-Liouville eigenvalue problem consisting of the differential equation

$$-u'' - \lambda u = 0 \tag{8.180}$$

and the boundary conditions

$$u(0) = 0, \tag{8.181}$$
$$\cos \beta u(1) + \sin \beta u'(1) = 0. \tag{8.182}$$

Without loss of generality, we assume that $-\frac{\pi}{2} < \beta \le \frac{\pi}{2}$.

This equation arises from (for instance) the separation of variables of a one-dimensional heat conduction problem modeling a long, thin rod that is insulated on the sides and with one end held at a fixed temperature and the other radiating heat.

Depending on the sign of λ we get the following two-parameter families of real solutions for (8.180):

$$y(x) = \begin{cases} A \cosh \sqrt{-\lambda}x + B \sinh \sqrt{-\lambda}x, & \lambda < 0 \\ A + Bx, & \lambda = 0 \\ A \cos \sqrt{\lambda}x + B \sin \sqrt{\lambda}x, & \lambda > 0. \end{cases} \tag{8.183}$$

Applying the boundary condition (8.181) implies

$$y(x) = \begin{cases} B \sinh \sqrt{-\lambda}x, & \lambda < 0 \\ Bx, & \lambda = 0 \\ B \sin \sqrt{\lambda}x, & \lambda > 0. \end{cases} \tag{8.184}$$

We now analyze the second boundary condition.

1. For $\lambda < 0$ we let $\omega = \sqrt{-\lambda}$. Then (8.182) becomes

$$B(\cos \beta \sinh \omega + \omega \sin \beta \cosh \omega) = 0. \tag{8.185}$$

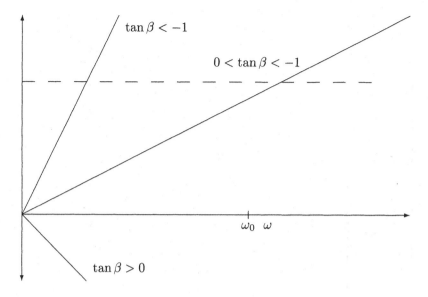

Figure 8.1. Solutions of $\tanh \omega = -\omega \tan \beta$.

Note that if $\beta = \pi/2$, there is no solution. Thus, there is a nontrivial solution if and only if there is an $\omega > 0$ satisfying

$$\tanh \omega = -\omega \tan \beta. \tag{8.186}$$

We see from Figure 8.1 that if $\tan \beta \in (-1, 0)$ (i.e., $\beta \in (-\pi/4, 0)$), there is exactly one solution ω_0, otherwise there is no solution. If a solution exists, it corresponds to the eigenvalue $\lambda_0 = -\omega_0^2$ and the eigenfunction $u_0(x) = \sinh \omega_0 x$.

2. For $\lambda = 0$, (8.182) becomes

$$B(\cos \beta + \sin \beta) = 0. \tag{8.187}$$

Thus, there is a nontrivial solution if and only if $\beta = -\pi/4$. In this case $\lambda_0 = 0$ is an eigenvalue and $u_0(x) = x$ is the corresponding eigenfunction.

3. For $\lambda > 0$ we let $\omega = \sqrt{\lambda}$. Then (8.182) becomes

$$B(\cos \beta \sin \omega + \omega \sin \beta \cos \omega) = 0. \tag{8.188}$$

For $\beta \in (-\pi/2, \pi/2)$ we have nontrivial solutions if and only if there is an $\omega > 0$ which satisfies

$$\tan \omega = -\omega \tan \beta. \tag{8.189}$$

From Figure 8.2 we see that for any $\beta \in (-\pi/2, \pi/2)$ this equation has an infinite family of solutions ω_n, $n = 1, 2, 3, \ldots$, such

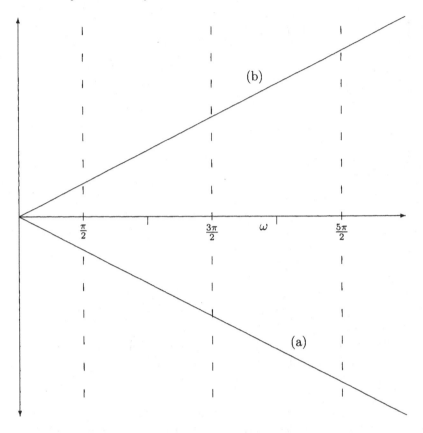

Figure 8.2. Solutions of $\tan \omega = -\omega \tan \beta$. (a) $\tan \beta > 0$. (b) $\tan \beta < 0$.

that $\lim_{n \to \infty} \omega_n = \infty$. Thus there is an infinite family of eigenvalues $\lambda_n = \omega_n^2$ with corresponding eigenfunctions $u_n(x) = \sin \omega_n x$, $n = 1, 2, 3, \ldots$.

If $\beta = \pi/2$, we solve (8.188) directly to get the eigenvalues $\lambda_n = (2n-1)^2 \pi^2/4$ with corresponding eigenfunctions $\sin[(2n-1)\pi x/2]$, $n = 1, 2, 3, \ldots$.

Problems

8.48. Find the eigenvalues and eigenfunctions for the following boundary-value problem:

$$-u'' - \lambda u = 0,$$

$$u'(0) = 0,$$

$$\cos \beta u(1) + \sin \beta u'(1) = 0.$$

8.49. Find the eigenvalues and eigenfunctions for the following boundary-value problem:

$$-(xu')' - \lambda xu = 0, \quad 0 < a < x < b;$$

$$u(a) = 0,$$
$$u(b) = 0.$$

8.50. Prove (8.156) for the case $\tan \alpha \in (-\infty, 0)$.

8.51. Consider the S-L problem (8.147) with Dirichlet boundary conditions. Let λ_n be the eigenvalues and let ϕ_n be the normalized eigenfunctions. Let $u \in L^2(a, b)$ have the expansion $u(x) = \sum_{n \in \mathbb{N}} \alpha_n \phi_n(x)$. Prove that

(a) $u \in H^2(a, b) \cap H_0^1(a, b)$ iff $\sum_{n \in \mathbb{N}} (1 + \lambda_n^2) |\alpha_n^2| < \infty$.

(b) $u \in H_0^1(a, b)$ iff $\sum_{n \in \mathbb{N}} (1 + |\lambda_n|) |\alpha_n|^2 < \infty$. Hint for (b): Consider the inner product $(u, Lu)_w$.

8.52. Let

$$\lambda = \min_{\substack{u \in H_0^1(a,b) \\ \|u\|_w = 1}} \int_a^b p(x)u'(x)^2 + q(x)u(x)^2 \, dx. \tag{8.190}$$

Prove that λ is the smallest eigenvalue of the S-L problem (8.147) with Dirichlet boundary conditions.

8.53. Obviously, the characterization of λ in the preceding problem can be used to derive upper and lower bounds for λ. Derive some such bounds.

8.7 The Fredholm Index

For many linear PDEs it is much easier to prove uniqueness than existence. For operators in a finite-dimensional vector space, it is well-known that uniqueness and existence are in fact equivalent; this is known as the Fredholm alternative. It is important to consider those operators in infinite dimensions for which a Fredholm alternative holds. We begin with a definition.

Definition 8.99. *Let X and Y be Banach spaces. We say that the operator $A \in \mathcal{L}(X, Y)$ is **semi-Fredholm** if $\mathcal{R}(A)$ is closed and if either $\mathcal{N}(A)$ is finite-dimensional or $\mathcal{R}(A)$ has a finite-dimensional complement in Y. If both are true, the operator is called **Fredholm**.*

We have restricted our definition to bounded operators. However, if A is unbounded, we can always regard it as a bounded operator defined on the Banach space $\mathcal{D}(A)$, where $\mathcal{D}(A)$ is equipped with the graph norm (cf. Problem 8.7).

The most important property of semi-Fredholm operators is a quantity called the Fredholm index.

Definition 8.100. *Let $A \in \mathcal{L}(X,Y)$ be semi-Fredholm. Then the dimension of $\mathcal{N}(A)$ is called the* **nullity** *of A and the dimension of the complement to $\mathcal{R}(A)$ is called the* **deficiency** *of A (If $\mathcal{R}(A)$ does not have a finite-dimensional complement, the deficiency is infinite.) The quantity*

$$\text{ind } A = \text{nul } A - \text{def } A \qquad (8.191)$$

is called the **(Fredholm) index** *of A.*

If A is Fredholm, the index is finite; otherwise it is either plus or minus infinity. The crucial theorem about semi-Fredholm operators is the following:

Theorem 8.101. *Let $A \in \mathcal{L}(X,Y)$ be semi-Fredholm. Then there exists $\epsilon > 0$ such that any $B \in \mathcal{L}(X,Y)$ with $\|B - A\| < \epsilon$ is also semi-Fredholm and, moreover,* ind $B =$ ind A.

The proof which we give works if either A is Fredholm or X and Y are Hilbert spaces. The difficulty in the general case is that a closed subspace of a Banach space does not necessarily have a closed complement; we refer to [Ka] for a proof in the general case. For the case when A is Fredholm, we note the following lemma:

Lemma 8.102. *Let X be a Banach space and assume that V is a closed subspace of X which is either finite-dimensional or of finite codimension. Then there is a closed subspace W of X such that $X = V \oplus W$.*

Proof. If V has finite codimension, we merely have to note that every finite-dimensional normed vector space is complete; hence every finite-dimensional subspace of X is closed. If V is finite-dimensional, let e_i, $i = 1, \ldots, n$, be a basis of V. By the Hahn-Banach theorem, we can construct linear functionals $x_i^* \in X^*$ such that $x_i^*(e_j) = \delta_{ij}$. We then define W to be the intersection of the nullspaces of the x_i^*. $\qquad \square$

In the following proof, we shall also have to use the fact that the direct sum of a closed subspace and a finite-dimensional subspace is closed; we leave the proof of this as an exercise (Problem 8.54).

We now proceed to the proof of the theorem assuming that either A is Fredholm or X and Y are Hilbert spaces.

Proof. Let V and W be subspaces of X and Y, respectively, so that $X = \mathcal{N}(A) \oplus V$, $Y = \mathcal{R}(A) \oplus W$. We define $\tilde{A} : V \times W \to Y$ by $\tilde{A}(v,w) = Av + w$. Then $\tilde{A}$ is bijective. Analogously, we define $\tilde{B}$ for some given operator $B \in \mathcal{L}(X,Y)$. If $\|B - A\|$ is sufficiently small, then the same is true for $\|\tilde{B} - \tilde{A}\|$, hence $\tilde{B}$ is bijective. In other words, the equation $Bv + w = y$ for given $y \in Y$ has a unique solution $v \in V$, $w \in W$. It follows immediately that $\mathcal{R}(B) + W = Y$, i.e., def $B \leq$ def A and that $\mathcal{N}(B) \cap V = \{0\}$, i.e., nul $B \leq$ nul A. Moreover, $B(V) = \{y \in Y \mid \tilde{B}^{-1}(y) \in V\}$ is closed and has finite codimension in $\mathcal{R}(B)$, since either $B(V)$ has finite codimension in Y or V has finite codimension in X. Hence $\mathcal{R}(B)$ is closed and B is semi-Fredholm.

Since either V has finite codimension in X or $B(V)$ has finite codimension in Y, we conclude that $V + \mathcal{N}(B)$ has finite codimension in X. Let now Z be a space such that $X = V \oplus \mathcal{N}(B) \oplus Z$. Then we have $Y = B(V) \oplus W$ and $\mathcal{R}(B) = B(V) \oplus B(Z)$, i.e., def $B = \dim W - \dim B(Z) = \dim W - \dim Z$ and nul $B = \dim \mathcal{N}(A) - \dim Z$; the equality of the indices follows immediately. $\qquad \square$

We have the following corollary of Theorem 8.101:

Corollary 8.103. *The set of all semi-Fredholm operators is open in $\mathcal{L}(X, Y)$. Moreover, the index is constant on each connected component.*

Hence, if $A(t) \in \mathcal{L}(X, Y)$ is semi-Fredholm for every t and depends continuously on t for $t \in [a, b]$, then the index is independent of t. This is often useful in applications, where the index of $A(a)$ may be easier to find than that of $A(b)$. One application of this approach is the following theorem.

Theorem 8.104. *Let $A \in \mathcal{L}(X, Y)$ be semi-Fredholm and let $k \in \mathcal{L}(X, Y)$ be compact. Then $A + k$ is semi-Fredholm and ind $A + k = $ ind A.*

Proof. We first consider the special case where $Y = X$ and $A = I$. We have to show that $I + k$ is Fredholm with index zero. Let $x_n \in \mathcal{N}(I + k)$ with $\|x_n\| \leq 1$. After taking a subsequence, we may assume that kx_n converges, and since $x_n + kx_n = 0$, we conclude that x_n converges. Hence the unit ball in $\mathcal{N}(I + k)$ is compact, which implies that $\mathcal{N}(I + k)$ is finite-dimensional.

We next show that $\mathcal{R}(I+k)$ is closed. Let V be a complement of $\mathcal{N}(I+k)$, then $I + k$ is a bijection from V to $\mathcal{R}(I + k)$. If we show that the inverse is bounded, then $\mathcal{R}(I + k)$ is isomorphic to V, hence complete. Suppose now that the inverse is not bounded, i.e., there is a sequence $x_n \in V$ such that $\|x_n\| = 1$, but $x_n + kx_n \to 0$. Then we may again take a subsequence and assume that kx_n converges, say to y. It follows that $x_n \to -y$ and that $y + ky = 0$, i.e., $y \in \mathcal{N}(I + k)$. But this is a contradiction, since, on the other hand, $y \in V$ and $\|y\| = 1$.

We have thus proved that $I + k$ is semi-Fredholm. That the index is zero follows by considering the family $\{I + tk \mid t \in [0, 1]\}$.

For the general case, we shall again assume that either A is Fredholm or that X and Y are Hilbert spaces. Let $X = V \oplus \mathcal{N}(A)$, $Y = W \oplus \mathcal{R}(A)$. Let $S \in \mathcal{L}(Y, X)$ be defined as follows: If $y \in \mathcal{R}(A)$, then Sy is the unique $x \in V$ such that $Ax = y$, and if $y \in W$, then $Sy = 0$. If W is finite-dimensional, then AS is a finite rank perturbation of the identity, i.e., $AS - I$ is compact. Hence $(A + k)S - I$ is also compact, i.e., $(A + k)S$ is Fredholm with index zero. Hence $\mathcal{R}((A+k)S)$ and consequently $\mathcal{R}(A+k)$ has finite codimension.

If on the other hand $\mathcal{N}(A)$ is finite-dimensional, then $SA - I$ is a finite rank operator and $S(A + k)$ is Fredholm with index zero. Hence the nullspace of $S(A + k)$ and a fortiori the nullspace of $A + k$ are finite-dimensional. We need to show that the range of $A + k$ is closed. Let U be a complement in X to the linear span of $\mathcal{N}(A)$ and $\mathcal{N}(A + k)$. We shall show that $A + k|_U$ has a bounded inverse. This implies that $(A + k)(U)$ is complete, and since this space has finite codimension in $\mathcal{R}(A + k)$, the closedness of $\mathcal{R}(A+k)$ follows. Assume now that $x_n \in U$ with $\|x_n\| = 1$ and $(A + k)x_n \to 0$. After taking a subsequence, we may assume that kx_n and hence Ax_n converges. But $A|_U$ has a bounded inverse, hence x_n converges, say to y. As above, we find a contradiction, since $(A + k)y = 0$, but on the other hand $y \in U$ with $\|y\| = 1$.

Hence we have shown that $A + k$ is semi-Fredholm and the statement about the index follows again by considering the family $\{A + tk \mid t \in [0, 1]\}$. $\square$

Now let A be a densely defined, closed symmetric operator in a Hilbert space H. If $\lambda \in \mathbb{C}\backslash\mathbb{R}$, then $A - \lambda I$ has a bounded inverse; hence it is injective and has closed range. By Corollary 8.103, the deficiency of $A - \lambda I$ is constant in the upper and lower half-planes. Thus the Fredholm index of $A - \lambda I$ for λ in the upper half-plane is equal to minus the deficiency index γ^+ of A, and the Fredholm index in the lower half-plane is equal to $-\gamma^-$ (cf. Definition 8.67). By Theorem 8.68, A is self-adjoint if and only if both deficiency indices are zero.

Problems

8.54. Let V be a closed subspace of X and let W be a finite-dimensional subspace. Prove that $V + W$ is closed in X.

8.55. Let $c \in C([0, 1])$ be such that $c(x) \geq 0$. Prove that the equation $y'' - c(x)y = f(x)$ subject to boundary conditions $y(0) = y(1) = 0$ has a solution in $C^2([0, 1])$ for every given $f \in C([0, 1])$.

8.56. Let $H = L^2(0, \infty)$ and let $Au = iu'$ with domain $H_0^1(0, \infty)$. Find the deficiency indices of A and its adjoint.

8.57. Let A be closed, densely defined and symmetric and such that the resolvent set of A contains at least one real number. Prove that A is self-adjoint. If in addition A is positive definite, show that the negative real axis belongs to the resolvent set of A.

9
Linear Elliptic Equations

9.1 Definitions

We will begin our study of linear elliptic PDEs by considering differential operators of the form $L(\mathbf{x}, D)u = 0$ defined in (2.9). Note that by Definition 2.8 a differential operator of order m is elliptic at $\mathbf{x}_0$ if and only if

$$L^p(\mathbf{x}_0, \boldsymbol{\xi}) = \sum_{|\alpha|=m} a_\alpha(\mathbf{x})\boldsymbol{\xi}^\alpha \neq 0 \quad \text{for every } \boldsymbol{\xi} \in \mathbb{R}^n \backslash \{\mathbf{0}\}. \tag{9.1}$$

In fact, we can show the following.

Lemma 9.1. *If a linear partial differential operator L of order m is elliptic at $\mathbf{x}_0 \in \mathbb{R}^n$, $n > 1$, then m is an even integer $(m = 2k)$ and $\boldsymbol{\xi} \mapsto L^p(\mathbf{x}_0, \boldsymbol{\xi})$ takes on only one sign on $\boldsymbol{\xi} \neq \mathbf{0}$.*

Proof. By definition, $\boldsymbol{\xi} \mapsto L^p(\mathbf{x}_0, \boldsymbol{\xi})$ is continuous and takes on the value 0 only at $\boldsymbol{\xi} = \mathbf{0}$. Suppose $L^p(\mathbf{x}_0, \boldsymbol{\xi}_1) < 0$ and $L^p(\mathbf{x}_0, \boldsymbol{\xi}_2) > 0$, and then connect $\boldsymbol{\xi}_1$ and $\boldsymbol{\xi}_2$ using a path not going through $\mathbf{0}$. As noted, $L^p(\mathbf{x}_0, \boldsymbol{\xi})$ must vary continuously along the path, taking on the value 0. This is a contradiction.

It now follows that, for any $\boldsymbol{\xi} \in \mathbb{R}^n$,

$$L^p(\mathbf{x}_0, \boldsymbol{\xi}) \quad \text{and} \quad L^p(\mathbf{x}_0, -\boldsymbol{\xi}) = (-1)^m L^p(\mathbf{x}_0, \boldsymbol{\xi})$$

must have the same sign. This implies that m is even. $\qquad \square$

In light of this result, we will use the following somewhat restricted definition of an elliptic operator for the remainder of the chapter.

Definition 9.2. Let $\Omega \subseteq \mathbb{R}^n$ be a domain. We say that a linear partial differential operator

$$L(\mathbf{x}, D) = \sum_{|\alpha| \le 2k} a_\alpha(\mathbf{x}) D^\alpha \tag{9.2}$$

is **elliptic** in Ω if

$$(-1)^k \sum_{|\alpha|=2k} a_\alpha(\mathbf{x}) \boldsymbol{\xi}^\alpha > 0 \quad \text{for every } \mathbf{x} \in \Omega, \ \boldsymbol{\xi} \in \mathbb{R}^n \backslash \{\mathbf{0}\}. \tag{9.3}$$

We say that L is **uniformly elliptic** in Ω if there exists a constant $\theta > 0$ such that

$$(-1)^k \sum_{|\alpha|=2k} a_\alpha(\mathbf{x}) \boldsymbol{\xi}^\alpha \ge \theta |\boldsymbol{\xi}|^{2k} \quad \text{for every } \mathbf{x} \in \Omega, \ \boldsymbol{\xi} \in \mathbb{R}^n \backslash \{\mathbf{0}\}. \tag{9.4}$$

Example 9.3. The reader should recall the calculations of Chapter 2 which showed that the negative of the Laplacian $-\Delta$ (which is of order 2) and the Biharmonic Δ^2 (order 4) are uniformly elliptic with $\theta = 1$.

Example 9.4. A second-order operator in n space dimensions of the form

$$Lu = a_{ij}(\mathbf{x}) \frac{\partial^2 u}{\partial x_i \partial x_j} + b_i(\mathbf{x}) \frac{\partial u}{\partial x_i} + c(\mathbf{x}) u \tag{9.5}$$

is uniformly elliptic on a domain Ω provided there exists a constant θ such that

$$\boldsymbol{\xi}^T \mathbf{A}(\mathbf{x}) \boldsymbol{\xi} > \theta |\boldsymbol{\xi}|^2 \tag{9.6}$$

for every $\mathbf{x} \in \Omega$. Here $\mathbf{A}(\mathbf{x})$ is the $n \times n$ matrix with components $-a_{ij}(\mathbf{x})$.

In our discussion of existence and regularity theory below, it is convenient to put our differential operators in a form which is amenable to integration by parts.

Definition 9.5. We say that an operator is in **divergence form** if there are functions $a_{\sigma\gamma} : \Omega \to \mathbb{R}$ such that

$$L(\mathbf{x}, D)u = \sum_{0 \le |\sigma|, |\gamma| \le k} (-1)^{|\sigma|} D^\sigma (a_{\sigma\gamma}(\mathbf{x}) D^\gamma u). \tag{9.7}$$

Remark 9.6. Note that an operator in divergence form is elliptic if and only if

$$\sum_{|\sigma|, |\gamma|=k} \boldsymbol{\xi}^\sigma a_{\sigma\gamma}(\mathbf{x}) \boldsymbol{\xi}^\gamma > 0 \quad \text{for every } \mathbf{x} \in \Omega, \ \boldsymbol{\xi} \in \mathbb{R}^n \backslash \{\mathbf{0}\}, \tag{9.8}$$

and uniformly elliptic if and only if there exists $\theta > 0$ such that

$$\sum_{|\sigma|, |\gamma|=k} \boldsymbol{\xi}^\sigma a_{\sigma\gamma}(\mathbf{x}) \boldsymbol{\xi}^\gamma > \theta |\boldsymbol{\xi}|^{2k} \quad \text{for every } \mathbf{x} \in \Omega, \ \boldsymbol{\xi} \in \mathbb{R}^n \backslash \{\mathbf{0}\}. \tag{9.9}$$

If our coefficients are smooth enough, we can put a general PDE into divergence form. We give conditions for doing so here which are sufficient, though by no means necessary.

Lemma 9.7. *Let*

$$a_\alpha \in C_b^{|\alpha|-k}(\overline{\Omega}) \quad \text{for } k < |\alpha| \le 2k \tag{9.10}$$

and

$$a_\alpha \in C_b(\overline{\Omega}) \quad \text{for } |\alpha| \le k. \tag{9.11}$$

Then there exist $a_{\sigma\gamma} \in C_b^{|\sigma|}(\overline{\Omega})$ such that for every $u \in C^{2k}(\Omega)$ we have

$$L(\mathbf{x}, D)u = \sum_{|\alpha| \le 2k} a_\alpha(\mathbf{x}) D^\alpha$$

$$= \sum_{0 \le |\sigma|, |\gamma| \le k} (-1)^{|\sigma|} D^\sigma (a_{\sigma\gamma}(\mathbf{x}) D^\gamma u).$$

Proof. We do the proof here for general k, but the notation is rather cumbersome, so the reader should work through the details for the case $k = 1$.

For every $|\alpha| \le 2k$ we choose σ_α and γ_α satisfying $|\sigma_\alpha|, |\gamma_\alpha| \le k$, $\sigma_\alpha + \gamma_\alpha = \alpha$. This choice is, of course, not unique.

Now, for any $u \in C^{2k}(\Omega)$ and $\phi \in \mathcal{D}(\Omega)$ we have

$$\int_\Omega L(\mathbf{x}, D)u\, \phi\, d\mathbf{x} = \sum_{|\alpha| \le 2k} \int_\Omega (D^\alpha u) a_\alpha \phi\, d\mathbf{x}$$

$$= \sum_{|\alpha| \le 2k} \int_\Omega (D^{\sigma_\alpha + \gamma_\alpha} u) a_{(\sigma_\alpha + \gamma_\alpha)} \phi\, d\mathbf{x}$$

$$= \sum_{|\alpha| \le 2k} (-1)^{|\sigma_\alpha|} \int_\Omega (D^{\gamma_\alpha} u)(D^{\sigma_\alpha}[a_{(\sigma_\alpha + \gamma_\alpha)}\phi])\, d\mathbf{x}$$

$$= \sum_{|\alpha| \le 2k} (-1)^{|\sigma_\alpha|} \int_\Omega D^{\gamma_\alpha} u \sum_{\rho \le \sigma_\alpha} \binom{\sigma_\alpha}{\rho} D^{\sigma_\alpha - \rho} a_{(\sigma_\alpha + \gamma_\alpha)} D^\rho \phi\, d\mathbf{x}$$

$$= \sum_{\substack{|\alpha| \le 2k \\ \rho \le \sigma_\alpha}} (-1)^{|\sigma_\alpha| + |\rho|} \int_\Omega D^\rho \left[\binom{\sigma_\alpha}{\rho} D^{\sigma_\alpha - \rho} a_{(\sigma_\alpha + \gamma_\alpha)} D^{\gamma_\alpha} u \right] \phi\, d\mathbf{x}$$

$$=: \sum_{0 \le |\sigma|, |\gamma| \le k} \int_\Omega (-1)^{|\sigma|} D^\sigma (a_{\sigma\gamma}(\mathbf{x}) D^\gamma u) \phi\, d\mathbf{x}.$$

(Note that the last equality is a definition.) Since this holds for all $\phi \in \mathcal{D}(\Omega)$ we have our result. $\qquad\square$

Remark 9.8. Unless explicitly stated otherwise, we shall assume that our coeficients satisfy the smoothness assumptions (9.10) and (9.11).

When dealing with systems of PDEs (as opposed to single equations) we will not, in general, focus on a more restrictive definition of ellipticity. That is, we will stick to the definition of an elliptic system as one with no real characteristics, and characteristics are to be determined by the principal part of the operator defined using appropriate weights. However, the reader should be aware of some particularly important examples of elliptic systems that arise in the calculus of variations. Let Ω be a domain in $\mathbb{R}^n$ and let $\mathbf{u} : \Omega \to \mathbb{R}^N$. Consider the system of N second-order differential equations in divergence form

$$\frac{\partial}{\partial x_k}\left(A_{kl}^{IJ}(\mathbf{x})\frac{\partial u_J}{\partial x_l}\right) + c^{IJ}(\mathbf{x})u_J(\mathbf{x}) = 0, \quad I = 1,\ldots,N. \qquad (9.12)$$

Here, the coefficients A_{kl}^{IJ} and c^{IJ}, $I, J = 1,\ldots,N$, $k, l = 1,\ldots,n$, are assumed to be sufficiently smooth, and the summation convention is assumed to hold from 1 to N for repeated uppercase indices and from 1 to n for repeated lowercase indices.

Note that if

$$M_k^I A_{kl}^{IJ} M_l^J > 0 \qquad (9.13)$$

for every $\mathbf{x} \in \Omega$ and for every nonzero $N \times n$ matrix $\mathbf{M}$, then we can show that the system is elliptic simply by taking the "obvious" principal part (giving weight 1 to each of the equations and dependent variables). Condition (9.13) or the uniform version

$$M_k^I A_{kl}^{IJ} M_l^J \geq \theta|\mathbf{M}|^2 \qquad (9.14)$$

for every $\mathbf{x} \in \Omega$ and every nonzero $N \times n$ matrix $\mathbf{M}$ for some $\theta > 0$ is often given as a definition of an elliptic system. However, such a definition does not fit such systems as the Stokes system.

Another important "ellipticity condition" is the **Legendre-Hadamard** condition

$$\eta^I \xi_k A_{kl}^{IJ} \eta^J \xi_l > 0 \qquad (9.15)$$

for every $\mathbf{x} \in \Omega$ and for every nonzero $\boldsymbol{\eta} \in \mathbb{R}^N$ and $\boldsymbol{\xi} \in \mathbb{R}^n$. The uniform version states that there exists $\theta > 0$ such that for every $\mathbf{x} \in \Omega$

$$\eta^I \xi_k A_{kl}^{IJ} \eta^J \xi_l > \theta|\boldsymbol{\eta}|^2|\boldsymbol{\xi}|^2 \qquad (9.16)$$

for every nonzero $\boldsymbol{\eta} \in \mathbb{R}^N$ and $\boldsymbol{\xi} \in \mathbb{R}^n$. These conditions turn out to be more physically reasonable than (9.13) or (9.14) for many problems in elasticity.

Note that (9.15) and (9.16) are much weaker than the corresponding conditions (9.13) and (9.14). (The inequalities have to hold only for rank-1 $N \times n$ matrices.) Despite this, (9.15) and (9.16) are sometimes referred to as *strong ellipticity* conditions. As this example shows, the reader should be forewarned that the nomenclature surrounding elliptic systems does not necessarily make sense. More importantly, there is not universal agreement regarding these definitions. In reading the literature one needs to be careful to note the definitions various authors use.

9.2 Existence and Uniqueness of Solutions of the Dirichlet Problem

9.2.1 The Dirichlet Problem—Types of Solutions

We begin with a statement of the classical Dirichlet problem.

Definition 9.9. Let $\Omega \subset \mathbb{R}^n$ be a bounded domain and suppose $f \in C_b(\Omega)$ is given. A function

$$u \in C_b^{2k}(\Omega) \cap C_b^{2k-1}(\overline{\Omega})$$

is a **classical solution** of the Dirichlet problem if

$$L(\mathbf{x}, D)u = \sum_{0 \leq |\sigma|, |\gamma| \leq k} (-1)^{|\sigma|} D^\sigma (a_{\sigma\gamma}(\mathbf{x}) D^\gamma u) = f \qquad (9.17)$$

in Ω; and

$$D^\alpha u = 0 \quad \text{for } |\alpha| \leq k - 1 \qquad (9.18)$$

on $\partial\Omega$.

One of the most important ideas of the modern analysis is that if you want to guarantee the existence of a solution to a problem, it is usually easier to do so in a "bigger" space of functions. This is clearly the case with the classical Dirichlet problem. Although we might expect a solution to have all of the smoothness suggested in the statement of the problem, we must relax the conditions on the solution at first so that we can use the methods of the last three chapters. The first step in relaxing the conditions on the solution is to state the problem in terms of Sobolev spaces.

Definition 9.10. Let $\Omega \subset \mathbb{R}^n$ be a bounded domain and suppose $f \in L^2(\Omega)$ is given. A function

$$u \in H^{2k}(\Omega) \cap H_0^k(\Omega)$$

is a **strong solution** of the Dirichlet problem if

$$L(\mathbf{x}, D)u = \sum_{0 \leq |\sigma|, |\gamma| \leq k} (-1)^{|\sigma|} D^\sigma (a_{\sigma\gamma}(\mathbf{x}) D^\gamma u) = f \qquad (9.19)$$

in Ω.

Note the following.

1. We have relaxed the conditions not only on the solution u, but on the data f. The space $L^2(\Omega)$ is certainly the obvious space for f

once we have relaxed the conditions on u, so the additional generality (which includes such physically reasonable situations as discontinuous forcing functions) will come along "for free." (In fact, we will be able to weaken the conditions on f each time we relax the conditions on the solution, as we shall see below.)

2. For classical solutions, the differential equation (9.17) is taken to hold in a pointwise sense. For strong solutions, the differential equation (9.19) is understood either in terms of equivalence classes (the right and left sides of the equation represent the same equivalence class of sequences in the $L^2(\Omega)$ norm) or in an "almost everywhere" sense (for those who have studied measure theory.)

3. Instead of imposing boundary conditions (9.18) explicitly as we did in the classical problem, we have incorporated them into the space $H_0^k(\Omega)$ in the new problem.

4. By combining the previous observations we see that the new problem is indeed a generalization of the classical problem; i.e., any classical solution of the Dirichlet problem is also a strong solution.

We now take a further step in weakening the conditions on solutions of the Dirichlet problem: we state the problem in variational form. This is the same process which was used in discussing weak solutions of conservation laws in Chapter 3. The first step is to create a bilinear form from the differential operator L using integration by parts. Let $\phi \in \mathcal{D}(\Omega)$ and $u \in H^{2k}(\Omega)$, then

$$
\int_\Omega \phi L u \, d\mathbf{x} = \sum_{0 \leq |\sigma|, |\gamma| \leq k} (-1)^{|\sigma|} \int_\Omega \phi D^\sigma (a_{\sigma\gamma}(\mathbf{x}) D^\gamma u) \, d\mathbf{x}
$$

$$
= \sum_{0 \leq |\sigma|, |\gamma| \leq k} \int_\Omega a_{\sigma\gamma}(\mathbf{x}) D^\gamma u D^\sigma \phi \, d\mathbf{x}.
$$

(9.20)

With this in mind we define

$$
B[v, u] := \sum_{0 \leq |\sigma|, |\gamma| \leq k} \int_\Omega a_{\sigma\gamma}(\mathbf{x}) D^\gamma u D^\sigma v \, d\mathbf{x}
$$

(9.21)

to be the **bilinear form** associated with the elliptic partial differential operator L. Note that $B[v, u]$ is well defined for u and v that are merely in $H^k(\Omega)$. With this in mind, we give the following definition of yet another type of solution of the Dirichlet problem.

Definition 9.11. Let $\Omega \subset \mathbb{R}^n$ be a bounded domain and suppose $f \in H^{-k}(\Omega)$ is given. A function

$$
u \in H_0^k(\Omega)
$$

is a **weak solution** of the Dirichlet problem if

$$B[v, u] = f(v) \tag{9.22}$$

for every $v \in H_0^k(\Omega)$.

Remark 9.12. We can extend the bilinear form B on the real Hilbert space $H_0^k(\Omega)$ to be a *sesquilinear form* on the complex Hilbert space (also denoted $H_0^k(\Omega)$) by letting

$$B[v, u] := \sum_{0 \leq |\sigma|, |\gamma| \leq k} \int_\Omega a_{\sigma\gamma}(\mathbf{x}) D^\gamma u \overline{D^\sigma v} \, d\mathbf{x}. \tag{9.23}$$

We will be rather sloppy about the distinction between complex-valued and real-valued functions, using the same notation for the spaces and the bilinear forms. Since we are mainly interested in discussing real solutions to partial differential equations, we will only go to complex spaces when forced to, such as when using Fourier transforms or discussing spectral theory.

Note that by using the calculations of (9.20) (though this time with a function $v \in H_0^k(\Omega)$ in place of $\phi \in \mathcal{D}(\Omega)$) we can see that any strong solution of the Dirichlet problem is automatically a weak solution. However, since we require so much less smoothness of weak solutions than strong ones, it will be far easier to show that weak solutions exist. Once we have done this, we will be able to show that if Ω, f and the coefficients $a_{\sigma\gamma}$ are sufficiently "nice," the weak solution is, in fact, a strong solution or a classical solution.

Example 9.13. An important series of classical examples of Dirichlet problems come from electrostatics. Without going into any of the physics, let us assume that we wish to know a scalar quantity u called the *electrostatic potential* or more commonly the *voltage* in a domain $\Omega \subset \mathbb{R}^3$. We will assume that the boundary of the domain is grounded; i.e.,

$$u(\mathbf{x}) = 0 \quad \text{for } \mathbf{x} \in \partial\Omega. \tag{9.24}$$

Within Ω there is a distribution of charge $\rho : \Omega \to \mathbb{R}$, and this charge "generates" a voltage through the formula

$$-\Delta u = \rho. \tag{9.25}$$

The solution of this class of problems is the subject of classical potential theory, and many of the techniques of the classical theory (eigenfunction expansion, Green's functions) are included in the modern theory as well. However, the classical and modern theory share the same approach only when we are looking for classical ($\rho \in C_b(\Omega)$) or strong ($\rho \in L^2(\Omega)$ or piecewise continuous) solutions.

The classical and modern theories take a very different approach in dealing with charge distributions that occur on surfaces; i.e., situations where $S \subset \Omega$ is a smooth surface and a charge distribution $\omega : S \to \mathbb{R}$ is defined.

(We assume that either $\omega \in C_b(S)$ or $\omega \in L^2(S)$.) Of course, since ω is defined only on a surface (a set of measure zero to readers who have had measure theory) the differential equation $-\Delta u = \omega$ does not make sense either classically or as the identification of equivalence classes of sequences in $L^2(\Omega)$. In the modern theory the situation is very clear: although ω is not in $L^2(\Omega)$, we can use it to define a perfectly nice functional in $H^{-1}(\Omega)$ through the formula

$$(\omega, \phi) = \int_S \omega(\mathbf{x})\phi(\mathbf{x}) \, da(\mathbf{x}), \tag{9.26}$$

for every $\phi \in H_0^1(\Omega)$. (Recall that the trace theorem implies that $\phi \in L^2(S)$.) Here $da(\mathbf{x})$ indicates differential area at $\mathbf{x} \in S$. Thus, the modern theory would simply have us look for a (weak) solution $u \in H_0^1(\Omega)$ of the variational problem

$$B[v, u] := \int_\Omega \nabla v \cdot \nabla u \, d\mathbf{x} = (\omega, v) \tag{9.27}$$

for all $v \in H_0^1(\Omega)$. As we shall see below, this problem is well-posed. The classical theory solves this problem (and some other similar ones) using the theory of single and double layer potentials: essentially integral operators defined using singular surface integrals. As is so often the case, the classical theory lacks much of the conceptual unity of the modern theory, but provides much more detailed information in special (though often the most important) cases. We will not go into the results of classical potential theory in this book, but the reader is encouraged to read *Foundations of Potential Theory* by O.D. Kellogg [Ke] as a good starting point for more information on this subject.

9.2.2   The Lax-Milgram Lemma

The first tool that we will develop for deriving the existence theory for elliptic equations is commonly known as the Lax-Milgram lemma; though because of its importance we designate it as a theorem. The result is simply a generalization of the Riesz Representation Theorem to bilinear forms that need not be symmetric.

Theorem 9.14 (Lax-Milgram). *Let H be a Hilbert space and let*

$$B : H \times H \to \mathbb{R} \tag{9.28}$$

be a bilinear mapping. Suppose there exist positive constants c_1 and c_2 such that

$$|B[x, y]| \le c_1 \|x\|_H \|y\|_H \quad \text{for all } x \text{ and } y \text{ in } H \tag{9.29}$$

and

$$B[x, x] \ge c_2 \|x\|_H^2 \quad \text{for all } x \in H. \tag{9.30}$$

Then for every $f \in H^*$ there exists a unique $y \in H$ such that

$$B[x, y] = f(x) \quad \text{for all } x \in H. \tag{9.31}$$

Furthermore, there exists a constant C, independent of f, such that

$$\|y\|_H \leq C\|f\|_{H^*}. \tag{9.32}$$

Remark 9.15. A mapping B satisfying (9.30) for some $c_2 > 0$ is called **coercive**. The inequality (9.30) can be thought of as an energy estimate. (The inequality says that the energy (the norm squared) can only blow up as fast as the bilinear form).

Remark 9.16. Note that by (9.29) and (9.30)

$$\|x\|_B := \sqrt{B[x, x]} \tag{9.33}$$

is equivalent to the original norm on H. Furthermore, if B is symmetric, i.e.,

$$B[x, y] = B[y, x] \quad \text{for all } x \text{ and } y \text{ in } H, \tag{9.34}$$

then $B[x, y]$ defines a new inner product on H. Thus, in this case, the Riesz Representation Theorem directly implies that for every $f \in H^*$ there exists a unique $y \in H$ such that (9.31) holds. Therefore, the significance of the Lax-Milgram lemma is that it does not require B to be symmetric.

We now prove the Lax-Milgram lemma.

Proof. For every fixed $y \in H$, the mapping

$$H \in x \mapsto B[x, y] \in \mathbb{R} \tag{9.35}$$

is bounded and linear, i.e., an element of H^*. Thus, by the Riesz Representation Theorem there exists a unique $z \in H$ such that

$$B[x, y] = (x, z) \quad \text{for all } x \in H. \tag{9.36}$$

Since a unique $z \in H$ can be derived for each fixed $y \in H$ we can define a mapping $A : H \to H$ by

$$z =: A(y). \tag{9.37}$$

The question of existence of a solution of (9.31) is now translated to the question of the invertibility of A. That is, for any $f \in H^*$ let $z \in H$ be the unique element such that

$$(x, z) = f(x) \quad \text{for all } x \in H. \tag{9.38}$$

Then if for every $z \in H$ can we find a unique solution of $y \in H$ of $A(y) = z$, then y is the unique solution of

$$B[x, y] = (x, A(y)) = (x, z) = f(x) \quad \text{for all } x \in H. \tag{9.39}$$

We now note some basic properties of A.

1. A is linear. To see this note that

$$
\begin{aligned}
(x, A(\alpha_1 y_1 + \alpha_2 y_2)) &:= B[x, \alpha_1 y_1 + \alpha_2 y_2] \\
&= \alpha_1 B[x, y_1] + \alpha_2 B[x, y_2] \\
&= \alpha_1 (x, A(y_1)) + \alpha_2 (x, A(y_2)) \\
&= (x, \alpha_1 A(y_1) + \alpha_2 A(y_2)).
\end{aligned}
$$

Since this holds for arbitrary x, α_i and y_i we have shown linearity.

2. A is bounded. Using (9.29) we get

$$
\|A(y)\|^2 = (A(y), A(y)) = B[A(y), y] \leq c_1 \|A(y)\| \|y\|. \tag{9.40}
$$

Canceling, we get

$$
\|A(y)\| \leq c_1 \|y\|. \tag{9.41}
$$

3. The range of A is dense in H. To see this we use (9.30) to note that if $y \in \mathcal{R}(A)^\perp$, then

$$
c_2 \|y\|^2 \leq B[y, y] = (y, A(y)) = 0. \tag{9.42}
$$

4. A is bounded below. Using (9.30) again (now for arbitrary $y \in H$) we get

$$
c_2 \|y\|^2 \leq B[y, y] = (y, A(y)) \leq \|y\| \|A(y)\| \tag{9.43}
$$

or

$$
\|A(y)\| \geq c_2 \|y\|. \tag{9.44}
$$

5. Combining this with Problem 8.2 implies that the range of A, $\mathcal{R}(A)$, is closed. Since $\mathcal{R}(A)$ is dense, A is surjective.

It follows that A is invertible, which gives us the existence of a unique solution y. Finally, the estimate (9.32) follows from the Riesz representation theorem and the fact that A is bounded below. $\qquad\square$

9.2.3 Gårding's Inequality

We now prove the basic energy or coercivity estimate for the elliptic Dirichlet problem.

Theorem 9.17 (Gårding's inequality). *Let Ω be a bounded domain with the k-extension property. Let $L(\mathbf{x}, D)$ be a linear partial differential operator in divergence form of order $2k$ such that for some $\theta > 0$ the uniform ellipticity condition (9.9) holds. Also suppose that*

$$
a_{\sigma\gamma} \in C_b(\overline{\Omega}) \quad \text{for all } |\sigma| = |\gamma| = k \tag{9.45}
$$

and

$$
a_{\sigma\gamma} \in L^\infty(\Omega) \quad \text{for all } |\sigma|, |\gamma| \leq k. \tag{9.46}
$$

Then there exist constants c_3 and $\lambda_G \geq 0$ such that

$$B[u,u] + \lambda\|u\|_{L^2(\Omega)}^2 \geq c_3\|u\|_{H^k(\Omega)}^2 \quad \text{for all } u \in H_0^k(\Omega). \tag{9.47}$$

Proof. Let $u \in H_0^k(\Omega)$. We begin by splitting $B[u,u]$ into principal part and lower-order terms; i.e., we let

$$B[u,u] = I_1 + I_2, \tag{9.48}$$

where

$$I_1 := \sum_{|\sigma|=|\gamma|=k} \int_\Omega a_{\sigma\gamma}(\mathbf{x}) D^\gamma u D^\sigma u \, dx, \tag{9.49}$$

$$I_2 := \sum_{\substack{0 \leq |\sigma|,|\gamma| \leq k \\ |\sigma|+|\gamma| < 2k}} \int_\Omega a_{\sigma\gamma}(\mathbf{x}) D^\gamma u D^\sigma u \, dx. \tag{9.50}$$

We begin by estimating the lower-order terms. Let $\epsilon > 0$ be given.

$$|I_2| \leq C \sum_{\substack{0 \leq |\sigma|,|\gamma| \leq k \\ |\sigma|+|\gamma| < 2k}} \int_\Omega |D^\gamma u|\,|D^\sigma u| \, d\mathbf{x}$$

$$\leq C \sum_{\substack{0 \leq |\sigma|,|\gamma| \leq k \\ |\sigma|+|\gamma| < 2k}} \|D^\gamma u\|_2 \|D^\sigma u\|_2 \tag{9.51}$$

$$\leq C\|u\|_{k,2}\|u\|_{k-1,2}$$

$$\leq \frac{\epsilon}{2}\|u\|_{k,2}^2 + C(\epsilon)\|u\|_{k-1,2}^2.$$

Here we have used Hölder's inequality and the elementary inequality

$$ab \leq \varepsilon a^2 + \frac{1}{4\varepsilon}b^2. \tag{9.52}$$

Now let $\delta > 0$. We use the abstract version of Ehrling's lemma (Theorem 7.30) and the previous estimate to get

$$|I_2| \leq \frac{\epsilon}{2}\|u\|_{k,2}^2 + C(\epsilon)\left[\delta\|u\|_{k,2}^2 + c(\delta)\|u\|_2^2\right]. \tag{9.53}$$

Now, for any $\epsilon > 0$, we can choose $\delta = \delta(\epsilon) > 0$ sufficiently small that

$$C(\epsilon)\delta \leq \epsilon/2. \tag{9.54}$$

Combining this with the previous inequality gives us the estimate:

$$|I_2| \leq \epsilon\|u\|_{k,2}^2 + C(\epsilon)\|u\|_2^2. \tag{9.55}$$

We now estimate the principal part. We assert the fact that each function $a_{\sigma\gamma}$ can be extended to be a continuous function on all of $\mathbb{R}^n$. (We already know this to be true for Lipschitz domains since they have the k-extension property for any k. In fact, by the Tietze extension theorem (consult a topology text), it holds for any domain Ω.)

Now let Ω' be any bounded open domain such that Ω is compactly contained in Ω'. Since each extended $a_{\sigma\gamma}$ ($|\sigma| = |\gamma| = k$) is uniformly continuous on Ω', there exists a nondecreasing *modulus of continuity* function $\omega : [0, \infty) \to [0, \infty)$ satisfying

$$0 = \omega(0) = \lim_{\delta \to 0^+} \omega(\delta) \qquad (9.56)$$

and

$$|a_{\sigma\gamma}(\mathbf{x}) - a_{\sigma\gamma}(\mathbf{y})| \leq \omega(|\mathbf{x} - \mathbf{y}|) \qquad (9.57)$$

for every $|\sigma| = |\gamma| = k$ and every $\mathbf{x}, \mathbf{y} \in \Omega'$.

Now let $B = B(\mathbf{x}_0, \delta)$ for some $\mathbf{x}_0 \in \Omega$. We will choose $\delta > 0$ later, but for now we assume only that it is sufficiently small so that $B \subset \Omega'$. The first step in our estimate of I_1 is to do an estimate in the case where $u \in H_0^k(B)$. In this case we have

$$I_1 = I_{11} + I_{12}, \qquad (9.58)$$

where

$$I_{11} := \sum_{|\sigma|=|\gamma|=k} \int_{\mathbb{R}^n} a_{\sigma\gamma}(\mathbf{x}_0) D^\gamma u D^\sigma u \, dx, \qquad (9.59)$$

$$I_{12} := \sum_{|\sigma|=|\gamma|=k} \int_B [a_{\sigma\gamma}(\mathbf{x}) - a_{\sigma\gamma}(\mathbf{x}_0)] D^\gamma u D^\sigma u \, dx. \qquad (9.60)$$

(Note that in the definition of I_{11} we have assumed u is extended by 0 to all of $\mathbb{R}^n$.)

We can use (9.57) and Hölder's inequality to get an easy estimate for I_{12}:

$$|I_{12}| \leq \omega(\delta)\|u\|_{k,2}^2. \qquad (9.61)$$

To estimate I_{11} we use Fourier transforms

$$
\begin{aligned}
I_{11} &= \sum_{|\sigma|=|\gamma|=k} a_{\sigma\gamma}(\mathbf{x}_0) \int_{\mathbb{R}^n} D^\gamma u(\mathbf{x}) \overline{D^\sigma u}(\mathbf{x}) \, d\mathbf{x} \\
&= \sum_{|\sigma|=|\gamma|=k} a_{\sigma\gamma}(\mathbf{x}_0) \int_{\mathbb{R}^n} \widehat{D^\gamma u}(\boldsymbol{\xi}) \overline{\widehat{D^\sigma u}}(\boldsymbol{\xi}) \, d\boldsymbol{\xi} \\
&= \sum_{|\sigma|=|\gamma|=k} \int_{\mathbb{R}^n} a_{\sigma\gamma}(\mathbf{x}_0)(i\boldsymbol{\xi})^\gamma (-i\boldsymbol{\xi})^\sigma |\hat{u}|^2 \, d\boldsymbol{\xi} \\
&= \sum_{|\sigma|=|\gamma|=k} \int_{\mathbb{R}^n} \boldsymbol{\xi}^\sigma a_{\sigma\gamma}(\mathbf{x}_0) \boldsymbol{\xi}^\gamma |\hat{u}|^2 \, d\boldsymbol{\xi} \\
&\geq \theta \int_{\mathbb{R}^n} |\boldsymbol{\xi}|^{2k} |\hat{u}|^2 \, d\boldsymbol{\xi}.
\end{aligned}
$$

In the last inequality we have used the uniform ellipticity condition. To continue, we use Theorem 7.12 to get

$$
\begin{aligned}
I_{11} &\geq \theta \int_{\mathbb{R}^n} (1+|\boldsymbol{\xi}|^{2k}) |\hat{u}|^2 \, d\boldsymbol{\xi} - \theta \int_{\mathbb{R}^n} |\hat{u}|^2 \, d\boldsymbol{\xi} \\
&\geq \bar{C} \|u\|_{k,2}^2 - \theta \|u\|_2^2
\end{aligned}
$$

for some $\bar{C} > 0$ which depends only on Ω'.

We now combine the estimates of I_{11} and I_{12} to get an estimate for I_1. At this time we assume that δ is sufficiently small so that $\omega(\delta) \leq \bar{C}/2$. Then we have

$$
\begin{aligned}
I_1 &\geq I_{11} - |I_{12}| \\
&\geq \bar{C} \|u\|_{k,2}^2 - \theta \|u\|_2^2 - \omega(\delta) \|u\|_{k,2}^2 \qquad (9.62) \\
&\geq \frac{\bar{C}}{2} \|u\|_{k,2}^2 - \theta \|u\|_2^2.
\end{aligned}
$$

We now continue with our estimate of I_1 in the case of a general $u \in H_0^k(\Omega)$. The basic idea is to break up u using a partition of unity, so that we can use the previous estimate.

We begin by covering $\overline{\Omega}$ with a finite collection of balls

$$
B_i := B(\mathbf{x}_i, \delta_i), \quad i = 1, \ldots, M, \qquad (9.63)
$$

with $\mathbf{x}_i \in \Omega$ and $\delta_i > 0$, selected as in the previous estimate so that $B_i \subset \Omega'$. Now let ψ_i be a partition of unity on $\overline{\Omega}$ subordinate to the covering B_i. We then set

$$
\phi_i(\mathbf{x}) := \left(\frac{\psi_i^2(\mathbf{x})}{\sum_{j=1}^M \psi_j^2(\mathbf{x})} \right)^{1/2}. \qquad (9.64)
$$

We then have

1. $0 \leq \phi_i(\mathbf{x}) \leq 1$,

2. $\phi_i \in C^\infty(B_i \cap \overline{\Omega})$,

3. $\sum_{i=1}^M \phi_i^2(\mathbf{x}) = 1$ for each $\mathbf{x} \in \Omega$, and

4. $u_i := u\phi_i \in H_0^k(B_i)$.

This can be used to write

$$
\begin{aligned}
I_1 &= \sum_{|\sigma|=|\gamma|=k} \int_\Omega a_{\sigma\gamma}(\mathbf{x}) D^\sigma u D^\gamma u \, d\mathbf{x} \\
&= \sum_{i=1}^M \sum_{|\sigma|=|\gamma|=k} \int_\Omega a_{\sigma\gamma}(\mathbf{x}) \phi_i D^\sigma u \phi_i D^\gamma u \, d\mathbf{x} \\
&= \sum_{i=1}^M \sum_{|\sigma|=|\gamma|=k} \int_\Omega a_{\sigma\gamma}(\mathbf{x}) D^\sigma(\phi_i u) D^\gamma(\phi_i u) \, d\mathbf{x} \\
&\quad + \sum_{i=1}^M \sum_{|\sigma|=|\gamma|=k} \int_\Omega a_{\sigma\gamma}(\mathbf{x})[\phi_i D^\sigma u - D^\sigma(\phi_i u)]\phi_i D^\gamma u \, d\mathbf{x} \\
&\quad + \sum_{i=1}^M \sum_{|\sigma|=|\gamma|=k} \int_\Omega a_{\sigma\gamma}(\mathbf{x})[\phi_i D^\gamma u - D^\gamma(\phi_i u)]D^\sigma(\phi_i u) \, d\mathbf{x} \\
&\geq \sum_{i=1}^M \sum_{|\sigma|=|\gamma|=k} \int_\Omega a_{\sigma\gamma}(\mathbf{x}) D^\sigma u_i D^\gamma u_i \, d\mathbf{x} \\
&\quad - C\|u\|_{k,2}\|u\|_{k-1,2}
\end{aligned}
$$

We can now use the previous estimate for each $u_i \in H_0^k(B_i)$ to get

$$
\begin{aligned}
I_1 &\geq \frac{\bar{C}}{2} \sum_{i=1}^M \|u_i\|_{k,2}^2 - C\|u\|_2^2 \\
&\quad - C\|u\|_{k,2}\|u\|_{k-1,2} \\
&= \frac{\bar{C}}{2} \sum_{i=1}^M \sum_{|\alpha|\leq k} \int_\Omega |D^\alpha(\phi_i u)|^2 \, d\mathbf{x} \\
&\quad - C[\|u\|_2^2 + \|u\|_{k,2}\|u\|_{k-1,2}] \\
&\geq \frac{\bar{C}}{2} \sum_{|\alpha|\leq k} \int_\Omega \sum_{i=1}^M (\phi_i^2) |D^\alpha u|^2 \, d\mathbf{x} \\
&\quad - C[\|u\|_2^2 + \|u\|_{k,2}\|u\|_{k-1,2}] \\
&= \frac{\bar{C}}{2}\|u\|_{k,2}^2 - C[\|u\|_2^2 + \|u\|_{k,2}\|u\|_{k-1,2}]
\end{aligned}
$$

Thus, using (9.52) and (7.14) we get

$$I_1 \geq \frac{\bar{C}}{4}\|u\|_{k,2}^2 - C\|u\|_2^2. \tag{9.65}$$

Finally we combine this estimate with (9.55) to get

$$
\begin{aligned}
B[u,u] &= I_1 + I_2 \\
&\geq \left(\frac{\bar{C}}{4} - \epsilon\right)\|u\|_{k,2}^2 - C(\epsilon)\|u\|_2^2 \\
&:= c_3\|u\|_{k,2}^2 - \lambda_G\|u\|_2^2,
\end{aligned}
$$

where in defining c_3 and λ_G we have taken ϵ to be sufficiently small, say $\epsilon = \bar{C}/8$. $\qquad\square$

Gårding's inequality is much easier to prove for second-order equations; i.e., in the case where $L(x, D)$ is a second-order differential operator of the form

$$L(x,D)u := \sum_{i,j=1}^n \frac{\partial}{\partial x_i} a_{ij}(\mathbf{x}) \frac{\partial u}{\partial x_j} + \sum_{i=1}^n b_i(\mathbf{x}) \frac{\partial u}{\partial x_i} + c(\mathbf{x})u, \tag{9.66}$$

with corresponding bilinear form

$$B[v,u] := -\sum_{i,j=1}^n \int_\Omega a_{ij}(\mathbf{x})u_{x_j}v_{x_i}\,d\mathbf{x} + \sum_{i=1}^n \int_\Omega b_i(\mathbf{x})u_{x_i}v\,d\mathbf{x} + \int_\Omega c(\mathbf{x})uv\,d\mathbf{x}. \tag{9.67}$$

In this case we do not need to use either Fourier transforms or the partition of unity technique, and the proof can be carried out under weaker hypotheses on the higher-order coefficients.

Theorem 9.18. *Let Ω be a bounded domain. Let $L(\mathbf{x}, D)$ be a second-order linear partial differential operator in divergence form of the form described in (9.66) such that for some $\theta > 0$ the uniform ellipticity condition (9.6) holds. Also suppose that a_{ij}, $b_k \in L^\infty(\Omega)$ for $i, j = 1, \ldots, n$, $k = 0, \ldots, n$. Then there exist constants c_3 and $\lambda_G \geq 0$ such that*

$$B[u,u] + \lambda_G\|u\|_{L^2(\Omega)}^2 \geq c_3\|u\|_{H^1(\Omega)}^2 \quad \text{for all } u \in H_0^1(\Omega), \tag{9.68}$$

where B is as defined in (9.67).

Proof. We start by using the uniform ellipticity condition and Hölder's inequality to get

$$
\begin{aligned}
B[u, u] \quad &:= \quad -\sum_{i,j=1}^{n} \int_{\Omega} a_{ij}(\mathbf{x}) u_{x_i} u_{x_j} \, d\mathbf{x} + \sum_{i=1}^{n} \int_{\Omega} b_i(\mathbf{x}) u_{x_i} u \, d\mathbf{x} \\
&\quad + \int_{\Omega} c(\mathbf{x}) u^2 \, d\mathbf{x} \\
&\geq \quad \theta \int_{\Omega} |\nabla u|^2 \, d\mathbf{x} - \max \|b_i\|_{L^\infty(\Omega)} \int_{\Omega} |\nabla u| \, |u| \, d\mathbf{x} \\
&\quad - \|c\|_{L^\infty(\Omega)} \int_{\Omega} |u|^2 \, d\mathbf{x}.
\end{aligned}
$$

We now use (9.52) and Poincaré's inequality (7.17) to get

$$
\begin{aligned}
B[u, u] \quad &\geq \quad \frac{\theta}{2} \int_{\Omega} |\nabla u|^2 \, d\mathbf{x} - C\|u\|_2^2 \\
&\geq \quad C_1 \|u\|_{1,2}^2 - \lambda_G \|u\|_2^2.
\end{aligned}
$$

This completes the proof. $\qquad\qquad\qquad\qquad\qquad\qquad\qquad\qquad\qquad\qquad$ $\square$

9.2.4 Existence of Weak Solutions

We are now in a position to prove our basic existence result for weak solutions.

Theorem 9.19. *Let $L(\mathbf{x}, D)$ be a linear partial differential operator in divergence form of order $2k$, satisfying the hypotheses of Theorem 9.17 (Gårding's inequality). Then there exists $\lambda_G \geq 0$ such that for any $\tilde{\lambda} \geq \lambda_G$, and any $f \in H^{-k}(\Omega)$, the Dirichlet problem for the operator*

$$
\tilde{L}(\mathbf{x}, D) := L(\mathbf{x}, D) + \tilde{\lambda} \tag{9.69}
$$

has a unique weak solution $u \in H_0^k(\Omega)$. Furthermore, this solution satisfies

$$
\|u\|_{k,2} \leq C\|f\|_{-k,2}. \tag{9.70}
$$

Proof. Theorem 9.17 guarantees the existence of $\lambda_G \geq 0$ such that (9.47) holds. Let $\tilde{\lambda} \geq \lambda_G$. Note that

$$
\tilde{B}[u, v] := B[u, v] + \tilde{\lambda}(u, v)_{L^2(\Omega)} \tag{9.71}
$$

is the bilinear form associated with the operator $\tilde{L}$ defined in (9.69). We now show that $\tilde{B}$ satisfies the hypotheses of the Lax-Milgram lemma.

Let $H = H_0^k(\Omega)$, and let $u, v \in H$. Then

$$
\begin{aligned}
|\tilde{B}[v, u]| &\leq |B[v, u]| + |\tilde{\lambda}|(u, v) \\
&\leq \sum_{0 \leq |\sigma|, |\gamma| \leq k} \int_\Omega |a_{\sigma\gamma}(\mathbf{x})| \, |D^\gamma u| \, |D^\sigma v| \, d\mathbf{x} + |\tilde{\lambda}|(u, v) \\
&\leq \max_{|\sigma|, |\gamma| \leq k} \|a_{\sigma\gamma}\|_{L^\infty(\Omega)} \sum_{0 \leq |\sigma|, |\gamma| \leq k} \int_\Omega |D^\gamma u| \, |D^\sigma v| \, d\mathbf{x} + |\tilde{\lambda}|(u, v) \\
&\leq C\|v\|_H \|u\|_H.
\end{aligned}
$$

Thus, $\tilde{B}$ satisfies (9.29).

Now by Gårding's inequality (9.47) we have

$$
\tilde{B}[u, u] = \tilde{\lambda}\|u\|_2 + B[u, u] \geq c_3 \|u\|_H^2. \tag{9.72}
$$

Thus, $\tilde{B}$ satisfies (9.30).

Thus, Lax-Milgram guarantees that for every $f \in H^{-k} = H^*$ there is a unique weak solution $u \in H$ of the Dirichlet problem, and that the solution satisfies the estimate (9.70). $\qquad\square$

Problems

9.1. Let D be the unit disk in the plane and let $\Omega = D\backslash\{0\}$. It is well-known that the Dirichlet problem $\Delta u = 1$ with $u = 0$ on $\partial\Omega$ has no classical solution. What is the weak "solution" given by Theorem 9.19? Hint: First characterize $H_0^1(\Omega)$.

9.2. Consider the ODE boundary-value problem $y''+p(x)y'+q(x)y = f(x)$, $y(0) = y(1) = 0$. Here $p \in C^1[0, 1]$, $q \in C[0, 1]$. Prove that a unique solution exists if $p' - 2q \geq 0$.

9.3. Let the double sequence a_{ij} be such that $\sum_{i,j=1}^\infty |a_{ij}|^2 < \infty$. Assume, moreover, that the matrix a_{ij}, $i, j = 1, \ldots, N$, is positive definite for every N. Prove that the equation

$$
u_n + \sum_{j=1}^\infty a_{nj}u_j = f_n \tag{9.73}
$$

has a unique solution $u \in \ell^2$ for every $f \in \ell^2$.

9.4. Consider a "weak" solution of the Dirichlet problem for the differential operator defined in (9.66) in a situation where the coefficients a_{ij}, b_i and c have discontinuities across a smooth surface. Assume you know that the solution is smooth on both sides of this interface. Determine the "matching conditions" which are satisfied across the interface.

9.3   Eigenfunction Expansions

Under suitable hypotheses on the elliptic operator L, Theorem 9.19 guarantees that there exists λ_G such that if $\tilde{\lambda} > \lambda_G$, then for any $f \in H^{-k}(\Omega)$ there exists a unique (weak) solution $u \in H_0^k(\Omega)$ of the Dirichlet problem for

$$\tilde{L}(\mathbf{x}, D)u := L(\mathbf{x}, D)u + \tilde{\lambda}u = f.$$

In this section we will apply some of the operator techniques developed in the previous chapter to this problem. This investigation will give us two basic improvements over the present existence theory. First, the Fredholm theorems will give us information on the existence and uniqueness of solutions for values of $\tilde{\lambda} < \lambda_G$. Second, if the operator L satisfies a symmetry condition, we can use the method of eigenfunction expansion to construct (or in real life approximate) solutions.

9.3.1   Fredholm Theory

In this section we consider the nonhomogeneous eigenvalue problem

$$L(\mathbf{x}, D)u + \lambda u = f \qquad (9.74)$$

for $f \in L^2(\Omega)$, where $L(\mathbf{x}, D)$ is the operator

$$L(\mathbf{x}, D)u = \sum_{0 \le |\sigma|, |\gamma| \le k} (-1)^{|\sigma|} D^\sigma (a_{\sigma\gamma}(\mathbf{x}) D^\gamma u),$$

and the bilinear form associated with L is

$$B[v, u] = \sum_{0 \le |\sigma|, |\gamma| \le k} \int_\Omega a_{\sigma\gamma}(\mathbf{x}) D^\gamma u D^\sigma v \, d\mathbf{x}.$$

Let us assume the hypotheses of Theorem 9.19 are satisfied and fix $\tilde{\lambda} > \lambda_G$ with $\tilde{\lambda} > 0$. Then for any $f \in L^2(\Omega)$ there is a unique solution $u \in H_0^k(\Omega)$ to the problem

$$B_{\tilde{\lambda}}[v, u] := B[v, u] + \tilde{\lambda}(v, u) = (v, f)_{L^2(\Omega)} \quad \text{for every } v \in H_0^k(\Omega). \quad (9.75)$$

We now define an operator $\overline{G} : L^2(\Omega) \to H_0^k(\Omega)$ as follows: for every $f \in L^2(\Omega)$ we define

$$\overline{G}(f) := \tilde{\lambda}u, \qquad (9.76)$$

where u is the unique (weak) solution of the Dirichlet problem for

$$L(\mathbf{x}, D)u + \tilde{\lambda}u = f; \qquad (9.77)$$

i.e., u solves (9.75). In other words, for every $f \in L^2(\Omega)$ and $v \in H_0^k(\Omega)$ we have

$$B_{\tilde{\lambda}}[v, \overline{G}(f)] = \tilde{\lambda}(v, f)_{L^2(\Omega)}. \qquad (9.78)$$

Formally, we have

$$\overline{G} = \tilde{\lambda}(L + \tilde{\lambda})^{-1}. \tag{9.79}$$

By (9.70), the operator $\overline{G}$ is bounded. We now define the operator $G :$ $L^2(\Omega) \to L^2(\Omega)$ by the composition of $\overline{G}$ and $\bar{I}$,

$$G := \bar{I}\overline{G}, \tag{9.80}$$

where $\bar{I}$ is the identity mapping from $H^k(\Omega)$ to $L^2(\Omega)$. We know from Theorem 7.29 that this operator is compact. Since the composition of a bounded operator and a compact operator is compact (cf. Problem 8.39) we have the following.

Lemma 9.20. *The solution operator $G : L^2(\Omega) \to L^2(\Omega)$ is compact.*

We now apply the Fredholm alternative theorem (Theorem 8.93) to the operator G to get the following.

Theorem 9.21. *Let $L(\mathbf{x}, D)$ be a uniformly elliptic differential operator of order $2k$ satisfying the hypotheses of Theorem 9.19. Then for every $\mu \in \mathbb{C}$ the Fredholm alternative holds; i.e., either*

1. *for every $f \in L^2(\Omega)$ there exists a unique weak solution $u \in H_0^k(\Omega)$ of the Dirichlet problem for the equation*

$$L(\mathbf{x}, D)u - \mu u = f, \tag{9.81}$$

i.e.,

$$B[v, u] - \mu(v, u)_{L^2(\Omega)} = (v, f)_{L^2(\Omega)} \tag{9.82}$$

for all $v \in H_0^k(\Omega)$, or

2. *there exists at most a finite linearly independent collection of functions $u_i \in H_0^k(\Omega)$, $i = 1, \ldots, N$, such that*

$$B[v, u_i] - \mu(v, u_i) = 0, \tag{9.83}$$

for all $v \in H_0^k(\Omega)$.

Furthermore, the set of values at which the second alternative holds forms an infinite discrete set with no finite accumulation point.

Proof. We first write the equation

$$Lu = \mu u \tag{9.84}$$

as

$$(L + \tilde{\lambda})u = (\tilde{\lambda} + \mu)u. \tag{9.85}$$

Then by a formal calculation in which we act on both sides of (9.85) with $G/\tilde{\lambda} = (L + \tilde{\lambda})^{-1}$ we see that (9.84) has a nontrivial solution u if and only if u solves

$$u = (L + \tilde{\lambda})^{-1}(L + \tilde{\lambda})u = \frac{\tilde{\lambda} + \mu}{\tilde{\lambda}}Gu = \frac{1}{\sigma}Gu. \qquad (9.86)$$

Thus, we see that $u \in L^2(\Omega)$ is an eigenfunction of G corresponding to the eigenvalue σ if and only if u is an eigenfunction of L corresponding to the eigenvalue μ where

$$\sigma = \frac{\tilde{\lambda}}{\tilde{\lambda} + \mu}, \qquad (9.87)$$

$$\mu = -\tilde{\lambda} + \frac{\tilde{\lambda}}{\sigma}. \qquad (9.88)$$

By the Fredholm alternative theorem, the nonzero eigenvalues of G are of finite multiplicity and thus the eigenvalues of L are as well. Also, the eigenvalues of G form a discrete set whose only possible accumulation point is zero, and since we have arranged it so that 0 is not an eigenvalue of G, G must have an infinite collection of eigenvalues. Thus, there must be an infinite collection of eigenvalues of L with no finite accumulation point.

When $\mu \neq -\tilde{\lambda}$ is not an eigenvalue of L, we note that $u \in H_0^k(\Omega)$ is a solution of (9.81) if and only if u is a solution of

$$G(u) - \frac{\tilde{\lambda}}{\mu + \tilde{\lambda}}u = -\frac{1}{\mu + \tilde{\lambda}}G(f). \qquad (9.89)$$

We leave it to the reader to supply the rigor necessary to shore up this formal argument. The only delicate points involve showing that functions u that are solutions of equations involving G (and are thus naturally thought of as being only in $L^2(\Omega)$) must actually be functions in $H_0^k(\Omega)$ imbedded into $L^2(\Omega)$ (and can thus work as weak solutions of equations involving L).   □

9.3.2   Eigenfunction Expansions

When the coefficients of $L(\mathbf{x}, D)$ satisfy the symmetry condition

$$a_{\sigma\gamma} = a_{\gamma\sigma}, \qquad (9.90)$$

then it is easy to show that L is symmetric. Moreover, by direct calculation we see that for every $u, v \in H_0^k(\Omega)$ we have

$$B[u, v] = B[v, u]. \qquad (9.91)$$

For any $f, g \in L^2(\Omega)$ this gives us

$$
\begin{aligned}
(G(f), g)_{L^2(\Omega)} &= (\bar{G}(f), g)_{L^2(\Omega)} \\
&= \frac{1}{\bar{\lambda}} B_{\bar{\lambda}}[\bar{G}(f), \bar{G}(g)] \\
&= \frac{1}{\bar{\lambda}} B_{\bar{\lambda}}[\bar{G}(g), \bar{G}(f)] \\
&= (\bar{G}(g), f)_{L^2(\Omega)} \\
&= (f, G(g))_{L^2(\Omega)}.
\end{aligned}
$$

So G is self-adjoint. Thus, we can use the Hilbert-Schmidt theorem to get the following.

Theorem 9.22. *If L is symmetric, then there is a sequence of real eigenvalues*

$$
\tilde{\lambda} \leq \lambda_1 \leq \lambda_2 \leq \cdots \leq \lambda_n \leq \cdots \tag{9.92}
$$

with no finite accumulation point and $\lim_{i \to \infty} \lambda_i = \infty$, and an orthonormal set of eigenfunctions $\{\phi_i\}_{i=1}^{\infty}$ such that

$$
L\phi_i = \lambda_i \phi_i \tag{9.93}
$$

(in the weak sense). Furthermore, if $\mu \neq \lambda_i$, $i = 1, 2, \ldots, \infty$, then for any $f \in L^2(\Omega)$ the unique weak solution of

$$
L(\mathbf{x}, D)u - \mu u = f \tag{9.94}
$$

is given by

$$
u = \sum_{i=1}^{\infty} \frac{(\phi_i, f)}{\lambda_i - \mu} \phi_i. \tag{9.95}
$$

If μ is an eigenvalue; i.e., $\mu = \lambda_j$ for j in some index set $J \subset \mathbb{N}$, then (9.94) is solvable if and only if

$$
(\phi_j, f) = 0, \quad j \in J. \tag{9.96}
$$

If so, there is a family of solutions given by

$$
u = \sum_{j \in J} \alpha_j \phi_j + \sum_{\mathbb{N} \setminus J} \frac{(\phi_i, f)}{\lambda_i - \mu} \phi_i. \tag{9.97}
$$

(Here the series (9.95) and (9.97) converge in $L^2(\Omega)$.)

The proof is left to the reader.

9.4 General Linear Elliptic Problems

So far in this chapter, we have discussed only Dirichlet boundary conditions for elliptic problems. In this section we shall discuss a few of the other boundary conditions that arise in physical and mathematical problems. Since physical problems present us with a wide variety of boundary conditions for consideration our discussion will not be exhaustive.

9.4.1   The Neumann Problem

After the Dirichlet problem, the second most common and important elliptic boundary-value problem is the Neumann problem.

Definition 9.23. Let $\Omega \subset \mathbb{R}^n$ be a bounded domain with C^1 boundary and suppose $f \in C_b(\Omega)$ is given. A function

$$u \in C_b^2(\Omega) \cap C_b^1(\overline{\Omega})$$

is a **classical solution** of the Neumann problem if

$$L(x, D)u := \sum_{i,j=1}^{n} \frac{\partial}{\partial x_i} a_{ij}(\mathbf{x}) \frac{\partial u}{\partial x_j} + \sum_{i=1}^{n} b_i(\mathbf{x}) \frac{\partial u}{\partial x_i} + c(\mathbf{x})u = f \qquad (9.98)$$

in Ω; and

$$\sum_{i,j=1}^{n} a_{ij}(\mathbf{x}) \frac{\partial u}{\partial x_j} \eta_i(\mathbf{x}) = 0 \qquad (9.99)$$

on $\partial\Omega$ where $\boldsymbol{\eta}(\mathbf{x})$ is the unit outward normal to $\partial\Omega$ at $\mathbf{x}$.

As with the Dirichlet problem, we can define a strong solution of the Neumann problem.

Definition 9.24. Let $\Omega \subset \mathbb{R}^n$ be a bounded domain with a C^1 boundary and suppose $f \in L^2(\Omega)$ is given. A function

$$u \in H^2(\Omega)$$

is a **strong solution** of the Neumann problem if (9.98) holds in $L^2(\Omega)$ and (9.99) holds in the sense of trace on $\partial\Omega$.

In order to state the Neumann problem in weak form we proceed as before and use integration by parts to create a bilinear form from the differential operator L and the boundary conditions. Note that for any ϕ and u in $H^2(\Omega)$ we have

$$\int_{\Omega} \phi L u \, d\mathbf{x} = B[\phi, u] + \sum_{i,j=1}^{n} \int_{\partial\Omega} a_{ij}(\mathbf{x}) \frac{\partial u}{\partial x_j}(\mathbf{x}) \eta_i(\mathbf{x}) \phi(\mathbf{x}) \, d\mathbf{x}, \qquad (9.100)$$

where the bilinear form B is defined in (9.67). Thus, if u satisfies the boundary condition (9.99), then we have

$$\int_{\Omega} \phi L u \, d\mathbf{x} = B[\phi, u] \qquad (9.101)$$

for every $\phi \in H^1(\Omega)$. In fact, we take this as the definition of a weak solution of the Neumann problem.

Definition 9.25. Let $\Omega \subset \mathbb{R}^n$ be a bounded domain and suppose $f \in L^2(\Omega)$ is given. A function

$$u \in H^1(\Omega)$$

is a **weak solution** of the Neumann problem if

$$B[v, u] = (f, v) \tag{9.102}$$

for every $v \in H^1(\Omega)$.

A few comments are in order.

1. Of course, we have constructed things so that every strong solution of the Neumann problem is also a weak solution.

2. In the construction of the weak form, the boundary conditions "disappear"; i.e., condition (9.99) does not appear explicitly in either the bilinear form or the space of admissible functions. Because of this, Neumann conditions are referred to as *natural boundary conditions*. In fact, since the trace theorem does not guarantee the existence of normal derivatives of $H^1(\Omega)$ functions, it does not necessarily make sense to evaluate the boundary condition (9.99) on a weak solution of (9.102).

3. We have assumed that the data f is in the space $L^2(\Omega)$. This can be weakened, for instance by taking $f \in L^2(S)$ where S is a smooth surface contained in Ω. However, we cannot take arbitrary data in $H^{-1}(\Omega)$ as we did for the Dirichlet problem.

As we have indicated above, the key to obtaining an existence theory is proving an energy estimate analogous to Gårding's inequality.

Theorem 9.26. *Let $L(\mathbf{x}, D)$ be a second-order linear partial differential operator in divergence form of the form described in (9.66) such that for some $\theta > 0$ the uniform ellipticity condition (9.6) holds. Also suppose that a_{ij}, $b_k \in L^\infty(\Omega)$ for $i, j = 1, \ldots, n$, $k = 0, \ldots, n$. Then there exist constants $\bar{c}$ and $\lambda_N \geq 0$ such that*

$$B[u, u] + \lambda_N \|u\|^2_{L^2(\Omega)} \geq \bar{c}\|u\|^2_{H^1(\Omega)} \quad \text{for all } u \in H^1(\Omega), \tag{9.103}$$

where B is as defined in (9.67).

Proof. The statement of this theorem is identical to that of Theorem 9.18 except that now we are trying to prove the theorem over the space $H^1(\Omega)$ rather than just $H^1_0(\Omega)$. Thus, the only difference in the proof of this result is that we no longer have Poincaré's inequality. However, as before we can get

$$B[u, u] \geq \frac{\theta}{2} \int_{\Omega} |\nabla u|^2 \, d\mathbf{x} - C\|u\|_2^2. \tag{9.104}$$

And instead of using Poincaré's inequality at this point we simply write

$$B[u, u] \geq \frac{\theta}{2}\|u\|_{1,2}^2 - (C + \frac{\theta}{2})\|u\|_2^2, \tag{9.105}$$

which completes the proof.   □

It is worth noting that the only real reason for using Poincaré in the proof of Theorem 9.18 was to get a sharper estimate on the constant λ_G. However, since we haven't been trying to specify optimal constants anyway, this effort was sort of wasted.

With our energy estimate in place to take care of the coercivity condition in the Lax-Milgram lemma, the existence of a weak solution of the Neumann problems follows with only minor modifications of the proof of Theorem 9.19.

Theorem 9.27. *Let $L(\mathbf{x}, D)$ be a second-order linear partial differential operator in divergence form satisfying the hypotheses of Theorem 9.26. Then there exists $\lambda_N \geq 0$ such that for any $\tilde{\lambda} \geq \lambda_N$ and for any $f \in L^2(\Omega)$, the Neumann problem for the operator*

$$\tilde{L}(\mathbf{x}, D) := L(\mathbf{x}, D) + \tilde{\lambda} \tag{9.106}$$

has a unique weak solution $u \in H^1(\Omega)$. Furthermore, this solution satisfies

$$\|u\|_{1,2} \leq C\|f\|_2. \tag{9.107}$$

9.4.2   The Complementing Condition for Elliptic Systems

For ODE boundary-value problems, it is well known that a "reasonable" boundary-value problem is obtained if the number of boundary conditions at each point equals half the order of the differential equation and the boundary conditions at each point are linearly independent. For elliptic PDEs, the picture is more complicated. Consider the problem

$$\Delta\Delta u = 0 \tag{9.108}$$

with boundary conditions

$$\Delta u = \frac{\partial}{\partial n}\Delta u = 0. \tag{9.109}$$

Although the two boundary conditions are independent of each other, we see that every harmonic function satisfies both the differential equation and the boundary conditions. There are infinitely many linearly independent harmonic functions even within the class of polynomials. This illustrates the need to formulate hypotheses which classify those boundary conditions leading to "good" problems. These hypotheses will not just express independence of the boundary conditions from each other, but also involve a relationship between the boundary conditions and the differential equation.

The complementing condition provides such a characterization of "good" boundary conditions. We shall state it for general elliptic systems. As in Chapter 2, let us consider a $k \times k$ system of equations

$$L_{ij}(\mathbf{x}, D)u_j = f_i(\mathbf{x}), \ \mathbf{x} \in \Omega, \ i = 1, \ldots, k. \tag{9.110}$$

As before, we assign the "weights" s_i to the ith equation and t_j to the jth independent variable in such a way that L_{ij} is at most of order $s_i + t_j$, and we denote the terms that are exactly of order $s_i + t_j$ by L_{ij}^p. The condition of ellipticity is that

$$\det \mathbf{L}^p(\mathbf{x}, \boldsymbol{\xi}) \neq 0 \quad \forall \boldsymbol{\xi} \in \mathbb{R}^n \backslash \{0\}. \tag{9.111}$$

As we remarked in Chapter 2, this condition can be interpreted as follows: Take the values of the coefficients at any fixed point $\mathbf{x}_0$ in Ω and consider the system $\mathbf{L}^p(\mathbf{x}_0, D)\mathbf{u} = 0$ on all of $\mathbb{R}^n$. Ellipticity means that this constant coefficient system has no nonconstant periodic solutions. The complementing condition will be an analogue of this for points at the boundary.

Let the order of the system, i.e., the order of the polynomial $\det \mathbf{L}^p(\mathbf{x}, \boldsymbol{\xi})$, be $2m$. We impose m boundary conditions

$$B_{lj}(\mathbf{x}, D)u_j = g_l(\mathbf{x}), \ \mathbf{x} \in \partial\Omega, \ l = 1, \ldots, m. \tag{9.112}$$

We now define weights r_l to each boundary condition $l = 1, \ldots, m$ so that the order of B_{lj} be bounded by $r_l + t_j$. Again the terms which are precisely of order $r_l + t_j$ will be considered the principal part. (If $r_l + t_j$ is negative, it is of course understood that $B_{lj} = 0$.) Let now $\mathbf{x}_0$ be a point on $\partial\Omega$ and let $\mathbf{n}$ be the outer normal to Ω. We consider the constant coefficient problem

$$L_{ij}^p(\mathbf{x}_0, D)u_j = 0, \ i = 1, \ldots, k, \tag{9.113}$$

on the half-space $(\mathbf{x} - \mathbf{x}_0) \cdot \mathbf{n} < 0$, with boundary conditions

$$B_{lj}^p(\mathbf{x}_0, D)u_j = 0, \ l = 1, \ldots, m, \tag{9.114}$$

on the boundary $(\mathbf{x} - \mathbf{x}_0) \cdot \mathbf{n} = 0$.

Definition 9.28. *We say that the* **complementing conditions** *holds at* $\mathbf{x}_0$, *if there are no nontrivial solutions of (9.113), (9.114) of the following form:*

$$\mathbf{u}(\mathbf{x}) = \exp(i\boldsymbol{\xi} \cdot (\mathbf{x} - \mathbf{x}_0))\mathbf{v}(\eta), \tag{9.115}$$

where $\boldsymbol{\xi}$ *is a nonzero real vector perpendicular to* $\mathbf{n}$, $\eta = (\mathbf{x} - \mathbf{x}_0) \cdot \mathbf{n}$, *and* $\mathbf{v}(\eta)$ *tends to* $\mathbf{0}$ *exponentially as* $\eta \to -\infty$.

Example 9.29. Consider the Stokes system $\Delta\mathbf{u} - \nabla p = \mathbf{0}$, div $\mathbf{u} = 0$, where $\mathbf{u} \in \mathbb{R}^3$, $p \in \mathbb{R}$, in the half-space $z > 0$, with the Dirichlet boundary condition $\mathbf{u} = \mathbf{0}$ on $z = 0$. As we showed in Chapter 2, the system is elliptic, and of order 6. Assume now that we have a solution $\mathbf{u}(x,y,z) = \exp(i\zeta_1 x + i\zeta_2 y)\mathbf{v}(z)$, $p(x,y,z) = \exp(i\zeta_1 x + i\zeta_2 y)q(z)$, where $\mathbf{v}$ and q tend to zero exponentially as $z \to \infty$. Let $\Sigma = (0,L_1) \times (0,L_2)$, where L_i is a multiple of $2\pi/\zeta_i$ if $\zeta_i \neq 0$ and arbitrary if $\zeta_i = 0$. We find

$$0 = \int_{\Sigma \times \mathbb{R}^+} (\Delta\mathbf{u} - \nabla p) \cdot \mathbf{u} \, dx \, dy \, dz = -\int_{\Sigma \times \mathbb{R}^+} |\nabla\mathbf{u}|^2 \, dx \, dy \, dz. \quad (9.116)$$

This implies that $\mathbf{u}$ is constant, which is compatible with our assumptions only if $\mathbf{u} = \mathbf{0}$. It easily follows that p is also zero. Hence the Stokes system with Dirichlet boundary conditions satisfies the complementing condition.

Example 9.30. Consider the biharmonic equation $\Delta\Delta u = 0$ in the half-plane $y > 0$ with boundary conditions $\Delta u = \frac{\partial}{\partial y}\Delta u = 0$ on the line $y = 0$. For every $\xi \in \mathbb{R}$, the function $u(x,y) = \exp(i\xi x - |\xi|y)$ is a solution. Hence the complementing condition does not hold.

The complementing condition or its failure is not always as easy to verify as in the preceding examples. However, it can always be reduced to a purely algebraic problem. If we insert the ansatz (9.115) into (9.113), we obtain a system of ODEs for $\mathbf{v}(\eta)$, which can as usual be solved by the ansatz $\mathbf{v}(\eta) = \exp(\lambda\eta)\mathbf{v}_0$. Ellipticity means that no roots λ are imaginary, and if the coefficients of our system are real, then an equal number of roots must have positive and negative real parts. Let $\lambda_l^+(\mathbf{x}_0, \boldsymbol{\xi})$ denote the roots with positive real part. Then one obtains m linearly independent solutions of (9.113) in the form $\exp(i\boldsymbol{\xi} \cdot (\mathbf{x} - \mathbf{x}_0) + \lambda_l^+\eta)\mathbf{u}_l$ (in the usual way, this may need to be modified by including powers of η if there are repeated roots). It remains to be checked if any linear combination of these solutions satisfies the boundary conditions, which is a purely algebraic problem. For equivalent algebraic characterizations of the complementing conditions we refer to the literature, see [ADN2].

What can we get out of the complementing condition? This question was answered in the work of Agmon, Douglis and Nirenberg [ADN2]. Before we state their results, let us introduce some notation. For $M \in \mathbb{N}$, let

$$X_M = \prod_{j=1}^{k} H^{M+t_j}(\Omega), \quad Y_M = \prod_{i=1}^{k} H^{M-s_i}(\Omega), \quad Z_M = \prod_{l=1}^{m} H^{M-r_l-1/2}(\partial\Omega).$$

$$(9.117)$$

We now consider the problem (9.110) with boundary condition (9.112). We write the equations in the compact form $\mathbf{Lu} = \mathbf{f}$ and $\mathbf{Bu} = \mathbf{g}$ and we denote by $\mathcal{A}$ the operator which maps $\mathbf{u}$ to $(\mathbf{Lu}, \mathbf{Bu}|_{\partial\Omega})$. In choosing weights, we shall now make the convention that $s_i \leq 0$ and $t_j \geq 0$ for all i and j; this can always be achieved by subtracting a constant from the s_i and r_l and adding the same constant to the t_j. Let $t' = \max_j t_j$, $M_1 = \max(0, \max_l r_l + 1)$. Then the following result holds.

Theorem 9.31 (Agmon, Douglas, Nirenberg). *Let $M \geq M_1$ be an integer. Assume that Ω is a bounded domain of class $C^{M+t'}$, that the coefficients of L_{ij} are of class $C^{M-s_i}(\overline{\Omega})$ and that the coefficients of B_{lj} are of class $C^{M-r_l}(\partial\Omega)$. Moreover, assume that ellipticity holds throughout $\overline{\Omega}$ and that the complementing condition holds everywhere on $\partial\Omega$. Assume that $\mathbf{f} \in Y_M$ and $\mathbf{g} \in Z_M$. Then the following hold:*

1. *Every solution $\mathbf{u} \in X_{M_1}$ is in fact in X_M.*

2. *There is a universal constant K, independent of $\mathbf{u}$, $\mathbf{f}$ and $\mathbf{g}$, such that, for every solution $\mathbf{u} \in X_M$, we have*

$$\|\mathbf{u}\|_{X_M} \leq K \left(\|\mathbf{f}\|_{Y_M} + \|\mathbf{g}\|_{Z_M} + \sum_{j=1}^{k} \|u_j\|_{L^2(\Omega)} \right). \tag{9.118}$$

If $\mathbf{u}$ is a unique solution, then the last term in (9.118) can be omitted.

The result thus consists of a regularity statement and an a priori estimate. Agmon, Douglas and Nirenberg actually prove more than we have stated; they establish similar results in L^p-based Sobolev spaces and also in Hölder spaces. We also note that some of the smoothness hypotheses on Ω and the coefficients can be weakened. We shall not pursue this point here. A proof of the theorem is beyond the scope of this introductory text. However, we refer to Sections 9.5 and 9.6 for a proof of a special case, namely, second-order elliptic PDEs with Dirichlet boundary condition.

We next derive an interesting corollary.

Corollary 9.32. *Let all assumptions be as in the preceding theorem. Assume in addition that $M + t_j > 0$ for every j. Then the operator $\mathcal{A} : X_M \to Y_M \times Z_M$ is semi-Fredholm.*

Proof. It easily follows from the smoothness hypotheses on the coefficients that $\mathcal{A}$ does indeed map X_M to $Y_M \times Z_M$. Let $N(\mathcal{A})$ be the nullspace of $\mathcal{A}$, and let B be the intersection of $N(\mathcal{A})$ with the unit ball in $(L^2(\Omega))^k$. By the theorem, B is bounded in the norm of X_M, hence precompact in $(L^2(\Omega))^k$. Since the unit ball in an infinite-dimensional space is never precompact, $N(\mathcal{A})$ must be finite-dimensional.

Next, we shall show that the range of $\mathcal{A}$ is closed. For that purpose, assume that $\mathbf{u}_N$ is a solution of $\mathbf{L}\mathbf{u}_N = \mathbf{f}_N$ with boundary conditions $\mathbf{B}\mathbf{u}_N = \mathbf{g}_N$, and that $\mathbf{f}_N$ and $\mathbf{g}_N$ converge in Y_M and Z_M to $\mathbf{f}$ and $\mathbf{g}$, respectively. Without loss of generality, we may assume that $\mathbf{u}_N$ is perpendicular to $N(\mathcal{A})$ in $(L^2(\Omega))^k$. We claim that $\mathbf{u}_N$ is then bounded in $(L^2(\Omega))^k$. Suppose not. After taking a subsequence, we may assume $\|\mathbf{u}_N\|_2 \to \infty$. Let $\mathbf{v}_N = \mathbf{u}_N/\|\mathbf{u}_N\|_2$. Then $\mathbf{v}_N$ solves the problem $\mathbf{L}\mathbf{v}_N = \mathbf{f}_N/\|\mathbf{u}_N\|_2$ with boundary conditions $\mathbf{B}\mathbf{v}_N = \mathbf{g}_N/\|\mathbf{u}_N\|_2$. It follows from (9.118) that the sequence $\mathbf{v}_N$ is bounded in X_M. Hence it has a subsequence which converges weakly in X_M, hence strongly in $(L^2(\Omega))^k$. Let $\mathbf{v}$ be the limit. Then $\mathbf{v}$ is in the nullspace of $\mathcal{A}$ and in its orthogonal complement, hence zero. But this is a contradiction, since $\|\mathbf{v}\|_2 = \lim_{n \to \infty} \|\mathbf{v}_N\|_2 = 1$. Since $\mathbf{u}_N$ is bounded in $(L^2(\Omega))^k$, (9.118) implies that it is also bounded in X_M. Hence, after taking a subsequence, $\mathbf{u}_N$ converges weakly in X_M and strongly in $(L^2(\Omega))^k$. Applying (9.118) again, we see that $\mathbf{u}_N$ actually converges strongly in X_M. The limit $\mathbf{u}$ is a solution of $\mathbf{Lu} = \mathbf{f}$ with boundary condition $\mathbf{Bu} = \mathbf{g}$.   □

The next interesting question is of course if the index of $\mathcal{A}$ is finite, and, more particularly, when it is zero. One of the standard methods in answering this question is to exploit the homotopy invariance of the Fredholm index. Consider for example a second-order elliptic operator

$$L(\mathbf{x}, D)u = a_{ij}(\mathbf{x})\frac{\partial^2 u}{\partial x_i \partial x_j} + b_i(\mathbf{x})\frac{\partial u}{\partial x_i} + c(\mathbf{x})u \qquad (9.119)$$

with Dirichlet boundary condition $B(\mathbf{x}, D)u = u$. We assume the matrix a_{ij} is symmetric and positive definite. We may then consider the one-parameter family of operators

$$L_t = (1 - t)\Delta + tL, \ B_t = B. \qquad (9.120)$$

If Ω and the coefficients satisfy the relevant smoothness assumptions, then the assumptions of Theorem 9.31 apply for every $t \in [0, 1]$; hence the Fredholm index for (L, B) is the same as for Laplace's equation. In Section 9.2, we proved that the problem $\Delta u = f$ with boundary condition $u = 0$ has a unique solution $u \in H^1(\Omega)$ for every $f \in H^{-1}(\Omega)$. Using the inverse trace theorem, we can trivially conclude that there is a unique solution $u \in H^1(\Omega)$ of the problem $\Delta u = f$, $u|_{\partial\Omega} = g$ for every $f \in H^{-1}(\Omega)$, $g \in H^{1/2}(\Omega)$. What we would now like to know is that if $f \in L^2(\Omega)$ and $g \in H^{3/2}(\Omega)$, then $u \in H^2(\Omega)$. This is a statement much along the lines of the first assertion of Theorem 9.31, but is not actually implied by Theorem 9.31. The reason is that for the Dirichlet problem of Laplace's equation, we would choose $s_1 = 0$, $t_1 = 2$ and $r_1 = -2$, making $M_1 = 0$ and $X_{M_1} = H^2(\Omega)$. Hence the theorem asserts higher regularity of H^2 solutions if the data are appropriate, but not H^2 regularity of H^1 solutions. Nevertheless, the regularity of weak solutions can be proved along very similar lines as Theorem 9.31 and Agmon, Douglis and Nirenberg actually state such results for scalar elliptic equations. For second-order equations with Dirichlet conditions, see Sections 9.5 and 9.6.

A natural question is now to ask for a class of problems to which the approach of Section 9.2, based on the Lax-Milgram lemma, can be extended. This will lead us to Agmon's condition, to be discussed in Subsection 9.4.4. The Lax-Milgram lemma will imply existence of a "weak" solution, and again the regularity of weak solutions has to be addressed before Theorem 9.31 is applicable.

Another interesting question is to characterize the orthogonal complement of the range of $\mathcal{A}$; i.e., what conditions must $\mathbf{f}$ and $\mathbf{g}$ satisfy so that the problem $\mathbf{L}u = \mathbf{f}$ with boundary conditions $\mathbf{B}u = \mathbf{g}$ is solvable? Usually, one can find a u satisfying $\mathbf{B}u = \mathbf{g}$ by an application of the inverse trace theorem (see next subsection); hence we are reduced to the case $\mathbf{g} = \mathbf{0}$. This leaves us with the question of characterizing those $\mathbf{v}$ for which $(\mathbf{v}, \mathbf{L}u) = 0$ for every u satisfying $\mathbf{B}u = \mathbf{0}$. By formally integrating by parts, one can obtain an elliptic boundary-value problem for $\mathbf{v}$, known as the adjoint boundary-value problem. We shall study adjoint boundary-value problems for scalar elliptic equations in the next subsection. Of course, a priori $\mathbf{v}$ will satisfy the adjoint boundary-value problem only in a "weak" or "generalized" sense. Hence the regularity of weak solutions becomes again an important issue. In particular, in order to show that the operator $\mathcal{A}$ is Fredholm, one has to show that the nullspace of the adjoint is finite-dimensional. Of course, one has to show this for weak solutions of the adjoint problem, not just for strong solutions. Indeed, it is possible to prove this. If the coefficients are smooth enough, it turns out that weak solutions of the adjoint problem are actually smooth.

9.4.3 The Adjoint Boundary-Value Problem

Throughout this subsection, let $L(\mathbf{x}, D)$ be a scalar elliptic differential operator of order $2m$ and let $B_j(\mathbf{x}, D)$, $j = 1, \ldots, m$, be m boundary operators which satisfy the complementing conditions. The general theory of adjoints requires rather stringent regularity assumptions on Ω and the coefficients; for simplicity we shall assume they are of class C^∞ and that Ω is bounded. We make these assumptions throughout. We shall make the additional assumption that the B_j are normal. This property is defined as follows.

Definition 9.33. *The boundary operators $B_j(\mathbf{x}, D)$ are called **normal**, if their orders m_j are different from each other and less than or equal to $2m - 1$ and if, moreover, the leading-order term in B_j contains a purely normal derivative, i.e., $B_j^p(\mathbf{x}, \mathbf{n}) \neq 0$ for every $\mathbf{x} \in \partial\Omega$ (here $\mathbf{n}$ is the unit normal to $\partial\Omega$).*

The orders of the B_j cover only half the values from 0 to $2m - 1$. We can add additional boundary operators S_j, $j = 1, \ldots, m$, to fill in the missing orders. Obviously, we can do this in such a way that the extended set of boundary operators still satisfies the conditions of normality; we merely have to take S_j to be the appropriate powers of $\partial/\partial n$. We make the following definition.

Definition 9.34. *The boundary operators $F_j(\mathbf{x}, D)$, $j = 1, \ldots, p$, are called a **Dirichlet system** of order p, if their orders m_j cover all values from zero to $p - 1$ and if, moreover, the leading-order term in F_j contains a purely normal derivative, i.e., $F_j^p(\mathbf{x}, \mathbf{n}) \neq 0$ for every $\mathbf{x} \in \partial\Omega$ (here $\mathbf{n}$ is the unit normal to $\partial\Omega$).*

We have the following lemma.

Lemma 9.35. *Let $F_i(\mathbf{x}, D)$, $i = 1, \ldots, p$, be a Dirichlet system, and suppose the order of F_i is $i-1$. Then there exist tangential differential operators $\Phi_{ij}(\mathbf{x}, D)$ and $\Psi_{ij}(\mathbf{x}, D)$, of order $i - j$, such that*

$$F_i(\mathbf{x}, D) = \sum_{j=0}^{i} \Phi_{ij}(\mathbf{x}, D) \frac{\partial^{j-1}}{\partial n^{j-1}},$$

$$\frac{\partial^{i-1}}{\partial n^{i-1}} = \sum_{j=0}^{i} \Psi_{ij}(\mathbf{x}, D) F_j(\mathbf{x}, D). \tag{9.121}$$

The existence of the Φ_{ij} is obvious from the definition. The Ψ_{ij} are then obtained by inverting the triangular matrix of the Φ_{ij}. We leave the details of the proof as an exercise; see Problem 9.7.

Corollary 9.36. *Let F_i, $i = 1, \ldots, 2m$, be a Dirichlet system, and let m_i denote the order of F_i. Let $g_i \in H^{2m+k-m_i-1/2}(\partial\Omega)$ be given. Then there exists $u \in H^{2m+k}(\Omega)$ such that $F_i u = g_i$ on $\partial\Omega$.*

The proof follows immediately from the previous lemma and Theorem 7.40.

We are now ready to state Green's formula.

Theorem 9.37. *Let $L(\mathbf{x}, D)$ be an elliptic operator of order $2m$ on $\overline{\Omega}$ and let $B_j(\mathbf{x}, D)$, $j = 1, \ldots, m$, be a set of normal boundary operators. Let $S_j(\mathbf{x}, D)$, $j = 1, \ldots, m$, be a set of boundary operators which complements the B_j to form a Dirichlet system. Then there exist boundary operators $C_j(\mathbf{x}, D)$, $T_j(\mathbf{x}, D)$, $j = 1, \ldots, m$, with the following properties:*

1. *ord $C_j = 2m - 1 - $ ord S_j, ord $T_j = 2m - 1 - $ ord B_j. (ord stands for the order of the operator.)*

2. *The C_j and T_j form a Dirichlet system.*

3. *For every $u, v \in H^{2m}(\Omega)$, we have*

$$\int_\Omega (Lu)v - u(L^*v) \, d\mathbf{x} = \sum_{j=1}^{m} \int_{\partial\Omega} (S_j u)(C_j v) - (B_j u)(T_j v) \, dS. \tag{9.122}$$

Here L^* is the formal adjoint of L; see Definition 5.53.

4. *If the B_j satisfy the complementing condition for L, the C_j satisfy the complementing condition for L^*.*

Proof. Integration by parts yields a formula of the form

$$\int_\Omega (Lu)v - u(L^*v) \, d\mathbf{x} = \sum_{\alpha, \beta} \int_{\partial\Omega} a_{\alpha\beta}(\mathbf{x}) D^\alpha u D^\beta v \, dS, \tag{9.123}$$

where the sum extends over α and β with $|\alpha| + |\beta| \leq 2m - 1$. We next integrate by parts on $\partial\Omega$ and move all tangential derivatives from u to v, so that only purely normal derivatives of u are left (carrying out this step requires a partition of unity and local coordinate charts on $\partial\Omega$). This leads to a formula of the form

$$\int_\Omega (Lu)v - u(L^*v)\ d\mathbf{x} = \sum_{j=0}^{2m-1} \int_{\partial\Omega} a_j(\mathbf{x})\frac{\partial^j u}{\partial n^j} E_j(\mathbf{x}, D)v\ dS, \qquad (9.124)$$

where E_j is a differential operator of order $2m - j - 1$. If L is elliptic (or, even more generally, if $\partial\Omega$ is noncharacteristic), then E_j contains a terms proportional to $\frac{\partial^{2m-j-1}}{\partial n^{2m-j-1}}$ with a nonzero coefficient; in other words, the E_j form a Dirichlet system. We next use Lemma 9.35 to find

$$\frac{\partial^j u}{\partial n^j} = \sum_{k=1}^m \Psi_{jk}(\mathbf{x}, D)B_k(\mathbf{x}, D)u + \sum_{k=1}^m \Psi'_{jk}(\mathbf{x}, D)S_k(\mathbf{x}, D)u. \qquad (9.125)$$

We substitute this into (9.124) and then integrate by parts on $\partial\Omega$ to move the tangential differential operators Ψ_{jk} and Ψ'_{jk} from u to v. This yields (9.122).

To verify the complementing condition, let Ω be the half-space $\{x_n > 0\}$ and let L have constant coefficients. Consider solutions of the form $\exp(i\boldsymbol\xi \cdot \mathbf{x})v(x_n)$, where $\xi_n = 0$ and $v(x_n) \to 0$ as $x_n \to \infty$. Green's formula (9.122) holds with Ω replaced by $\Sigma \times \mathbb{R}^+$, where Σ is a parallelepiped in $\mathbb{R}^{n-1}$ corresponding to one period. Moreover, L now becomes an ordinary differential operator. For such operators, a Fredholm alternative holds, i.e., the initial-value problem

$$L(\mathbf{x}, D)u = 0, \ B_j(\mathbf{x}, D)u = 0 \quad \text{for } x_n = 0 \qquad (9.126)$$

has only the trivial solution if and only if the problem

$$L(\mathbf{x}, D)u = f, \ B_j(\mathbf{x}, D)u = g_j \quad \text{for } x_n = 0 \qquad (9.127)$$

is solvable for all f and g_j. By Green's formula, the latter condition implies that the initial-value problem

$$L^*(\mathbf{x}, D)v = 0, \ C_j(\mathbf{x}, D)v = 0 \quad \text{for } x_n = 0 \qquad (9.128)$$

has only the trivial solution, i.e., that the C_j satisfy the complementing condition for L^*. Note that if v were a nontrivial solution of (9.128), then f and g_j in (9.127) would have to satisfy

$$\int_\Omega fv\ d\mathbf{x} = -\sum_{j=1}^m \int_{\partial\Omega} g_j T_j v\ d\mathbf{x}. \qquad (9.129)$$

This completes the proof. $\qquad\qquad\qquad\qquad\qquad\qquad\qquad\qquad\qquad\quad\Box$

Suppose now that $v \in L^2(\Omega)$ is such that

$$\int_\Omega (Lu)v\ d\mathbf{x} = 0 \qquad (9.130)$$

for every $u \in H^{2m}(\Omega)$ such that $B_j u = 0$ on $\partial\Omega$ for $j = 1, \ldots, m$. If we actually knew that $v \in H^{2m}(\Omega)$, then we could use Green's formula to conclude that

$$\int_\Omega u(L^* v) \, dx = -\sum_{j=1}^m \int_{\partial\Omega} S_j u C_j v \, dS. \qquad (9.131)$$

Since the $B_j u$ and $S_j u$ can be chosen arbitrarily and independently, (9.131) implies that $L^* v = 0$ and $C_j v = 0$ on the boundary. Even without the assumption that $v \in H^{2m}(\Omega)$, we find $L^* v = 0$ in the sense of distributions by restricting u to $\mathcal{D}(\Omega)$. What now remains to be done is to show that any $v \in L^2(\Omega)$ satisfying (9.130) actually is in $H^{2m}(\Omega)$.

Let

$$N^* = \{v \in H^{2m}(\Omega) \mid L^* v = 0, \ C_1 v = \cdots = C_m v = 0 \text{ on } \partial\Omega\}, \qquad (9.132)$$

let $(N^*)^\perp$ be the orthogonal complement of N^* in $L^2(\Omega)$ and let $M^* = (N^*)^\perp \cap H^{2m}(\Omega)$. On M^*, we consider the quadratic form

$$[u, v] = \int_\Omega L^* u L^* v \, dx + \sum_{j=1}^m (C_j u, C_j v)_{2m-m_j-1/2}, \qquad (9.133)$$

where m_j is the order of C_j and $(\cdot, \cdot)_s$ denotes the inner product in $H^s(\partial\Omega)$. We shall show that this quadratic form is coercive on the space M^*.

Lemma 9.38. *There exists a constant C such that, for all $u \in M^*$, we have*

$$\|u\|_{2m}^2 \leq C[u, u]. \qquad (9.134)$$

Proof. Since the C_j boundary operators satisfy the complementing condition for L^*, we can use the Agmon-Douglis-Nirenberg theorem to get

$$\|u\|_{2m}^2 \leq C([u, u] + \|u\|_0^2) \qquad (9.135)$$

for any $u \in H^{2m}(\Omega)$.

Suppose now for the sake of contradiction that $u_n \in M^*$ and $[u_n, u_n] \to 0$, whereas $\|u_n\|_{2m} = 1$. Then a subsequence of the u_n converges weakly in $H^{2m}(\Omega)$ and strongly in $L^2(\Omega)$; let u be the limit. It now follows from (9.135) that the subsequence actually converges to u strongly in $H^{2m}(\Omega)$. Since $[u, u] = 0$, we have $u \in N^*$, and since $u_n \in M^*$, we also have $u \in M^*$. Since N^* and M^* are orthogonal in $L^2(\Omega)$, this implies $u = 0$. But that is a contradiction, since $\|u\|_{2m} = 1$.   □

The Lax-Milgram lemma now implies that the equation $[u, v] = (f, v)$ for all $v \in M^*$ has a unique solution $u \in M^*$. If $f \in (N^*)^\perp$, we actually have $[u, v] = (f, v)$ for all $v \in H^{2m}(\Omega)$. We can summarize this as follows.

Lemma 9.39. *The equation*

$$[u, v] = \int_\Omega fv \ d\mathbf{x} = (f, v) \ \forall v \in H^{2m}(\Omega) \qquad (9.136)$$

has a solution $u \in H^{2m}(\Omega)$ *if and only if* $f \in (N^*)^\perp$. u *is unique up to addition of an arbitrary element of* N^*.

The following is a regularity theorem along the lines of the Agmon-Douglis-Nirenberg result.

Theorem 9.40. *Let* u *be a solution of (9.136). Then* $u \in H^{4m}(\Omega)$.

The proof is hard and tedious and will not be given. We are now ready to state the main result of this subsection.

Theorem 9.41. *The boundary-value problem*

$$Lu = f, \ B_j u = 0 \text{ on } \partial\Omega \quad \text{for } j = 1, \ldots, m, \qquad (9.137)$$

with $f \in L^2(\Omega)$ *has a solution* $u \in H^{2m}(\Omega)$ *if and only if* $f \in (N^*)^\perp$.

Proof. It is obvious from Green's formula that the condition $f \in (N^*)^\perp$ is necessary. Let now $f \in (N^*)^\perp$. Then we can find $g \in M^* \cap H^{4m}(\Omega)$ such that $[g, v] = (f, v)$ for every $v \in H^{2m}(\Omega)$. Now let $u = L^*g \in H^{2m}(\Omega)$. For every $v \in H^{2m}(\Omega)$ which satisfies the boundary conditions $C_j v = 0$, we conclude

$$\int_\Omega u(L^*v) \ d\mathbf{x} = \int_\Omega fv \ d\mathbf{x} \qquad (9.138)$$

and, after integration by parts,

$$\int_\Omega (Lu)v \ d\mathbf{x} + \sum_{j=1}^m \int_{\partial\Omega} B_j u T_j v \ dS = \int_\Omega fv \ d\mathbf{x}. \qquad (9.139)$$

It follows readily that u satisfies $Lu = f$ and $B_j u = 0$. $\square$

Remark 9.42. The adjoint boundary operators C_j are generally not unique. However, it is clear from the last theorem that the space N^* is uniquely determined. Hence different sets of adjoint boundary conditions are equivalent in the sense that they determine the same nullspace.

9.4.4 Agmon's Condition and Coercive Problems

We consider a scalar elliptic operator of order $2m$, given in divergence form as in (9.7):

$$L(\mathbf{x}, D)u = \sum_{|\alpha|,|\beta| \le m} (-1)^{|\alpha|} D^\alpha(a_{\alpha\beta}(\mathbf{x})D^\beta u), \qquad (9.140)$$

where the $a_{\alpha\beta}$ are continuous on $\overline{\Omega}$ and the ellipticity condition

$$\sum_{|\alpha|=|\beta|=m} \xi^\alpha a_{\alpha\beta}(\mathbf{x})\xi^\beta > 0 \qquad (9.141)$$

holds throughout $\overline{\Omega}$. Moreover, we consider p normal boundary-value operators $B_j(\mathbf{x}, D)$, with coefficients of class $C^{m-m_j}(\partial\Omega)$, where $m_j < m$ is the order of B_j. In general, p can take any value between 0 and m. We define

$$V = \{u \in H^m(\Omega) \mid B_j(\mathbf{x}, D)u = 0 \text{ on } \partial\Omega, \ j = 1,\ldots,p\}. \qquad (9.142)$$

We consider the quadratic form

$$a(u, v) = \int_\Omega \sum_{|\alpha|\le m, |\beta|\le m} a_{\alpha\beta}(\mathbf{x})D^\beta u D^\alpha v \ dx, \qquad (9.143)$$

and we ask for conditions under which this form is coercive on V:

$$a(u, u) \ge c_1\|u\|_m^2 - c_2\|u\|_0^2 \quad \forall\, u \in V. \qquad (9.144)$$

If the form is coercive, we can apply the Lax-Milgram lemma to conclude that, for λ large enough, the equation

$$a(u, v) + \lambda(u, v) = (f, v) \quad \forall v \in V \qquad (9.145)$$

has a unique solution $u \in V$ for every $f \in V'$. It is then clear that $L(\mathbf{x}, D)u + \lambda u = f$ in the sense of distributions, and that $B_j(\mathbf{x}, D)u = 0$ on the boundary. In addition u will satisfy $m - p$ "natural" boundary conditions, which arise in a similar way as the Neumann boundary condition in Section 9.4.1.

The condition guaranteeing coercivity is known as Agmon's condition. Consider a point $\mathbf{x}_0 \in \partial\Omega$; we may orient our coordinate system in such a way that $\mathbf{x}_0$ is the origin and the inner normal points in the x_n direction. We then consider the constant coefficient problem $L^p(\mathbf{0}, D)u = 0$ in the half-space $x_n > 0$ with boundary conditions $B_j^p(\mathbf{0}, D)u = 0$ for $j = 1,\ldots,p$. We shall use the notation $\mathbf{x} = (\mathbf{x}', x_n)$, where $\mathbf{x}' \in \mathbb{R}^{n-1}$, and correspondingly we write $\alpha = (\alpha', \alpha_n)$ for a multi-index α. We now pick any $\boldsymbol{\xi}' \in \mathbb{R}^{n-1}\backslash\{0\}$ and consider the ODE

$$L^p\left(\mathbf{0}, i\boldsymbol{\xi}', \frac{d}{dt}\right)v(t) = 0, \ t > 0, \qquad (9.146)$$

with initial conditions

$$B_j^p\left(\mathbf{0}, i\boldsymbol{\xi}', \frac{d}{dt}\right)v(0) = 0, \ j = 1,\ldots,p. \qquad (9.147)$$

Definition 9.43. *We say that* **Agmon's condition** *holds if for any $\boldsymbol{\xi}' \in \mathbb{R}^{n-1}\backslash\{0\}$, and any nonzero solution $v(t)$ of (9.146) and (9.147) such that* *v tends to zero exponentially as* $t \to \infty$, *we have the inequality*

$$\int_0^\infty \sum_{\substack{|\alpha'|+k=m \\ |\beta'|+l=m}} a_{(\alpha',k)(\beta',l)}(\mathbf{0})(\boldsymbol{\xi}')^{\alpha'}(\boldsymbol{\xi}')^{\beta'} \frac{d^k v(t)}{dt^k} \frac{d^l v(t)}{dt^l} \, dt > 0. \tag{9.148}$$

Remark 9.44. If $p = m$ and the complementing condition holds, then Agmon's condition is vacuously true. Indeed, if $p = m$, then, by Lemma 9.35, the boundary conditions are equivalent to Dirichlet conditions. In fact, Dirichlet conditions always satisfy the complementing condition; see Problem 9.6.

The following result generalizes Gårding's inequality.

Theorem 9.45. *Let* L, B_j *and* a *be as above. Assume that Agmon's condition holds at each point of* $\partial\Omega$. *Then there exist constants* c_1 *and* c_2 *such that (9.144) holds.*

We now address the question how (9.145) is to be interpreted as an elliptic boundary-value problem. For this, we first need a regularity statement.

Theorem 9.46. *Assume that* Ω *and the coefficients of* L *and the* B_j *are sufficiently smooth. Assume in addition that* $f \in L^2(\Omega)$. *Then the solution* u *of (9.145) lies in* $H^{2m}(\Omega)$.

Next, we need a Green's formula.

Theorem 9.47. *Let* L *and* a *be as above. Let* $B_i(\mathbf{x}, D)$, $i = 1, \ldots, m$, *be a Dirichlet system of order* m. *Assume that* Ω *and the coefficients of the operators involved are sufficiently smooth. Then there exist normal boundary-value operators* C_i, *of order* $2m - 1 - \text{ord } B_i$, *such that, for all* $u, v \in H^{2m}(\Omega)$, *we have*

$$a(u, v) = \int_\Omega (Lu)v \, d\mathbf{x} - \sum_{i=1}^m \int_\Omega (C_i u)(B_i v) \, dS. \tag{9.149}$$

The proof is completely analogueous to that of Theorem 9.37. For $u \in H^{2m}(\Omega)$ and $f \in L^2(\Omega)$, equation (9.145) now assumes the form

$$\int_\Omega (Lu + \lambda u)v \, d\mathbf{x} - \sum_{j=p+1}^m \int_{\partial\Omega} (C_j u)(B_j v) \, dS = \int_\Omega fv \, d\mathbf{x}. \tag{9.150}$$

This identifies (9.145) as the weak form of the elliptic boundary-value problem

$$Lu + \lambda u = f, \quad B_j u = 0, \ j = 1, \ldots, p, \quad C_j u = 0, \ j = p+1, \ldots, m. \tag{9.151}$$

The first set of boundary conditions is called essential; they are directly imposed on u in the weak formulation of the problem. The second set of boundary conditions is called "natural"; they are not imposed explicitly, but arise from an integration by parts just like Neumann's condition in Section 9.4.1.

Problems

9.5. Assume that Ω is bounded, connected, and has the 1-extension property. Let

$$V = \left\{ u \in H^1(\Omega) \mid \int_\Omega u(\mathbf{x})\, d\mathbf{x} = 0 \right\}.$$

(a) Show that for each $f \in L^2(\Omega)$ there is a unique $u \in V$ such that

$$\int_\Omega \nabla u \cdot \nabla v = \int_\Omega fv \quad \text{for all } v \in V. \tag{9.152}$$

(See Problem 7.15.)

(b) Explain why it is appropriate to regard (9.152) as a weak form of the Neumann problem

$$\begin{aligned} -\Delta u &= f & \text{in } \Omega \\ \frac{\partial u}{\partial \mathbf{n}} &= 0 & \text{on } \partial\Omega \end{aligned} \tag{9.153}$$

if $\int_\Omega f = 0$.

(c) If $\int_\Omega f \neq 0$, is it still reasonable to call the solution of (9.152) a solution of (9.153)? Explain.

9.6. Show that Dirichlet boundary conditions for scalar elliptic PDEs always satisfy the complementing condition.

9.7. Fill in the details for the proof of Lemma 9.35.

9.8. Suppose that Agmon's condition holds. Show that the complementing condition is satisfied for (9.151). Hint: Apply (9.149) on a half-space.

9.9. Formulate a weak form of (9.151) when the boundary conditions are allowed to be inhomogeneous.

9.10. Show that the "traction boundary conditions" $(\nabla \mathbf{u} + (\nabla \mathbf{u})^T)) \cdot \mathbf{n} - p\mathbf{n} = \mathbf{0}$ satisfy the complementing condition for the Stokes system.

9.11. Show that a scalar elliptic operator with Dirichlet conditions has Fredholm index 0. Hint: Show that the adjoint problem also has Dirichlet conditions.

9.5   Interior Regularity

In Section 9.2, we have shown the existence of weak solutions $u \in H^k(\Omega)$ of the Dirichlet problem for elliptic operators of order $2k$. We now wish to show that under suitable hypotheses on the smoothness of the coefficients $a_{\sigma\gamma}$, the forcing function f and the boundary of Ω, our weak solution is, in fact, a strong solution or a classical solution. In order to give some idea of how we plan to go about this, we make a couple of formal calculations.

For our first calculation let us assume that Ω has a smooth boundary $\partial\Omega$ with unit outward normal $\boldsymbol{\eta} = (\eta_1, \ldots, \eta_n)$ and that u is a classical solution of

$$-\Delta u = f \tag{9.154}$$

in Ω, and

$$u = 0 \tag{9.155}$$

on $\partial\Omega$. Our goal is to show that (weak) solutions of elliptic problems such as the one above are actually in a "better" space than $H_0^1(\Omega)$. In order to prepare for this, we will now estimate the $L^2(\Omega)$ norm of the matrix of second partials of u in terms of the $H^1(\Omega)$ norm. Since this is simply a formal calculation, we will proceed as if we already know that u is as smooth as we like.

$$
\begin{aligned}
\int_\Omega |\Delta u|^2 \, d\mathbf{x} &= \sum_{i,j=1}^n \int_\Omega u_{x_i x_i} u_{x_j x_j} \, d\mathbf{x} \\
&= -\sum_{i,j=1}^n \int_\Omega u_{x_i} u_{x_j x_i x_j} \, d\mathbf{x} \\
&\quad + \sum_{i,j=1}^n \int_{\partial\Omega} u_{x_i} u_{x_j x_j} \eta_i \, dS \\
&= \sum_{i,j=1}^n \int_\Omega u_{x_i x_j} u_{x_j x_i} \, d\mathbf{x} \\
&\quad + \sum_{i,j=1}^n \int_{\partial\Omega} u_{x_i} u_{x_j x_j} \eta_i - u_{x_i} u_{x_i x_j} \eta_j \, dS.
\end{aligned}
$$

We also have

$$\int_\Omega |\Delta u|^2 \, d\mathbf{x} = \int_\Omega |f|^2 \, d\mathbf{x}. \tag{9.156}$$

Combining these two results gives us

$$\int_\Omega |\nabla^2 u|^2 \, d\mathbf{x} \leq \int_\Omega |f|^2 \, d\mathbf{x} + |\text{boundary terms}|. \tag{9.157}$$

Thus, if we had some additional information on the boundary terms, we could derive an a priori estimate on the $H^2(\Omega)$ norm of a solution u in terms of the data f.

Unfortunately, estimates on boundary terms are rather delicate, so we will put off this subject until the next section. In the meantime, we will concentrate on *interior* estimates of higher-order derivatives. For example, let Ω' be any domain such that $\Omega' \subset\subset \Omega$. (The notation $\Omega' \subset\subset \Omega$ means that Ω' is *compactly contained* in the open set Ω; i.e., $\overline{\Omega'}$ is compact and $\overline{\Omega'} \subset \Omega$.) We now choose a cutoff function $\zeta \in \mathcal{D}(\Omega)$ such that $0 \le \zeta \le 1$ and $\zeta \equiv 1$ on Ω'. We can now make some calculations very similar to those above, but without any boundary terms getting in the way.

$$
\begin{aligned}
\int_\Omega \zeta^2 |\Delta u|^2 \, d\mathbf{x} &= \sum_{i,j=1}^n \int_\Omega u_{x_i x_i} u_{x_j x_j} \zeta^2 \, d\mathbf{x} \\
&= -\sum_{i,j=1}^n \int_\Omega u_{x_i} u_{x_j x_i x_j} \zeta^2 \, d\mathbf{x} \\
&\quad -\sum_{i,j=1}^n \int_\Omega u_{x_i} u_{x_j x_j} 2\zeta\zeta_{x_i} \, d\mathbf{x} \\
&= \sum_{i,j=1}^n \int_\Omega u_{x_i x_j} u_{x_j x_i} \zeta^2 \, d\mathbf{x} \\
&\quad + \sum_{i,j=1}^n \int_\Omega u_{x_i} u_{x_j x_i} 2\zeta\zeta_{x_j} - u_{x_i} u_{x_j x_j} 2\zeta\zeta_{x_i} \, d\mathbf{x}.
\end{aligned}
$$

We now use this with inequalities of the form

$$
|u_{x_i} u_{x_j x_i} 2\zeta\zeta_{x_j}| \le \epsilon u_{x_i x_j}^2 \zeta^2 + \frac{1}{\epsilon} u_{x_i}^2 \zeta_{x_j}^2 \tag{9.158}
$$

and

$$
\int_\Omega |\Delta u|^2 \zeta^2 \, d\mathbf{x} = \int_\Omega |f|^2 \zeta^2 \, d\mathbf{x} \le \int_\Omega |f|^2 \, d\mathbf{x} \tag{9.159}
$$

to get

$$
\int_\Omega |\nabla^2 u|^2 \zeta^2 \, d\mathbf{x} \le \int_\Omega \{|f|^2 + \epsilon |\nabla^2 u|^2 \zeta^2 + C(\epsilon)|\nabla u|^2 |\nabla \zeta|^2\} \, d\mathbf{x}. \tag{9.160}
$$

We now let $\epsilon = 1/2$ and use the fact that $\zeta \equiv 1$ on Ω' to get

$$
\int_{\Omega'} |\nabla^2 u|^2 \, d\mathbf{x} \le \int_\Omega \zeta^2 |\nabla^2 u|^2 \, d\mathbf{x} \le C \left(\|f\|_2^2 + \|\nabla u\|_2^2 \right). \tag{9.161}
$$

Thus, we have an estimate on the $H^2(\Omega')$ norm of a solution u for any $\Omega' \subset\subset \Omega$ in terms of the $L^2(\Omega)$ of the data f and the $H^1(\Omega)$ norm of u.

Of course, one of the major objections to the calculations performed above is that we needed to make unwarranted assumptions about the smoothness of the solution u in order to perform the integrations by parts involved. In the rigorous versions of these calculations below, these operations are replaced by analogous techniques involving difference quotients. Because the technique of using difference quotients is so important in this section, we present the following short digression on this topic.

9.5.1 Difference Quotients

Let $\Omega \subset \mathbb{R}^n$ and let $\{\mathbf{e}_1, \ldots, \mathbf{e}_n\}$ be the standard orthonormal basis for $\mathbb{R}^n$. For any function $u \in L^p(\Omega)$ we can formally define the difference quotient in the direction $\mathbf{e}_i$ to be

$$D_i^h u(\mathbf{x}) := \frac{u(\mathbf{x} + h\mathbf{e}_i) - u(\mathbf{x})}{h}. \tag{9.162}$$

Of course, since $\mathbf{x} + h\mathbf{e}_i$ might extend beyond Ω for $\mathbf{x}$ near the boundary, this function might not be well defined for all $\mathbf{x} \in \Omega$. However, we can get the following result.

Lemma 9.48. *Let $u \in W^{1,p}(\Omega)$, $1 \le p \le \infty$. Then for any $\Omega' \subset\subset \Omega$ and any $h < dist(\Omega', \partial\Omega)$, we have $D_i^h u \in L^p(\Omega')$ and*

$$\|D_i^h u\|_{L^p(\Omega')} \le \|u_{x_i}\|_{L^p(\Omega)}. \tag{9.163}$$

Proof. Let Ω' and h satisfy the hypotheses of the lemma. For any $\xi \in [0, h]$ we define

$$\Omega'_{\xi,i} := \{\mathbf{x} \in \Omega \mid \mathbf{x} = \bar{\mathbf{x}} + \xi\mathbf{e}_i, ; \quad \bar{\mathbf{x}} \in \Omega'\}. \tag{9.164}$$

For $p \in [1, \infty)$ we first consider the case where $u \in C_b^1(\Omega) \cap W^{1,p}(\Omega)$. Then for any $\mathbf{x} \in \Omega'$, we can use the fundamental theorem of calculus to write

$$\begin{aligned} D_i^h u(\mathbf{x}) &= \frac{u(\mathbf{x} + h\mathbf{e}_i) - u(\mathbf{x})}{h} \\ &= \frac{1}{h} \int_0^h u_{x_i}(\mathbf{x} + \xi\mathbf{e}_i) \, d\xi. \end{aligned} \tag{9.165}$$

Thus, using Hölder's inequality, and switching orders of integration, we get

$$\begin{aligned} \int_{\Omega'} |D_i^h u(\mathbf{x})|^p \, d\mathbf{x} &\le \int_{\Omega'} \frac{1}{h} \int_0^h |u_{x_i}(\mathbf{x} + \xi\mathbf{e}_i)|^p \, d\xi \, d\mathbf{x} \\ &\le \frac{1}{h} \int_0^h \int_{\Omega'_{\xi,i}} |u_{x_i}(\mathbf{x})|^p \, d\mathbf{x} \, d\xi \\ &\le \int_{\Omega} |u_{x_i}|^p \, d\mathbf{x}. \end{aligned}$$

By Theorem 7.48, this inequality extends to the whole space by taking limits.

For $p = \infty$ we note that (9.165) holds in the sense of distributions. Since the test functions are dense in L^1, the inequality follows from manipulating and estimating $D_i^h u$ and u_{x_i} as linear functionals on sets of test functions. We leave this to the reader. $\qquad\square$

Of course, it is not at all surprising that if a function is in $W^{1,p}(\Omega)$, its difference quotients obey some bound in terms of its partial derivatives. The following result is more substantial; it says that if we start out knowing that u is in $L^p(\Omega)$ and can obtain a bound on its difference quotients that is independent of h, then we can deduce that u is in the space $W^{1,p}(\Omega)$.

Lemma 9.49. *Let* $u \in L^p(\Omega)$, $1 < p \leq \infty$, *and suppose there exists a constant* $\bar{C}$ *such that for any* $\Omega' \subset\subset \Omega$ *and* $h < \mathrm{dist}(\Omega', \partial\Omega)$ *we have* $D_i^h u \in L^p(\Omega')$ *and*

$$\|D_i^h u\|_{L^p(\Omega')} \leq \bar{C}. \tag{9.166}$$

Then u_{x_i} *(which is a priori well defined as a distribution) is in fact in* $L^p(\Omega)$ *and satisfies*

$$\|u_{x_i}\|_{L^p(\Omega)} \leq \bar{C}. \tag{9.167}$$

Proof. Recall that at the end of Section 5.5.1 we showed the distributional derivative of a function could be obtained as the limit of difference quotients. In terms of the present problem we have

$$(u_{x_i}, \phi) = \lim_{h \to 0} (D_i^h u, \phi) \tag{9.168}$$

for each $\phi \in \mathcal{D}(\Omega)$.

We now note that there exists a function $v \in L^p(\Omega)$ with $\|v\|_{L^p(\Omega)} \leq \bar{C}$ and a sequence $h_m \to 0$ such that for every $\phi \in \mathcal{D}(\Omega)$

$$\lim_{h_m \to 0} \int_\Omega \phi D_i^{h_m} u \, d\mathbf{x} = \int_\Omega \phi v \, d\mathbf{x}. \tag{9.169}$$

This follows from the weak compactness of the bounded set $D_i^h u$ in $L^p(\Omega')$ for every $\Omega' \subset\subset \Omega$ (cf. Theorem 6.64) and the fact that Ω can be covered with a countable collection of closed subsets $\Omega_j \subset\subset \Omega$ such that each $\Omega' \subset\subset \Omega$ intersects at most a finite number of the Ω_j.

Thus,

$$(u_{x_i}, \phi) = \int_\Omega \phi v \, d\mathbf{x}; \tag{9.170}$$

i.e., the distributional derivative of u is given (uniquely) by the function $v \in L^p(\Omega)$. $\qquad\square$

We also need to develop a few important tools using difference operators; namely, the analogues of the product rule and integration by parts in differential calculus.

Lemma 9.50. *Suppose* $\Omega' \subset\subset \Omega$ *and* $h < \mathrm{dist}(\Omega', \partial\Omega)$. *Then for any* $u \in L^p(\Omega)$, $1 < p < \infty$, *and any* $v \in C_b(\Omega)$ *with* supp $v \subseteq \Omega'$ *we have*

$$D_i^h(uv)(\mathbf{x}) = u(\mathbf{x})D_i^h v(\mathbf{x}) + D_i^h u(\mathbf{x})v(\mathbf{x} + h\mathbf{e}_i) \tag{9.171}$$

and

$$\int_\Omega D_i^h u(\mathbf{x})v(\mathbf{x}) \, d\mathbf{x} = - \int_\Omega u(\mathbf{x})D_i^{-h} v(\mathbf{x}) \, d\mathbf{x}. \tag{9.172}$$

The proof is left to the reader.

9.5.2   Second-Order Scalar Equations

In order to eliminate many technical details, we will give a proof of an interior regularity result only in the case of a second-order scalar equation. We already gave statements of results for higher-order equations and systems in Section 9.4 above.

Theorem 9.51 (Interior regularity). *Let L be a uniformly elliptic second-order operator of the form*

$$L(x, D)u := \sum_{i,j=1}^{n} \frac{\partial}{\partial x_i} a_{ij}(\mathbf{x}) \frac{\partial u}{\partial x_j} + \sum_{i=1}^{n} b_i(\mathbf{x}) \frac{\partial u}{\partial x_i} + c(\mathbf{x})u,$$

with corresponding bilinear form

$$B[v, u] := - \sum_{i,j=1}^{n} \int_{\Omega} a_{ij}(\mathbf{x}) u_{x_j} v_{x_i}\, dx + \sum_{i=1}^{n} \int_{\Omega} b_i(\mathbf{x}) u_{x_i} v\, dx + \int_{\Omega} c(\mathbf{x}) uv\, dx.$$

Suppose the coefficients satisfy $a_{ij} \in W^{1,\infty}(\Omega)$, $b_i, c \in L^\infty(\Omega)$ and that $f \in L^2(\Omega)$. Let $u \in H_0^1(\Omega)$ be a weak solution of the Dirichlet problem for $L(D, \mathbf{x})u = f$. Then $u \in H^2(\Omega')$ for every $\Omega' \subset\subset \Omega$, and

$$\|u\|_{H^2(\Omega')} \le C(\|u\|_{H^1(\Omega)} + \|f\|_{L^2(\Omega)}). \tag{9.173}$$

Proof. We begin with the identity

$$\int_{\Omega} a_{ij} u_{x_j} v_{x_i}\, dx = \int_{\Omega} gv\, dx \tag{9.174}$$

for all $v \in H_0^1(\Omega)$, where $g \in L^2(\Omega)$ is given by

$$g := b_i u_{x_i} + cu - f. \tag{9.175}$$

Now suppose $v \in H^1(\Omega)$ and $\operatorname{supp} v \subset\subset \Omega$ and let

$$|2h| < \operatorname{dist}(\operatorname{supp} v, \partial\Omega).$$

Then we have (for any $k = 1, \ldots, n$) $D_k^{-h} v \in H_0^1(\Omega)$, and thus we can use (9.174) and the "differencing by parts formula" (9.172) to get

$$\int_{\Omega} D_k^h(a_{ij} u_{x_j}) v_{x_i}\, dx = - \int_{\Omega} a_{ij} u_{x_j} D_k^{-h} v_{x_i}\, dx$$

$$= - \int_{\Omega} g D_k^{-h} v\, dx.$$

We can now use this and the product rule for difference quotients (9.171) to get

$$\int_{\Omega} a_{ij}(\mathbf{x} + h\mathbf{e}_k)(D_k^h u_{x_j}) v_{x_i}\, dx = - \int_{\Omega} (D_k^h a_{ij}) u_{x_j} v_{x_i} + g D_k^{-h} v\, dx. \tag{9.176}$$

We can now use Lemma 9.48 to estimate this by

$$\left| \int_{\Omega} a_{ij}(\mathbf{x} + h\mathbf{e}_k)(D_k^h u_{x_j}) v_{x_i}\, dx \right| \le C(\|u\|_{1,2} + \|f\|_2) \|\nabla v\|_2. \tag{9.177}$$

Let $\Omega' \subset\subset \Omega$. We now choose a cutoff function $\zeta \in \mathcal{D}(\Omega)$ with the following properties.

1. $\zeta \equiv 1$ on Ω'.

2. $|\nabla\zeta| < 2/d$ where $d = \text{dist}(\Omega', \partial\Omega)$.

We now use the ellipticity condition (9.6), elementary inequalities, and the previous estimate (with $v = \zeta^2 D_k^h u$) to obtain

$$\theta \int_\Omega |\zeta \nabla D_k^h u|^2 \, d\mathbf{x} \le - \int_\Omega \zeta^2 a_{ij}(\mathbf{x} + h\mathbf{e}_k) D_k^h u_{x_i} D_k^h u_{x_j} \, d\mathbf{x}$$

$$= - \int_\Omega a_{ij}(\mathbf{x} + h\mathbf{e}_k) D_k^h u_{x_i} \left[(\zeta^2 D_k^h u)_{x_j} - 2 D_k^h u \zeta \zeta_{x_j} \right] \, d\mathbf{x}$$

$$\le C[(\|u\|_{1,2} + \|f\|_2)(\|\zeta \nabla D_k^h u\|_2 + 2\|D_k^h u \nabla \zeta\|_2)$$
$$+ \sup_{i,j} \|a_{ij}\|_\infty \|\zeta \nabla D_k^h u\|_2 \|D_k^h u \nabla \zeta\|_2]$$

$$\le C(\epsilon)[(\|u\|_{1,2} + \|f\|_2 + \|D_k^h u \nabla \zeta\|_2)^2 + \epsilon\|\zeta \nabla D_k^h u\|_2^2]$$

for any $\epsilon > 0$.

Thus, after making an appropriate choice of ϵ and rearranging, we have

$$\|D_k^h \nabla u\|_{L^2(\Omega')} \le \|\zeta D_k^h \nabla u\|_{L^2(\Omega)}$$
$$\le C(\|u\|_{H^1(\Omega)} + \|f\|_{L^2(\Omega)} + \|D_k^h u \nabla \zeta\|)$$
$$\le C(\|u\|_{H^1(\Omega)} + \|f\|_{L^2(\Omega)}).$$

Here we have used Lemma 9.48 and our pointwise bound on $|\nabla\zeta|$.

Finally, we can use this estimate on the difference quotients of ∇u and Lemma 9.49 to deduce that $\nabla u \in H^1(\Omega')$. The estimate (9.173) follows immediately. $\qquad\square$

Problems

9.12. Show that if Ω is bounded, $f : \mathbb{R} \to \mathbb{R}$ is uniformly Lipschitz, and $u \in W^{1,p}(\Omega)$ with $1 < p \le \infty$, then the composite $f \circ u$ belongs to $W^{1,p}(\Omega)$.

9.13. Give an example to show that Lemma 9.49 fails for $p = 1$.

9.6 Boundary Regularity

In the previous section we showed that if the data and coefficients are sufficiently smooth, then weak solutions of elliptic problems are "as smooth as one could expect" in the *interior* of the domain on which the problem is posed. In the following example we see that a solution is not necessarily smooth up to the boundary of the domain on which the problem is posed if that boundary is not sufficiently smooth. (The reader should also note that this is not some sort of weird "cooked up" counterexample; the domain in question simply has a corner.)

Example 9.52. Let **i** and **j** give an orthonormal basis for $\mathbb{R}^2$, and let (r, θ) be the standard polar coordinates for $\mathbb{R}^2$ defined by the map

$$\hat{\mathbf{x}}(r, \theta) := r\mathbf{e}_1(\theta), \tag{9.178}$$

where

$$\mathbf{e}_1(\theta) := \cos\theta\mathbf{i} + \sin\theta\mathbf{j}, \tag{9.179}$$
$$\mathbf{e}_2(\theta) := -\sin\theta\mathbf{i} + \cos\theta\mathbf{j}. \tag{9.180}$$

Recall that for a real-valued function $f(r, \theta)$ we can calculate the gradient and Laplacian as follows:

$$\nabla f = f_r\mathbf{e}_1 + \frac{1}{r}f_\theta\mathbf{e}_2, \tag{9.181}$$

$$\Delta f = \frac{1}{r}\frac{\partial}{\partial r}\left(r\frac{\partial f}{\partial r}\right) + \frac{1}{r^2}\frac{\partial^2 f}{\partial\theta^2}. \tag{9.182}$$

In addition, the Hessian or second gradient matrix is given by

$$\begin{pmatrix} f_{rr} & \frac{1}{r}f_{r\theta} - \frac{1}{r^2}f_\theta \\ \frac{1}{r}f_{r\theta} - \frac{1}{r^2}f_\theta & \frac{1}{r}f_r + \frac{1}{r^2}f_{\theta\theta} \end{pmatrix}. \tag{9.183}$$

We now consider the following problem for Laplace's equation with nonhomogeneous boundary conditions. Let $0 < \beta < 2\pi$ and define

$$\Omega_\beta := \{\mathbf{x} \in \mathbb{R}^2 \mid \mathbf{x} = \hat{\mathbf{x}}(r, \theta), \quad 0 < r < 1, \quad 0 < \theta < \beta\}. \tag{9.184}$$

We seek $u : \Omega_\beta \to \mathbb{R}$ satisfying

$$\Delta u = 0 \quad \text{in } \Omega_\beta, \tag{9.185}$$

and the boundary conditions

$$u(r, 0) = 0, \quad 0 < r < 1$$
$$u(r, \beta) = 0, \quad 0 < r < 1$$
$$u(1, \theta) = \sin\frac{\pi\theta}{\beta}, \quad 0 < \theta < \beta.$$

Using separation of variables, we can find the solution of the problem to be

$$u(r, \theta) = r^{\pi/\beta}\sin\frac{\pi\theta}{\beta}. \tag{9.186}$$

This function (as the results of the previous section assure us) is in $C^\infty(\Omega')$ for any Ω' compactly contained in Ω_β. However, note that if $\beta > \pi$, then our solution decays at the origin like r^α with $\alpha < 1$. To see the impact of this we calculate the gradient of u

$$\nabla u = \frac{\pi}{\beta} r^{\pi/\beta - 1} \left(\sin \frac{\pi\theta}{\beta} \mathbf{e}_1(\theta) + \cos \frac{\pi\theta}{\beta} \mathbf{e}_2(\theta) \right) \tag{9.187}$$

and its $L^2(\Omega_\beta)$ norm

$$\int_{\Omega_\beta} |\nabla u|^2 \, dx = C \int_0^\beta \int_0^1 r^{2(\pi/\beta - 1)} r \, dr \, d\theta \leq \infty. \tag{9.188}$$

We see that (as our basic existence theorem for weak solutions implies) our solution u is in $H^1(\Omega_\beta)$. However, by calculating the second gradient and computing its norm, we see that if $\beta > \pi$, then u is **not** in $H^2(\Omega_\beta)$. Thus, despite the fact that we have all of the interior regularity guaranteed by the results of the previous section, *we do not have regularity up to the boundary.* The culprit here is the lack of smoothness of the boundary.

As the example above indicates, we will need to assume that the boundary has some smoothness properties in order to get a boundary regularity result (also called a global regularity result). In order to emphasize the most important techniques in the proof (breaking up the domain using a partition of unity and mapping the pieces containing portions of the boundary to a half-space) we will give the proof only for second-order scalar equations and in the proof we will ignore lower-order terms.

Theorem 9.53 (Global regularity). *Suppose that the hypotheses of Theorem 9.51 hold and that in addition $\partial\Omega$ is of class C^2. Then $u \in H^2(\Omega)$ and*

$$\|u\|_{H^2(\Omega)} \leq C(\|u\|_{L^2(\Omega)} + \|f\|_{L^2(\Omega)}). \tag{9.189}$$

The proof of this result is rather long and involved, so we will break it up by proving a number of preliminary lemmas. One of our basic techniques is to decompose the domain into pieces using a partition of unity and "flattening out" any portion of the boundary. As we see in our first lemma (which is essentially a version of the main result in the case where the boundary is already flat) a flat boundary allows us to use difference quotients to our advantage.

Lemma 9.54. *Let $R > 0$, $\lambda \in (0,1)$, and define*

$$D^+ \quad := \quad B_R(\mathbf{0}) \cap \{\mathbf{x} \in \mathbb{R}^n \mid x_n > 0\}, \tag{9.190}$$
$$Q^+ \quad := \quad B_{\lambda R}(\mathbf{0}) \cap \{\mathbf{x} \in \mathbb{R}^n \mid x_n > 0\}. \tag{9.191}$$
$$\tag{9.192}$$

Let L be a uniformly elliptic second-order operator of the form

$$L(x, D)u := \sum_{i,j=1}^n \frac{\partial}{\partial x_i} a_{ij}(\mathbf{x}) \frac{\partial u}{\partial x_j} + \sum_{i=1}^n b_i(\mathbf{x}) \frac{\partial u}{\partial x_i} + c(\mathbf{x})u$$

with corresponding bilinear form

$$B[v, u] := -\sum_{i,j=1}^{n} \int_{D^+} a_{ij}(\mathbf{x}) u_{x_j} v_{x_i} \, d\mathbf{x} + \sum_{i=1}^{n} \int_{D^+} b_i(\mathbf{x}) u_{x_i} v \, d\mathbf{x}$$
$$+ \int_{D^+} c(\mathbf{x}) uv \, d\mathbf{x}.$$

Suppose the coefficients satisfy $a_{ij} \in W^{1,\infty}(D^+)$, $b_i, c \in L^\infty(D^+)$ and that $f \in L^2(D^+)$. Suppose $u \in H^1(D^+)$ satisfies the variational equation

$$B[v, u] = (v, f) \qquad (9.193)$$

for all $v \in H_0^1(D^+)$ and that $u \equiv 0$ in the sense of trace on $\{\mathbf{x} \in \mathbb{R}^n \mid x_n = 0\}$. Then $u \in H^2(Q^+)$ and there exists a constant C depending on R such that

$$\|u\|_{H^2(Q^+)} \le C(\|f\|_{L^2(D^+)} + \|u\|_{L^2(D^+)}). \qquad (9.194)$$

Proof. Let $h \in (0, R(1-\lambda)/2)$ and fix an index $k = 1, \ldots, n-1$ (i.e., $k \ne n$). Now choose $\zeta \in C_b^\infty(D^+)$ such that

1. $0 \le \zeta \le 1$,

2. $\zeta \equiv 1$ on Q^+,

3. $U := \operatorname{supp} \zeta \subset B_{R(1+\lambda)/2}(0)$.

Note: The function ζ in not in $\mathcal{D}(D^+)$ since $\zeta \ne 0$ on the flat part of the boundary of D^+: $\{x_n \equiv 0\}$.

Now define

$$v := -D_k^{-h}(\zeta^2 D_k^h u). \qquad (9.195)$$

After some manipulations using the definition of the difference quotients, we get the following identity.

$$v(\mathbf{x}) = -\frac{1}{h^2}(\zeta^2(\mathbf{x})[u(\mathbf{x} + h\mathbf{e}_k) - u(\mathbf{x})] + \zeta^2(\mathbf{x} - h\mathbf{e}_k)[u(\mathbf{x} - h\mathbf{e}_k) - u(\mathbf{x})]).$$
$$(9.196)$$

Note that in constructing v we have used translations only in directions tangential to the plane $x_n = 0$. The key idea is that we can "slide the support of u" along the plane $x_n = 0$ without destroying the boundary conditions. Also note that none of the translations moves the support of ζ outside of $\overline{D^+}$. These facts ensure that

1. v is well defined on D^+,

2. $v \in H^1(D^+)$ (since $u \in H^1(D^+)$),

3. $v \in H_0^1(D^+)$ (since ζ is zero on the curved part of the boundary of D^+ and u is zero on the flat part of the boundary (and the same goes for any of the translations of ζ and u)).

We define

$$w := \zeta D_k^h u, \tag{9.197}$$

so that $v = -D_k^{-h}(\zeta w)$.

Now, we follow a procedure similar to the derivation of the estimate (9.177) of Theorem 9.51, but using v as defined above. We get

$$
\begin{aligned}
-\int_{D+} g D_k^{-h}(\zeta w)\, d\mathbf{x} &= \int_{D+} a_{ij} u_{x_j} v_{x_i}\, d\mathbf{x} \\
&= \int_{D+} D_k^h(a_{ij} u_{x_j})(\zeta w)_{x_i}\, d\mathbf{x} \\
&= \int_{D+} a_{ij}(\mathbf{x} + h\mathbf{e}_k) D_k^h u_{x_j}(\zeta w)_{x_i}\, d\mathbf{x} \\
&\quad + \int_{D+} D_k^h a_{ij} u_{x_j}(\zeta w)_{x_i}\, d\mathbf{x} \\
&= \int_{D+} a_{ij}(\mathbf{x} + h\mathbf{e}_k) w_{x_j} w_{x_i}\, d\mathbf{x} \\
&\quad - \int_{D+} a_{ij}(\mathbf{x} + h\mathbf{e}_k)\zeta_{x_j} D_k^h u w_{x_i}\, d\mathbf{x} \\
&\quad + \int_{D+} a_{ij}(\mathbf{x} + h\mathbf{e}_k) D_k^h u_{x_j} \zeta_{x_i} w\, d\mathbf{x} \\
&\quad + \int_{D+} D_k^h a_{ij} u_{x_j}(\zeta w)_{x_i}\, d\mathbf{x}.
\end{aligned}
$$

Here g is defined as in (9.175). Rearranging, we get

$$\int_{D+} a_{ij}(\mathbf{x} + h\mathbf{e}_k) w_{x_j} w_{x_i}\, d\mathbf{x} = I_1 + I_2 + I_3 + I_4, \tag{9.198}$$

where

$$I_1 := -\int_{D+} g D_k^{-h}(\zeta w)\, d\mathbf{x}, \tag{9.199}$$

$$I_2 := \int_{D+} a_{ij}(\mathbf{x} + h\mathbf{e}_k)\zeta_{x_j} D_k^h u w_{x_i}\, d\mathbf{x}, \tag{9.200}$$

$$I_3 := -\int_{D+} a_{ij}(\mathbf{x} + h\mathbf{e}_k) D_k^h u_{x_i} \zeta_{x_i} w\, d\mathbf{x}, \tag{9.201}$$

$$I_4 := -\int_{D+} D_k^h a_{ij} u_{x_j}(\zeta w)_{x_i}\, d\mathbf{x}. \tag{9.202}$$

We estimate these terms using techniques which should be familiar to the reader from the proof of the Theorem 9.51 (basically Hölder's inequality and Lemma 9.48). We will make use of the estimate

$$\|v\|_2 = \|D_k^{-h}(\zeta w)\|_2 \le 2\|\zeta\|_{1,\infty}\|w\|_{1,2}. \tag{9.203}$$

(Here, $\|\cdot\|_2 = \|\cdot\|_{L^2(D^+)}$, etc.) We get the following:

$$
\begin{aligned}
|I_1| &\leq \sum_i \|b_i\|_\infty \|\nabla u\|_2 \|v\|_2 + \|c\|_\infty \|u\|_2 \|v\|_2 + \|f\|_2 \|v\|_2 \\
&\leq C(\|u\|_{1,2} + \|f\|_2) \|w\|_{1,2}.
\end{aligned}
$$

$$
\begin{aligned}
|I_2| &\leq \sum_{i,j} \|a_{ij}\|_\infty \|D_k^h u\|_2 \|w_{x_i}\|_2 \|\nabla \zeta\|_\infty \\
&\leq C\|u\|_{1,2} \|w\|_{1,2}.
\end{aligned}
$$

$$
\begin{aligned}
|I_3| &\leq \sum_{i,j} \|a_{ij}\|_\infty \|\zeta_{x_i}\|_\infty \|D_k^h u\|_2 \|w\|_2 \\
&\leq C\|u\|_{1,2} \|w\|_{1,2}.
\end{aligned}
$$

$$
\begin{aligned}
|I_4| &\leq \sum_{i,j} \|a_{ij}\|_{1,\infty} \|\zeta\|_{1,\infty} \|u\|_{1,2} \|w\|_{1,2} \\
&\leq C\|u\|_{1,2} \|w\|_{1,2}.
\end{aligned}
$$

Combining these estimates and the uniform ellipticity condition gives us

$$
\theta \int_{D^+} |\nabla w|^2 \, d\mathbf{x} \leq - \int_{D^+} a_{ij}(\mathbf{x} + h\mathbf{e}_k) w_{x_i} w_{x_j} \, d\mathbf{x} \leq C(\|u\|_{1,2} + \|f\|_2) \|w\|_{1,2}. \tag{9.204}
$$

It follows that

$$
\int_{D^+} |\nabla w|^2 \leq C(\|f\|_2^2 + \|u\|_{1,2}^2). \tag{9.205}
$$

We can use this and the fact that $\zeta \equiv 1$ on Q^+ to show that

$$
\begin{aligned}
\int_{Q^+} |D_k^h \nabla u|^2 \, d\mathbf{x} &= \int_{Q^+} |\nabla w|^2 \, d\mathbf{x} \\
&\leq \int_{D^+} |\nabla w|^2 \, d\mathbf{x} \\
&\leq C(\|f\|_2^2 + \|u\|_{1,2}^2).
\end{aligned}
$$

Thus, using Lemma 9.48, we can get an estimate for all second-order mixed partial derivatives except for $u_{x_n x_n}$; i.e., we have

$$
\sum_{\substack{i,j=1 \\ i+j<2n}}^n \int_{Q^+} |u_{x_i x_j}|^2 \, d\mathbf{x} \leq C(\|f\|_2^2 + \|u\|_{1,2}^2). \tag{9.206}
$$

To estimate $u_{x_n x_n}$ we proceed as follows. From the interior regularity result we know that the differential equation

$$
L(D, \mathbf{x})u = f
$$

is satisfied in the strong or L^2 sense. Since $a_{nn} \in W^{1,\infty}(\Omega)$, the Sobolev embedding theorem ensures that it is continuous. Furthermore, the strong ellipticity condition ensures that $a_{nn} \geq \theta > 0$. Thus, we can divide the PDE by a_{nn} and rearrange to get

$$u_{x_n x_n} = -\frac{1}{a_{nn}} \left[\sum_{\substack{i,j=1 \\ i+j<2n}}^{n} a_{ij} u_{x_i x_j} - g \right] \tag{9.207}$$

Note that we have shown that the right-hand side of this equation is in $L^2(Q^+)$; and in addition, we have the estimate

$$\int_{Q^+} |u_{x_n x_n}|^2 \, dx \leq C(\|f\|_2^2 + \|u\|_{1,2}^2). \tag{9.208}$$

Combining this with our previous results we get our final estimate (9.194). $\square$

We now state without proof the "higher-order" version of Lemma 9.54.

Lemma 9.55. *Let the hypotheses of Lemma 9.54 hold. In addition, assume that the coefficients satisfy $a_{ij} \in W^{k+1,\infty}(D^+)$, $b_i, c \in W^{k,\infty}(D^+)$ and that $f \in H^k(D^+)$. Then $u \in H^{k+2}(Q^+)$ and there exists a constant C depending on R such that*

$$\|u\|_{H^{k+2}(Q^+)} \leq C(\|f\|_{H^k(D^+)} + \|u\|_{H^1(D^+)}). \tag{9.209}$$

To prove Theorem 9.53 we must consider a general domain Ω. Of course, since we did all the work of proving Lemma 9.54 on a half-space, our next task is to "straighten out" the boundary of Ω. Since $\partial\Omega$ is a C^2 surface, we know that for each $\bar{x} \in \partial\Omega$ there is an $R > 0$ and a $C^2(\mathbb{R}^{n-1})$ function ψ such that (after a possible renumbering and reorientation of coordinates)

$$\partial\Omega \cap B_R(\bar{x}) = \{x \in B_R(\bar{x}) \mid x_n = \psi(x_1, x_2, \ldots, x_{n-1})\}, \tag{9.210}$$
$$\Omega \cap B_R(\bar{x}) = \{x \in B_R(\bar{x}) \mid x_n > \psi(x_1, x_2, \ldots, x_{n-1})\}; \tag{9.211}$$

and moreover, the mapping

$$B_R(\bar{x}) \ni x \mapsto y = \Psi(x) \in \mathbb{R}^n \tag{9.212}$$

defined by

$$y_i := x_i - \bar{x}_i, \quad i = 1, \ldots, n-1,$$
$$y_n := x_n - \psi(x_1, \ldots, x_{n-1})$$

is one-to one. Define $\Phi := \Psi^{-1}$. Note that Φ is a C^2 function which transforms the set $\Omega' := \Omega \cap B_R(\bar{x})$ (in what we refer to as x space) into a set Ω'' in the half-space $y_n > 0$ (of y space). Note also that the point $\bar{x}$ is mapped to the origin of y space (cf. Figure 9.1).

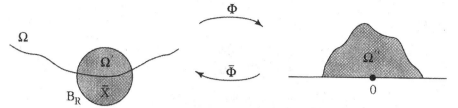

Figure 9.1. Straightening out the boundary.

Our task now is obvious (and obviously unpleasant). We must change the differential equation $L(\mathbf{x}, D)u = f$ into $\mathbf{y}$ coordinates. To facilitate this task we define the following notation: for any function

$$v : \Omega' \to \mathbb{R},$$

we define

$$\tilde{v} : \Omega'' \to \mathbb{R} \tag{9.213}$$

by

$$\tilde{v}(\mathbf{y}) := v(\Phi(\mathbf{y})). \tag{9.214}$$

Note that for any function $v \in L^2(\Omega)$ there are constants c_1 and c_2 such that

$$c_1 \|v\|_{L^2(\Omega')} \le \|\tilde{v}\|_{L^2(\Omega'')} \le c_2 \|v\|_{L^2(\Omega')}. \tag{9.215}$$

The action of the change of variables on our partial differential operator is described by the following lemma.

Lemma 9.56. *Let $u \in H^1(\Omega')$ satisfy $u \equiv 0$ (in the sense of trace) on $\partial\Omega \cap \partial\Omega'$ and let u be a solution of the variational equation*

$$B[v, u] = (f, v), \tag{9.216}$$

for all $v \in H_0^1(\Omega')$. Then $\tilde{u} \in H^1(\Omega'')$ satisfies $\tilde{u} \equiv 0$ on $\partial\Omega'' \cap \{\mathbf{y} \mid y_n = 0\}$ and $\tilde{u}$ is a solution of the variational equation

$$\tilde{B}[\tilde{v}, \tilde{u}] = (\tilde{f}, \tilde{v}), \tag{9.217}$$

for every $\tilde{v} \in H_0^1(\Omega'')$. Here

$$\tilde{B}[\tilde{v}, \tilde{w}] := -\sum_{k,l=1}^{n} \int_{\Omega''} \bar{a}_{kl}(\mathbf{y}) \tilde{w}_{y_l} \tilde{v}_{y_k} \, d\mathbf{y} + \sum_{k=1}^{n} \int_{\Omega''} \bar{b}_k(\mathbf{y}) \tilde{w}_{y_k} \tilde{v} \, d\mathbf{y}$$

$$quad + \int_{\Omega''} \bar{c}(\mathbf{y}) \tilde{v} \tilde{w} \, d\mathbf{y},$$

$$\tag{9.218}$$

with

$$\bar{a}_{kl} := \sum_{i,j=1}^{n} \tilde{a}_{ij}(\mathbf{y}) \frac{\partial \Psi_k}{\partial x_i}(\Phi(\mathbf{y})) \frac{\partial \Psi_l}{\partial x_j}(\Phi(\mathbf{y})), \tag{9.219}$$

$$\bar{b}_k := \sum_{i=1}^{n} \tilde{b}_i(\mathbf{y}) \frac{\partial \Psi_k}{\partial x_i}(\Phi(\mathbf{y})), \tag{9.220}$$

$$\bar{c}(\mathbf{y}) := \tilde{c}(\mathbf{y}). \tag{9.221}$$

The proof uses standard techniques and is left to the reader.

Before applying Lemma 9.54 we need to show that the transformed differential operator is uniformly elliptic.

Lemma 9.57. *The operator $\tilde{L}$ defined by*

$$\tilde{L}(\mathbf{y}, D)v := \sum_{k,l=1}^{n} \left(\bar{a}_{kl}(\mathbf{y}) v_{y_l} \right)_{y_k} + \sum_{k=1}^{n} \bar{b}_k(\mathbf{y}) v_{y_k} + \bar{c}(\mathbf{y})v \tag{9.222}$$

is uniformly elliptic in Ω''.

Proof. We must show that there exists a constant $\tilde{\theta} > 0$ such that

$$-\sum_{k,l=1}^{n} \bar{a}_{kl}(\mathbf{y}) \xi_k \xi_l \geq \tilde{\theta} |\boldsymbol{\xi}|^2 \tag{9.223}$$

for every $\boldsymbol{\xi} \in \mathbb{R}^n$ and every $\mathbf{y} \in \Omega''$.

For any $\boldsymbol{\xi} \in \mathbb{R}^n$ let $\boldsymbol{\eta} := \mathbf{A}\boldsymbol{\xi}$ where

$$\mathbf{A}(\mathbf{y}) := \nabla \Psi(\Phi(\mathbf{y}))^T$$
$$= \begin{pmatrix} 1 & 0 & \cdots & 0 & -\psi_{x_1} \\ 0 & 1 & \cdots & 0 & -\psi_{x_2} \\ \vdots & \vdots & \ddots & \vdots & \vdots \\ 0 & 0 & \cdots & 1 & -\psi_{x_{n-1}} \\ 0 & 0 & \cdots & 0 & 1 \end{pmatrix}.$$

Note that $\mathbf{A}(\mathbf{y})$ is invertible. Let

$$\tilde{C} := \sup_{\mathbf{y} \in \Omega''} |\mathbf{A}^{-1}(\mathbf{y})|. \tag{9.224}$$

Then

$$|\boldsymbol{\xi}| \leq \tilde{C} |\boldsymbol{\eta}|. \tag{9.225}$$

Now, using (9.225) and the uniform ellipticity of L we get

$$-\sum_{k,l=1}^{n} \bar{a}_{kl}\xi_k\xi_l = -\sum_{i,j=1}^{n} a_{ij}(\phi(\mathbf{y}))\sum_{k,l=1}^{n}\left(\frac{\partial\Psi_k}{\partial x_i}(\Phi(\mathbf{y}))\xi_k\right)\left(\frac{\partial\Psi_l}{\partial x_j}(\Phi(\mathbf{y}))\xi_l\right)$$

$$= -\sum_{i,j=1}^{n} a_{ij}(\phi(\mathbf{y}))\eta_i\eta_j$$

$$\geq \theta|\boldsymbol{\eta}|^2$$

$$\geq \frac{\theta}{\bar{C}^2}|\boldsymbol{\xi}|^2.$$

Thus, $\tilde{L}$ is uniformly elliptic with constant $\tilde{\theta} := \theta/\bar{C}^2$. $\square$

We can now put the previous lemmas together to get the following result.

Lemma 9.58. *Let the hypotheses of Theorem 9.53 be satisfied. Then for each $\bar{\mathbf{x}} \in \partial\Omega$ there exists an open set $\tilde{Q} \subset \mathbb{R}^n$ containing $\bar{\mathbf{x}}$ such that $u \in H^2(\tilde{Q} \cap \Omega)$, and furthermore*

$$\|u\|_{H^2(\tilde{Q}\cap\Omega)} \leq C(\|f\|_{L^2(\Omega)} + \|u\|_{H^1(\Omega)}). \tag{9.226}$$

Proof. For each $\bar{x} \in \partial\Omega$ we let the sets Ω' in $\mathbf{x}$ space, Ω'' in $\mathbf{y}$ space and the maps $\Psi : \Omega' \to \Omega''$ and $\Phi : \Omega'' \to \Omega'$ be defined as above (cf. Figure 9.1). Let $\bar{R}$ be such that $B_{\bar{R}}(\mathbf{0}) \cap \{\mathbf{y} \mid y_n > 0\} \subset \Omega''$ and define

$$Q^+ := B_{\bar{R}}(\mathbf{0}) \cap \{\mathbf{y} \mid y_n > 0\}, \tag{9.227}$$

$$\tilde{Q} := \Phi(B_{\bar{R}}(\mathbf{0})), \tag{9.228}$$

$$\tilde{Q}^+ := \Phi(Q^+). \tag{9.229}$$

Now, we can use Lemmas 9.54 and 9.57 to get

$$\|\tilde{u}\|_{H^2(Q^+)} \leq C(\|\tilde{f}\|_{L^2(\Omega'')} + \|\tilde{u}\|_{H^1(\Omega'')}). \tag{9.230}$$

From inequalities such as (9.215) we get

$$\|u\|_{H^2(\tilde{Q}^+)} \leq C(\|f\|_{L^2(\Omega')} + \|u\|_{H^1(\Omega')}), \tag{9.231}$$

which leads immediately to (9.226). $\square$

We now prove Theorem 9.53.

Proof. It is now a simple matter to put together the proof of the global regularity theorem. We simply provide an open cover for $\overline{\Omega}$ using the neighborhoods $\tilde{Q}$ constructed in Lemma 9.58 for each point $\bar{\mathbf{x}} \in \partial\Omega$ and one additional set $\Omega_0 \subset\subset \Omega$ to cover the interior. Since $\overline{\Omega}$ is compact, there is a finite subcover (in which we assume Ω_0 is included and which we label $\{\Omega_i\}_{i=0}^{N}$) such that

$$\Omega \subset \bigcup_{i=0}^{N} \Omega_i. \tag{9.232}$$

Now, using the interior regularity result (Theorem 9.51) for Ω_0 and Lemma 9.58 for each of the other sets we get

$$\|u\|_{H^2(\Omega)} \leq \sum_{i=0}^{N} \|u\|_{H^2(\Omega_i)} \leq C(\|f\|_{L^2(\Omega)} + \|u\|_{H^1(\Omega)}). \qquad (9.233)$$

A standard application of Ehrling's lemma gives us the final result.   □

10

Nonlinear Elliptic Equations

In this chapter we shall discuss nonlinear elliptic equations from three perspectives: the implicit function theorem, the calculus of variations, and nonlinear operator theory. This is the only chapter of the book in which we assume that the reader is familiar with the basic results of measure theory. In particular, we shall assume that the reader understands the following concepts and results.

- The definition of a set of measure zero and the idea of functions agreeing "almost everywhere."

- The idea of Lebesgue measurable functions and the definition of the L^p spaces as equivalence classes of functions that agree almost everywhere.

- The equivalence of the "measure theoretic" definition of the L^p spaces and the "completion" definiton used in the rest of this book.

- The idea of almost everywhere convergence of sequences of functions, the interrelationship between various types of convergence. This includes an understanding of such results as Fatou's lemma and the Lebesgue dominated convergence theorem.

10.1   Perturbation Results

Many results on differential equations say that a nonlinear equation behaves essentially like its linearization as long as one considers solutions which are small enough so that the linear terms dominate over the nonlinear ones. In Chapter 1, we stated the implicit function theorem from classical calculus, which provides such a result for finite-dimensional systems of equations. In this section, we shall generalize the implicit function theorem to a Banach space setting and then consider applications to elliptic PDEs.

10.1.1   The Banach Contraction Principle and the Implicit Function Theorem

The Banach contraction principle is one of the most used techniques for finding solutions of nonlinear equations. It consists of the following theorem.

Theorem 10.1 (Banach contraction). *Let (X, d) be a complete metric space. Assume that X is not empty and let $T : X \to X$ be a contraction, i.e., a mapping with the property that there exists $\theta \in [0, 1)$ with the property that $d(T(x), T(y)) \leq \theta d(x, y)$ for every $x, y \in X$. Then T has a unique fixed point.*

Proof. Suppose there were two fixed points x and y. Then

$$d(x, y) = d(T(x), T(y)) \leq \theta d(x, y), \qquad (10.1)$$

which is possible only if $x = y$.

Now consider any point $x_0 \in X$ and define recursively $x_{n+1} = T(x_n)$. Then we have $d(x_{n+1}, x_n) \leq \theta d(x_n, x_{n-1})$, from which it easily follows that the sequence x_n is Cauchy. Let x be the limit. Then we have

$$T(x) = T(\lim_{n \to \infty} x_n) = \lim_{n \to \infty} T(x_n) = \lim_{n \to \infty} x_{n+1} = x. \qquad (10.2)$$

This completes the proof.   □

In the following, we want to use the Banach contraction principle to prove an implicit function theorem for functions between Banach spaces. In order to do so, we first need to define the concept of a derivative.

Definition 10.2. *Let X and Y be Banach spaces and let x_0 be a point in X. Let F be a mapping from a neighborhood of x_0 into Y. Then F is called* **differentiable** *at x_0 if there exists a linear operator $A \in \mathcal{L}(X, Y)$ with the property that*

$$F(x) = F(x_0) + A(x - x_0) + G(x), \qquad (10.3)$$

where

$$\lim_{x \to x_0} \|G(x)\|_Y / \|x - x_0\|_X = 0. \qquad (10.4)$$

If such an A exists, we call it the **(Fréchet) derivative** *of F at x_0.*

Naturally, we shall call a function differentiable if it is differentiable at all points of its domain. We shall use familiar notations from calculus, such as $F'(x_0)$ or $DF(x_0)$ to denote Fréchet derivatives. We call a function continuously differentiable, if $F'(x)$ (as an element of $\mathcal{L}(X, Y)$) depends continuously on x. Clearly, we can use the definition recursively to define higher-order derivatives. For example, $F''(x)$, if it exists, is an element of $\mathcal{L}(X, \mathcal{L}(X, Y))$. Alternatively, we can regard $F''(x)$ as a bilinear mapping from X to Y, see Problem 10.2. Many of the elementary properties of derivatives can be generalized to Fréchet derivatives in a straightforward manner; see, e.g., Problem 10.3.

Example 10.3. Let f be continuously differentiable function from $\mathbb{R}$ to $\mathbb{R}$ and let $X = Y = C([0, 1])$. The mapping f induces a mapping from X to Y by pointwise action: $\phi(t) \to f(\phi(t))$. The mapping f is differentiable as a mapping from X to Y and its derivative at ϕ is the linear mapping which takes the function $\psi(t)$ to $f'(\phi(t))\psi(t)$. See Problem 10.4.

The crucial result concerning the solvability of equations is the inverse function theorem, which says that locally an equation is uniquely solvable if its linearization is.

Theorem 10.4 (Inverse function theorem). *Let X and Y be Banach spaces and let U be an open neighborhood of the origin in X. Let $F : U \to Y$ be continuously differentiable and assume that $F'(0) \in \mathcal{L}(X, Y)$ is one-to-one and onto. Then there exists an open neighborhood V of $F(0)$ in Y and a continuously differentiable mapping $G : V \to X$ with the property that $F(G(y)) = y$. Moreover, $G(y)$ is the only sufficiently small solution x of the equation $F(x) = y$.*

Proof. We consider the mapping T_y defined by

$$T_y(x) = F'(0)^{-1}(y - F(0)) - F'(0)^{-1}(F(x) - F(0) - F'(0)x). \quad (10.5)$$

Let $M = \|F'(0)^{-1}\|$. We claim: If ϵ is chosen small enough and $\|y - F(0)\| < \epsilon/(2M)$, then T_y is a contraction mapping the closed ϵ-ball in X into itself. To see this, we note that

$$\|T_y(x)\| \leq M\|y - F(0)\| + M\|F(x) - F(0) - F'(0)x\|. \quad (10.6)$$

The first term on the right-hand side is less than $\epsilon/2$, and the second term is $o(\epsilon)$ by the definition of the derivative. Hence we find $\|T_y(x)\| < \epsilon$ if ϵ is sufficiently small. Moreover, note that

$$T_y'(x) = -F'(0)^{-1}(F'(x) - F'(0)) \quad (10.7)$$

has norm less than, say, $1/2$ if ϵ is small. The contraction property now follows from the equation

$$T_y(x) - T_y(z) = \int_0^1 T_y'(z + \theta(x - z))(x - z)\, d\theta. \quad (10.8)$$

By the Banach contraction principle, T_y has a unique fixed point within the ball of radius ϵ. We call this fixed point $G(y)$. It is easy to check that the equation $T_y(x) = x$ is equivalent to $F(x) = y$.

It remains to be shown that G is in fact continuously differentiable. We first show continuity. We have $G(y) = T_y(G(y))$ and hence

$$\|G(y+h) - G(y)\| \le \|T_{y+h}(G(y+h)) - T_y(G(y+h))\|$$
$$+ \|T_y(G(y+h)) - T_y(G(y))\| \tag{10.9}$$
$$\le M\|h\| + \frac{1}{2}\|G(y+h) - G(y)\|,$$

where the latter estimate follows from the contraction property of T_y. From (10.9), it is immediate that G is Lipschitz continuous.

Let now $k = G(y+h) - G(y)$, then it is immediate from (10.9) that $\|k\| \le 2M\|h\|$. We now have

$$0 = F(G(y+h)) - F(G(y)) - h = F'(G(y))k + R(k) - h, \tag{10.10}$$

where $\|R(k)\|/\|k\| \to 0$ as $\|k\| \to 0$. It follows that

$$k = F'(G(y))^{-1}h - F'(G(y))^{-1}R(k), \tag{10.11}$$

and since $\|k\| \le 2M\|h\|$, we have that $\|R(k)\|/\|h\| \to 0$ as $\|h\| \to 0$. This proves that G is differentiable and the derivative is $F'(G(y))^{-1}$. $\square$

Remark 10.5. If F is k times continuously differentiable, then so is G (Problem 10.5).

The inverse function theorem is often used in the following form, known as the implicit function theorem.

Theorem 10.6 (Implicit function theorem). *Let X, Y and Z be Banach spaces, and let U, V be neighborhoods of the origin in X and Y, respectively. Let F be continuously differentiable from $U \times V$ into Z, and assume that $D_y F(0,0) \in \mathcal{L}(Y, Z)$ is one-to-one and onto. Assume, moreover, that $F(0,0) = 0$. Then there exists a neighborhood W of the origin in X and a continuously differentiable mapping $f : W \to Y$ such that $F(x, f(x)) = 0$. Moreover, for small x and y, $f(x)$ is the only solution y of the equation $F(x, y) = 0$.*

Proof. For the proof, we simply consider the mapping $\tilde{F} : U \times V \to X \times Z$ defined by $\tilde{F}(x, y) = (x, F(x, y))$ and apply the previous theorem to solve the equation $\tilde{F}(x, y) = (x, 0)$. $\square$

Although the implicit function theorem asserts the existence of a unique solution, it is often also useful in situations where uniqueness fails. Let us consider an equation $F(x, \lambda) = 0$, where F is a continuously differentiable function from $X \times \mathbb{R}$ to Y. We assume that $F(0, \lambda) = 0$ for every λ. If $D_x F(0, \lambda)$ is one-to-one and onto, then the implicit function theorem tells us that $x = 0$ is the only small solution of $F(x, \lambda) = 0$. It is of interest to consider those values of λ where this assumption fails. This leads to the subject of bifurcation theory, on which there is an extensive literature. We shall consider only the simplest case of a bifurcation. Specifically, we shall assume that F is twice continuously differentiable, that $D_x F(0,0)$ has a one-dimensional nullspace spanned by x_0 and that its range has codimension one, and that $D_{x\lambda}F(0,0)x_0$ is not in the range of $D_x F(0,0)$. Let now Z be a subspace of X which complements the span of x_0; we substitute $x = \epsilon(x_0 + z)$, where $z \in Z$. We then define $G(\epsilon, z, \lambda) = \frac{1}{\epsilon}F(\epsilon(x_0 + z), \lambda)$, with the obvious definition $G(0, z, \lambda) = D_x F(0, \lambda)(x_0+z)$ in the limit $\epsilon = 0$. Since F was assumed of class C^2, we find that G is still of class C^1. We can now use the implicit function theorem to solve the equation $G(\epsilon, z, \lambda)$ for z and λ as functions of ϵ. To see this, we simply note that the derivatives are given by $D_z G(0,0,0) = D_x F(0,0)|_Z$ and $D_\lambda G(0,0,0) = D_{x\lambda}(0,0)x_0$. The assumption that $D_{x\lambda}(0,0)x_0$ is not in the range of $D_x F(0,0)$ is precisely what is needed to guarantee the invertibility of the linearization. We thus obtain a bifurcating branch of nontrivial solutions of the equation $F(x, \lambda) = 0$. This branch of nontrivial solutions is parameterized by the "amplitude factor" ϵ.

10.1.2 Applications to Elliptic PDEs

We want to apply the results of the previous subsection to nonlinear elliptic PDEs. In order to do this, we have to set up such problems as abstract nonlinear equations in Banach space. The following lemma is crucial for this.

Lemma 10.7. *Let Ω be a bounded domain in $\mathbb{R}^n$ with smooth boundary. Moreover, let $f : \mathbb{R}^k \to \mathbb{R}$ be of class C^{m+1}, where $m > n/2$. Then f, interpreted pointwise, induces a continuously differentiable mapping from $(H^m(\Omega))^k$ into $H^m(\Omega)$.*

Proof. By the Sobolev embedding theorem, we have $H^m(\Omega) \subset C(\overline{\Omega})$. Hence we have $f(\mathbf{u}) \in C(\overline{\Omega}) \subset L^2(\Omega)$ for every $\mathbf{u} \in (H^m(\Omega))^k$. Moreover, if $\mathbf{u}$ is in $(C^m(\overline{\Omega}))^k$, we can obtain the derivatives of $f(\mathbf{u})$ by the chain rule, and in the general case, we can use approximation by smooth functions. For example, let $\mathbf{u} = \lim_{n\to\infty} \mathbf{u}_n$ in the topology of $(H^m(\Omega))^k$, where the $\mathbf{u}_n$ are smooth. We find

$$\left(f(\mathbf{u}), \frac{\partial \phi}{\partial x_i} \right) = \lim_{n\to\infty} \left(f(\mathbf{u}_n), \frac{\partial \phi}{\partial x_i} \right)$$
$$= -\lim_{n\to\infty} \left(\nabla f(\mathbf{u}_n)\frac{\partial \mathbf{u}_n}{\partial x_i}, \phi \right) \qquad (10.12)$$
$$= -\left(\nabla f(\mathbf{u})\frac{\partial \mathbf{u}}{\partial x_i}, \phi \right),$$

i.e., the chain rule applies for differentiating $f(\mathbf{u})$. In a similar fashion, we can deal with higher derivatives. Note that all derivatives of $f(\mathbf{u})$ have the form of a product involving a derivative of f and derivatives of $\mathbf{u}$. The first factor is in $C(\overline{\Omega})$, while any lth derivative of $\mathbf{u}$ lies in $H^{m-l}(\Omega)$, which imbeds into $L^{2n/(n-2(m-l))}(\Omega)$ if $m-l < n/2$. We can use this fact and Hölder's inequality to show that all derivatives of $f(\mathbf{u})$ up to order m are in $L^2(\Omega)$; moreover, it is clear from this argument that f is actually continuous from $(H^m(\Omega))^k$ into $H^m(\Omega)$. The differentiability of f follows along similar lines by exploiting the relation

$$f(\mathbf{u}) - f(\mathbf{v}) = \int_0^1 \nabla f(\mathbf{v} + \theta(\mathbf{u}-\mathbf{v})) \cdot (\mathbf{u}-\mathbf{v}) \, d\theta. \tag{10.13}$$

We leave it to the reader to fill in some more details of the proof (Problem 10.6). □

It is now easy to give applications of the implicit function theorem to elliptic PDEs. Consider the equation

$$\Delta u + f(u) = g(\mathbf{x}) \tag{10.14}$$

on a bounded smooth domain in $\mathbb{R}^3$, subject to Dirichlet boundary conditions. Assume that f is of class C^3 with $f(0) = f'(0) = 0$. Then it follows from Lemma 10.7 that the mapping $u \to f(u)$ is continuously differentiable from $H^2(\Omega)$ into itself. We can now apply the inverse function theorem to conclude that, for sufficiently small $g \in L^2(\Omega)$, the equation has a solution $u \in H^2(\Omega) \cap H_0^1(\Omega)$. Moreover, this solution is unique among solutions of small norm. (We note that actually we only need the mapping $u \to f(u)$ to be continuously differentiable as a mapping from $H^2(\Omega)$ to $L^2(\Omega)$. For this, it would suffice that f be of class C^1.) It is clear how to formulate and prove more general results involving nonlinearities which depend on derivatives of u and/or nonlinear boundary conditions.

As an example of a bifurcation problem, consider the equation

$$\Delta u + \lambda u = f(u), \tag{10.15}$$

again with Dirichlet boundary condition and with Ω and f as before. The zero solution is the only small solution as long as λ is not an eigenvalue of $-\Delta$. If λ_0 is a simple eigenvalue, the assumptions of the previous section apply, since the range of $\Delta + \lambda$ is precisely the orthogonal complement of the nullspace, and, with u_0 denoting the eigenfunction, and $F(u,\lambda) = \Delta u + \lambda u - f(u)$, we find $D_{u\lambda}F(0,\lambda_0)u_0 = u_0$. It turns out that the first eigenvalue of the Dirichlet problem for the Laplacian is always simple.

Lemma 10.8. *Let Ω be a bounded domain. Then the first eigenvalue of the Dirichlet problem for Laplace's equation is simple.*

Proof. Let λ_0 be the smallest eigenvalue; then we have

$$\lambda_0 = \min_{u \in H_0^1(\Omega), \|u\|_2 = 1} \|u\|_{1,2}^2. \tag{10.16}$$

Recall Problem 8.52, where the analogous result was established for Sturm-Liouville problems. Let u_0 be a minimizer, and assume that u_0 has zeros inside Ω. We can now construct a sequence of piecewise polynomials p_n, which converges to u_0 in $H_0^1(\Omega)$, and uniformly on compact subsets of Ω. Clearly, we have

$$\lambda_0 = \lim_{n \to \infty} \|p_n\|_{1,2}^2 / \|p_n\|_2^2. \tag{10.17}$$

Now $|p_n|$ has the same H^1- and L^2-norms as p_n itself; in particular, the H^1-norms of $|p_n|$ are bounded as $n \to \infty$. Hence there is a subsequence of $|p_n|$ which converges weakly in $H_0^1(\Omega)$, and strongly in $L^2(\Omega)$. Since the p_n converge to u_0 uniformly on compact subsets of Ω, it also follows that $|p_n| \to |u_0|$. We thus find that

$$\| \, |u_0| \, \|_{1,2}^2 / \| \, |u_0| \, \|_2^2 \leq \lambda_0, \tag{10.18}$$

i.e., $|u_0|$ must also be an eigenfunction. But $|u_0|$ is non-negative and has zeros inside Ω, which contradicts the maximum principle.

Hence our assumption of an eigenfunction that changes sign must have been wrong, and every eigenfunction is either positive or negative. No two such functions can be orthogonal to each other, and hence the eigenvalue is simple. $\square$

Problems

10.1. Apply the Banach contraction principle to prove the Picard-Lindelöf existence theorem for ODEs (Theorem 1.1).

10.2. Show that there is a natural correspondence between continuous bilinear mappings from X to Y and linear mappings from X to $\mathcal{L}(X, Y)$.

10.3. Establish the chain rule for Fréchet derivatives.

10.4. Verify the claim in Example 10.3.

10.5. Verify Remark 10.5.

10.6. Complete the details in the proof of Lemma 10.7.

10.7. Use the implicit function theorem to obtain existence results for fully nonlinear second-order elliptic PDEs with nonlinear boundary conditions.

10.8. Consider the composition mapping

$$L^2(0,1) \ni u(\cdot) \mapsto \sin(u(\cdot)) \in L^2(0,1).$$

Show that this mapping is nowhere differentiable.

10.2 Nonlinear Variational Problems

10.2.1 Convex problems

An important class of nonlinear elliptic PDEs arise from problems in the calculus of variations. We will focus on a particularly vital set of representatives of this class: problems describing the equilibrium of *elastic* materials. In discussing this class of problems we will be able to examine not only many of the classical techniques for elliptic problems from the calculus of variations, but also some important techniques which have been developed over the last twenty years.

We begin our brief discussion of nonlinear elasticity by describing the *kinematics* or geometry of deformation of three-dimensional bodies. We let a domain $\Omega \subset \mathbb{R}^3$ represent the *reference configuration* of a material body. We assume that Ω is bounded with Lipschitz boundary. (It is usually convenient to think of the reference configuration Ω as the "rest" or "unstressed" configuration of the body, but we will not restrict ourselves to this case.) A *deformation* of the body is simply a mapping of the form

$$\mathbb{R}^3 \supset \Omega \ni \mathbf{x} \mapsto \mathbf{p}(\mathbf{x}) \in \mathbb{R}^3. \tag{10.19}$$

It will ease our computations somewhat to spell out all of our vectors and matrices in terms of components. Thus, we let $\mathbf{e}_1$, $\mathbf{e}_2$, $\mathbf{e}_3$ constitute a fixed orthonormal basis for $\mathbb{R}^3$, and for a given vector $\mathbf{v}$, we define $v_i := \mathbf{v} \cdot \mathbf{e}_i$.

We will assume a certain amount of smoothness of the deformation: We either make the classical assumption that $\mathbf{p} \in C_b^1(\overline{\Omega})$ or use a Sobolev space $\mathbf{p} \in W^{1,p}(\Omega)$. In either case, we are able to define the *deformation gradient*

$$\mathbf{F}(\mathbf{x}) := \frac{\partial \mathbf{p}}{\partial \mathbf{x}}(\mathbf{x}). \tag{10.20}$$

In terms of components we have

$$F_{ij}(\mathbf{x}) := \frac{\partial p_i}{\partial x_j}(\mathbf{x}). \tag{10.21}$$

We would like to restrict the deformations to be one-to-one mappings so that the material does not interpenetrate or overlap when it deforms. However, such a constraint is hard to treat analytically so we will ignore it. We will, however, assume that the material *preserves orientation* (i.e., no "mirror-image" deformations). If $\mathbf{p} \in C_b^1(\overline{\Omega})$ this constraint can be expressed by the pointwise inequality

$$\det \mathbf{F}(\mathbf{x}) > 0. \tag{10.22}$$

If we are taking $\mathbf{p} \in W^{1,p}(\Omega)$, then we will assume that the inequality (10.22) is satisfied almost everywhere.

We will specify *displacement boundary conditions* on our deformations. The most natural way to state such conditions is to specify a continuous function $\mathbf{b} : \partial\Omega \to \mathbb{R}^3$ and to require that

$$\mathbf{p}(\mathbf{x}) = \mathbf{b}(\mathbf{x}) \quad \text{for all } \mathbf{x} \in \partial\Omega. \tag{10.23}$$

If we wish to consider $\mathbf{p} \in W^{1,p}(\Omega)$ we have to require that (10.23) holds in the sense of trace. However, when working in Sobolev spaces, it is more convenient to enforce boundary conditions by specifying a function $\mathbf{g} : \Omega \to \mathbb{R}^3$ with $\mathbf{g} \in W^{1,p}(\Omega)$ and requiring

$$\mathbf{p}_0 := \mathbf{p} - \mathbf{g} \in W_0^{1,p}(\Omega). \tag{10.24}$$

This ensures that $\mathbf{p}$ and $\mathbf{g}$ have the same trace on the boundary of Ω. If we are working with deformations in the space $H^1(\Omega)$ and $\partial\Omega$ is of class C^1, then Theorem 7.37 (the inverse trace theorem) ensures that if $\mathbf{b} \in H^{1/2}(\partial\Omega)$, then there exists $\mathbf{g} \in H^1(\Omega)$ such that the trace of $\mathbf{g}$ on $\partial\Omega$ is $\mathbf{b}$. Thus, by using this $\mathbf{g}$, the two boundary conditions (10.23) and (10.24) can be made to coincide. We can achieve a similar result for the space $W^{1,p}(\Omega)$ with $p \neq 2$ by using a more general version of Theorem 7.37, but instead of trying to find the most general inverse trace theorem (on the roughest possible boundary) we will content ourselves with boundary conditions of the form (10.24). In addition, we will be able to assume such conditions as

$$\det\left(\frac{\partial\mathbf{g}}{\partial\mathbf{x}}\right) > 0 \tag{10.25}$$

almost everywhere. (The existence of such a $\mathbf{g}$ is difficult to address in a general inverse trace theorem.) To sum up, we assume that $\mathbf{g} \in W^{1,p}(\Omega)$ satisfying (10.25) exists and take the domain of our elasticity boundary-value problem to be

$$\mathcal{D}_{\mathcal{E}} := \{\mathbf{p} \in W^{1,p}(\Omega) \mid \det \nabla\mathbf{p} > 0 \text{ a.e., and } \mathbf{p} - \mathbf{g} \in W_0^{1,p}(\Omega)\}. \tag{10.26}$$

We now pose a mathematical problem whose solutions will describe the equilibrium configurations of the body. A material whose equilibria are described by such a problem is said to be *elastic*. We begin by defining an *energy functional*:

$$\mathcal{E}(\mathbf{p}) := \int_\Omega \mathcal{W}(\mathbf{x}, \mathbf{F}(\mathbf{x})) \, d\mathbf{x} + \int_\Omega \Psi(\mathbf{p}(\mathbf{x})) \, d\mathbf{x}. \tag{10.27}$$

Here $\mathbf{F} := \nabla\mathbf{p}$ and $\mathcal{W} : \Omega \times Q \to R$ is the *stored energy density*, where Q is the set of 3×3 matrices with positive determinant. The function $\Psi : \mathbb{R}^3 \to \mathbb{R}$ is the *potential energy density*. The first term in the energy is called the *stored energy functional*, and this describes the energy stored from mechanical deformation within the material. The second term is a *potential energy functional* and it is the energy from exterior forces (assumed to be conservative).

We make the following assumptions about the density functionals.

1. We assume that $\mathcal{W} \in C^2(\Omega \times Q)$ and $\Psi \in C^1(\mathbb{R}^3)$.

2. For some $p > 3$, there exists a constant $k > 0$ and a function $\omega \in C_b(\Omega)$ such that

$$\mathcal{W}(\mathbf{x}, \mathbf{F}) \geq \omega(\mathbf{x}) + k|\mathbf{F}|^p. \tag{10.28}$$

Note: From now on this p is to be used in the definition of $\mathcal{D}_\mathcal{E}$.

3. There exists a constant C such that for every $\mathbf{p} \in \mathcal{D}_\mathcal{E}$ we have

$$\int_\Omega \Psi(\mathbf{p}(\mathbf{x}))\, d\mathbf{x} \geq C. \tag{10.29}$$

4. There exists $\mathbf{p} \in \mathcal{D}_\mathcal{E}$ such that $\mathcal{E}(\mathbf{p}) < \infty$.

5. For every $\mathbf{x} \in \Omega$, we have $\mathcal{W}(\mathbf{x}, \mathbf{F}) \to \infty$ as $\det F \to 0$.

We now define the equilibrium configurations of the body to be those that minimize the energy $\mathcal{E}$; i.e., we say that $\bar{\mathbf{p}} \in \mathcal{D}_\mathcal{E}$ is an *equilibrium state* or a *minimizer* if

$$\mathcal{E}(\bar{\mathbf{p}}) \leq \mathcal{E}(\mathbf{p}) \quad \text{for all } \mathbf{p} \in \mathcal{D}_\mathcal{E}. \tag{10.30}$$

The primary goal of this section is to examine the question of existence of solutions of the minimization problem (10.30). But before doing so, we wish to expose the connection of this problem to elliptic partial differential equations. We do so in the following result.

Theorem 10.9 (Euler-Lagrange equations). *Suppose $\bar{\mathbf{p}} \in \mathcal{D}_\mathcal{E} \cap C^2(\overline{\Omega})$ solves (10.30) and that (10.22) holds almost everywhere in Ω. Then at every $\mathbf{x} \in \Omega$, $\bar{\mathbf{p}}$ must satisfy the* Euler-Lagrange equations

$$-\sum_j \frac{\partial}{\partial x_j} \mathcal{A}_{ij}(\mathbf{x}) + \frac{\partial \Psi}{\partial p_i}(\mathbf{p}(\mathbf{x})) = 0, \quad i = 1, 2, 3, \tag{10.31}$$

where

$$\mathcal{A}_{ij}(\mathbf{x}) := \frac{\partial \hat{\mathcal{W}}}{\partial F_{ij}}(\mathbf{x}, \mathbf{F}(\mathbf{x})). \tag{10.32}$$

In a homogeneous material: $\mathcal{W} = \hat{\mathcal{W}}(\mathbf{F})$, we have

$$\sum_j \frac{\partial}{\partial x_j} \mathcal{A}_{ij}(\mathbf{x}) = \sum_{j,k,l} a_{ijkl}(\mathbf{x}) \frac{\partial^2 p_k}{\partial x_j \partial x_l}, \tag{10.33}$$

where

$$a_{ijkl}(\mathbf{x}) := \frac{\partial^2 \mathcal{W}}{\partial F_{ij} \partial F_{kl}}(\mathbf{F}(\mathbf{x})). \tag{10.34}$$

Proof. Since $\bar{\mathbf{p}} \in \mathcal{D}_\mathcal{E} \cap C^2(\overline{\Omega})$, we see that for any $\phi = \sum \phi_i \mathbf{e}_i \in [\mathcal{D}(\Omega)]^3$ we have

$$\bar{\mathbf{p}} + \epsilon\phi \in \mathcal{D}_\mathcal{E} \tag{10.35}$$

for ϵ sufficiently small. Thus, for any ϕ, the real-valued function

$$f(\epsilon) := \mathcal{E}(\bar{\mathbf{p}} + \epsilon\phi) \tag{10.36}$$

is well defined in an interval about $\epsilon = 0$. Furthermore, f is minimized at $\epsilon = 0$ since by hypothesis $\mathcal{E}$ is minimized at $\bar{\mathbf{p}}$. Now, using standard results on uniform convergence to take the derivative under the integral, we see that f is differentiable. We get

$$f'(\epsilon) = \int_\Omega \sum_{i,j} \frac{\partial W}{\partial F_{ij}}(\mathbf{x}, \bar{\mathbf{F}}(\mathbf{x}) + \epsilon\nabla\phi(\mathbf{x})) \frac{\partial \phi_i}{\partial x_j} + \sum_i \frac{\partial \Psi}{\partial p_i}(\bar{\mathbf{p}} + \epsilon\phi)\phi_i \, d\mathbf{x}. \tag{10.37}$$

Setting the derivative equal to zero at $\epsilon = 0$ and integrating by parts gives us

$$\int_\Omega \sum_i \left[-\sum_j \frac{\partial}{\partial x_j} \frac{\partial W}{\partial F_{ij}}(\mathbf{x}, \bar{\mathbf{F}}(\mathbf{x})) + \frac{\partial \Psi}{\partial p_i}(\bar{\mathbf{p}}(\mathbf{x})) \right] \phi_i(\mathbf{x}) \, d\mathbf{x} = 0. \tag{10.38}$$

Since ϕ is arbitrary, this implies that (10.31) is satisfied in the sense of distributions. Since each term in the equation is continuous, it must, in fact, be satisfied pointwise. $\qquad\square$

We now return to the question of existence. We will use what is called a *direct method* in the calculus of variations. (We will try to minimize the energy directly rather than solve, for instance, the Euler-Lagrange equations.) The following lemma provides a first important step.

Lemma 10.10. *There exists a minimizing sequence* $\{\mathbf{p}_n\} \in \mathcal{D}_\mathcal{E}$ *such that*

$$\lim_{n\to\infty} \mathcal{E}(\mathbf{p}_n) \le \mathcal{E}(\mathbf{p}) \quad \text{for all } \mathbf{p} \in \mathcal{D}_\mathcal{E} \tag{10.39}$$

and with the additional property that $\mathbf{p}_n$ *is weakly convergent; i.e., there exists* $\tilde{\mathbf{p}} \in W^{1,p}(\Omega)$ *such that*

$$\mathbf{p}_n \rightharpoonup \tilde{\mathbf{p}} \text{ in } W^{1,p}(\Omega). \tag{10.40}$$

Proof. We first note that $\mathcal{E}$ is bounded below on $\mathcal{D}_\mathcal{E}$; i.e., for any $\mathbf{p} \in \mathcal{D}_\mathcal{E}$ we can use (10.28) and (10.29)

$$\mathcal{E}(\mathbf{p}) \ge \int_\Omega \omega(\mathbf{x}) + \Psi(\mathbf{p}(\mathbf{x})) \, d\mathbf{x} \ge C - \|\omega\|_\infty |\Omega|, \tag{10.41}$$

where $|\Omega|$ is the volume of Ω. Since $\mathcal{E}(\mathbf{p})$ is bounded below it must have a greatest lower bound L, and hence there must be a sequence $\mathbf{p}_n \in \mathcal{D}_\mathcal{E}$ such that

$$\lim_{n\to\infty} \mathcal{E}(\mathbf{p}_n) = L. \tag{10.42}$$

Since $\mathcal{E}(\mathbf{p}_n)$ is a convergent sequence in $\mathbb{R}$ it must be bounded, say, by a constant K. Using this and (10.28) we get

$$K \geq |\mathcal{E}(\mathbf{p}_n)| \geq k \int_\Omega |\nabla \mathbf{p}_n(\mathbf{x})|^p \, d\mathbf{x} - \|\omega\|_\infty |\Omega| - C. \tag{10.43}$$

Rearranging and combining this with Poincaré's inequality (on $\mathbf{p}_n - \mathbf{g}$) gives us

$$\|\mathbf{p}_n\|_{1,p} \leq \tilde{K}, \tag{10.44}$$

for some constant $\tilde{K}$ independent of n.

Since $\mathbf{p}_n$ is bounded, Theorem 6.64 implies that it has a weakly convergent subsequence (which we also label $\mathbf{p}_n$). Defining $\tilde{\mathbf{p}}$ to be the weak limit of this sequence gives us (10.40). Since the original sequence $\mathcal{E}(\mathbf{p}_n)$ converges to L, so does any subsequence; this gives us (10.39) and completes the proof.   □

Of course, since $\mathbf{p}_n$ is a minimizing sequence, our first guess is that its "limit," $\tilde{\mathbf{p}}$, is a solution of our problem. However, two questions remain.

1. Is $\tilde{\mathbf{p}} \in \mathcal{D}_\mathcal{E}$; i.e., is the constraint $\det \nabla \tilde{\mathbf{p}} > 0$ satisfied?

2. If so, is $\tilde{\mathbf{p}}$ actually a minimizer; i.e., is it true that

$$\mathcal{E}(\tilde{\mathbf{p}}) = \lim_{n\to\infty} \mathcal{E}(\mathbf{p}_n) = L \leq \mathcal{E}(\mathbf{p}) \quad \text{for all } \mathbf{p} \in \mathcal{D}_\mathcal{E}? \tag{10.45}$$

In order to answer the first question, we extend the domain of definition of $\mathcal{E}$ to functions which do not satisfy the constraint (10.22). We define the function

$$\Omega \times \mathcal{M}^{3\times3} \ni (\mathbf{x}, \mathbf{f}) \mapsto \bar{\mathcal{W}}(\mathbf{x}, \mathbf{F}) \in \bar{\mathbb{R}} := \mathbb{R} \cup \{-\infty, \infty\} \tag{10.46}$$

by

$$\bar{\mathcal{W}}(\mathbf{x}, \mathbf{F}) := \begin{cases} \mathcal{W}(\mathbf{x}, \mathbf{F}), & \det \mathbf{F} > 0 \\ \infty, & \det \mathbf{F} \leq 0. \end{cases} \tag{10.47}$$

We can now extend the domain of definition of our total energy to

$$W_\mathbf{g}^{1,p}(\Omega) := \{\mathbf{p} \in W^{1,p}(\Omega) \mid \mathbf{p} - \mathbf{g} \in W_0^{1,p}(\Omega)\}, \tag{10.48}$$

by defining

$$\bar{\mathcal{E}}(\mathbf{p}) := \int_\Omega \bar{\mathcal{W}}(\mathbf{x}, \mathbf{F}(\mathbf{x})) \, d\mathbf{x} + \int_\Omega \Psi(\mathbf{p}(\mathbf{x})) \, d\mathbf{x}. \tag{10.49}$$

The following result is then immediate.

Lemma 10.11. *If $\bar{\mathcal{E}}(\mathbf{p}) < \infty$, then $\mathbf{p} \in \mathcal{D}_\mathcal{E}$.*

This means that if we can indeed answer the second question in the affirmative; i.e., if $\tilde{\mathbf{p}}$ is a minimizer, then $\tilde{\mathbf{p}} \in \mathcal{D}_\mathcal{E}$. Thus, we focus on the second question and identify conditions that will ensure the weak limit is indeed a minimizer.

Definition 10.12. Let X be a Banach space. We say that a nonlinear mapping $\mathcal{F} : X \to \mathbb{R}$ is **sequentially weakly lower semicontinuous (wlsc)** if whenever

$$v_n \rightharpoonup \bar{v} \text{ in } X, \tag{10.50}$$

we have

$$\mathcal{F}(\bar{v}) \le \liminf_{n \to \infty} \mathcal{F}(v_n). \tag{10.51}$$

We say that $\mathcal{F} : X \to \mathbb{R}$ is **sequentially weak-star lower semicontinuous (wslsc)** if whenever

$$v_n \overset{*}{\rightharpoonup} \bar{v} \text{ in } X, \tag{10.52}$$

it follows that (10.51) holds.

A mapping $\mathcal{F}$ is **sequentially weakly (weak-star) continuous** if

$$\mathcal{F}(\bar{v}) = \lim_{n \to \infty} \mathcal{F}(v_n) \tag{10.53}$$

whenever $v_n \rightharpoonup \bar{v}$ $(v_n \overset{*}{\rightharpoonup} \bar{v})$ in X.

Remark 10.13. We will drop the use of the word "sequentially" from here on. (A more thorough study of functional analysis would highlight the differences between the sequential and topological notions of continuity and lower semicontinuity.)

Remark 10.14. The reader already knows a basic lower semicontinuity result: *Fatou's lemma*, which implies (among other things) that the L^1 norm is lower semicontinuous.

We state a theorem (usually attributed to Tonelli) which is the fundamental result on weak lower semicontinuity. However, before doing this we give a definition of a convex function whose domain is in a general Banach space X and whose range is the extended real line $\overline{\mathbb{R}}$.

Definition 10.15. Let $K \subset X$ be a convex set. Then we say a mapping $G : K \to \overline{\mathbb{R}}$ is **convex** if for every $u, v \in K$, we have

$$G(\lambda u + (1 - \lambda)v) \le \lambda G(u) + (1 - \lambda)G(v) \quad \text{for all } \lambda \in [0, 1], \tag{10.54}$$

whenever the right-hand side of the inequality is well defined.

Theorem 10.16 (Tonelli). *For functions* $\mathbf{u} : \Omega \to \mathbb{R}^m$, *define the nonlinear function*

$$\mathcal{F}(\mathbf{u}) := \int_\Omega f(\mathbf{u}(x)) \, d\mathbf{x}, \tag{10.55}$$

where $f : \mathbb{R}^m \to \overline{\mathbb{R}}$ *is continuous. Then the function* $\mathcal{F}$ *is sequentially weakly lower semicontinuous on the space* $L^p(\Omega)$ *for* $1 < p < \infty$ *and weak-star lower semicontinuous on* $L^\infty(\Omega)$ *if and only if* $\mathbb{R}^m \ni \mathbf{u} \mapsto f(\mathbf{u}) \in \overline{\mathbb{R}}$ *is convex.*

Proof. The proof of this theorem is found in many texts on convex analysis or the calculus of variations. We give only a sketch of the proof here. We begin by showing that weak-star lower semicontinuity of $\mathcal{F}$ in $L^\infty(\Omega)$ implies that f is convex. We first prove a lemma which highlights one of the most important types of weak convergence: a wildly oscillating sequence which converges to its average value.

Lemma 10.17. *For any* $\mathbf{a}, \mathbf{b} \in \mathbb{R}^m$ *and* $\theta \in (0,1)$ *define* $\mathbf{u} : \mathbb{R} \to \mathbb{R}^m$ *to be a function of period one such that*

$$\mathbf{u}(x) := \begin{cases} \mathbf{a}, & x \in [0, \theta) \\ \mathbf{b}, & x \in [\theta, 1). \end{cases} \tag{10.56}$$

Let the sequence of functions $\mathbf{u}_n : [0,1] \to \mathbb{R}^m$ *be given by*

$$\mathbf{u}_n(x) := \mathbf{u}(nx), \quad x \in [0,1]. \tag{10.57}$$

Then

$$\mathbf{u}_n \rightharpoonup \bar{\mathbf{u}} \tag{10.58}$$

in $L^p(0,1)$ *for* $1 < p < \infty$ *and*

$$\mathbf{u}_n \overset{*}{\rightharpoonup} \bar{\mathbf{u}} \tag{10.59}$$

in $L^\infty(0,1)$, *where* $\bar{\mathbf{u}} := \theta\mathbf{a} + (1-\theta)\mathbf{b}$ *is a constant function.*

Proof. Here as well, we just give a sketch of the proof. We wish to show that

$$\lim_{n\to\infty} \int_0^1 \mathbf{u}_n(x)\phi(x)\,dx = [\theta\mathbf{a} + (1-\theta)\mathbf{b}] \int_0^1 \phi(x)\,dx, \tag{10.60}$$

for any $\phi \in L^p(0,1)$, $1 \leq p < \infty$. The reader should fill in the following steps.

1. The assertion is easy to prove for functions of the form

$$\phi(x) = \begin{cases} c, & x \in I \\ 0, & x \notin I, \end{cases} \tag{10.61}$$

   where $I \subset [0,1]$ is an interval. (For large n the interval I will contain a large integral number of periods of $\mathbf{u}_n$ plus some small "slop" at the ends.)

2. One can use the previous observation to show that (10.60) holds if ϕ is a *simple function;* i.e., if ϕ is piecewise constant with a finite number of jump discontinuities.

3. One can then show that for any $\phi \in L^p(0,1)$ $p \in [1,\infty)$, and any $\epsilon > 0$, we can find a simple function ϕ_s such that

$$\|\phi - \phi_s\|_1 < \epsilon. \tag{10.62}$$

(By the definition of L^1 given in this book, we can approximate ϕ arbitrarily closely by a bounded, continuous function. The reader should verify that a bounded continuous function can be approximated arbitrarily closely in the L^1 norm by a simple function.)

The combination of these observations completes the proof.   □

Remark 10.18. The choice of the interval $[0,1]$ was arbitrary. With only trivial modifications of the proof, one can show that

$$\mathbf{u}_n \rightharpoonup \bar{\mathbf{u}} \qquad (10.63)$$

in $L^p(I)$, for any compact interval $I \subset \mathbb{R}$. Furthermore, we proved this result for functions whose domains is a subset of $\mathbb{R}$ only for clarity; an analogous construction can be created using functions $\mathbf{u}_n$ defined on the domain $\Omega \subset \mathbb{R}^n$ by letting the functions oscillate in a single coordinate direction.

We now return to the proof of Tonelli's theorem. We assume that $\mathcal{F}$ is weak-star lower semicontinuous. Let $C \subset \Omega \subset \mathbb{R}^n$ be a hypercube. For any $\mathbf{a}, \mathbf{b} \in \mathbb{R}^m$, $\theta \in (0,1)$, let $\mathbf{u}_n$ be a sequence of oscillating functions with support on C such that

$$\mathbf{u}_n \rightharpoonup \bar{\mathbf{u}} = \theta \mathbf{a} + (1 - \theta)\mathbf{b} \qquad (10.64)$$

in $L^\infty(\Omega)$ The key observation here is that the sequence of composite functions $f(\mathbf{u}_n)$ oscillate between the values $f(\mathbf{a})$ and $f(\mathbf{b})$ with volume fractions θ and $1 - \theta$, respectively. Hence, it follows from the arguments in the proof of the lemma above that

$$f(\mathbf{u}_n) \overset{*}{\rightharpoonup} \theta f(\mathbf{a}) + (1 - \theta)f(\mathbf{b}) \qquad (10.65)$$

in $L^\infty(\Omega)$. Combining this with the weak-star lower semicontinuity of $\mathcal{F}$, we have

$$
\begin{aligned}
|\Omega| f(\theta \mathbf{a} + (1 - \theta)\mathbf{b}) &= \mathcal{F}(\bar{\mathbf{u}}) \\
&\leq \liminf_{n \to \infty} \mathcal{F}(\mathbf{u}_n) \\
&= \lim_{n \to \infty} \int_\Omega f(\mathbf{u}_n) \\
&= |\Omega| \left\{ \theta f(\mathbf{a}) + (1 - \theta)f(\mathbf{b}) \right\}.
\end{aligned}
$$

It follows that f is convex.

We now assume that f is convex and show that $\mathcal{F}$ is weak-star lower semicontinuous. We let $\mathbf{u}_n \rightharpoonup \bar{\mathbf{u}}$ be an arbitrary weakly convergent sequence and let

$$L = \liminf \mathcal{F}(\mathbf{u}_n). \qquad (10.66)$$

By taking a subsequence, we can actually assume that

$$L = \lim \mathcal{F}(\mathbf{u}_n). \qquad (10.67)$$

To complete the proof of the theorem we will use Mazur's lemma which we state without proof. (The proof uses elementary convex analysis and is found in many texts (cf., e.g., [ET]).)

Lemma 10.19 (Mazur). *Let X be a Banach space and suppose*

$$u_n \rightharpoonup \bar{u} \tag{10.68}$$

in X. Then there exists a function $N : \mathbb{N} \to \mathbb{N}$, and a sequence of sets of real numbers $\{\alpha(n)_k\}_{k=n}^{N(n)}$ such that $\alpha(n)_k \geq 0$ and $\sum_{k=n}^{N(n)} \alpha(n)_k = 1$ such that the sequence

$$v_n := \sum_{k=n}^{N(n)} \alpha(n)_k u_k \tag{10.69}$$

*converges **strongly** to $\bar{u}$ in X.*

Remark 10.20. We say that v_n as defined above is a *convex combination* of elements of the set $\{u_k\}_{k=n}^{N(n)}$. The set of all possible convex combinations of elements of a set $\mathcal{S}$ is called the *convex hull* of $\mathcal{S}$.

We can use Mazur's lemma to construct a sequence such that $\mathbf{v}_n \to \bar{\mathbf{u}}$ strongly in $L^p(\Omega)$ for every $p \in [1, \infty)$. Thus, at least for a subsequence we have $\mathbf{v}_n \to \bar{\mathbf{u}}$ almost everywhere. There are two steps to the remainder of the proof of Theorem 10.16.

1. Since $f(\mathbf{v}_n) \to f(\bar{\mathbf{u}})$ almost everywhere, it follows from Fatou's lemma and the convexity of f that

$$\mathcal{F}(\bar{\mathbf{u}}) \leq \liminf_{n \to \infty} \mathcal{F}(\mathbf{v}_n). \tag{10.70}$$

(If f is bounded below, as it is in most applications, we can use Fatou's lemma without using convexity. Otherwise, one can use the fact that any convex function is bounded below by an affine function.)

2. We now let $\epsilon > 0$ be given and choose N sufficiently large so that $\mathcal{F}(\bar{\mathbf{u}}) \leq \mathcal{F}(\mathbf{v}_k) + \epsilon/2$ and $\mathcal{F}(\mathbf{u}_k) \leq L + \epsilon/2$ for $k \geq N$. Then for $n \geq N$ we have

$$
\begin{aligned}
\mathcal{F}(\bar{\mathbf{u}}) &\leq \mathcal{F}(\mathbf{v}_n) + \epsilon/2 \\
&= \int_\Omega f\left(\sum_{k=n}^{N(n)} \alpha(n)_k \mathbf{u}_k \right) d\mathbf{x} + \epsilon/2 \\
&\leq \sum_{k=n}^{N(n)} \alpha(n)_k \int_\Omega f(\mathbf{u}_k) \, d\mathbf{x} + \epsilon/2 \\
&\leq L + \epsilon.
\end{aligned}
$$

Here we have used the fact that $\alpha(n)_k \geq 0$ and $\sum_{k=n}^{N(n)} \alpha(n)_k = 1$. Since ϵ was arbitrary, the proof is complete.

□

By applying the previous theorem to $\mathcal{F}$ and $-\mathcal{F}$ we immediately get the following result.

Corollary 10.21. *The mapping $\mathcal{F}$ is weakly (weak-star) continuous if and only if $\mathbf{u} \mapsto f(\mathbf{u})$ is affine; i.e.,*

$$f(\mathbf{u}) = \alpha + \mathbf{b} \cdot \mathbf{u}, \tag{10.71}$$

for some $\alpha \in \mathbb{R}$ and some $\mathbf{b} \in \mathbb{R}^n$.

(The reader should verify (or already know) that if both f and $-f$ are convex, then f is affine.)

Tonelli's theorem can now be applied directly to energy functional $\bar{\mathcal{E}}$.

Corollary 10.22. *If the function $\mathbf{F} \mapsto \bar{W}(\mathbf{x}, \mathbf{F})$ is convex, then $\bar{\mathcal{E}}$ is weakly lower semicontinuous. Furthermore, there exists an equilibrium state $\bar{\mathbf{p}}$.*

Proof. The theorem applies directly to the minimizing sequence and the stored energy W. To take care of the potential energy term we use compact imbedding to get $\mathbf{p}_n \to \bar{\mathbf{p}}$ strongly in $L^p(\Omega)$. Then, using Fatou's lemma once again, we get

$$\int_\Omega \Psi(\bar{\mathbf{p}}) \, dx \leq \liminf_{n \to \infty} \int_\Omega \Psi(\mathbf{p}_n) \, dx \tag{10.72}$$

This completes the proof.

□

The result above is useful in some situations (e.g., linear elasticity). However, in problems in nonlinear elasticity (where we really want to apply this theory in the first place) there are good reasons why the assumption of convexity of the energy density is physically unreasonable. We will not discuss these reasons here; for an introductory discussion of these issues the reader could consult [Ba].

If we weaken the convexity assumption on W, Tonelli's theorem would seem to imply that we will not be able to show that the energy $\mathcal{E}$ is weakly lower semicontinuous. However, this is not the case. We know more about sequences $\mathbf{F}_n \rightharpoonup \bar{\mathbf{F}}$ than the fact that they are weakly convergent. We know that each element of the sequence is a gradient; i.e., $\mathbf{F}_n = \nabla \mathbf{p}_n$.

How can we identify which matrix-valued functions $\mathbf{F} : \mathbb{R}^3 \to Q$ are given by gradients? Using the equality of mixed partial derivatives, we see that

$$\frac{\partial F_{ij}}{\partial x_k} = \frac{\partial^2 p_i}{\partial x_j \partial x_k} = \frac{\partial^2 p_i}{\partial x_k \partial x_j} = \frac{\partial F_{ik}}{\partial x_j}. \tag{10.73}$$

(This is clear if $\mathbf{p}$ is smooth. It holds in the space $W^{-1,p}(\Omega)$ for general $\mathbf{p} \in W^{1,p}(\Omega)$.) This is simply a complicated version of the condition that the curl of a gradient be zero. With this in mind, it is natural to study weak lower semicontinuity under the assumption of *differential constraints*. A very important example of a result of this type is the following: often called the *div-curl lemma*.

Theorem 10.23 (Div-Curl). *Let $\Omega \subseteq \mathbb{R}^n$ be a domain. Suppose $\mathbf{u}^k : \Omega \to \mathbb{R}^n$ and $\mathbf{v}^k : \Omega \to \mathbb{R}^n$ are sequences of vector valued functions satisfying*

$$\mathbf{u}^k \;\rightharpoonup\; \bar{\mathbf{u}} \;\; in\; L^2(\Omega), \tag{10.74}$$

$$\mathbf{v}^k \;\rightharpoonup\; \bar{\mathbf{v}} \;\; in\; L^2(\Omega), \tag{10.75}$$

and suppose the sequences

$$div\,\mathbf{u}^k \;:=\; \sum_{i=1}^{n} \frac{\partial u_i^k}{\partial x_i}, \tag{10.76}$$

$$curl\,\mathbf{v}^k \;:=\; \frac{\partial v_j^k}{\partial x_i} - \frac{\partial v_i^k}{\partial x_j}, \quad i,j = 1,\dots,n, \tag{10.77}$$

lie in a compact set in $H_{\mathrm{loc}}^{-1}(\Omega)$. Then

$$\mathbf{u}^k \cdot \mathbf{v}^k \to \bar{\mathbf{u}} \cdot \bar{\mathbf{v}} \;\; in\; \mathcal{D}'(\Omega). \tag{10.78}$$

Here, $H_{\mathrm{loc}}^{-1}(\Omega)$ is the set of functions f such that for any test function $\phi \in \mathcal{D}(\Omega)$ we have $\phi f \in H^{-1}(\Omega)$.

Since we won't use this lemma below we will skip the proof. However, we leave the proof of an easier version as an exercise (cf. Problem 10.9).

A more useful result from the point of view of our problems in elasticity is the following theorem on the weak continuity of subdeterminants of gradients.

Theorem 10.24. *Let $\Omega \subset \mathbb{R}^n$ be a bounded domain. Suppose $n < p < \infty$ and suppose the sequence of functions $\mathbf{p}_k : \Omega \to \mathbb{R}^n$ satisfy*

$$\mathbf{p}_k \rightharpoonup \bar{\mathbf{p}} \;\; in\; W^{1,p}(\Omega). \tag{10.79}$$

Let $m \le n$ and let M_k be the sequence of $m \times m$ subdeterminants obtained by taking a fixed m rows and m columns of $\nabla \mathbf{p}_k$, and let $\bar{M}$ be the corresponding subdeterminant of $\nabla \bar{\mathbf{p}}$. Then

$$M_k \rightharpoonup \bar{M} \;\; in\; L^{p/m}(\Omega). \tag{10.80}$$

Proof. The proof proceeds by induction, and the first step is left to the reader (Problem 10.10).

Without loss of generality, we can complete the proof by showing that

$$\det(\nabla \mathbf{p}_k) \rightharpoonup \det(\nabla \bar{\mathbf{p}}) \;\; in\; L^{p/n}(\Omega) \tag{10.81}$$

under the assumption that any $(n-1) \times (n-1)$ subdeterminant M_k satisfies (10.80) (with $m = n-1$). However, under this assumption we have

$$\mathrm{cof}\,\nabla \mathbf{p}_k \rightharpoonup \mathrm{cof}\,\nabla \bar{\mathbf{p}} \;\; in\; L^{p/(n-1)}(\Omega). \tag{10.82}$$

Here we have used cof $\mathbf{A}$ to denote the *cofactor matrix* of $\mathbf{A}$. (The (i,j)th component of the cofactor matrix is $(-1)^{i+j}$ times the (i,j)th minor.)

We now use the fact that for smooth functions $\mathbf{p}$

$$\det \nabla \mathbf{p} = \sum_j \frac{\partial p_i}{\partial x_j} (\text{cof} \nabla \mathbf{p})_{ij}, \tag{10.83}$$

for any $i = 1, \ldots, n$. In addition, we use the identity

$$\sum_j \frac{\partial}{\partial x_j} (\text{cof} \nabla \mathbf{p})_{ij} = 0, \quad i = 1, \ldots, n. \tag{10.84}$$

(This identity can be verified by direct calculation (cf. Problem 10.11).) Therefore, we can derive the formula

$$\det \nabla \mathbf{p} = \sum_j \frac{\partial}{\partial x_j} \left(p_i (\text{cof} \nabla \mathbf{p})_{ij} \right). \tag{10.85}$$

Thus, after approximating our sequence $\mathbf{p}_k$ by smooth functions, we have for any $\phi \in \mathcal{D}(\Omega)$

$$\int_\Omega \phi \det \nabla \mathbf{p}_k \, dx \;=\; -\int_\Omega \sum_j \frac{\partial \phi}{\partial x_j} p_i^k (\text{cof} \nabla \mathbf{p}_k)_{ij} \, dx$$

$$\rightarrow \;-\int_\Omega \sum_j \frac{\partial \phi}{\partial x_j} \bar{p}_i (\text{cof} \nabla \bar{\mathbf{p}})_{ij} \, dx$$

$$=\; \int_\Omega \phi \det \nabla \bar{\mathbf{p}} \, dx.$$

In taking the limit above we have used the fact that by compact imbedding $\mathbf{p}^k \to \bar{\mathbf{p}}$ (strongly) in $L^p(\Omega)$, and hence (since Ω is bounded) $\mathbf{p}^k \to \bar{\mathbf{p}}$ (strongly) in $L^q(\Omega)$ where $q = p/(1 + p - n)$ is the conjugate exponent of $p/(n-1)$. We can use this and Problem 6.33 to take the limit.

Thus, $\det \nabla \mathbf{p}_k \to \det \nabla \bar{\mathbf{p}}$ in $\mathcal{D}'(\Omega)$. To complete the proof, we use the fact that $\det \nabla \mathbf{p}_k$ is bounded in $L^{p/n}(\Omega)$ and a density argument (cf. Problem 6.32). $\qquad\square$

We can use this result coupled with the following definition in our study of variational problems of nonlinear elasticity.

Definition 10.25. Let $\mathcal{M}^{m \times n}$ be the set of $m \times n$ matrices. A function $G : \mathcal{M}^{m \times n} \to \mathbb{R}$ is said to be **polyconvex** if $\mathbf{A} \mapsto G(\mathbf{A})$ can be represented as a convex function of the subdeterminants of $\mathbf{A}$.

In the particular case of three-dimensional elasticity, we say that the stored energy density $\mathcal{W}(\mathbf{x}, \mathbf{F})$ is polyconvex if there exists a function

$$\Omega \times \mathcal{M}^{3 \times 3} \times \mathcal{M}^{3 \times 3} \times (0, \infty) \ni (\mathbf{x}, \mathbf{A}, \mathbf{B}, d) \mapsto g(\mathbf{x}, \mathbf{A}, \mathbf{B}, d) \in \mathbb{R}, \tag{10.86}$$

such that for every $\mathbf{x} \in \Omega$, $(\mathbf{A}, \mathbf{B}, d) \mapsto g(\mathbf{x}, \mathbf{A}, \mathbf{B}, d)$ is convex and

$$\mathcal{W}(\mathbf{x}, \mathbf{F}) = g(\mathbf{x}, \mathbf{F}, \text{cof} \mathbf{F}, \det \mathbf{F}). \tag{10.87}$$

Remark 10.26. Polyconvexity is indeed a weaker assumption than convexity. In particular, such functions as

$$G(\mathbf{F}) = \begin{cases} \frac{1}{\det \mathbf{F}}, & \det \mathbf{F} > 0 \\ \infty, & \det \mathbf{F} \le 0, \end{cases} \tag{10.88}$$

are polyconvex but not convex.

Remark 10.27. In elasticity, where $\mathbf{F}(\mathbf{x})$ is the deformation gradient at $\mathbf{x} \in \Omega$, the components of $\mathbf{F}(\mathbf{x})$ reflect local changes under the deformation in the length of curves going through $\mathbf{x}$, the components of $\mathrm{cof}\mathbf{F}(\mathbf{x})$ reflect changes in the areas of variously oriented surfaces and $\det \mathbf{F}(\mathbf{x})$ reflects the local change in volume.

Remark 10.28. The physical objections raised in the literature for convex stored energy functions do not apply to polyconvex functions. Polyconvexity is widely accepted as an assumption which is both extremely general and physically reasonable.

We now state our basic existence result for elastic materials with polyconvex energies.

Theorem 10.29. *Let $\bar{W}$ satisfy assumptions 1-5 on page 343 and, in addition, be polyconvex. Then there exists a minimizer $\tilde{\mathbf{p}} \in \mathcal{D}_{\mathcal{E}}$ of the energy $\bar{\mathcal{E}}$.*

Proof. By Lemma 10.10 there exists a minimizing sequence $\mathbf{p}_n \in \mathcal{D}_{\mathcal{E}}$ and a function $\tilde{\mathbf{p}} \in W_{\mathbf{g}}^{1,p}(\Omega)$ such that $\mathbf{p}_n \rightharpoonup \tilde{\mathbf{p}}$ in $W^{1,p}(\Omega)$. Now let $\mathbf{F}_n := \nabla \mathbf{p}_n$, $\tilde{\mathbf{F}} := \nabla \tilde{\mathbf{p}}$. Then Theorem 10.24 gives us

$$\mathrm{cof}\mathbf{F}_n \rightharpoonup \mathrm{cof}\tilde{\mathbf{F}} \quad \text{in } L^{p/2}(\Omega), \tag{10.89}$$

$$\det \mathbf{F}_n \rightharpoonup \det \tilde{\mathbf{F}} \quad \text{in } L^{p/3}(\Omega). \tag{10.90}$$

Now, we use Theorem 10.16 to get

$$\begin{aligned}
\bar{\mathcal{E}}(\tilde{\mathbf{p}}) &= \int_\Omega \bar{W}(\mathbf{x}, \tilde{\mathbf{F}}(\mathbf{x})) + \Psi(\tilde{\mathbf{p}}(\mathbf{x})) \, d\mathbf{x} \\
&= \int_\Omega g(\mathbf{x}, \tilde{\mathbf{F}}(\mathbf{x}), \mathrm{cof}\tilde{\mathbf{F}}(\mathbf{x}), \det \tilde{\mathbf{F}}(\mathbf{x})) + \Psi(\tilde{\mathbf{p}}(\mathbf{x})) \, d\mathbf{x} \\
&\le \liminf_{n\to\infty} \int_\Omega g(\mathbf{x}, \mathbf{F}_n(\mathbf{x}), \mathrm{cof}\mathbf{F}_n(\mathbf{x}), \det \mathbf{F}_n(\mathbf{x})) + \Psi(\mathbf{p}_n(\mathbf{x})) \, d\mathbf{x} \\
&= \lim_{n\to\infty} \int_\Omega \bar{W}(\mathbf{x}, \mathbf{F}_n(\mathbf{x})) + \Psi(\mathbf{p}_n(\mathbf{x})) \, d\mathbf{x} \\
&= L.
\end{aligned}$$

Where L is the greatest lower bound of $\bar{\mathcal{E}}$ over $\mathcal{D}_{\mathcal{E}}$. Thus, $\tilde{\mathbf{p}}$ is indeed a minimizer and this completes the proof. $\qquad\square$

10.2.2 Nonconvex Problems

The study of nonconvex variational problems has occupied a significant fraction of the PDE community over the last twenty years. Such problems arise in PDEs which model phase transitions , e.g. the liquid-gas transition, phase transitions is crystalline solids, ferromagnetism and superconductivity. We discuss the main issues which arise in a simple one-dimensional example.

A nonconvex model problem

Let us try to minimize

$$I(u) = \int_0^1 u(x)^2 + (u'(x)^2 - 1)^2 \, dx, \qquad (10.91)$$

subject to the boundary conditions $u(0) = u(1) = 0$. Note that to do so we must meet two conflicting demands. The second term under the integral is a "nice" convex function of u. To make it small we must look for functions u that are as close to zero as possible. The second term is nonconvex. It is minimized when u' is either one or minus one. In between these values the term achieves a local maximum when $u' = 0$.

With this in mind minimizing sequence is easily found. For each $i = 0, 1, ..., n - 1$, let

$$u_n(x) = \begin{cases} x - \frac{i}{n}, & \frac{i}{n} \le x \le \frac{i}{n} + \frac{1}{2n}, \\ \frac{i+1}{n} - x, & \frac{i}{n} + \frac{1}{2n} \le x \le \frac{i+1}{n}. \end{cases} \qquad (10.92)$$

That is, u_n is a sequence of piecewise linear functions that "zigzag" between derivatives ± 1. The maximum height of the function is $1/2n$. Since the second term in I is identically zero, we can easily calculate that $I(u_n) = 1/(12n^2)$. Thus

$$\lim_{n \to \infty} I(u_n) = 0,$$

and since I is always nonnegative, the sequence approaches the minimum value of I.

However, because of the nonconvex term, I is not continuous in the way we might expect. That is even though the sequence of functions u_n converges to zero uniformly, we have

$$I(\lim_{n \to \infty} u_n) = I(0) = 1 \ne \lim_{n \to \infty} I(u_n).$$

Indeed, it is easy to see that no function can possibly attain the minimum of I. Any function u with $I(u) = 0$ would have to satisfy $u = 0$ and $u' = \pm 1$, but there is no such function.

The behavior found in this example turns out to be typical of nonconvex variational problems. There are two possible strategies to cope with this:

1. Identify the problem satisfied by the limits of minimizing sequences; obviously this will be a problem different from the original one.

2. Actually define a "function" which is equal to zero but has derivative ± 1.

In this section, we shall give a brief outline of each of these approaches.

Convexification

Definition 10.30. For a function f(p), defined on an interval, we define the lower convex envelope by

$$Cf = \sup\{g \mid g \text{ convex, } g \leq f\}. \tag{10.93}$$

There are several other equivalent characterizations, see for instance [Dac]. For variational problems of the form

$$\min \int_a^b f(x, u, u') \, dx, \tag{10.94}$$

it is possible to show under quite general hypotheses that the limits of minimizing sequences satisfy the problem

$$\min \int_a^b Cf(x, u, u') \, dx, \tag{10.95}$$

where the lower convex envelope is with respect to the third variables. For instance, in the example above, we have (see Problem 10.12)

$$f(x, u, u') = ((u')^2 - 1)^2 + u^2, \tag{10.96}$$

and

$$Cf(x, u, u') = \begin{cases} ((u')^2 - 1)^2 + u^2, & |u'| \geq 1, \\ u^2, |u'| < 1. \end{cases} \tag{10.97}$$

Obviously, $u = 0$ is indeed a minimizer of the modified problem.

In the multidimensional case, however, convexity is not the right notion, as we already saw in the previous section. The appropriate notion is quasiconvexity.

Definition 10.31. The function $f : \mathbb{R}^{nm} \to \mathbb{R}$ is called *quasiconvex* if

$$\frac{1}{m(D)} \int_D f(\mathbf{A} + \nabla \mathbf{u}(\mathbf{x})) \, d\mathbf{x} \geq f(\mathbf{A}) \tag{10.98}$$

for every bounded domain $D \subset \mathbb{R}^n$, every $\mathbf{A} \in \mathbb{R}^{nm}$ and every $\mathbf{u} \in (\mathcal{D}(D))^m$. Here $m(D)$ denotes the volume of D.

Quasiconvexity is linked to lower semicontinuity in the multidimensional case in the same way as convexity is in the one-dimensional case. If appropriate technical hypotheses are satisfied, it is therefore possible to show that limits of minimizing sequences for the problem

$$\min \int_\Omega f(\mathbf{x}, \mathbf{u}, \nabla \mathbf{u}) \, d\mathbf{x} \qquad (10.99)$$

are minimizers for the functional

$$\min \int_\Omega Qf(\mathbf{x}, \mathbf{u}, \nabla \mathbf{u}) \, d\mathbf{x}, \qquad (10.100)$$

where Qf denotes the lower quasiconvex (with respect to the third variable) envelope:

$$Qf(\mathbf{x}, \mathbf{u}, \mathbf{A}) = \sup\{g \,|\, g \text{ quasiconvex in } \mathbf{A}, \ g \leq f\}. \qquad (10.101)$$

Unfortunately, quasiconvexity is not a simple pointwise condition on f as convexity is and is difficult to verify. Polyconvexity as defined in the preceding section is a sufficient condition. A necessary condition is rank one convexity.

Definition 10.32. The function $f : \mathbb{R}^{nm} \to \mathbb{R}$ is called *rank one convex* if

$$f(\lambda \mathbf{A} + (1-\lambda)\mathbf{B}) \leq \lambda f(\mathbf{A}) + (1-\lambda)f(\mathbf{B}) \qquad (10.102)$$

for every $\lambda \in (0,1)$ and every $\mathbf{A}$ and $\mathbf{B}$ such that $\text{rank}(\mathbf{A} - \mathbf{B}) = 1$.

We can define a lower polyconvex envelope Pf and a lower rank one convex envelope Rf in an analogous fashion as the lower convex envelope Cf and lower quasiconvex envelope Qf. Since it can be shown that

$$\text{convexity} \quad \Rightarrow \quad \text{polyconvexity} \Rightarrow \text{quasiconvexity}$$
$$\Rightarrow \quad \text{rank one convexity}, \qquad (10.103)$$

we have

$$Cf \leq Pf \leq Qf \leq Rf \leq f. \qquad (10.104)$$

Hence we have a characterization of Qf if we can show that $Pf = Rf$.

Generalized functions

How can one define a "function" which is equal to zero but has derivative ± 1? The solution to this dilemma is to assign probabilities rather than values to the derivative. The classical derivative is then the average of the probability distribution. This allows the function 0 to have derivatives ± 1, as long as both values occur with equal probability. The development of this theory, originally introduced by L.C. Young, rests on the following result (see [Dac]).

Theorem 10.33. Let $K \subset \mathbb{R}^m$, $\Omega \subset \mathbb{R}^n$ be bounded, open sets, and let $\mathbf{v}_n \in (L^\infty(\Omega))^m$ be such that $\mathbf{v}_n$ has values in K. Then there exists a family of probability measures ν_x, $x \in \Omega$, on $\bar{K}$ and a subsequence $\mathbf{v}_{n_k}$ such that, *in the sense of weak-∗ convergence in $(L^\infty(\Omega))^m$, we have*

$$\lim_{k\to\infty} f(\mathbf{v}_{n_k}) = \int_{\bar K} \nu_x(\mathbf{p}) f(\mathbf{p})\, d\mathbf{p} \qquad (10.105)$$

for every continuous function $f : \bar K \to \mathbb{R}$.

For instance, if we take v_n to be the derivative of the zigzag function given by (10.92), we have

$$\lim f(v_n) = \frac{1}{2}(f(1) + f(-1)); \qquad (10.106)$$

in particular this yields $\lim v_n = 0$, and $\lim(v_n^2 - 1)^2 = 0$.

Problems

10.9. Prove the following version of the div-curl lemma. Let $\Omega \subset \mathbb{R}^n$ be a domain. Let $\mathbf{u}^k : \Omega \to \mathbb{R}^n$ be a sequence of vector-valued functions and $w^k : \Omega \to \mathbb{R}$ be a sequence of scalar-valued functions satisfying

$$\mathbf{u}^k \rightharpoonup \bar{\mathbf{u}} \quad \text{in } L^2(\Omega), \qquad (10.107)$$

$$\operatorname{div}\mathbf{u}^k \quad \text{bounded in } L^2(\Omega), \qquad (10.108)$$

$$w^k \rightharpoonup \bar w \quad \text{in } H^1(\Omega). \qquad (10.109)$$

Show that

$$\mathbf{u}^k \cdot \nabla w^k \to \bar{\mathbf{u}} \cdot \nabla \bar w \quad \text{in } \mathcal{D}'(\Omega). \qquad (10.110)$$

10.10. Do the first step of the induction in the proof of Theorem 10.24; i.e., show that for any indices i, j, l, m from 1 to n we have

$$\lim_{k\to\infty} \frac{\partial p_i^k}{\partial x_l}\frac{\partial p_j^k}{\partial x_m} - \frac{\partial p_i^k}{\partial x_m}\frac{\partial p_j^k}{\partial x_l} = \frac{\partial \bar p_i}{\partial x_l}\frac{\partial \bar p_j}{\partial x_m} - \frac{\partial \bar p_i}{\partial x_m}\frac{\partial \bar p_j}{\partial x_l}.$$

10.11. Let $\mathcal{M}^{n\times n}$ be the set of $n \times n$ matrices and let

$$\mathcal{M} \ni \mathbf{F} \mapsto g(\mathbf{F}) \in \mathbb{R}$$

be given by an $m \times m$ subdeterminant of $\mathbf{F}$. Let $\Omega \in \mathbb{R}^n$ be a domain. Consider functions $\mathbf{p} : \Omega \to \mathbb{R}^n$. Compute the Euler-Lagrange equations for the energy functional

$$\mathcal{E}(\mathbf{p}) := \int_\Omega g(\nabla\mathbf{p}(\mathbf{x}))\, d\mathbf{x}. \qquad (10.111)$$

Explain why subdeterminant functions like g are called *null Lagrangians*. Can you think of any other null Lagrangians?

10.12. Let $f(p) = (p^2 - 1)^2$ and let

$$F(p) = \begin{cases} f(p), & |p| \geq 1, \\ 0, & |p| < 1. \end{cases} \qquad (10.112)$$

a. Show that F is convex.

b. Prove that if $x = \lambda y + (1 - \lambda)z$, where $0 \leq \lambda \leq 1$, and if g is convex with $g \leq f$, then $g(x) \leq \lambda f(y) + (1 - \lambda)f(z)$.

c. Show that if g is convex with $g \leq f$, then $g \leq F$.

10.13. Show that convexity implies quasiconvexity and that the converse holds if $n = m = 1$.

10.14. Formulate the Euler-Lagrange equations for the problem

$$\min \int_{\Omega} f(\mathbf{x}, \mathbf{u}, \nabla \mathbf{u}) \, d\mathbf{x}. \tag{10.113}$$

Explore what rank one convexity of f implies for these equations.

10.15. Let $u_n(x) = \sin(nx)$. Explicitly find a function $r(x)$ such that

$$\lim_{n \to \infty} f(u_n(x)) = \int_{-1}^{1} f(y) r(y) \, dy \tag{10.114}$$

in the weak sense.

10.3 Nonlinear Operator Theory Methods

In this section we give a number of results on nonlinear mappings from a Banach space X to its dual:

$$X \ni v \mapsto T(v) \in X^*. \tag{10.115}$$

We shall usually use the term *mapping* to refer to a nonlinear mapping and reserve the term *operator* for when we are assuming a mapping is linear. However, the reader should be warned that this is an affectation adopted for this book which is not used throughout the literature. Authors use the term "operator" for both linear and nonlinear mappings, the assumption that an operator is linear is often left unspecified in contexts where only linear operators are studied. (We do this in Chapter 8.)

10.3.1 Mappings on Finite-Dimensional Spaces

In this section we study mappings $\mathbf{f} : \mathbb{R}^n \to \mathbb{R}^n$. Our goal is a better understanding of the problem of "n equations in n unknowns":

$$\mathbf{f}(\mathbf{u}) = \mathbf{a}. \tag{10.116}$$

We already have discussed a "local" result for the problem, the inverse function theorem, in Chapter 1; but we desire something stronger. As an example of what we expect, consider the following result when $n = 1$.

Theorem 10.34. *Suppose* $f : \mathbb{R} \to \mathbb{R}$ *is continuous and satisfies*

$$f(u) \to \pm\infty \quad \text{as } u \to \pm\infty. \tag{10.117}$$

Then, for every $a \in \mathbb{R}$ *the equation*

$$f(u) = a \tag{10.118}$$

has a solution $u \in \mathbb{R}$. *Furthermore, if* f *is monotone increasing, then, for each* a, *the set of solutions forms a closed interval; if* f *is strictly monotone, the solution is unique.*

The proof is an elementary application of the intermediate value theorem and the definition of monotone and continuous functions.

In order to get an analogue of this result for mappings from $\mathbb{R}^n$ to $\mathbb{R}^n$ we must generalize both the growth condition (10.117) and the definition of monotone increasing (order preserving).

Definition 10.35. We say that a function $\mathbf{f} : \mathbb{R}^n \to \mathbb{R}^n$ is **coercive** if

$$\frac{\mathbf{f}(\mathbf{u}) \cdot \mathbf{u}}{|\mathbf{u}|} \to \infty \quad \text{as } |\mathbf{u}| \to \infty. \tag{10.119}$$

Definition 10.36. We say that a function $\mathbf{f} : \mathbb{R}^n \to \mathbb{R}^n$ is **monotone** if

$$(\mathbf{f}(\mathbf{u}) - \mathbf{f}(\mathbf{v})) \cdot (\mathbf{u} - \mathbf{v}) \geq 0 \quad \text{for all } \mathbf{u}, \mathbf{v} \in \mathbb{R}^n. \tag{10.120}$$

We say that $\mathbf{f}$ is **strictly monotone** if the inequality in (10.120) is strict whenever $\mathbf{u} \neq \mathbf{v}$.

We first study the implications of monotonicity. We begin with the following immediate consequence of the definition.

Theorem 10.37. *If* $\mathbf{f}$ *is strictly monotone, then the equation*

$$\mathbf{f}(\mathbf{u}) = \mathbf{a} \tag{10.121}$$

has at most one solution.

To get a result when $\mathbf{f}$ is only monotone we first prove the following lemma.

Lemma 10.38. *Let* $\mathbf{a} \in \mathbb{R}^n$ *be given and let* $\mathbf{f} : \mathbb{R}^n \to \mathbb{R}^n$ *be continuous. Then if* $\mathbf{u} \in \mathbb{R}^n$ *is a solution of the variational inequality*

$$(\mathbf{f}(\mathbf{v}) - \mathbf{a}) \cdot (\mathbf{v} - \mathbf{u}) \geq 0 \quad \text{for every } \mathbf{v} \in \mathbb{R}^n, \tag{10.122}$$

it follows that $\mathbf{u}$ *is also a solution of equation (10.121).*

In addition, if $\mathbf{f}$ *is monotone, then every solution of equation (10.121) is also a solution of the variational inequality (10.122).*

Proof. Suppose $\mathbf{f}$ is continuous and (10.122) holds. Then, for any $\mathbf{w} \in \mathbb{R}^n$, let $\mathbf{v} = \mathbf{u} + t\mathbf{w}$ for $t > 0$. This gives us (after dividing (10.122) by t)

$$(\mathbf{f}(\mathbf{u} + t\mathbf{w}) - \mathbf{a}) \cdot \mathbf{w} \geq 0 \quad \text{for all } \mathbf{w} \in \mathbb{R}^n. \tag{10.123}$$

Letting $t \to 0$ gives us

$$(\mathbf{f}(\mathbf{u}) - \mathbf{a}) \cdot \mathbf{w} \geq 0 \quad \text{for all } \mathbf{w} \in \mathbb{R}^n. \tag{10.124}$$

Using $\mathbf{w} = \pm(\mathbf{f}(\mathbf{u}) - \mathbf{a})$ gives us $\mathbf{f}(\mathbf{u}) = \mathbf{a}$.

Now, suppose $\mathbf{f}$ is monotone and (10.121) holds. Then

$$(\mathbf{f}(\mathbf{v}) - \mathbf{a}) \cdot (\mathbf{v} - \mathbf{u}) = (\mathbf{f}(\mathbf{v}) - \mathbf{f}(\mathbf{u})) \cdot (\mathbf{v} - \mathbf{u}) \geq 0 \quad \text{for every } \mathbf{v} \in \mathbb{R}^n, \tag{10.125}$$

by the definition of monotonicity. □

As an immediate consequence of this we get the following.

Theorem 10.39. *Let* $\mathbf{a} \in \mathbb{R}^n$ *and let* $\mathbf{f}$ *be continuous and monotone. Then the set* $\mathcal{K}$ *of solutions of* $\mathbf{f}(\mathbf{u}) = \mathbf{a}$ *is closed and convex.*

Proof. For any fixed $\mathbf{v} \in \mathbb{R}^n$ let

$$\mathcal{S}_\mathbf{v} = \{\mathbf{u} \in \mathbb{R}^n \mid (\mathbf{f}(\mathbf{v}) - \mathbf{a}) \cdot (\mathbf{v} - \mathbf{u}) \geq 0\}. \tag{10.126}$$

Note that each set $\mathcal{S}_\mathbf{v}$ is a closed half-space (and is hence convex). By our Lemma 10.38 we can write the solution set as

$$\mathcal{K} = \bigcap_{\mathbf{v} \in \mathbb{R}^n} \mathcal{S}_\mathbf{v}. \tag{10.127}$$

Since the arbitrary intersection of closed sets is closed and the arbitrary intersection of convex sets is convex, our theorem is proved. □

We now examine the consequences of coercivity.

Theorem 10.40. *Let* $\mathbf{f} : \mathbb{R}^n \to \mathbb{R}^n$ *be continuous and coercive. Then for every* $\mathbf{a} \in \mathbb{R}^n$, *equation (10.121) has a solution* $\mathbf{u} \in \mathbb{R}^n$.

In order to prove this we need the following important result.

Theorem 10.41 (Brouwer fixed point theorem). *Let* C *be a compact, convex, nonempty subset of* $\mathbb{R}^n$, *and suppose* $\mathbf{f}$ *is a continuous function that maps* C *into* C. *Then* $\mathbf{f}$ *has a fixed point in* C; *i.e., there exists* $\mathbf{u} \in C$ *such that*

$$\mathbf{f}(\mathbf{u}) = \mathbf{u}. \tag{10.128}$$

Proof. For the general case, we refer the reader to the literature (cf. [DS]), but when $n = 1$ the proof is easy to see graphically. In this case the hypotheses of the theorem state that we have a continuous function f which maps an interval $[a, b]$ into itself. The graph of the function must lie in the box $[a, b] \times [a, b]$ in the (x, y)-plane. There are only three possibilities: $f(a) = a$, $f(b) = b$ or the graph of f starts to the left of the line $x = y$ and ends on the right. Since the function is continuous the graph must cross the line $x = y$ somewhere. Any crossing gives us a fixed point. □

We now prove Theorem 10.40.

Proof. We begin by noting that it is sufficient to show that $\mathbf{f}(\mathbf{u}) = \mathbf{0}$ has a solution. To see this let an arbitrary $\mathbf{a} \in \mathbb{R}^n$ be given, and define

$$\mathbf{f_a}(\mathbf{u}) := \mathbf{f}(\mathbf{u}) - \mathbf{a}. \tag{10.129}$$

We now note that $\mathbf{f_a}$ is coercive if and only if $\mathbf{f}$ is, since

$$
\begin{aligned}
\frac{\mathbf{f_a}(\mathbf{u}) \cdot \mathbf{u}}{|\mathbf{u}|} &= \frac{\mathbf{f}(\mathbf{u}) \cdot \mathbf{u}}{|\mathbf{u}|} - \frac{\mathbf{a} \cdot \mathbf{u}}{|\mathbf{u}|} \\
&\geq \frac{\mathbf{f}(\mathbf{u}) \cdot \mathbf{u}}{|\mathbf{u}|} - \frac{|\mathbf{a}||\mathbf{u}|}{|\mathbf{u}|} \\
&= \frac{\mathbf{f}(\mathbf{u}) \cdot \mathbf{u}}{|\mathbf{u}|} - |\mathbf{a}|.
\end{aligned}
$$

Thus, if we can show that for every coercive function, $\mathbf{f}(\mathbf{u}) = \mathbf{0}$ has a solution, it immediately follows that $\mathbf{f}(\mathbf{u}) = \mathbf{a}$ has one for every $\mathbf{a} \in \mathbb{R}^n$.

We now convert $\mathbf{f}(\mathbf{u}) = \mathbf{0}$ to a fixed point problem. As a first step, we define $\mathbf{g} : \mathbb{R}^n \to \mathbb{R}^n$ by

$$\mathbf{g}(\mathbf{u}) := \mathbf{u} - \mathbf{f}(\mathbf{u}). \tag{10.130}$$

Now note that

$$\mathbf{g}(\mathbf{u}) \cdot \mathbf{u} = |\mathbf{u}|^2 - \mathbf{f}(\mathbf{u}) \cdot \mathbf{u}. \tag{10.131}$$

Since $\mathbf{f}$ is coercive, there exists $R > 0$ such that

$$\mathbf{g}(\mathbf{u}) \cdot \mathbf{u} < |\mathbf{u}|^2 \quad \text{whenever } |\mathbf{u}| \geq R. \tag{10.132}$$

We now define

$$\mathbf{r}(\mathbf{v}) := \begin{cases} \mathbf{v}, & |\mathbf{v}| \leq R \\ \frac{R\mathbf{v}}{|\mathbf{v}|}, & |\mathbf{v}| > R. \end{cases} \tag{10.133}$$

Finally, we let

$$\mathbf{h}(\mathbf{u}) := \mathbf{r}(\mathbf{g}(\mathbf{u})). \tag{10.134}$$

We note that $\mathbf{h}$, as the composition of continuous functions, is continuous. Furthermore, if we let

$$B := \{\mathbf{u} \in \mathbb{R}^n \mid |\mathbf{u}| < R\}, \tag{10.135}$$

then $\mathbf{h}$ maps the closed ball $\overline{B}$ to itself. Thus, by the Brouwer fixed point theorem, $\mathbf{h}$ has a fixed point $\mathbf{u} \in \overline{B}$.

If $\mathbf{u} \in \partial B$, then

$$R = |\mathbf{u}| = |\mathbf{h}(\mathbf{u})| = |\mathbf{r}(\mathbf{g}(\mathbf{u}))|. \tag{10.136}$$

Thus, by the definition of $\mathbf{r}$, we must have

$$\frac{R}{|\mathbf{g}(\mathbf{u})|} \leq 1. \tag{10.137}$$

But we now use (10.132) to get

$$|\mathbf{u}|^2 = \mathbf{u} \cdot \mathbf{u} = \mathbf{h}(\mathbf{u}) \cdot \mathbf{u} = \frac{R}{|\mathbf{g}(\mathbf{u})|} \mathbf{g}(\mathbf{u}) \cdot \mathbf{u} < \frac{R}{|\mathbf{g}(\mathbf{u})|} |\mathbf{u}|^2 < |\mathbf{u}|^2. \qquad (10.138)$$

This is a contradiction. Thus, we must have

$$R > |\mathbf{u}| = |\mathbf{h}(\mathbf{u})| = |\mathbf{r}(\mathbf{g}(\mathbf{u}))|. \qquad (10.139)$$

Hence, $\mathbf{r}(\mathbf{g}(\mathbf{u})) = \mathbf{g}(\mathbf{u})$ and so

$$\mathbf{u} = \mathbf{h}(\mathbf{u}) = \mathbf{r}(\mathbf{g}(\mathbf{u})) = \mathbf{g}(\mathbf{u}) = \mathbf{u} - \mathbf{f}(\mathbf{u}). \qquad (10.140)$$

Finally, this gives us

$$\mathbf{f}(\mathbf{u}) = \mathbf{0}, \qquad (10.141)$$

and completes the proof. $\qquad\qquad\qquad\qquad\qquad\qquad\qquad\qquad\square$

Combining the results of this section gives us

Corollary 10.42. *If* $\mathbf{f} : \mathbb{R}^n \to \mathbb{R}^n$ *is continuous, strictly monotone and coercive, then for every* $\mathbf{a} \in \mathbb{R}^n$ *there exists a unique* $\mathbf{u} \in \mathbb{R}^n$ *such that*

$$\mathbf{f}(\mathbf{u}) = \mathbf{a}.$$

10.3.2 Monotone Mappings on Banach Spaces

In this section we will see that we can expand some of the ideas of the previous section to infinite-dimensional spaces. We will assume X is a real, reflexive Banach space and we study mappings $T : X \to X^*$. For an element $g \in X^*$ we will use the standard inner product notation and write (g, v) for $g(v)$. *In the remaining sections of this chapter, we will, in order to make the text easier to follow, give proofs only in the case where X and X^* are separable.*

We begin by giving a string of definitions which generalize concepts from functions on finite-dimensional spaces to mappings on reflexive Banach spaces.

Definition 10.43. We say that a mapping $T : X \to X^*$ is **bounded** if it maps bounded sets in X to bounded sets in X^*. The mapping is **continuous** if for every $u \in X$ we have

$$\|T(u) - T(v)\|_{X^*} \to 0 \quad \text{whenever } \|u - v\|_X \to 0. \qquad (10.142)$$

Definition 10.44. We say that a mapping $T : X \to X^*$ is **monotone** if

$$(T(u) - T(v), u - v) \geq 0 \quad \text{for all } u, v \in X, \qquad (10.143)$$

and **strictly monotone** if this inequality is strict whenever $u \neq v$.

Definition 10.45. We say that a mapping $T : X \to X^*$ is **coercive** if

$$\frac{(T(u), u)}{\|u\|} \to \infty \quad \text{as } \|u\| \to \infty. \qquad (10.144)$$

Remark 10.46. Using the definitions above, the results of the previous section can easily be extended to mappings T from an n-dimensional Banach space X to another n-dimensional Banach space Y. The only real trick here is in defining the appropriate bilinear form

$$X \times Y \ni x, y \mapsto \langle x, y \rangle \in \mathbb{R} \tag{10.145}$$

to replace the "inner product"

$$X \times X^* \ni x, y \mapsto (x, y) \in \mathbb{R} \tag{10.146}$$

in the definitions in this section and the dot product in the previous section. The details are left to the reader.

The proofs of Lemma 10.38 and Theorem 10.39 were based on monotonicity and were in no way dependent on the dimension of the spaces involved. Thus, we can get the following two analogous results using only minor changes of notation in the previous proofs.

Lemma 10.47. *Let $g \in X^*$ be given and let $T : X \to X^*$ be continuous. Then if $u \in X$ is a solution of the variational inequality*

$$(T(v) - g, v - u) \geq 0 \quad \text{for every } v \in X, \tag{10.147}$$

then u satisfies the equation

$$T(u) = g. \tag{10.148}$$

In addition, if T is monotone, then every solution of the equation (10.148) is also a solution of the variational inequality (10.147).

Theorem 10.48. *Let $g \in X^*$ and let $T : X \to X^*$ be continuous and monotone. Then the set $\mathcal{K}$ of solutions of $T(u) = g$ is closed and convex.*

Our first infinite-dimensional analogue of Theorem 10.40 is the following.

Theorem 10.49 (Browder-Minty). *Let X be a real, reflexive Banach space and let $T : X \to X^*$ be bounded, continuous, coercive and monotone. Then for any $g \in X^*$ there exists a solution u of the equation*

$$T(u) = g; \tag{10.149}$$

i.e., $T(X) = X^*$.

Proof. Let $\{x_i\}_{i=1}^\infty$ be a set whose span is dense in X. (Recall that we are giving proofs only in the case where X and X^* are separable.) Now let

$$X \supset X_n := \text{span}\{x_1, x_2, \ldots, x_n\}. \tag{10.150}$$

Our plan is to approximate the problem (10.149) by a finite-dimensional problem

$$T_n(u_n) = g_n, \tag{10.151}$$

where $u_n \in X_n$, $g_n \in X_n^*$ and $T_n : X_n \to X_n^*$. The idea of approximating infinite-dimensional by finite-dimensional problems is known as *Galerkin's method*. It is well-known as a device for doing numerical calculations, but is equally useful as a theoretical tool (as we use it here).

We assume $g \in X^*$ is given and define $g_n \in X_n^*$ to be the restriction

$$g_n := g|_{X_n}. \qquad (10.152)$$

Similarly, we define the operator $T_n : X_n \to X_n^*$ by

$$T_n(u) := T(u)|_{X_n} \quad \text{for all } u \in X_n. \qquad (10.153)$$

Another way of saying this is that for any $u, v \in X_n$ we have

$$(g - g_n, v) = (T(u) - T_n(u), v) = 0. \qquad (10.154)$$

The continuity, boundedness and coercivity of T_n are inherited directly from properties of T. Thus, using Theorem 10.40, we see that there exists a sequence $\{u_n\}_{n=1}^{\infty}$ such that

$$T_n(u_n) = g_n. \qquad (10.155)$$

We now use (10.154) to note that

$$\frac{(T(u_n), u_n)}{\|u_n\|} = \frac{(T_n(u_n), u_n)}{\|u_n\|} = \frac{(g_n, u_n)}{\|u_n\|} = \frac{(g, u_n)}{\|u_n\|} \leq \frac{\|g\| \, \|u_n\|}{\|u_n\|} = \|g\|. \qquad (10.156)$$

Thus, since T is coercive, we must have $\|u_n\|$ bounded. Also, since T is bounded we must have $\|T(u_n)\|$ bounded. Thus, using the weak compactness theorem 6.64, the reflexivity of X and the separability of X^*, we see that there exists $u \in X$, $\tilde{g} \in X^*$ and a subsequence (also labeled u_n) such that

$$u_n \rightharpoonup u \quad \text{in } X \qquad (10.157)$$

and

$$T(u_n) \rightharpoonup \tilde{g} \quad \text{in } X^*. \qquad (10.158)$$

We now show that $\tilde{g} = g$ and that $T(u) = g$.

We use (10.154) again to note that for every basis vector x_i,

$$(\tilde{g} - g, x_i) = \lim_{n \to \infty} (T(u_n) - g, x_i) = 0. \qquad (10.159)$$

Thus, $\tilde{g} = g$.

The key result here is that, once again using (10.154), we can get

$$\begin{aligned}
(T(u_n), u_n) &= (T_n(u_n), u_n) \\
&= (g_n, u_n) \\
&= (g, u_n) \\
&\to (g, u).
\end{aligned} \qquad (10.160)$$

(Since the left-hand side is the product of two *weakly* convergent sequences, this result is nontrivial.) We now use this and the monotonicity of T to get that for every $v \in X$

$$0 \leq (T(u_n) - T(v), u_n - v) \to (T(v) - g, v - u). \tag{10.161}$$

The theorem now follows immediately from Lemma 10.47.   □

10.3.3 Applications of Monotone Operators to Nonlinear PDEs

In this section we apply the theory of monotone operators to scalar second-order quasilinear PDEs in divergence form. Thus, we assume $\Omega \subset \mathbb{R}^n$ is a domain, and for sufficiently smooth functions $u : \Omega \to \mathbb{R}$, we set

$$A(u) = \sum_{|\alpha| \leq 1} (-1)^{|\alpha|} D^\alpha \mathcal{A}_\alpha(\mathbf{x}, \delta_1(u(\mathbf{x}))), \tag{10.162}$$

where

$$\delta_1(u(\mathbf{x})) := \{D^\alpha u(\mathbf{x}) \mid |\alpha| \leq 1\}. \tag{10.163}$$

We will consider the Dirichlet problem for

$$A(u) = f \tag{10.164}$$

with $f \in W^{-1,q}(\Omega)$, where $p \in (1, \infty)$ is given and $\frac{1}{p} + \frac{1}{q} = 1$. (Here, in analogy to Section 7.3, we define $W^{-1,q}(\Omega)$ to be the dual space of $W_0^{1,p}(\Omega)$.) The "bivariate form" corresponding to A is

$$B(u,v) := \sum_{|\alpha| \leq 1} \int_\Omega \mathcal{A}_\alpha(\mathbf{x}, \delta_1(u(\mathbf{x}))) D^\alpha v(\mathbf{x}) \, d\mathbf{x}. \tag{10.165}$$

Definition 10.50. We say that $u \in W_0^{1,p}(\Omega)$ is a **weak solution** of the Dirichlet problem for the quasilinear PDE (10.164) if

$$B(u,v) = (f,v) \quad \text{for all } v \in W_0^{1,p}(\Omega). \tag{10.166}$$

In order to get an existence theorem we make the following assumptions about the functions $\mathcal{A}_\alpha$.

H-1. For each $|\alpha| \leq 1$,

$$\mathbf{x} \mapsto \mathcal{A}_\alpha(\mathbf{x}, \delta_1) \tag{10.167}$$

is measurable for every fixed $\delta_1 = (\mathbf{x} i_\alpha)_{|\alpha| \leq 1} \in \mathbb{R}^{n+1}$.

H-2. For each $|\alpha| \leq 1$,

$$\delta_1 \mapsto \mathcal{A}_\alpha(\mathbf{x}, \delta_1) \tag{10.168}$$

is in $C(\mathbb{R}^{n+1})$ for almost every $\mathbf{x} \in \Omega$.

H-3. For every $\delta_1^1 = (\mathbf{x}i_\alpha^1)_{|\alpha|\leq 1} \in \mathbb{R}^{n+1}$, $\delta_1^2 = (\mathbf{x}i_\alpha^2)_{|\alpha|\leq 1} \in \mathbb{R}^{n+1}$ and every $\mathbf{x} \in \Omega$, we have

$$\sum_{|\alpha|\leq 1} [\mathcal{A}_\alpha(\mathbf{x}, \delta_1^1) - \mathcal{A}_\alpha(\mathbf{x}, \delta_1^2)](\mathbf{x}i_\alpha^1 - \mathbf{x}i_\alpha^2) \geq 0. \qquad (10.169)$$

H-4. There exists $p \in (1,\infty)$, a constant $c_0 > 0$, a function $h \in L^1(\Omega)$ such that for every $\mathbf{x} \in \Omega$ and every $\delta_1 = (\mathbf{x}i_\alpha)_{|\alpha|\leq 1} \in \mathbb{R}^{n+1}$ we have

$$\sum_{|\alpha|\leq 1} \mathcal{A}_\alpha(\mathbf{x}, \delta_1)\mathbf{x}i_\alpha \geq c_0|\delta_1|^p - h(\mathbf{x}). \qquad (10.170)$$

H-5. There exists a constant $c_1 > 0$ and a function $g \in L^q(\Omega)$, ($q := p/(p-1)$) such that for every $\mathbf{x} \in \Omega$ and every $\delta_1 = (\mathbf{x}i_\alpha)_{|\alpha|\leq 1} \in \mathbb{R}^{n+1}$ we have

$$|\mathcal{A}_\alpha(\mathbf{x}, \delta_1)| \leq c_1|\delta_1|^{p-1} + g(\mathbf{x}). \qquad (10.171)$$

Under these assumptions we get the following result.

Theorem 10.51. *Let $\mathcal{A}_\alpha$ satisfy hypotheses H-1–H-5, and let p and q be as defined in hypotheses H-4 and H-5, respectively. Then, for every $f \in W^{-1,q}(\Omega)$, there exists a weak solution $u \in W_0^{1,p}(\Omega)$ of the Dirichlet problem for the quasilinear PDE (10.164).*

We break the proof of this theorem into a series of lemmas, the first of which is the following.

Lemma 10.52. *For each fixed $u \in W_0^{1,p}(\Omega)$, the mapping*

$$W_0^{1,p}(\Omega) \ni v \mapsto B(u,v) \in \mathbb{R} \qquad (10.172)$$

is a bounded linear functional. Thus, there exists a mapping

$$X = W_0^{1,p}(\Omega) \ni u \mapsto T(u) \in W^{-1,q}(\Omega) = X^* \qquad (10.173)$$

such that

$$B(u,v) = (T(u),v), \qquad (10.174)$$

for all u, v in $W_0^{1,p}(\Omega)$. Furthermore, the nonlinear mapping is bounded.

Proof. The linearity of the mapping $v \mapsto B(u,v)$ is obvious. If $u, v \in W_0^{1,p}(\Omega)$ then, using H-5 and Hölder's inequality, we get the estimate:

$$
\begin{aligned}
|B(u,v)| &\leq \sum_{|\alpha| \leq 1} \int_\Omega |\mathcal{A}_\alpha(x, \delta_1(u(\mathbf{x}))) D^\alpha v(\mathbf{x})| \, d\mathbf{x} \\
&\leq \sum_{|\alpha| \leq 1} \int_\Omega \left[c_1 |\delta_1(u(\mathbf{x}))|^{p-1} + g(\mathbf{x}) \right] |D^\alpha v(\mathbf{x})| \, d\mathbf{x} \\
&\leq \sum_{|\alpha| \leq 1} c_1 \left\{ \int_\Omega |\delta_1(u(\mathbf{x}))|^{q(p-1)} \right\}^{1/q} \left\{ \int_\Omega |D^\alpha v|^p \, d\mathbf{x} \right\}^{1/p} \\
&\quad + \sum_{|\alpha| \leq 1} \left\{ \int_\Omega |g|^q \, d\mathbf{x} \right\}^{1/q} \left\{ \int_\Omega |D^\alpha v|^p \, d\mathbf{x} \right\}^{1/p} \\
&\leq C(\|u\|_{1,p}^{p-1} + 1) \|v\|_{1,p}.
\end{aligned}
$$

This shows that for each $u \in W_0^{1,p}(\Omega)$, the mapping $v \mapsto B(u,v)$ is bounded. This implies the existence of the mapping T.

Furthermore, the same estimate shows that T is bounded since

$$\|T(u)\|_{-1,q} = \sup_{\substack{v \in W_0^{1,p}(\Omega) \\ \|v\|_{1,p}=1}} |(T(u), v)| \leq C(\|u\|_{1,p}^{p-1} + 1). \qquad (10.175)$$

Thus, T maps bounded sets in $W_0^{1,p}(\Omega)$ into bounded sets in $W^{-1,q}(\Omega)$. This completes the proof. $\qquad \square$

Lemma 10.53. *The operator T defined in Lemma 10.52 is monotone and coercive.*

Proof. To prove monotonicity we use hypothesis H-3 to get

$$
\begin{aligned}
(T(u) - T(v), u - v) &= B(u, u - v) - B(v, u - v) \\
&= \sum_{|\alpha| \leq 1} \int_\Omega [\mathcal{A}_\alpha(x, \delta_1(u)) - \mathcal{A}_\alpha(x, \delta_1(v))] [D^\alpha u - D^\alpha v] \, dx \\
&\geq 0.
\end{aligned}
$$

To prove coercivity we use hypothesis H-4 to get

$$
\begin{aligned}
\frac{(T(u), u)}{\|u\|_{1,p}} &= \frac{B(u,u)}{\|u\|_{1,p}} \\
&= \|u\|_{1,p}^{-1} \sum_{|\alpha| \leq 1} \int_\Omega \mathcal{A}_\alpha(x, \delta_1(u)) D^\alpha u \, dx \\
&\geq \|u\|_{1,p}^{-1} \left[c_0 \int_\Omega |\delta_1(u)|^p \, dx - \int_\Omega h(\mathbf{x}) \, dx \right] \\
&\geq \|u\|_{1,p}^{-1} C[\|u\|_{1,p}^p - 1].
\end{aligned}
$$

Since $p > 1$ we have

$$\frac{(T(u), u)}{\|u\|_{1,p}} \to \infty \quad \text{as } \|u\|_{1,p} \to \infty. \tag{10.176}$$

This completes the proof. $\qquad\qquad\qquad\qquad\qquad\qquad\qquad\qquad\qquad\square$

Thus, to apply the Browder-Minty theorem to the mapping T and complete the proof of Theorem 10.51 we need only show the following.

Lemma 10.54. *The mapping $T : W_0^{1,p}(\Omega) \to W^{-1,q}(\Omega)$ is continuous.*

In the next section we describe a tool called *Nemytskii operators* which we can use to prove this lemma.

Example 10.55. Consider the second-order nonlinear partial differential operator

$$A(u) := -\sum_{i=1}^{n} \frac{\partial}{\partial x_i} \left(\left| \frac{\partial u}{\partial x_i} \right|^{p-2} \frac{\partial u}{\partial x_i} \right) + |u|^{p-2} u, \tag{10.177}$$

where $p \in (1, \infty)$. Note that the case $p = 2$ is simply the Laplacian plus a lower-order term which we have already considered in our material on linear problems. Here,

$$\mathcal{A}_{(0,0,\ldots,0)} = a_0(\mathbf{x}, u, u_{x_1}, \ldots, u_{x_n}) = |u|^{p-2} u,$$
$$\mathcal{A}_{(0,\ldots,0,1,0,\ldots,0)} = a_i(\mathbf{x}, u, u_{x_1}, \ldots, u_{x_n}) = |u_{x_i}|^{p-2} u_{x_i}, \quad i = 1, \ldots, n.$$

We wish to verify that these $\mathcal{A}_\alpha$ satisfy the hypotheses H-1 to H-5. Hypotheses H-1 and H-2 obviously hold. To verify H-3 we let $\delta_1^1 = (\mathbf{x}i_0^1, \mathbf{x}i_1^1, \ldots, \mathbf{x}i_n^1)$ and $\delta_1^2 = (\mathbf{x}i_0^2, \mathbf{x}i_1^2, \ldots, \mathbf{x}i_n^2)$ and calculate

$$\sum_{|\alpha| \leq 1} [\mathcal{A}_\alpha(\mathbf{x}, \delta_1^1) - \mathcal{A}_\alpha(\mathbf{x}, \delta_1^2)](\mathbf{x}i_\alpha^1 - \mathbf{x}i_\alpha^2)$$
$$= \sum_{i=0}^{n} \left(|\mathbf{x}i_i^1|^{p-2} \mathbf{x}i_i^1 - |\mathbf{x}i_i^2|^{p-2} \mathbf{x}i_i^2 \right) \left(\mathbf{x}i_i^1 - \mathbf{x}i_i^2 \right)$$
$$\geq 0.$$

To verify H-4 we let $\delta_1 = (\mathbf{x}i_0, \mathbf{x}i_1, \ldots, \mathbf{x}i_n)$ and get

$$\sum_{|\alpha| \leq 1} \mathcal{A}_\alpha(\mathbf{x}, \delta_1) \mathbf{x}i_\alpha = \sum_{i=0}^{n} |\mathbf{x}i_i|^{p-2} \mathbf{x}i_i \mathbf{x}i_i = \sum_{i=1}^{n} |\mathbf{x}i_i|^p \geq c_0 |\delta_i|^p \tag{10.178}$$

We see that H-5 holds since

$$|\mathcal{A}_\alpha(\mathbf{x}, \delta_1)| = |a_i(\mathbf{x}, \delta_1)| = |\mathbf{x}i_i|^{p-1} \leq |\delta_1|^{p-1}. \tag{10.179}$$

Thus the following existence result follows immediately from Theorem 10.51.

Theorem 10.56. *Let the nonlinear second order partial differential operator A be defined by (10.177). Then for every $f \in W^{-1,q}(\Omega)$ there exists a weak solution $u \in W_0^{1,p}$ of the equation*

$$A(u) = f. \tag{10.180}$$

10.3.4 Nemytskii Operators

In the following section we state without proof some important results on the composition of $L^p(\Omega)$ with nonlinear functions. For a more detailed treatment, the reader could consult [Li].

Definition 10.57. Let $\Omega \subset \mathbb{R}^n$ be a domain. We say that a function

$$\Omega \times \mathbb{R}^m \ni (\mathbf{x}, \mathbf{u}) \mapsto f(\mathbf{x}, \mathbf{u}) \in \mathbb{R} \tag{10.181}$$

satisfies the **Carathéodory conditions** if

$$\mathbf{u} \mapsto f(\mathbf{x}, \mathbf{u}) \text{ is continuous for almost every } \mathbf{x} \in \Omega \tag{10.182}$$

and

$$\mathbf{x} \mapsto f(\mathbf{x}, \mathbf{u}) \text{ is measurable for every } \mathbf{u} \in \Omega. \tag{10.183}$$

Given any f satisfying the Carathéodory conditions and a function $\mathbf{u} : \Omega \to \mathbb{R}^m$, we can define another function by composition

$$\mathcal{F}(\mathbf{u})(\mathbf{x}) := f(\mathbf{x}, \mathbf{u}(\mathbf{x})). \tag{10.184}$$

The composition operator $\mathcal{F}$ is called a *Nemytskii operator*. Our main theorem is on the boundedness and continuity of these operators from $L^p(\Omega)$ to $L^q(\Omega)$.

Theorem 10.58. *Let $\Omega \subset \mathbb{R}^n$ be a domain, and let*

$$\Omega \times \mathbb{R}^m \ni (\mathbf{x}, \mathbf{u}) \mapsto f(\mathbf{x}, \mathbf{u}) \in \mathbb{R} \tag{10.185}$$

*satisfy the **Carathéodory conditions**. In addition, let $p \in (1, \infty)$ and $g \in L^q(\Omega)$ (where $\frac{1}{p} + \frac{1}{q} = 1$) be given, and let f satisfy*

$$|f(\mathbf{x}, \mathbf{u})| \le C|\mathbf{u}|^{p-1} + g(\mathbf{x}). \tag{10.186}$$

Then the Nemytskii operator $\mathcal{F}$ defined by (10.184) is a bounded and continuous map from $L^p(\Omega)$ to $L^q(\Omega)$.

Remark 10.59. Lemma 10.54 follows as a corollary to this theorem. To see this we simply need to apply hypotheses H-1, H-2 and H-5 to see that each $\mathcal{A}_\alpha$ can be used as a Nemytskii operator satisfying the appropriate growth conditions. The continuity of T from $W^{1,p}(\Omega)$ to $W^{-1,q}(\Omega)$ follows from the continuity of $\delta_1(\mathbf{x}) \mapsto \mathcal{A}_\alpha(\mathbf{x}, \delta_1(\mathbf{x}))$ as a map from $L^p(\Omega)$ to $L^q(\Omega)$.

10.3.5 Pseudo-monotone Operators

In this section we examine a somewhat more general class of nonlinear mappings, called *pseudo-monotone operators*. In applications, it often occurs that the hypotheses imposed in the previous section are unnecessarily strong. In particular, the monotonicity assumption H-3 involves both the first-order derivatives and the function itself. As we shall see in this chapter, it is really only necessary to have a monotonicity assumption on the highest-order derivatives: Compactness will take care of the lower-order terms.

Definition 10.60. Let X be a reflexive Banach space. An operator $T : X \to X^*$ is called **pseudo-monotone** if T is bounded and if whenever

$$u_j \rightharpoonup \bar{u} \quad \text{in } X \tag{10.187}$$

and

$$\limsup_{j \to \infty}(T(u_j), u_j - \bar{u}) \leq 0, \tag{10.188}$$

it follows that

$$\liminf_{j \to \infty}(T(u_j), u_j - v) \geq (T(\bar{u}), \bar{u} - v) \quad \text{for all } v \in X. \tag{10.189}$$

The following can be proved using only a slight modification of the proof of the Browder-Minty theorem.

Theorem 10.61. *Let X be a real reflexive Banach space and suppose $T : X \to X^*$ is continuous, coercive and pseudo-monotone. Then for every $g \in X^*$ there exists a solution $u \in X$ of the equation*

$$T(u) = g. \tag{10.190}$$

The proof is left to the reader (Problem 10.17).

In practice, the following condition is easier to verify than pseudo-monotonicity.

Definition 10.62. Let X be a reflexive Banach space. An operator $T : X \to X^*$ is said to be of the **calculus of variations type** if it is bounded, and it has the representation

$$T(u) = \hat{T}(u, u) \tag{10.191}$$

where the mapping

$$X \times X \ni (u, v) \mapsto \hat{T}(u, v) \in X^* \tag{10.192}$$

satisfies the following hypotheses.

CV-1. For each $u \in X$, the mapping $v \mapsto \hat{T}(u, v)$ is bounded and continuous from X to X^*, and

$$(\hat{T}(u, u) - \hat{T}(u, v), u - v) \geq 0 \quad \text{for all } v \in X. \tag{10.193}$$

CV-2. For each $v \in X$, the mapping $u \mapsto \hat{T}(u, v)$ is bounded and continuous from X to X^*.

CV-3. If

$$u_j \rightharpoonup \bar{u} \text{ in } X \qquad (10.194)$$

and

$$(\hat{T}(u_j, u_j) - \hat{T}(u_j, \bar{u}), u_j - \bar{u}) \to 0, \qquad (10.195)$$

then for every $v \in X$

$$\hat{T}(u_j, v) \rightharpoonup \hat{T}(\bar{u}, v) \text{ in } X^*. \qquad (10.196)$$

CV-4. If

$$u_j \rightharpoonup \bar{u} \text{ in } X \qquad (10.197)$$

and

$$\hat{T}(u_j, v) \rightharpoonup \psi \text{ in } X^*, \qquad (10.198)$$

then

$$(\hat{T}(u_j, v), u_j) \to (\psi, \bar{u}). \qquad (10.199)$$

As we indicated above, we have the following.

Theorem 10.63. *If T is of the calculus of variations type, then T is pseudo-monotone.*

Proof. Let $u_j \rightharpoonup \bar{u}$ in X and suppose

$$\limsup_{j \to \infty}(T(u_j), u_j - \bar{u}) \leq 0. \qquad (10.200)$$

We wish to show that

$$\liminf_{j \to \infty}(T(u_j), u_j - v) \geq (T(\bar{u}), \bar{u} - v) \quad \text{for every } v \in X. \qquad (10.201)$$

Since $\hat{T}(u_j, \bar{u})$ is bounded in X^*, we can extract a subsequence u_j such that

$$\hat{T}(u_j, \bar{u}) \rightharpoonup \psi \quad \text{in } X^*, \qquad (10.202)$$

for some $\psi \in X^*$. We now use CV-4 to get

$$\lim_{j \to \infty}(\hat{T}(u_j, \bar{u}), u_j) = (\psi, \bar{u}). \qquad (10.203)$$

Thus, if we define

$$x_j := (\hat{T}(u_j, u_j) - \hat{T}(u_j, \bar{u}), u_j - \bar{u}) \in \mathbb{R}, \qquad (10.204)$$

we have

$$\limsup_{j \to \infty} x_j = \limsup_{j \to \infty}[(T(u_j), u_j - \bar{u}) - (\hat{T}(u_j, \bar{u}), u_j) + (\hat{T}(u_j, \bar{u}), \bar{u})] \leq 0.$$
$$(10.205)$$

Here we have used (10.200), (10.202) and (10.203). Since CV-1 implies that $x_j \geq 0$ we get

$$\lim_{j \to \infty} x_j = 0. \tag{10.206}$$

Thus, we can use CV-3 to get

$$\hat{T}(u_j, v) \rightharpoonup \hat{T}(\bar{u}, v) \text{ in } X^* \quad \text{for all } v \in X. \tag{10.207}$$

Hence, we can use CV-4 again to get

$$(\hat{T}(u_j, v), u_j) \to (\hat{T}(\bar{u}, v), \bar{u})$$

or

$$(\hat{T}(u_j, v), u_j - \bar{u}) \to 0 \quad \text{for all } v \in X. \tag{10.208}$$

We now use this and the fact that $x_j \geq 0$ to get

$$(T(u_j), u_j - \bar{u}) \geq (\hat{T}(u_j, \bar{u}), u_j - \bar{u}) \to 0. \tag{10.209}$$

Together with (10.200) this gives us

$$(T(u_j), u_j - \bar{u}) \to 0. \tag{10.210}$$

We now take the inequality

$$(T(u_j) - \hat{T}(u_j, w), u_j - w) \geq 0 \quad \text{for all } w \in X$$

from CV-1, and plug in

$$w = (1 - \theta)\bar{u} + \theta v, \tag{10.211}$$

for $\theta \in (0, 1)$. This yields

$$\theta(T(u_j), \bar{u} - v) \geq -(T(u_j), u_j - \bar{u}) + (\hat{T}(u_j, w), u_j - \bar{u}) + \theta(\hat{T}(u_j, w), \bar{u} - v). \tag{10.212}$$

Dividing this by θ and using (10.208) and (10.210) we get

$$\begin{aligned}
\liminf_{j \to \infty}(T(u_j), u_j - v) &= \liminf_{j \to \infty}(T(u_j), u_j - \bar{u}) + \liminf_{j \to \infty}(T(u_j), \bar{u} - v) \\
&\geq \liminf_{j \to \infty}(\hat{T}(u_j, w), \bar{u} - v) \\
&= (\hat{T}(\bar{u}, w), \bar{u} - v) \\
&= (\hat{T}(\bar{u}, (1 - \theta)\bar{u} + \theta v), \bar{u} - v).
\end{aligned}$$

Letting $\theta \searrow 0$ we get

$$\liminf_{j \to \infty}(T(u_j), u_j - v) \geq (T(\bar{u}), \bar{u} - v) \quad \text{for all } v \in X. \tag{10.213}$$

Since this argument holds for any subsequence of the original sequence, the inequality (10.213) holds for the entire original sequence. This completes the proof. $\qquad \square$

The following is immediate from the preceding results.

Corollary 10.64. *Let X be a real reflexive Banach space and suppose $T : X \to X^*$ is continuous, coercive and of the calculus of variations type. Then for every $g \in X^*$ there exists a solution $u \in X$ of the equation*

$$T(u) = g. \tag{10.214}$$

10.3.6 Application to PDEs

Let $\Omega \subset \mathbb{R}^n$ be a bounded domain with smooth boundary. We consider quasilinear second-order differential operators having the form

$$\tilde{A}(u)(x) = -\sum_{i=1}^{n} \frac{\partial}{\partial x_i} a_i(x, u(x), \nabla u(x)) + a_0(x, u(x), \nabla u(x)). \tag{10.215}$$

Our goal is to solve the Dirichlet problem for

$$\tilde{A}(u) = f \tag{10.216}$$

for appropriate f. Formally, we define the bivariate form

$$\tilde{B}(u, v) := \int_{\Omega} \sum_{i=1}^{n} a_i(x, u(x), \nabla u(x)) \frac{\partial v(x)}{\partial x_i} + a_0(x, u(x), \nabla u(x))v(x) \, dx. \tag{10.217}$$

We make the following hypotheses on the functions

$$\Omega \times \mathbb{R} \times \mathbb{R}^n \ni (x, \eta, xii) \mapsto a_i(x, \eta, xii) \in \mathbb{R}, \quad i = 0, \dots, n. \tag{10.218}$$

HP-1. For each $i = 0, \dots, n$,

$$x \mapsto a_i(x, \eta, xii) \tag{10.219}$$

is in $C_b(\overline{\Omega})$ for every fixed $(\eta, xii) \in \mathbb{R}^{n+1}$.

HP-2. For each $i = 0, \dots, n$,

$$(\eta, xii) \mapsto a_i(x, \eta, xii) \tag{10.220}$$

is in $C(\mathbb{R}^{n+1})$ for every $x \in \Omega$.

HP-3. There exists $p \in (1, \infty)$, a constant $c_0 > 0$, a function $k \in L^q(\Omega)$ ($\frac{1}{p} + \frac{1}{q} = 1$) such that for every $x \in \Omega$ and every $(\eta, xii) \in \mathbb{R}^{n+1}$ we have

$$|a_i(x, \eta, xii)| \le c[|\eta|^{p-1} + |xii|^{p-1} + k(x)], \tag{10.221}$$

for each $i = 0, \dots, n$.

HP-4. For every $xii \in \mathbb{R}^n$ and $xii^* \in \mathbb{R}^n$ such that $xii \ne xii^*$, and every $\eta \in \mathbb{R}$ and $x \in \Omega$ we have

$$\sum_{i=1}^{n} [a_i(x, \eta, xii) - a_i(x, \eta, xii^*)](xi_i - xi_i^*) > 0. \tag{10.222}$$

HP-5.

$$\frac{|\tilde{\mathcal{B}}(v,v)|}{\|v\|_{1,p}} \to \infty \quad \text{as } \|v\|_{1,p} \to \infty. \qquad (10.223)$$

HP-6. For every $\mathbf{x} \in \Omega$ and uniformly for $|\eta|$ in bounded sets, we have

$$\sum_{i=1}^{n} a_i(\mathbf{x}, \eta, \mathbf{x}ii)\mathbf{x}i_i \frac{1}{|\mathbf{x}ii| + |\mathbf{x}ii|^{p-1}} \to \infty \quad \text{as } |\mathbf{x}ii| \to \infty.$$
$$(10.224)$$

Note that by hypotheses HP-1, HP-2 and HP-3, the bivariate form $\tilde{\mathcal{B}}(u,v)$ is well defined for u, v in $W^{1,p}(\Omega)$ where $p \in (1,\infty)$ is given as in HP-3. Let $f \in W^{-1,q}(\Omega)$. As above, we say that $u \in W_0^{1,p}(\Omega)$ is a weak solution of the Dirichlet problem for (10.216) if

$$\tilde{\mathcal{B}}(u,v) = (f,v) \quad \text{for every } v \in W_0^{1,p}(\Omega). \qquad (10.225)$$

Hypotheses HP-1, HP-2 and HP-3 also imply that ɪor each fixed $u \in W_0^{1,p}(\Omega)$, the mapping

$$W_0^{1,p}(\Omega) \ni w \mapsto \tilde{\mathcal{B}}(u,w) \in \mathbb{R} \qquad (10.226)$$

is a bounded linear functional. Thus, there exists a mapping

$$W_0^{1,p}(\Omega) \ni u \mapsto \bar{T}(u) \in W^{-1,q}(\Omega) \qquad (10.227)$$

such that

$$\tilde{\mathcal{B}}(u,w) = (\bar{T}(u), w) \quad \text{for all } w \in W_0^{1,p}(\Omega). \qquad (10.228)$$

We now show the following.

Theorem 10.65. *Let a_i satisfy hypotheses HP-1–HP-6. Then the operator $\bar{T}$ is of the calculus of variations type.*

Proof. Of course, one of our goals here is to separate the effects of higher and lower derivatives; thus we define

$$\bar{\mathcal{B}}(u,v,w) := \mathcal{B}_1(u,v,w) + \mathcal{B}_0(u,w), \qquad (10.229)$$

where

$$\mathcal{B}_1(u,v,w) := \sum_{i=1}^{n} \int_{\Omega} a_i(\mathbf{x}, u(\mathbf{x}), \nabla v(\mathbf{x})) \frac{\partial w}{\partial x_i}(\mathbf{x}) \, d\mathbf{x}, \qquad (10.230)$$

$$\mathcal{B}_0(u,w) := \int_{\Omega} a_0(\mathbf{x}, u(\mathbf{x}), \nabla u(\mathbf{x})) w(\mathbf{x}) \, d\mathbf{x}. \qquad (10.231)$$

Using the same argument as above, we see that there exists a mapping

$$W_0^{1,p}(\Omega) \times W_0^{1,p}(\Omega) \ni (u,v) \mapsto \hat{T}(u,v) \in W^{-1,q}(\Omega) \qquad (10.232)$$

such that

$$\bar{B}(u, v, w) = (\hat{T}(u, v), w) \quad \text{for all } w \in W_0^{1,p}(\Omega). \tag{10.233}$$

Furthermore, we have

$$\bar{T}(u) = \hat{T}(u, u). \tag{10.234}$$

We now note that our results on Nemytskii operators and hypotheses HP-1, HP-2 and HP-3 immediately imply the following lemma.

Lemma 10.66. *For each $i = 0, \ldots, n$ the mapping*

$$W^{1,p}(\Omega) \ni u \mapsto a_i(\mathbf{x}, u(\mathbf{x}), \nabla u(\mathbf{x})) \in L^q(\Omega) \tag{10.235}$$

is bounded and continuous. Furthermore, for fixed $v \in L^p(\Omega)$ the mapping

$$W^{1,p}(\Omega) \ni u \mapsto a_i(\mathbf{x}, v(\mathbf{x}), \nabla u(\mathbf{x})) \in L^q(\Omega) \tag{10.236}$$

is bounded and continuous; and for fixed $w \in W^{1,p}(\Omega)$ the mapping

$$L^p(\Omega) \ni v \mapsto a_i(\mathbf{x}, v(\mathbf{x}), \nabla w(\mathbf{x})) \in L^q(\Omega) \tag{10.237}$$

is bounded and continuous.

This gives us the following corollary.

Corollary 10.67. *The following hold.*

1. *The operator $\bar{T}$ is bounded and continuous.*

2. *For each $u \in W_0^{1,p}(\Omega)$ the mapping*

$$W_0^{1,p}(\Omega) \ni v \mapsto \hat{T}(u, v) \in W^{-1,q}(\Omega) \tag{10.238}$$

   is bounded and continuous.

3. *For each $v \in W_0^{1,p}(\Omega)$ the mapping*

$$W_0^{1,p}(\Omega) \ni u \mapsto \hat{T}(u, v) \in W^{-1,q}(\Omega) \tag{10.239}$$

   is bounded and continuous.

This and hypothesis HP-4 show that conditions CV-1 and CV-2 are satisfied. Thus, to show that $\bar{T}$ is of the calculus of variations type and complete the proof of Theorem 10.65, we need only verify that conditions CV-3 and CV-4 are satisfied. To check condition CV-3 we assume that $u_j \rightharpoonup \bar{u}$ in $W_0^{1,p}(\Omega)$ and that

$$(\hat{T}(u_j, u_j) - \hat{T}(u_j, \bar{u}), u_j - \bar{u})$$

$$= \mathcal{B}_1(u_j, u_j, u_j - \bar{u}) - \mathcal{B}_1(u_j, \bar{u}, u_j - \bar{u}) \tag{10.240}$$

$$= \int_\Omega \sum_{i=1}^n [a_i(\mathbf{x}, u_j, \nabla u_j) - a_i(\mathbf{x}, u_j, \nabla \bar{u})] \frac{\partial}{\partial x_i}(u_j - \bar{u}) \, d\mathbf{x}$$

$$\to 0,$$

and we need to show that

$$(T(u_j, v), w) = \mathcal{B}_1(u_j, v, w) + \mathcal{B}_0(u_j, w)$$
$$\rightarrow \mathcal{B}_1(\bar{u}, v, w) + \mathcal{B}_0(\bar{u}, w) \qquad (10.241)$$
$$= (T(\bar{u}, v), w)$$

for all $w \in W_0^{1,p}(\Omega)$. Now by compact imbedding $u_j \rightarrow u$ (strongly) in $L^p(\Omega)$. Thus, even without using hypothesis (10.240), we have

$$\mathcal{B}_1(u_j, v, w) = \int_\Omega \sum_{i=1}^n a_i(\mathbf{x}, u_j(\mathbf{x}), \nabla v(\mathbf{x})) \frac{\partial w}{\partial x_i}(\mathbf{x}) \, d\mathbf{x}$$
$$\rightarrow \mathcal{B}_1(\bar{u}, v, w).$$

We use the results on Nemytskii operators here. Thus, we need only show that

$$\int_\Omega a_0(\mathbf{x}, u_j(\mathbf{x}), \nabla u_j(\mathbf{x})) w(\mathbf{x}) \, d\mathbf{x} \rightarrow \int_\Omega a_0(\mathbf{x}, \bar{u}(\mathbf{x}), \nabla \bar{u}(\mathbf{x})) w(\mathbf{x}) \, d\mathbf{x}.$$
$$(10.242)$$

The proof of this in the general case is an exercise in measure theory which we shall skip. We refer the reader to [Li, Lemma 2.2, p. 184]. In our example below we use a lower-order term of the form

$$a_0(\mathbf{x}, u(\mathbf{x}), \nabla u(\mathbf{x})) = \mathbf{b}_1(\mathbf{x}) \cdot \nabla u(\mathbf{x}) + \alpha_0(\mathbf{x}) |u(\mathbf{x})|^{p-2} u(\mathbf{x}), \qquad (10.243)$$

where $\mathbf{b}_1$ and α_0 are bounded continuous functions. For such a term, condition (10.242) can be verified directly (that is, we need not use (10.240), though this condition is essential in the general proof).

To check condition CV-4 we assume that $u_j \rightharpoonup \bar{u}$ in $W_0^{1,p}(\Omega)$ and $\hat{T}(u_j, v) \rightharpoonup \psi$ in $W^{-1,q}(\Omega)$; i.e.,

$$(\hat{T}(u_j, v), w) = \mathcal{B}_1(u_j, v, w) + \mathcal{B}_0(u_j, w) \rightarrow (\psi, w) \qquad (10.244)$$

for every $w \in W_0^{1,p}(\Omega)$. We need to show that

$$(\hat{T}(u_j, v), u_j) = \mathcal{B}_1(u_j, v, u_j) + \mathcal{B}_0(u_j, u_j) \rightarrow (\psi, \bar{u}). \qquad (10.245)$$

To do this, we write

$$(\hat{T}(u_j, v), u_j) = (\hat{T}(u_j, v), \bar{u}) + (\hat{T}(u_j, v), u_j - \bar{u}). \qquad (10.246)$$

Thus, by (10.244) we need only show that

$$\lim_{j \rightarrow \infty} (\hat{T}(u_j, v), u_j - \bar{u}) = 0. \qquad (10.247)$$

The proof of this is left to the reader (Problem 10.23). This shows that our operator is of calculus of variations type and completes the proof.   □

There is one more lemma to prove to get our main result on partial differential equations.

Lemma 10.68. *The operator $\bar{T}$ is coercive.*

The proof is left to the reader (Problem 10.24). The culmination of the previous results is the following existence theorem for quasilinear elliptic partial differential equations.

Theorem 10.69. *Let a_i satisfy hypotheses (HP-1–HP-6), and let p and q be as defined in hypothesis HP-3. Then for every $f \in W^{-1,q}(\Omega)$, there exists a weak solution $u \in W_0^{1,p}(\Omega)$ of the Dirichlet problem for the quasilinear PDE (10.216).*

Example 10.70. Consider the second-order nonlinear partial differential operator

$$\tilde{A}(u) := -\sum_{i=1}^{n} \frac{\partial}{\partial x_i}\left(\left|\frac{\partial u}{\partial x_i}\right|^{p-2}\frac{\partial u}{\partial x_i}\right) + \mathbf{b}_1(\mathbf{x}) \cdot \nabla u(\mathbf{x}) + \alpha_0(\mathbf{x})|u(\mathbf{x})|^{p-2}u(\mathbf{x}),$$

$$(10.248)$$

where $p \in (1, \infty)$, and where $\mathbf{b}_1$ and α_0 are bounded and continuous. Note that this is the same as the operator defined in (10.177) except for the difference in the lower-order terms. The reader is asked to show that there exists a constant $\tilde{C} > 0$ such that if

$$\alpha_0(\mathbf{x}) > -\tilde{C}, \quad \mathbf{x} \in \Omega, \qquad (10.249)$$

then hypotheses HP-1–HP-6 hold (Problem 10.25). By Theorem 10.69 we have the following existence result.

Theorem 10.71. *Let the nonlinear second-order partial differential operator $\tilde{A}$ be defined by (10.248). Then for every $f \in W^{-1,q}(\Omega)$ there exists a weak solution $u \in W_0^{1,p}$ of the equation*

$$\tilde{A}(u) = f. \qquad (10.250)$$

Problems

10.16. We say that a mapping $T : X \to X^*$ is **hemicontinuous** at $u \in X$ if

$$\mathbb{R} \ni t \mapsto (T(u + tv), w) \in \mathbb{R} \qquad (10.251)$$

is continuous for every $v, w \in X$.

Find a function $f : \mathbb{R}^2 \to \mathbb{R}^2$ which is hemicontinuous at the origin but not continuous.

10.17. Prove Theorem 10.61.

10.18. Show that Theorem 10.49 still holds if the hypothesis of continuity is replaced by hemicontinuity.

10.19. Show that Theorem 10.61 still holds if the hypothesis of continuity is replaced by hemicontinuity.

10.20. Show that a bounded, monotone operator is pseudo-monotone.

10.21. Show that if T is a pseudo-monotone operator and $u_j \to \bar{u}$ (strongly) in X, then $T(u_j) \rightharpoonup T(\bar{u})$ in X^*.

10.22. Show that an operator of the calculus of variations type is hemicontinuous. (Thus, by Problem 10.19, we can drop the hypothesis of continuity in Corollary 10.64 and the conclusion still holds.)

10.23. Assume $u_j \rightharpoonup \bar{u}$ in $W_0^{1,p}(\Omega)$ and $\hat{T}(u_j, v) \rightharpoonup \psi$ in $W^{-1,q}(\Omega)$. Verify (10.247).

10.24. Prove Lemma 10.68.

10.25. Show that there is a $\tilde{C} > 0$ such that if (10.249) holds, then hypotheses HP-1–HP-6 are satisfied for the quasilinear differential operator $\tilde{A}$ defined in (10.248). Identify which of the hypotheses H-1–H-5 do *not* hold for this operator.

11

Energy Methods for Evolution Problems

11.1 Parabolic Equations

In this section, we shall consider evolution problems of the form

$$u_t = A(t)u,$$

where u depends on $t \in [0, T]$ and $\mathbf{x} \in \Omega \subset \mathbb{R}^n$, and $A(t)$ is some elliptic differential operator. We shall formulate such problems as abstract evolution problems in a Hilbert space, such as $L^2(\Omega)$. In order to do so, we must first introduce spaces of functions whose values are in a Banach space.

11.1.1 Banach Space Valued Functions and Distributions

Let X be a Banach space, and let I be an interval (more generally, I could also be a set in $\mathbb{R}^n$). We define $C(I, X)$ to be the bounded continuous functions of the form

$$\mathbb{R} \supset I \ni t \mapsto u(t) \in X. \tag{11.1}$$

We equip this space with the norm

$$\|u\| = \sup_{t \in I} \|u(t)\|_X. \tag{11.2}$$

The space $C^n(I, X)$ contains functions whose derivatives (in I) up to order n are in $C(I, X)$.

Example 11.1. What we have in mind here is letting functions of both space and time,

$$\mathbb{R} \times \mathbb{R}^n \supset I \times \Omega \ni (t, \mathbf{x}) \mapsto u(t, \mathbf{x}) \in \mathbb{R},$$

be thought of as a collection of functions of space parameterized by time. For instance, the function described above might be of the form

$$\mathbb{R} \supset I \ni t \mapsto u(t, \cdot) \in L^2(\Omega).$$

Note that a function in, say $C([0, 1], L^2(\Omega))$ need not be continuous in $\mathbf{x}$. It needs only be true that any two "snapshots" of the function at nearby times be close in $L^2(\Omega)$. For example, if $v \in L^2(\Omega)$ and $g \in C^n(I)$, then

$$u(t, \mathbf{x}) := g(t)v(\mathbf{x})$$

is in $C^n(I, L^2(\Omega))$ no matter how many discontinuities v has.

We now let I be an open interval and define $\mathcal{D}(I, X)$ to be the space of all C^∞-functions from I to X which have compact support in I. A notion of convergence in $\mathcal{D}(I, X)$ is defined analogously as in Chapter 5; i.e., a sequence converges if the supports are contained in a common compact subset of I and all derivatives converge uniformly.

Let X^* be the dual space of X. Then we denote the set of continuous linear mappings from $\mathcal{D}(I, X)$ to the field of scalars (i.e., $\mathbb{R}$ or $\mathbb{C}$) by $\mathcal{D}'(I, X^*)$. We refer to the elements of $\mathcal{D}'(I, X^*)$ as X^*-valued distributions. It is clear that $C(I, X^*)$ is contained in $\mathcal{D}'(I, X^*)$. Moreover, the definitions of distributional derivatives are easily extended to Banach space valued distributions.

We can now define $L^p(I, X)$ to be the completion of $C(I, X)$ with respect to the norm

$$\|u\| = \left(\int_I \|u(t)\|_X^p \, dt \right)^{1/p}. \tag{11.3}$$

Clearly, the elements of $L^p(I, X)$ are X^{**}-valued distributions. Also, we can define Sobolev spaces of X-valued functions just as before. In most applications, X will be a Hilbert space. In this case, the density, extension, imbedding (except for compactness of imbeddings) and trace theorems can be established the same way as for scalar-valued functions, and we shall use them without restating and proving those theorems.

For a reflexive Banach space X, we shall use the notation $L^\infty(I, X)$ to denote the dual space of $L^1(I, X^*)$.

Example 11.2. Let $\Omega \subset \mathbb{R}^n$ be a domain and let $T > 0$ be given. The space $C([0, T], L^2(\Omega))$ has the norm

$$\|u\| = \sup_{t \in [0, T]} \left\{ \int_\Omega |u(\mathbf{x}, t)|^2 \, d\mathbf{x} \right\}^{1/2}. \tag{11.4}$$

The space $L^2((0,T), L^2(\Omega))$ has the norm

$$\|u\| = \left\{ \int_0^T \left[\int_\Omega |u(\mathbf{x},t)|^2 \, d\mathbf{x} \right] dt \right\}^{1/2}. \tag{11.5}$$

The space $H^1((0,T), L^2(\Omega))$ has the norm

$$\|u\| = \left\{ \int_0^T \left[\int_\Omega |u(\mathbf{x},t)|^2 + |u_t(\mathbf{x},t)|^2 \, d\mathbf{x} \right] dt \right\}^{1/2}. \tag{11.6}$$

The space $L^2((0,T), H^{-1}(\Omega))$ has the norm

$$\|u\| = \left\{ \int_0^T \left[\sup_{\substack{\phi \in H_0^1(\Omega) \\ \|\phi\|_{1,2}=1}} \int_\Omega u(\mathbf{x},t)\phi(\mathbf{x}) \, d\mathbf{x} \right]^2 dt \right\}^{1/2}. \tag{11.7}$$

11.1.2 Abstract Parabolic Initial-Value Problems

We consider a separable real Hilbert space H and another separable Hilbert space V, which is continuously and densely imbedded in H. We identify H with its own dual space; the dual of V is denoted by V^*. Thus we have $V \subset H \subset V^*$ with continuous and dense imbeddings. (For example, we could take $H_0^1(\Omega) \subset L^2(\Omega) \subset H^{-1}(\Omega)$.) We shall use the same notation $(\cdot, \cdot)$ for the inner product in H and for the pairing between V^* and V. We assume that $A(t) \in \mathcal{L}(V, V^*)$ depends continuously on $t \in [0,T]$. With $A(t)$, we can associate the parameterized quadratic form

$$a(t, u, v) = -(A(t)u, v) \tag{11.8}$$

defined on $\mathbb{R} \times V \times V$. We assume that this form satisfies the coercivity condition

$$a(t, u, u) \geq a\|u\|_V^2 - b\|u\|_H^2, \tag{11.9}$$

with positive constants a and b which are independent of $t \in [0,T]$.

We now consider the evolution problem

$$\frac{du}{dt} = A(t)u + f(t), \quad u(0) = u_0. \tag{11.10}$$

We shall establish the following result.

Theorem 11.3. *Let H, V and $A(t)$ be as above. Assume that the functions $f \in L^2((0,T), V^*)$ and $u_0 \in H$ are given. Then (11.10) has a unique solution $u \in L^2((0,T), V) \cap H^1((0,T), V^*)$.*

In this result, the differential equation in (11.10) is of course interpreted in the sense of V^*-valued distributions. Moreover, by the Sobolev imbedding theorem, we have $u \in C([0,T], V^*)$, which allows us to interpret the initial condition. Indeed, we can say more.

Lemma 11.4. *Suppose that $u \in L^2((0,T), V) \cap H^1((0,T), V^*)$. Then, in fact, $u \in C([0,T], H)$.*

This shows that Theorem 11.3 is optimal; i.e., if we want a solution with the regularity guaranteed by the theorem, then the assumptions which we made on f and u_0 are necessary.

We now prove the lemma.

Proof. First, let u be in $C^1([0,T], H)$. We then obtain the estimate

$$\|u(t)\|_H^2 = \|u(t^*)\|^2 + 2\int_{t^*}^t (\dot{u}(s), u(s)) \, ds. \tag{11.11}$$

We now choose t^* in such a way that $\|u(t^*)\|^2$ is equal to the mean value of $\|u(t)\|^2$; moreover, we estimate $(\dot{u}, u)$ by $\|\dot{u}\|_{V^*}\|u\|_V$. In this fashion, we obtain

$$\|u(t)\|_H^2 \leq \frac{1}{T}\int_0^T \|u(t)\|_H^2 \, dt + 2\int_0^T \|\dot{u}\|_{V^*}\|u\|_V \, dt. \tag{11.12}$$

Using Cauchy-Schwarz, we conclude

$$\max_{t\in[0,T]} \|u(t)\|_H^2 \leq \frac{1}{T}\|u\|_{L^2((0,T),H)}^2 + 2\|u\|_{H^1((0,T),V^*)}\|u\|_{L^2((0,T),V)}. \tag{11.13}$$

The rest follows by a density argument. $\qquad\square$

We now turn to the proof of the theorem. Without loss of generality, we assume that the constant b in (11.9) is zero; we can always achieve this by the substitution $u = v\exp(bt)$.

We first prove uniqueness. Let u be a solution. Using (11.10), we take the inner product with u and integrate from 0 to T. This yields

$$\frac{1}{2}(\|u(T)\|_H^2 - \|u_0\|_H^2) + \int_0^T a(t,u,u) \, dt = \int_0^T (f,u) \, dt. \tag{11.14}$$

Combining this with condition (11.10) leads to an a priori estimate of the form

$$\|u\|_{L^2((0,T),V)} \leq C(\|f\|_{L^2((0,T),V^*)} + \|u_0\|_H). \tag{11.15}$$

From this and linearity, uniqueness of solutions is obvious.

The realization that a priori estimates like (11.15) can indeed be used as a foundation of existence proofs rather than just uniqueness was one of the milestones in the modern theory of PDEs. We have already encountered this idea (in the form of Galerkin's method) in the proof of the Browder-Minty theorem in Chapter 10. More generally, the technique proceeds as follows. One first constructs a family of approximate problems for which an a priori estimate analogous to (11.15) holds, but which are easily shown to have solutions. This yields a sequence of approximate solutions, for which one has uniform bounds. Uniform bounds imply the existence of a weakly convergent subsequence. One then shows that the weak limit is the solution we seek.

To carry out this program for the abstract parabolic problem above, we need a set $\{\phi_n \mid n \in \mathbb{N}\}$ of linearly independent elements of V such that the linear span of the ϕ_n is dense in V. Let V_n be the span of $\phi_1, \phi_2, \ldots, \phi_n$ and let P_n be the orthogonal projection from H (not V!) onto V_n. Let now $u_n(t) = \sum_{i=1}^{n} \alpha_i(t)\phi_i$ be the solution of the following problem:

$$\left(\frac{du_n}{dt}, \phi_i\right) = (A(t)u_n, \phi_i) + (f(t), \phi_i), \ i = 1, \ldots, n,$$

$$u_n(0) = P_n u_0. \tag{11.16}$$

The system (11.16) is simply a system of linear ODEs for the coefficients $\alpha_i(t)$, which clearly has a unique solution. In complete analogy to (11.14), we obtain

$$\frac{1}{2}(\|u_n(T)\|_H^2 - \|P_n u_0\|_H^2) + \int_0^T a(t, u_n, u_n) \, dt = \int_0^T (f, u_n) \, dt. \tag{11.17}$$

From this, we obtain an a priori bound (independent of n) for the norm of u_n in $L^2((0,T), V)$. Hence a subsequence converges, weakly in $L^2((0,T), V)$, to a limit u. Let $\phi \in \mathcal{D}((0,T), V)$ be of the form

$$\phi(t) = \sum_{i=1}^{N} \beta_i(t)\phi_i \tag{11.18}$$

for some N, where $\beta_i \in \mathcal{D}((0,T), \mathbb{R})$. For $n \geq N$, we have

$$\left(\frac{du_n}{dt}, \phi\right) = (A(t)u_n, \phi) + (f, \phi); \tag{11.19}$$

integrating in time and passing to the limit we find

$$\int_0^T \left(\frac{du}{dt}, \phi\right) \, dt = \int_0^T (A(t)u, \phi) + (f, \phi) \, dt. \tag{11.20}$$

Since test functions of the form (11.18) are dense in $\mathcal{D}((0,T), V)$, it follows that (11.10) holds in the sense of V^*-valued distributions. In particular, this implies that $u \in H^1((0,T), V^*)$ and hence $u \in C([0,T], H)$.

Consider now, more generally, $\phi \in H^1((0,T), V)$ with the property that $\phi(T) = 0$. Again, functions of the form (11.18) are dense in this space of functions. Moreover, if ϕ has the form (11.18) and $n \geq N$, then

$$-\int_0^T (u_n(t), \dot{\phi}(t)) \, dt - (u_n(0), \phi(0)) = \int_0^T (A(t)u_n(t) + f(t), \phi(t)) \, dt. \tag{11.21}$$

In the limit we find

$$-\int_0^T (u(t), \dot{\phi}(t))\, dt - (u_0, \phi(0)) = \int_0^T (A(t)u(t) + f(t), \phi(t))\, dt. \quad (11.22)$$

If, on the other hand, we multiply (11.10) by ϕ and integrate, we find

$$-\int_0^T (u(t), \dot{\phi}(t))\, dt - (u(0), \phi(0)) = \int_0^T (A(t)u(t) + f(t), \phi(t))\, dt. \quad (11.23)$$

By comparing (11.22) and (11.23), we conclude that $u(0) = u_0$.

11.1.3 Applications

Example 11.5. Let $H = L^2(\Omega)$, $V = H_0^1(\Omega)$ and

$$A(t)u = \frac{\partial}{\partial x_j}\left(a_{ij}(\mathbf{x}, t)\frac{\partial u}{\partial x_i}\right) + b_i(\mathbf{x}, t)\frac{\partial u}{\partial x_i} + c(\mathbf{x}, t)u. \quad (11.24)$$

If the coefficients are continuous and the matrix a_{ij} is strictly positive definite, then the assumptions above apply (cf. Theorem 9.17). This yields an existence result for the initial/boundary-value problem

$$\begin{aligned}
\frac{\partial u}{\partial t} &= \frac{\partial}{\partial x_j}\left(a_{ij}(\mathbf{x}, t)\frac{\partial u}{\partial x_i}\right) + b_i(\mathbf{x}, t)\frac{\partial u}{\partial x_i} + c(\mathbf{x}, t)u + f(\mathbf{x}, t), \\
&\qquad\qquad\qquad\qquad \mathbf{x} \in \Omega,\ t \in (0, T), \qquad\qquad (11.25) \\
u(\mathbf{x}, t) &= 0,\ \mathbf{x} \in \partial\Omega,\ t \in (0, T), \\
u(\mathbf{x}, 0) &= u_0(\mathbf{x}),\ \mathbf{x} \in \Omega.
\end{aligned}$$

Here we have to assume $f \in L^2((0, T), H^{-1}(\Omega))$, $u_0 \in L^2(\Omega)$.

Example 11.6. Let $H = L^2(\Omega)$, $V = H_0^2(\Omega)$ and $Au = -\Delta\Delta u$. Then the associated quadratic form is $a(u, u) = (\Delta u, \Delta u)$. By using the elliptic regularity results for Laplace's equation (see Chapter 9), it can be shown that this quadratic form is equivalent to the inner product in $H_0^2(\Omega)$, provided $\partial\Omega$ is sufficiently smooth (say of class C^2) and Ω is bounded. Again the result above is applicable, yielding an existence result for the problem

$$\begin{aligned}
u_t &= -\Delta\Delta u,\ \mathbf{x} \in \Omega,\ t \in (0, T), \\
u &= \frac{\partial u}{\partial n} = 0,\ \mathbf{x} \in \partial\Omega,\ t \in (0, T), \qquad\qquad (11.26) \\
u(\mathbf{x}, 0) &= u_0(\mathbf{x}),\ \mathbf{x} \in \Omega.
\end{aligned}$$

Example 11.7. Let

$$a(t, u, v) = \int_\Omega a_{ij}(\mathbf{x}, t)\frac{\partial u}{\partial x_i}\frac{\partial v}{\partial x_j} - b_i(\mathbf{x}, t)\frac{\partial u}{\partial x_i}v - c(\mathbf{x}, t)uv\ d\mathbf{x}. \quad (11.27)$$

We assume that the coefficients are continuous on $\overline{\Omega} \times [0, T]$ and that the matrix a_{ij} is strictly positive definite. We choose $V = H^1(\Omega)$ and $H = L^2(\Omega)$. Let $A(t)$ be the operator from V to V^* defined by $(A(t)u, v) = -a(t, u, v)$. Again the assumptions of the theorem above are satisfied. That is, for every $u_0 \in L^2(\Omega)$ and every $f \in L^2((0, T), V^*)$ we have a unique solution of the problem $\dot{u} = A(t)u + f$ with initial condition $u(0) = u_0$. Since, however, V^* is not a space of distributions on Ω, we have to think a little about the interpretation of this equation. If Ω is smooth enough, then every function in $H^1(\Omega)$ has a trace on $\partial\Omega$, which lies in $H^{1/2}(\partial\Omega)$. If, for instance, we take $g \in L^2(\Omega)$ and $h \in H^{-1/2}(\partial\Omega)$, then the functional $g \oplus h$ defined by

$$(g \oplus h, v) = (g, v) + \int_{\partial\Omega} h(\mathbf{x})v(\mathbf{x}) \, dS \tag{11.28}$$

is certainly in V^*. Let us assume that $f(t)$ has this form. Then, formally, we have

$$
\begin{aligned}
(A(t)u + f(t), v) &= -a(t, u, v) + (g, v) + \int_{\partial\Omega} h(\mathbf{x})v(\mathbf{x}) \, dS \\
&= -\int_{\Omega} a_{ij}(\mathbf{x}, t) \frac{\partial u}{\partial x_i} \frac{\partial v}{\partial x_j} - b_i(\mathbf{x}, t) \frac{\partial u}{\partial x_i} v - c(\mathbf{x}, t)uv \, d\mathbf{x} \\
&\quad + (g, v) + \int_{\partial\Omega} h(\mathbf{x})v(\mathbf{x}) \, dS \\
&= \int_{\Omega} \left(\frac{\partial}{\partial x_j}\left(a_{ij}(\mathbf{x}, t) \frac{\partial u}{\partial x_i} \right) + b_i(\mathbf{x}, t) \frac{\partial u}{\partial x_i} + c(\mathbf{x}, t)u + g(\mathbf{x}, t) \right) v \, d\mathbf{x} \\
&\quad + \int_{\partial\Omega} \left(-a_{ij}(\mathbf{x}, t)n_j(\mathbf{x}) \frac{\partial u}{\partial x_i} + h(\mathbf{x}, t) \right) v(\mathbf{x}) \, dS.
\end{aligned}
$$

In a formal or "generalized" sense, u is therefore a solution of the PDE

$$\frac{\partial u}{\partial t} = \frac{\partial}{\partial x_j}\left(a_{ij}(\mathbf{x}, t) \frac{\partial u}{\partial x_i} \right) + b_i(\mathbf{x}, t) \frac{\partial u}{\partial x_i} + c(\mathbf{x}, t)u + g(\mathbf{x}, t) \tag{11.29}$$

with boundary condition

$$a_{ij}(\mathbf{x}, t)n_j(\mathbf{x}) \frac{\partial u}{\partial x_i} = h(\mathbf{x}, t). \tag{11.30}$$

A stricter interpretation of the boundary condition requires higher regularity, since the expression in (11.30) does not exist in the sense of trace if only $u \in H^1$ is assumed.

11.1.4 Regularity of Solutions

The usual way of proving regularity for solutions of evolution problems such as (11.10) is to first establish temporal regularity and then use elliptic estimates for the operator A to show spatial regularity. Let

$\mathcal{D}(A)$ denote the domain of A as an unbounded operator in H, i.e., $\mathcal{D}(A) = \{u \in V \mid Au \in H\}$. Assume now that $u_0 \in \mathcal{D}(A(0))$ and that $f \in H^1((0,T), V^*) \cap L^2((0,T), V)$. Moreover, let us assume that $A \in C^1([0,T], \mathcal{L}(V, V^*))$. We can then formally differentiate equation (11.10) with respect to time. This yields

$$\ddot{u} = A(t)\dot{u} + \dot{A}(t)u + \dot{f}(t), \quad \dot{u}(0) = A(0)u_0 + f(0). \tag{11.31}$$

We now consider $\dot{u}$ as a new variable v and consider the evolution problem

$$\dot{v} = A(t)v + (\dot{A}(t)u + \dot{f}(t)), \quad v(0) = A(0)u_0 + f(0). \tag{11.32}$$

We now have $\dot{A}(t)u + \dot{f}(t) \in L^2((0,T), V^*)$ and $A(0)u_0 + f(0) \in H$ according to our assumptions. Hence Theorem 11.3 is applicable and (11.32) has a solution $v \in H^1((0,T), V^*) \cap L^2((0,T), V)$. Below we shall prove that actually $v = \dot{u}$. Once this is known, it follows that $u \in H^1((0,T), V) \cap H^2((0,T), V^*)$. Moreover, the equation $\dot{u} = A(t)u + f(t)$ implies that $A(t)u \in H^1((0,T), V^*) \cap L^2((0,T), V)$. In concrete examples, where A is an elliptic operator, this implies further spatial regularity of u (see Problem 11.4).

It remains to give a rigorous justification that v is really equal to $\dot{u}$. For this, set

$$z(t) = u_0 + \int_0^t v(\tau) \, d\tau. \tag{11.33}$$

We conclude that $\dot{z} = v$ and

$$
\begin{aligned}
A(t)z(t) &= A(t)u_0 + \int_0^t A(t)v(\tau) \, d\tau \\
&= \int_0^t A(\tau)v(\tau) \, d\tau + \int_0^t (A(t) - A(\tau))v(\tau) \, d\tau + A(t)u_0 \\
&= v(t) - v(0) - f(t) + f(0) - \int_0^t \dot{A}(\tau)u(\tau) \, d\tau \\
&\quad + \int_0^t \dot{A}(\tau)z(\tau) \, d\tau - (A(t) - A(0))u_0 + A(t)u_0 \\
&= v(t) - f(t) + \int_0^t \dot{A}(\tau)(z(\tau) - u(\tau)) \, d\tau.
\end{aligned}
$$

With w denoting $z - u$, we thus obtain

$$\dot{w} = A(t)w - \int_0^t \dot{A}(\tau)w(\tau) \, d\tau, \quad w(0) = 0. \tag{11.34}$$

We take the inner product with w and integrate. This yields

$$\frac{1}{2}(w(t), w(t)) + \int_0^t a(s, w(s), w(s)) \, ds = -\int_0^t \left(\int_0^s \dot{A}(\tau)w(\tau) \, d\tau, w(s) \right) ds \tag{11.35}$$

for every $t \in [0,T]$. From this it is easy to show that $w = 0$; see Problem 11.5.

Problems

11.1. As mentioned in the introduction to this section, the main results on Sobolev spaces can be generalized to Hilbert space valued functions with essentially the same proofs. Where do problems arise when one wants to consider Banach space valued functions?

11.2. Let A be a self-adjoint, strictly negative definite operator in a Hilbert space K. Assume that A has a compact resolvent and let $-\lambda_n$ be the eigenvalues of A and ϕ_n the corresponding eigenfunctions. We define $(-A)^{1/2}$ as follows:

$$(-A)^{1/2} \sum_{i=1}^{\infty} \alpha_i \phi_i = \sum_{i=1}^{\infty} \alpha_i \lambda_i^{1/2} \phi_i. \tag{11.36}$$

Formulate and prove existence results for the problem

$$\dot{u} = Au + f(t), \quad u(0) = u_0,$$

based on the following choices: (a) $H = K$, $V = \mathcal{D}((-A)^{1/2})$, (b) $H = \mathcal{D}((-A)^{1/2})$, $V = \mathcal{D}(A)$.

11.3. In Example 11.6, consider the choice $V = H_0^1(\Omega) \cap H^2(\Omega)$ instead of $H_0^2(\Omega)$. To which boundary conditions does this correspond? Can one also choose $V = H^2(\Omega)$?

11.4. Apply the regularity results of Section 11.1.4 to the examples in Section 11.1.3.

11.5. Use (11.35) to show that $w = 0$.

11.2 Hyperbolic Evolution Problems

11.2.1 Abstract Second-Order Evolution Problems

Let H be a separable real Hilbert space, and let V be another separable Hilbert space, which is continuously and densely embedded in H. Moreover, let $A \in C^1([0,T], \mathcal{L}(V, V^*))$, and let $a(t,u,v) = -(A(t)u,v)$ be the associated quadratic form. We assume that a is symmetric:

$$a(t,u,v) = a(t,v,u), \tag{11.37}$$

and that (11.9) holds; i.e., there are positive constants a and b such that

$$a(t,u,u) \geq a\|u\|_V^2 - b\|u\|_H^2. \tag{11.38}$$

We consider the evolution problem

$$\ddot{u}(t) = A(t)u + f(t), \ u(0) = u_0, \ \dot{u}(0) = u_1. \tag{11.39}$$

Our goal is the following result.

Theorem 11.8. *Assume that* $f \in L^1((0,T), H)$, $u_0 \in V$, $u_1 \in H$ *and that A is as described above. Then there exists a unique weak solution* $u \in C([0,T], V) \cap C^1([0,T], H)$ *of the evolution problem (11.39).*

For the proof, we shall need to proceed in several steps. We shall first give a proof of existence. This proof is similar in spirit to the one given in the previous section for parabolic problems. We first derive an energy equation which yields an a priori estimate for a solution. Then we use an approximation scheme and uniform energy estimates to obtain a solution as a weak limit. The energy estimate yields uniqueness only if more regularity of the solution is assumed; we shall therefore need a separate argument to prove uniqueness. Finally, the existence proof we give will only show that the solution is in $L^\infty((0,T), V) \cap W^{1,\infty}((0,T), H)$, and a separate argument is needed to establish continuity.

We remark that if u has the regularity guaranteed by the theorem, it does not follow that $f \in L^1((0,T), H)$. Hyperbolic problems differ from parabolic problems in the fact that they lack coercive properties; solutions are not "as smooth as the data will allow." This makes these problems in many respects harder than parabolic equations.

11.2.2   Existence of a Solution

We begin with a formal energy estimate. In (11.39), we take the inner product with $\dot{u}$. (This is of course not justified; even if we already knew that the theorem is true, $\dot{u}(t)$ would not be in V!). This yields

$$(\ddot{u}, \dot{u}) + a(t, u, \dot{u}) = (f, \dot{u}). \tag{11.40}$$

We integrate from 0 to t, and obtain

$$a(t, u(t), u(t)) + \|\dot{u}(t)\|_H^2 = a(0, u_0, u_0) + \|u_1\|_H^2$$
$$+ \int_0^t a'(s, u(s), u(s)) + 2(f(s), \dot{u}(s)) \, ds. \tag{11.41}$$

Here we have set

$$a'(t, u, v) = \frac{\partial}{\partial t} a(t, u, v) = -(\dot{A}(t)u, v). \tag{11.42}$$

From (11.41), one easily derives an estimate of the form

$$\|u(t)\|_V + \|\dot{u}(t)\|_H \le C(\|u_0\|_V + \|u_1\|_H + \|f\|_{L^1((0,T), H)}). \tag{11.43}$$

We leave the details of this argument as an exercise (Problem 11.6).

Next, we construct a sequence of approximate problems for which an analogue of (11.41) holds. Let $\{\phi_n \mid n \in \mathbb{N}\}$ be a set of linearly independent vectors in V such that the span of the ϕ_n is dense. Let V_n be the span of $\phi_1, \phi_2, \ldots, \phi_n$, let P_n be the orthogonal projection from H onto V_n and let Π_n be the orthogonal projection from V onto V_n. We now seek

$$u_n(t) = \sum_{i=1}^{n} \alpha_i(t)\phi_i \tag{11.44}$$

satisfying

$$
\begin{aligned}
(\ddot{u}_n, \phi_i) &= (A(t)u_n, \phi_i) + (f(t), \phi_i), \quad i = 1, \ldots, n, \\
u_n(0) &= \Pi_n u_0, \\
\dot{u}_n(0) &= P_n u_1.
\end{aligned}
\tag{11.45}
$$

This is a system of second-order ODEs for the $\alpha_i(t)$, which has a unique solution. The energy equation (11.41) holds for u_n (with the same derivation as above), and hence one obtains uniform bounds for $u_n \in L^\infty((0,T), V)$ and $\dot{u}_n \in L^\infty((0,T), H)$. We can extract a weakly-$*$ convergent subsequence, which has a limit $u \in L^\infty((0,T), V) \cap W^{1,\infty}((0,T), H)$. For simplicity, we shall again use the notation u_n to denote the weakly-$*$ convergent subsequence.

It remains to be shown that u is a solution. Let

$$X = \{\psi \in C^1([0,T], V) \mid \psi(T) = 0\}. \tag{11.46}$$

Functions of the form $\psi = \sum_{i=1}^n \alpha_i(t)\phi_i$, where $n \in \mathbb{N}$, are dense in X. If ψ is any function of this form, we obtain, for $m \geq n$,

$$\int_0^T (\ddot{u}_m, \psi) \, dt = \int_0^T (A(t)u_m, \psi) + (f(t), \psi(t)) \, dt, \tag{11.47}$$

which yields after an integration by parts

$$\int_0^T -(\dot{u}_m, \dot{\psi}) + a(t, u_m, \psi) \, dt = \int_0^T (f, \psi) \, dt + (\dot{u}_m(0), \psi(0)). \tag{11.48}$$

Here we can take the limit, which yields

$$\int_0^T -(\dot{u}, \dot{\psi}) + a(t, u, \psi) \, dt = \int_0^T (f, \psi) \, dt + (u_1, \psi(0)). \tag{11.49}$$

By density, this identity actually holds for every $\psi \in X$. If we restrict ψ to test functions, it follows that the differential equation in (11.39) holds in the sense of V^*-valued distributions.

We now note that weak-$*$ convergence in $L^\infty((0,T), H)$ implies weak convergence in $L^2((0,T), H)$. Thus, it follows that u_n converges weakly in $H^1((0,T), H)$. By the Sobolev imbedding theorem, the mapping $u \to u(0)$ is continuous from $H^1((0,T), H)$ to H; hence $u_n(0)$ converges weakly in H

to $u(0)$. Since, on the other hand, $u_n(0)$ converges strongly in V to u_0, it follows that $u(0) = u_0$.

It follows from the differential equation that $\ddot{u} \in L^1((0,T), V^*)$ so that $\dot{u} \in C([0,T], V^*)$ (cf. Problem 11.10). Hence $\dot{u}(0)$ is meaningful as an element of V^*. Repeating the derivation which led to (11.48) (with u in place of u_m), we find

$$\int_0^T -(\dot{u}, \dot{\psi}) + a(t, u, \psi) \, dt = \int_0^T (f, \psi) \, dt + (\dot{u}(0), \psi(0)). \qquad (11.50)$$

By comparing with (11.49), we conclude that $\dot{u}(0) = u_1$.

11.2.3 Uniqueness of the Solution

It would be easy to conclude uniqueness from (11.41). The problem is, however, that the derivation of (11.41) requires more smoothness of the solution than we have. We can circumvent this difficulty by deriving an energy equation for time-integrated quantities. Let u be a solution of (11.39) for $f = 0$, $u_0 = u_1 = 0$. Fix $s \in (0, T)$ and let

$$v(t) = \begin{cases} -\int_t^s u(r) \, dr & t < s \\ 0 & t \geq s. \end{cases} \qquad (11.51)$$

We multiply (11.39) by v and integrate. This yields

$$0 = \int_0^T (\ddot{u} - A(t)u, v) \, dt = \int_0^T a(t, u, v) - (\dot{u}, \dot{v}) \, dt; \qquad (11.52)$$

note that this integration by parts is permissible and the boundary terms vanish since $\dot{u}(0) = v(T) = 0$. Using the definition of v, we conclude

$$\int_0^s a(t, v, \dot{v}) - (\dot{u}, u) \, dt = 0. \qquad (11.53)$$

Carrying out the integration, we find

$$a(0, v(0), v(0)) + \|u(s)\|_H^2 = -\int_0^s a'(t, v, v) \, dt. \qquad (11.54)$$

We conclude an estimate of the form

$$\|v(0)\|_V^2 + \|u(s)\|_H^2 \leq C\left(\int_0^s \|v(t)\|_V^2 \, dt + \|v(0)\|_H^2\right). \qquad (11.55)$$

Setting $w(t) = \int_0^t u(r) \, dr = v(t) - v(0)$, we conclude

$$\|w(s)\|_V^2 + \|u(s)\|_H^2 \leq C\left(\int_0^s \|w(t) - w(s)\|_V^2 \, dt + \|w(s)\|_H^2\right). \qquad (11.56)$$

We now use the inequalities

$$\|w(t) - w(s)\|_V^2 \leq 2(\|w(t)\|_V^2 + \|w(s)\|_V^2) \qquad (11.57)$$

and

$$\|w(s)\|_H^2 \leq s \int_0^s \|u(t)\|_H^2 \, dt. \tag{11.58}$$

(The latter follows from the definition of w and Cauchy-Schwarz.) By using these two inequalities in (11.56), we find

$$(1 - 2Cs)\|w(s)\|_V^2 + \|u(s)\|_H^2 \leq K \int_0^s \|w(t)\|_V^2 + \|u(t)\|_H^2 \, dt \tag{11.59}$$

with some new constant K. From this, one easily concludes that $w = u = 0$ as long as $s < 1/(2C)$ (cf. Gronwall's inequality). Since the constant C is independent of the starting time, we can use a stepping argument to conclude that $w = u = 0$ everywhere in $[0, T]$.

11.2.4   Continuity of the Solution

We already know that $u \in C([0, T], H)$ and $\dot{u} \in C([0, T], V^*)$. The following lemma allows us to draw a further conclusion from this:

Lemma 11.9. *Let V and H be Hilbert spaces such that V is continuously and densely embedded in H. Assume that $u \in L^\infty((0, T), V) \cap C([0, T], H)$. Then $u(t) \in V$ for every $t \in [0, T]$ and $u(t)$ is weakly continuous; i.e., $(f, u(t))$ is a continuous function of t for every $f \in V^*$.*

Proof. We shall establish that for each $t \in (0, T)$

$$\|u(t)\|_V \leq \|u\|_{L^\infty((0,T),V)}. \tag{11.60}$$

Suppose not. Since H is dense in V^*, there exists $f \in H$ such that

$$(f, u(t)) > \|f\|_{V^*} \|u\|_{L^\infty((0,T),V)}. \tag{11.61}$$

Since $u \in C([0, T], H)$, we conclude that

$$(f, u(s)) > \|f\|_{V^*} \|u\|_{L^\infty((0,T),V)} \tag{11.62}$$

for s in some neighborhood of t, say for $|s - t| < \epsilon$. Define now $g(s) = f$ for $|s - t| < \epsilon$ and $g(s) = 0$ otherwise. Then, we find

$$\int_0^T (g(s), u(s)) \, ds > \|g\|_{L^1((0,T),V^*)} \|u\|_{L^\infty((0,T),V)}. \tag{11.63}$$

This is a contradiction of Hölder's inequality. Hence $u(t)$ is a bounded function taking values in V.

Consider now $f \in V^*$. Then there exists a sequence $f_n \in H$ such that $f_n \to f$ in V^*. It follows that $(f_n, u(t))$ converges uniformly to $(f, u(t))$. Since $(f_n, u(t))$ is continuous, $(f, u(t))$ is continuous. $\square$

Using the lemma, we conclude that the solution u of (11.9) is weakly continuous with values in V and $\dot{u}$ is weakly continuous with values in H.

Let us now recall the construction of u in Section 11.2.2. The solution u was the limit of a sequence u_n, and for each u_n, we have

$$a(t, u_n(t), u_n(t)) + \|\dot{u}_n(t)\|_H^2$$
$$= \quad a(0, \Pi_n u_0, \Pi_n u_0) + \|P_n u_1\|_H^2 \tag{11.64}$$
$$+ \int_0^t a'(s, u_n(s), u_n(s)) + 2(f(s), \dot{u}_n(s)) \, ds.$$

Consider now any fixed $s \in (0, T]$. The quantity

$$\sup_{t \in [0,s]} [a(t, u, u) + \|\dot{u}(t)\|_H^2] \tag{11.65}$$

is equivalent to the square of the norm in $L^\infty((0, s), V) \times L^\infty((0, s), H)$. Since balls are weak-*-compact and hence weak-*-closed, we conclude from (11.64) that, in the limit $n \to \infty$:

$$\sup_{t \in [0,s]} a(t, u(t), u(t)) + \|\dot{u}(t)\|_H^2$$
$$\leq \quad a(0, u_0, u_0) + \|u_1\|_H^2 \tag{11.66}$$
$$+ \limsup_{n \to \infty} \int_0^s |a'(s, u_n(s), u_n(s)) + 2(f(s), \dot{u}_n(s))| \, ds.$$

By letting s tend to zero, we find

$$\limsup_{t \to 0+} a(t, u(t), u(t)) + \|\dot{u}(t)\|_H^2 \leq a(0, u_0, u_0) + \|u_1\|_H^2. \tag{11.67}$$

Since, on the other hand, $u(t) \in V$ and $\dot{u}(t) \in H$ are weakly continuous, we have

$$\liminf_{t \to 0+} a(t, u(t), u(t)) + \|\dot{u}(t)\|_H^2 \geq a(0, u_0, u_0) + \|u_1\|_H^2. \tag{11.68}$$

It follows that the quantity $a(t, u(t), u(t)) + \|\dot{u}(t)\|_H^2$ is right continuous at $t = 0$. However, we might just as well have taken any other time as the initial time; hence we have right continuity everywhere. Finally, since our equation is invariant under time reversal, all our results apply to the time-reversed problem as well. Hence $a(t, u(t), u(t)) + \|\dot{u}(t)\|_H^2$ is also left continuous.

We now note that

$$a(t, u(t) - u(s), u(t) - u(s)) + \|\dot{u}(t) - \dot{u}(s)\|_H^2$$
$$= a(t, u(t), u(t)) - 2a(t, u(t), u(s)) + a(s, u(s), u(s))$$
$$+ a(t, u(s), u(s)) - a(s, u(s), u(s)) \tag{11.69}$$
$$+ (\dot{u}(t), \dot{u}(t)) - 2(\dot{u}(t), \dot{u}(s)) + (\dot{u}(s), \dot{u}(s)).$$

By exploiting the weak continuity of u and $\dot{u}$ and the continuity of $a(t, u, u) + \|\dot{u}\|^2$, we see that the right-hand side of (11.69) tends to zero as $s \to t$. Hence the left-hand side also tends to zero; this implies continuity of $u(t) \in V$ and $\dot{u}(t) \in H$.

Problems

11.6. Use (11.41) to derive (11.43).

11.7. Discuss higher regularity of solutions in analogy to Section 11.1.4.

11.8. Establish an existence and uniqueness result for the following perturbation of (11.39): $\ddot{u} = A(t)u + B(t)u$, where A is as above and B is a bounded operator from V to H.

11.9. Apply the result of this section to the second-order analogues of the examples discussed in Section 11.1.3.

11.10. Let I be an open interval. Show that $W^{1,1}(I) \subset C(\bar{I})$.

11.11. In our construction of the solution, we took a subsequence of u_n. Show that in fact the whole sequence u_n converges.

12
Semigroup Methods

Roughly speaking, the semigroup approach is a point of view which regards time-dependent PDEs as ODEs on a function space. Consider for example, the following initial/boundary-value problem for the heat equation:

$$
\begin{aligned}
u_t &= u_{xx}, & x \in (0,1),\ t > 0, \\
u(0,t) &= 0, & t > 0, \\
u(1,t) &= 0, & t > 0, \\
u(x,0) &= u_0(x), & x \in (0,1).
\end{aligned}
\tag{12.1}
$$

Let X be the function space $L^2(0,1)$ and let $A = d^2/dx^2$ be the second-derivative operator with domain $\mathcal{D}(A) = \{u \in H^2(0,1) \mid u(0) = u(1) = 0\}$. Then we can think of (12.1) as an initial-value problem for an ODE in X:

$$
\dot{u} = Au,\ u(0) = u_0.
\tag{12.2}
$$

Any self-respecting physicist "knows" that the solution to (12.2) is $\exp(At)u_0$. For mathematicians, life is not as simple: we have to face the annoying issue of giving a meaning to $\exp(At)$. The theory of semigroups of linear operators generalizes the notion of the exponential matrix (or, for nonautonomous problems, the fundamental matrix) to problems in infinite-dimensional spaces involving unbounded operators. As a preparation, we review various ways that the exponential matrix can be defined in finite dimensions and we tentatively assess their potential of being generalizable to infinite dimensions.

1. The power series method: The conventional way of defining the exponential of a matrix A is by the power series

$$e^A = \sum_{n=0}^{\infty} \frac{A^n}{n!}. \tag{12.3}$$

It should be obvious that such a definition is virtually useless for unbounded operators. For example, if A is the second-derivative operator in the example above, then u has to be of class C^{∞} and derivatives of all even orders have to vanish at the endpoints in order for $A^n u$ to be defined for all n. The requirement of convergence of the exponential series would restrict u even further.

2. The spectral method: The way textbooks tell us to compute an exponential matrix is to diagonalize A first (or, more generally, transform it to Jordan canonical form). If $A = T^{-1}DT$ with D diagonal, then it easily follows from (12.3) that

$$e^A = T^{-1}e^D T. \tag{12.4}$$

The exponential of a diagonal matrix is of course trivial to compute. In infinite dimensions, even for bounded operators, there is in general no analogue of diagonalization or Jordan canonical forms. However, for restricted classes of operators, such as self-adjoint operators in Hilbert space, the spectral theorem can be viewed as a diagonalization, and (12.4) can indeed be used to define the exponential.

3. Another way to define the exponential of A is

$$e^A = \lim_{n \to \infty} \left(I + \frac{A}{n} \right)^n. \tag{12.5}$$

This approach clearly suffers from the same defects as the power series when unbounded operators are concerned. However, we may modify it as follows:

$$e^A = \lim_{n \to \infty} \left(I - \frac{A}{n} \right)^{-n}. \tag{12.6}$$

Now, instead of taking powers of A, we are taking powers of the resolvent. We shall see that formula (12.6) can indeed be used.

4. Representation by Cauchy's formula: Let C be a closed, simple, rectifiable, positively oriented curve enclosing all eigenvalues of A. Then we obtain

$$e^A = \frac{1}{2\pi i} \int_C e^\lambda (\lambda I - A)^{-1} \, d\lambda. \tag{12.7}$$

In PDE applications, the spectrum of A is usually unbounded, and we cannot find a curve enclosing it. However, one may think of using appopriate unbounded curves for C. Indeed, we shall return to this idea in the context of analytic semigroups.

5. Laplace transforms: We find

$$\int_0^\infty e^{At} e^{-\lambda t}\, dt = (\lambda I - A)^{-1};\tag{12.8}$$

i.e., the resolvent is the Laplace transform of the exponential matrix. Inverting the transform, we obtain

$$e^{At} = \frac{1}{2\pi i} \int_{\gamma - i\infty}^{\gamma + i\infty} e^{\lambda t}(\lambda I - A)^{-1}\, d\lambda.\tag{12.9}$$

Here γ must be taken larger than the real part of any eigenvalue of A. Modulo a contour deformation, (12.9) is actually the same as (12.7).

In this chapter, we shall only consider linear autonomous evolution problems of the form $\dot{u} = Au + f(t)$; we emphasize, however, that the methods discussed here have been extended to nonautonomous and nonlinear equations. We refer to the literature for such results.

Problem

12.1. Verify that the various definitions of the matrix exponential discussed above are indeed equivalent.

12.1   Semigroups and Infinitesimal Generators

12.1.1   Strongly Continuous Semigroups

If A is an $n \times n$ matrix, we have $\exp(A(t+s)) = \exp(At)\exp(As)$, i.e., the matrices $\exp(At)$, $t \geq 0$, form a *semigroup*. We shall consider families of bounded linear operators with the same property.

Definition 12.1. *Let X be a Banach space. A family $\{T(t)\}$, $t \geq 0$, of bounded linear operators in X is called a* **strongly continuous semigroup** *or* C_0**-semigroup,** *if it satisfies the following properties:*

1. $T(t+s) = T(t)T(s)$, $t, s \geq 0$,

2. $T(0) = I$,

3. *for every $x \in X$,*

$$[0,\infty) \ni t \mapsto T(t)x \in X$$

is continuous.

Remark 12.2. It can be shown that instead of condition 3, it is sufficient to require continuity of $T(t)x$ at $t = 0$. The uniform boundedness principle can then be used to show that $\|T(t)\|$ is bounded in a neighborhood of zero. Having this, the semigroup property yields that $\|T(t)\|$ is bounded on any finite interval. Now continuity of $T(t)x$ from the right follows immediately from the semigroup property, and continuity from the left can be shown by using the identity $T(t-h) - T(t) = T(t-h)(I - T(h))$. We leave the details of the argument to the reader; see Problem 12.2.

Example 12.3. If A is a bounded operator in X, then $\exp(At)$ can be defined using, for example, the power series. It is easy to see that the operators $\exp(At)$ form a strongly continuous semigroup. In fact, $\exp(At)$ is even continuous in the norm topology, not just strongly; i.e.,

$$[0, \infty) \ni t \mapsto \exp(At) \in \mathcal{L}(X)$$

is continuous. Such semigroups are referred to as uniformly continuous. It can be shown that every uniformly continuous semigroup is of the form $\exp(At)$, where A is a bounded operator. Since this result has little interest for PDEs, we shall not prove it.

Example 12.4. Let $X = L^2(\mathbb{R})$, and define $(T(t)u)(x) = u(x+t)$. Then $\{T(t)\}$, $t \geq 0$, is a strongly continuous semigroup. It is not uniformly continuous; indeed, we have $\|T(t) - T(s)\| = 2$ whenever $t \neq s$.

Example 12.5. Let $X = C_b(\mathbb{R})$, and let $T(t)$ be as in the previous example. Then $T(t)$ is not strongly continuous; the convergence of $T(t)u$ to u as $t \to 0$ need not be uniform on $\mathbb{R}$. (The reader should construct a counterexample: a function $u(x)$ for which one does not have $\lim_{t\to 0} T(t)u = 0$. Try a function consisting of a sequence of increasingly narrow "bumps" of unit height.)

We conclude this subsection with a result on the growth of $\|T(t)\|$ for strongly continuous semigroups. As a preparation we need a lemma.

Lemma 12.6. Let $\omega : [0, \infty) \to \mathbb{R}$ be bounded above on every finite interval and subadditive (i.e., $\omega(t_1 + t_2) \leq \omega(t_1) + \omega(t_2)$). Then

$$\inf_{t>0} \omega(t)/t = \lim_{t\to\infty} \omega(t)/t;$$

(it is understood that both sides in this equation may be $-\infty$).

Proof. Let $\omega_0 = \inf_{t>0} \omega(t)/t$ and let $\gamma > \omega_0$. Then there exists a $t_0 > 0$ with $\omega(t_0)/t_0 < \gamma$. Any $t \geq 0$ can be represented as a multiple of t_0 plus a remainder: $t = nt_0 + r$, $n \in \mathbb{N} \cup \{0\}$, $r \in [0, t_0)$. Subadditivity yields

$$\frac{\omega(t)}{t} \leq \frac{n\omega(t_0) + \omega(r)}{t}. \qquad (12.10)$$

As $t \to \infty$, n/t tends to $1/t_0$, and $\omega(r)/t$ is less than or equal to $\sup_{s\in[0,t_0)} \omega(s)/t$. Hence (12.10) yields

$$\limsup_{t\to\infty} \frac{\omega(t)}{t} \leq \frac{\omega(t_0)}{t_0} < \gamma. \qquad (12.11)$$

From this the lemma is immediate. $\qquad\qquad\qquad\qquad\qquad\qquad\square$

Theorem 12.7. *Let $\{T(t)\}$, $t \geq 0$ be a strongly continuous semigroup of bounded linear operators on a Banach space X. Then the limit*

$$\omega_0 = \lim_{t \to \infty} (\log \|T(t)\|)/t \qquad (12.12)$$

exists (with the understanding that its value may be $-\infty$). For every $\gamma > \omega_0$, there is a constant M_γ such that $\|T(t)\| \leq M_\gamma \exp(\gamma t)$.

Proof. It is an immediate consequence of the semigroup property that $\log \|T(t)\|$ is a subadditive function. Moreover, because of strong continuity, $\|T(t)x\|$ is bounded on any finite time interval, and by the uniform boundedness principle $\|T(t)\|$ is bounded on finite intervals. By the previous lemma, ω_0 as given by (12.12) exists. For any $\gamma > \omega_0$, there is a t_0 such that $\log \|T(t)\|/t < \gamma$ for $t \geq t_0$, i.e., $\|T(t)\| \leq \exp(\gamma t)$ for $t \geq t_0$. The theorem now follows with

$$M_\gamma = \max\Big(1, \sup_{t \in [0,t_0]} \|T(t)\| e^{-\gamma t}\Big). \qquad (12.13)$$

This completes the proof. $\qquad\qquad\qquad\qquad\qquad\qquad\qquad\qquad\square$

Definition 12.8. *The number ω_0 as given by the last theorem is called the* **type** *of the semigroup.*

Example 12.9. Let $X = L^2(0,1)$ and let

$$(T(t)u)(x) = \begin{cases} u(x+t) & x+t < 1 \\ 0 & \text{otherwise.} \end{cases} \qquad (12.14)$$

Then $T(t) = 0$ for $t \geq 1$. This example illustrates that the type of a semigroup may well be $-\infty$.

12.1.2   The Infinitesimal Generator

Our motivation for studying semigroups of operators was to generalize the matrix exponential. This naturally raises the question whether every strongly continuous semigroup is, in a sense to be made precise, given as $\exp(At)$ for some operator A. We shall answer this question affirmatively. We begin with the following definition.

Definition 12.10. *Let $\{T(t)\}$, $t \geq 0$, be a strongly continuous semigroup of bounded linear operators on a Banach space X. The* **infinitesimal generator** *of the semigroup is the operator A defined by*

$$Ax = \lim_{h \to 0+} \frac{T(h)x - x}{h}, \qquad (12.15)$$

and the domain of A is the set of all vectors $x \in X$ for which this limit exists.

Of course it is not clear from this definition that $\mathcal{D}(A)$ contains anything but 0. The notion of infinitesimal generator would be of little interest unless $\mathcal{D}(A)$ is suitably large. We shall see that $\mathcal{D}(A)$ is actually dense in X. As a preparation, we shall establish a number of facts which will also be useful for other purposes.

Lemma 12.11. *Let A be the infinitesimal generator of the strongly continuous semigroup $T(t)$. Then the following hold.*

1. *For $x \in X$,*

$$\lim_{h \to 0} \frac{1}{h} \int_t^{t+h} T(s)x \ ds = T(t)x. \qquad (12.16)$$

2. *For $x \in X$ and any $t > 0$, $\int_0^t T(s)x \ ds \in \mathcal{D}(A)$ and*

$$A\left(\int_0^t T(s)x \ ds \right) = T(t)x - x. \qquad (12.17)$$

3. *For $x \in \mathcal{D}(A)$, we have $T(t)x \in \mathcal{D}(A)$. Moreover, the function*

$$[0, \infty) \ni t \mapsto T(t)x \in X$$

is differentiable. (This means that difference quotients have a limit in the sense of norm convergence in X.) In fact,

$$\frac{d}{dt}T(t)x = AT(t)x = T(t)Ax. \qquad (12.18)$$

4. *For $x \in \mathcal{D}(A)$,*

$$T(t)x - T(s)x = \int_s^t T(\tau)Ax \ d\tau = \int_s^t AT(\tau)x \ d\tau. \qquad (12.19)$$

Proof. Part 1 is a straightforward consequence of strong continuity of the semigroup. For part 2, we choose $h > 0$, and obtain

$$\frac{T(h) - I}{h} \int_0^t T(s)x \ ds = \frac{1}{h} \int_0^t (T(s+h) - T(s))x \ ds$$

$$= \frac{1}{h} \int_t^{t+h} T(s)x \ ds - \frac{1}{h} \int_0^h T(s)x \ ds.$$

According to part 1, the right-hand side tends to $T(t)x - x$. This proves part 2. To prove part 3, we consider the identities

$$\frac{T(t+h)x - T(t)x}{h} = \frac{T(h) - I}{h}T(t)x = T(t)\frac{T(h)x - x}{h} \qquad (12.20)$$

and

$$\frac{T(t)x - T(t-h)x}{h} = T(t-h)\frac{T(h)x - x}{h}, \qquad (12.21)$$

and take the limit $h \to 0+$. Finally, part 4 follows from part 3 by the fundamental theorem of calculus. $\qquad \square$

Theorem 12.12. *Let A be the infinitesimal generator of a C_0-semigroup. Then $\mathcal{D}(A)$ is dense and A is closed.*

Proof. We have

$$x = \lim_{h \to 0+} \frac{1}{h} \int_0^h T(s)x \, ds, \tag{12.22}$$

and by the preceding lemma the right-hand side is in $\mathcal{D}(A)$. Hence $\mathcal{D}(A)$ is dense. Assume now that $x_n \in \mathcal{D}(A)$, $x_n \to x$ and $Ax_n \to y$. Then

$$T(h)x_n - x_n = \int_0^h T(s)Ax_n \, ds \tag{12.23}$$

by part 4 of the preceding lemma. Letting $n \to \infty$, we find

$$T(h)x - x = \int_0^h T(s)y \, ds. \tag{12.24}$$

It remains to divide by h and let $h \to 0+$. $\qquad\qquad\square$

Example 12.13. Let $X = L^2(\mathbb{R})$ and $(T(t)u)(x) = u(x + t)$. Then the infinitesimal generator is $A = d/dx$ and its domain is $H^1(\mathbb{R})$.

12.1.3 Abstract ODEs

Throughout this section, $T(t)$ is a strongly continuous semigroup of operators in X and A is the infinitesimal generator. We are interested in solutions of the initial-value problem

$$\dot{u} = Au + f(t), \quad u(0) = u_0. \tag{12.25}$$

Our notion of a solution is classical; we assume that $u_0 \in \mathcal{D}(A)$ and $f \in C([0, T]; X)$ and we seek a solution $u \in C^1([0, T]; X) \cap C([0, T]; \mathcal{D}(A))$. (Here we think of $\mathcal{D}(A)$ as a Banach space equiped with the graph norm.) The following is the abstract version of the well-known variation of constants formula for ODEs.

Theorem 12.14. *Let u be a classical solution of (12.25). Then u is represented by the formula*

$$u(t) = T(t)u_0 + \int_0^t T(t - s)f(s) \, ds. \tag{12.26}$$

Proof. Let $g(s) = T(t - s)u(s)$. Then

$$\begin{aligned}
\frac{dg}{ds} &= -AT(t - s)u(s) + T(t - s)\dot{u}(s) \\
&= -AT(t - s)u(s) + T(t - s)(Au(s) + f(s)) = T(t - s)f(s).
\end{aligned} \tag{12.27}$$

Hence we find

$$g(t) - g(0) = u(t) - T(t)u_0 = \int_0^t T(t - s)f(s) \, ds. \tag{12.28}$$

This completes the proof.   □

Clearly (12.26) makes sense under much weaker assumptions. For example, (12.26) represents a continuous function of t if we only assume that $u_0 \in X$ and $f \in L^1([0,T];X)$. This motivates the following definition.

Definition 12.15. *Assume that $u_0 \in X$ and $f \in L^1([0,T];X)$. Then $u(t)$ as given by (12.26) is called a* **mild solution** *of (12.25).*

Naturally, we shall henceforth consider the mild solution a "solution," regardless of whether or not it is indeed a classical solution. However, regularity of solutions is of some interest, and we shall now look into the question of when we can assert that the solution is classical. This question is answered by the following theorem.

Theorem 12.16. *Assume that $u_0 \in \mathcal{D}(A)$, $f \in C([0,T];X)$ and that in addition either $f \in W^{1,1}([0,T];X)$ or $f \in L^1([0,T];\mathcal{D}(A))$. Then the mild solution of (12.25) is a classical solution.*

Proof. The term $T(t)u_0$ is easily dealt with using Lemma 12.11. We can hence focus attention on the term

$$v(t) := \int_0^t T(t-s)f(s)\, ds. \tag{12.29}$$

We need to show that $v \in C^1([0,T];X) \cap C([0,T];\mathcal{D}(A))$ and that $\dot{v} = Av + f$. For this, we first note the identity

$$\frac{T(h)-I}{h}v(t) = \frac{v(t+h)-v(t)}{h} - \frac{1}{h}\int_t^{t+h} T(t+h-s)f(s)\, ds. \tag{12.30}$$

Using the continuity of f and the strong continuity of the semigroup, we find that the last term on the right tends to $-f(t)$ as $h \to 0+$. If v is differentiable with respect to t, then the right-hand side of (12.30) has a limit as $h \to 0+$, hence so does the left-hand side, i.e., $Av(t)$ exists. Conversely, if $Av(t)$ exists and is continuous, then the right derivative $D^+v(t)$ exists and is continuous. By substituting $t-h$ for t in (12.30), we also obtain the existence of the left derivative. Hence $v \in C^1([0,T];X)$ if and only if $v \in C([0,T];\mathcal{D}(A))$ and in either case we have $\dot{v} = Av + f$. If $f \in L^1([0,T];\mathcal{D}(A))$, then it is clear from (12.29) that $v \in C([0,T];\mathcal{D}(A))$. If, on the other hand, $f \in W^{1,1}([0,T];X)$, we rewrite v as

$$v(t) = \int_0^t T(s)f(t-s)\, ds, \tag{12.31}$$

from which we find

$$\dot{v}(t) = T(t)f(0) + \int_0^t T(s)\dot{f}(t-s)\, ds. \tag{12.32}$$

Hence $v \in C^1([0,T];X)$.   □

Problems

12.2. Fill in the details for the argument in Remark 12.2.

12.3. Let $C_0(\mathbb{R})$ be the space of all continuous functions on $\mathbb{R}$ which tend to zero at infinity and let $T(t)$ be as in Example 12.5. Show that $T(t)$ is strongly continuous.

12.4. Assume that there is some $t_0 > 0$ such that $T(t_0)$ has a nonzero eigenvalue. Show that the type of the semigroup cannot be $-\infty$.

12.5. Suppose two C_0-semigroups $T(t)$ and $S(t)$ have the same infinitesimal generator A. Show that the semigroups are equal. Hint: Consider $\frac{d}{ds}T(t-s)S(s)x$ for $x \in \mathcal{D}(A)$.

12.6. Let $T(t)$ be a C_0-semigroup with infinitesimal generator A. Let $\phi \in \mathcal{D}(0,\infty)$. Show that $\int_0^\infty \phi(s)T(s)x$ is in the domain of A^n for every $n \in \mathbb{N}$. Use this to show that $\bigcap_{n\in\mathbb{N}} \mathcal{D}(A^n)$ is dense in X.

12.7. Characterize the infinitesimal generator for the semigroup in Example 12.9.

12.8. Show by a counterexample that continuity of f alone is not sufficient for the conclusions of Theorem 12.16 to hold. Hint: Consider the equation $\dot{u} = Au + T(t)x$, $u(0) = 0$ for some fixed x. The solution is $u(t) = tT(t)x$.

12.2   The Hille-Yosida Theorem

12.2.1   The Hille-Yosida Theorem

In the preceding section, we showed that every strongly continuous semigroup has an infinitesimal generator and also how the semigroup can be used to solve initial-value problems. The most important question from the PDE point of view is how to recognize those operators which are infinitesimal generators of a C_0-semigroup. The Hille-Yosida theorem provides an answer to that question.

Theorem 12.17 (Hille-Yosida). *Let A be an operator in the Banach space X. Then A is the infinitesimal generator of a C_0-semigroup $T(t)$ satisfying $\|T(t)\| \le M\exp(\omega t)$ if and only if the following two conditions hold:*

1. *$\mathcal{D}(A)$ is dense and A is closed.*

2. *Every real number $\lambda > \omega$ is in the resolvent set of A and $\|R_\lambda(A)^n\| \le M/(\lambda - \omega)^n$ for every $n \in \mathbb{N}$.*

It will be clear from the proof below that actually every complex number with $\operatorname{Re}\lambda > \omega$ is in the resolvent set and that $\|R_\lambda(A)^n\| \le M/(\operatorname{Re}\lambda-\omega)^n$.

Proof. We already know that the first condition is necessary. To see the necessity of the second, recall that for any C_0-semigroup there are constants M and ω such that $\|T(t)\| \le M \exp(\omega t)$. Let us now consider the integral

$$I_n(\lambda)x = \frac{1}{(n-1)!} \int_0^\infty t^{n-1} e^{-\lambda t} T(t)x \, dt. \tag{12.33}$$

Clearly, $I_n(\lambda)$ is well defined for Re $\lambda > \omega$ and

$$\|I_n(\lambda)\| \le \frac{1}{(n-1)!} \int_0^\infty t^{n-1} e^{-\text{Re }\lambda t} M e^{\omega t} \, dt = \frac{M}{(\text{Re }\lambda - \omega)^n}. \tag{12.34}$$

If $x \in \mathcal{D}(A)$, we find, for $n > 1$,

$$\begin{aligned}
I_n(\lambda)Ax &= \frac{1}{(n-1)!} \int_0^\infty t^{n-1} e^{-\lambda t} T(t) Ax \, dt \\
&= \frac{1}{(n-1)!} \int_0^\infty t^{n-1} e^{-\lambda t} \frac{d}{dt}(T(t)x) \, dt \\
&= -\frac{1}{(n-2)!} \int_0^\infty t^{n-2} e^{-\lambda t} T(t)x \, dt \\
&\quad + \frac{\lambda}{(n-1)!} \int_0^\infty t^{n-1} e^{-\lambda t} T(t)x \, dx \\
&= -I_{n-1}(\lambda)x + \lambda I_n(\lambda)x.
\end{aligned}$$

For $n = 1$, we find instead $I_1(\lambda)Ax = -x + \lambda I_1(\lambda)x$. We can also apply $(T(h) - I)/h$ to (12.33) and go through a similar calculation involving "differencing by parts." We find, in the limit $h \to 0$, that

$$\begin{aligned}
AI_n(\lambda)x &= -I_{n-1}(\lambda)x + \lambda I_n(\lambda)x, \quad x \in X, \ n > 1, \\
AI_1(\lambda)x &= -x + \lambda I_1(\lambda)x, \quad x \in X.
\end{aligned}$$

In summary, we find that $I_n(\lambda)(A - \lambda I) = (A - \lambda I)I_n(\lambda) = -I_{n-1}(\lambda)$ for $n > 1$ and $I_1(\lambda)(A - \lambda I) = (A - \lambda I)I_1(\lambda) = -I$. This clearly implies that $R_\lambda(A)$ exists and $I_n(\lambda) = (-R_\lambda(A))^n$.

We now turn to the proof of sufficiency. For this, we consider the operators

$$U_n(t) = \left(I - \frac{t}{n}A\right)^{-n}. \tag{12.35}$$

It follows from condition 2 that $U_n(t)$ is well defined for $n/t > \omega$, i.e., in particular for sufficiently large n. Moreover, the operators $U_n(t)$ are uniformly bounded as $n \to \infty$. For $t > 0$, differentiation with respect to t yields

$$\dot{U}_n(t) = A\left(I - \frac{t}{n}A\right)^{-n-1} \tag{12.36}$$

in the sense of the operator norm topology. As $t \to 0$, we claim that $U_n(t) \to I$ strongly. For this, it suffices to show that $(I - \frac{t}{n}A)^{-1}$ converges to I

strongly; see Problem 12.10. Since we know that $(I - \frac{t}{n}A)^{-1}$ is uniformly bounded as $t \to 0$, it suffices to show that $(I - \frac{t}{n}A)^{-1}u$ converges for u in a dense set. We now note that, for $u \in \mathcal{D}(A)$,

$$\left\|\left(I - \frac{t}{n}A\right)^{-1} u - u\right\| = \frac{t}{n}\left\|\left(I - \frac{t}{n}A\right)^{-1} Au\right\| \leq Ct\|Au\|, \qquad (12.37)$$

which tends to zero as $t \to 0$.

We claim that $U_n(t)$ has a strong limit as $n \to \infty$. For that, we estimate $U_n(t)u - U_m(t)u$ for u in the dense set $\mathcal{D}(A^2) = (A - \lambda)^{-1}\mathcal{D}(A)$, Re $\lambda > \omega$. We note that

$$U_n(t)u - U_m(t)u = \lim_{\epsilon \to 0} \int_\epsilon^{t-\epsilon} \frac{d}{ds}\left[U_m(t - s)U_n(s)u\right]\, ds$$

$$= \lim_{\epsilon \to 0} \int_\epsilon^{t-\epsilon} \left[-\dot{U}_m(t - s)U_n(s)u + U_m(t - s)\dot{U}_n(s)u\right]\, ds.$$

Using (12.36), one finds, after some algebra, that

$$U_n(t)u - U_m(t)u = \int_0^t \left(\frac{s}{n} - \frac{t-s}{m}\right)\left(I - \frac{t-s}{m}A\right)^{-m-1}\left(I - \frac{s}{n}A\right)^{-n-1} A^2 u\, ds. \qquad (12.38)$$

It follows that

$$\|U_n(t)u - U_m(t)u\| \leq C\|A^2u\| \int_0^t \frac{s}{n} + \frac{t-s}{m}\, ds = \frac{Ct^2}{2}\left(\frac{1}{n} + \frac{1}{m}\right)\|A^2u\|. \qquad (12.39)$$

Hence the strong limit of $U_n(t)$ as $n \to \infty$ exists, and we shall call this limit $T(t)$. Since

$$\|U_n(t)\| \leq M\left(1 - \frac{t}{n}\omega\right)^{-n}, \qquad (12.40)$$

we find that $\|T(t)\| \leq M\exp(\omega t)$ in the limit $n \to \infty$.

It remains to be shown that $T(t)$ is indeed a semigroup and that A is its infinitesimal generator. It is easy to see from (12.39) that the convergence of $U_n(t)u$ to $T(t)u$ is indeed uniform for t is finite intervals. Since $U_n(t)u$ is continuous for every n, it follows that $T(t)u$ is continuous for every $u \in \mathcal{D}(A^2)$. Since $\mathcal{D}(A^2)$ is dense and $\|T(t)\|$ is bounded on finite intervals, it follows that $T(t)u$ is continuous for every $u \in X$. Hence $T(t)$ is strongly continuous. It is also obvious that $T(0) = I$, since $U_n(0) = I$ for every n. Moreover, we note that the same calculation which led to (12.38) yields

$$\frac{d}{ds}[U_n(t-s)U_n(s)u] = \frac{2s-t}{n}\left(I - \frac{t-s}{n}A\right)^{-n-1}\left(I - \frac{s}{n}A\right)^{-n-1} A^2 u \qquad (12.41)$$

for $u \in \mathcal{D}(A^2)$. Letting $n \to \infty$, we find that $T(t-s)T(s)u$ is independent of s for $u \in \mathcal{D}(A^2)$ and, by density, for $u \in X$. This is the semigroup property.

By letting $n \to \infty$ in (12.36) we find that

$$\left.\frac{d}{dt}T(t)u\right|_{t=0} = Au \qquad (12.42)$$

for every $u \in \mathcal{D}(A)$; hence the infinitesimal generator of the semigroup is an extension of A. This extension cannot be proper, because according to the growth estimate for $\|T(t)\|$ and the necessity part of this proof, the infinitesimal generator has a resolvent for $\lambda > \omega$. This resolvent must agree with the resolvent of A, since a bounded operator defined on the whole space cannot be properly extended. □

The above proof not only yields the existence of a semigroup; it also yields a practical way of approximating the semigroup. We have

$$T(t) = \lim_{n \to \infty} \left(I - \frac{t}{n}A\right)^{-n}. \qquad (12.43)$$

This corresponds to the difference scheme

$$\frac{u(t+h) - u(t)}{h} = Au(t+h) \qquad (12.44)$$

for solving the equation $\dot{u} = Au$, which is known as the implicit Euler scheme. Also, it corresponds to formula (12.6) for the matrix exponential. We now feel fully justified in dispensing with separate notations for a semigroup and its infinitesimal generator.

Definition 12.18. *Let A be the infinitesimal generator of a C_0-semigroup of bounded linear operators on X. Then $\exp(At)$, $t \geq 0$ denotes the semigroup generated by A.*

We also introduce the following convenient notation.

Definition 12.19. *We say that a linear operator A in X is in $\mathcal{G}(M,\omega)$ if it satisfies the hypotheses of Theorem 12.17.*

12.2.2 The Lumer-Phillips Theorem

The conditions of the Hille-Yosida theorem are still not easy to verify in applications since they require bounds on all powers of the resolvent. However, the situation is much simpler if $M = 1$: If $\|R_\lambda(A)\| \leq (\lambda - \omega)^{-1}$, then obviously $\|R_\lambda(A)^n\| \leq (\lambda - \omega)^{-n}$. Because of the importance of this case, we make the following definition.

Definition 12.20. *A C_0-semigroup $T(t)$ is called a* **quasicontraction semigroup** *if $\|T(t)\| \leq \exp(\omega t)$ for some ω. It is called a* **contraction semigroup** *if $\|T(t)\| \leq 1$.*

It is obvious that if A generates a quasicontraction semigroup, then $A - \omega I$ generates a contraction semigroup. Actually, every semigroup can be made into a quasicontraction semigroup.

Theorem 12.21. *Let $T(t)$ be a C_0-semigroup on X satisfying*

$$\|T(t)\| \leq M \exp(\omega t). \tag{12.45}$$

Then there is an equivalent norm on X such that, in the operator norm corresponding to this new norm on X, we have $\|T(t)\| \leq \exp(\omega t)$.

We omit the proof of this theorem; it can be found, e.g., in [Pa]. In many applications, it is more profitable to seek an appropriate equivalent norm than to try to verify the assumptions of the Hille-Yosida theorem directly. A practical criterium for generators of quasicontraction semigroups is given by the following result, known as the Lumer-Phillips theorem.

Theorem 12.22 (Lumer-Phillips). *Let H be a Hilbert space and let A be a linear operator in H satisfying the following conditions:*

1. $\mathcal{D}(A)$ is dense.

2. $\mathrm{Re}(x, Ax) \leq \omega(x, x)$ for every $x \in \mathcal{D}(A)$.

3. There exists a $\lambda_0 > \omega$ such that $A - \lambda_0 I$ is onto.

Then A generates a quasicontraction semigroup and $\|\exp(At)\| \leq \exp(\omega t)$.

Proof. We find, for $\lambda > \omega$,

$$\|(A - \lambda I)x\|\|x\| \geq \mathrm{Re}(x, (\lambda I - A)x) \geq (\lambda - \omega)(x, x), \tag{12.46}$$

i.e., we have $\|(A - \lambda I)x\| \geq (\lambda - \omega)\|x\|$. If we can show that $A - \lambda I$ is onto for every $\lambda > \omega$, then this implies that $R_\lambda(A)$ exists and $\|R_\lambda(A)\| \leq (\lambda - \omega)^{-1}$. In particular, we have $\lambda_0 \in \rho(A)$, and hence A is closed. For $\lambda > \omega$, $A - \lambda I$ has a bounded inverse, and hence its range must be closed. But this means that $A - \lambda I$ is semi-Fredholm and its index must be constant on (ω, ∞). $\square$

Example 12.23. Any self-adjoint operator whose spectrum is bounded from above generates a quasicontraction semigroup. Any skew-adjoint operator generates a contraction semigroup.

Example 12.24. Let $H = L^2(0,1)$ and let $Au = u'$ with domain $\mathcal{D}(A) = \{u \in H^1(0,1) \mid u(1) = 0\}$. Clearly $\mathcal{D}(A)$ is dense and it is easy to show that the spectrum of A is empty. Moreover, we have

$$(u, Au) = \int_0^1 u(x)u'(x)\, dx = -\frac{1}{2}(u(0))^2 \leq 0 \tag{12.47}$$

for every $u \in \mathcal{D}(A)$; hence A generates a contraction semigroup. Indeed, the semigroup generated by A is the one given in Example 12.9; cf. Problem 12.7.

Definition 12.25. *An operator in a Hilbert space which satisfies condition 2 of Theorem 12.22 is called* **quasidissipative**; *it is called* **dissipative** *if $\omega = 0$. An operator which satisfies conditions 2 and 3 is called* **quasi-m-dissipative**.

Problems

12.9. Prove the existence of a solution of the heat equation

$$u_t = -Lu + f(t), \quad u(0) = u_0.$$

where L is the second-order elliptic operator defined in (9.66) with Dirichlet boundary conditions. Identify the spaces that the data and solution should occupy.

12.10. Let $A_k \in \mathcal{L}(X, Y)$, $B_k \in \mathcal{L}(Y, Z)$ and assume that $A_k \to A$, $B_k \to B$ as $k \to \infty$ strongly. Show that $B_k A_k \to BA$ strongly.

12.11. Let $A \in \mathcal{G}(1, \omega)$ and let B be bounded. Show that $A + B \in \mathcal{G}(1, \omega + \|B\|)$.

12.12. In Example 12.24, replace the boundary condition $u(1) = 0$ by the condition $u(0) = 0$. Show that the resulting operator does not generate a C_0-semigroup.

12.13. If both A and $-A$ satisfy the hypotheses of the Hille-Yosida theorem, show that $\exp(At)$ and $\exp((-A)t)$ are inverse to each other. It is then of course natural to write $\exp(-At)$ for $\exp((-A)t)$. Show that the operators $\exp(At)$, $t \in \mathbb{R}$ form a group.

12.14. Let A be skew-adjoint. Show that $\exp(At)$ is unitary.

12.3   Applications to PDEs

12.3.1   Symmetric Hyperbolic Systems

Let $\mathbf{A}^i(\mathbf{x})$ be symmetric $p \times p$ matrices defined for $\mathbf{x} \in \mathbb{R}^m$. We consider the operator

$$\mathbf{A}\mathbf{u}(\mathbf{x}) = \mathbf{A}^i(\mathbf{x}) \frac{\partial \mathbf{u}}{\partial x_i} \tag{12.48}$$

in the function space $L^2(\mathbb{R}^m)$. We assume that $\mathbf{A}^i \in C_b^1(\mathbb{R}^m)$. It is no simple matter to characterize the domain for $\mathbf{A}$. Instead, we take a copout. We first define $\mathbf{A}$ on $H^1(\mathbb{R}^m)$ and then take the closure. We claim

Theorem 12.26. $\mathbf{A}$ is the infinitesimal generator of a C_0-semigroup.

Proof. We shall use the Lumer-Phillips theorem. Since the domain of $\mathbf{A}$ includes $H^1(\mathbb{R}^m)$, it is clear that $\mathbf{A}$ is densely defined. Moreover, for $\mathbf{u} \in H^1$, an integration by parts yields

$$
\begin{aligned}
(\mathbf{u}, \mathbf{A}\mathbf{u}) &= \int_{\mathbb{R}^m} \mathbf{u}(\mathbf{x}) \cdot \mathbf{A}^i(\mathbf{x}) \frac{\partial \mathbf{u}}{\partial x_i} \, d\mathbf{x} \\
&= -\frac{1}{2} \int_{\mathbb{R}^m} \mathbf{u} \cdot \frac{\partial \mathbf{A}^i}{\partial x_i} \mathbf{u} \, d\mathbf{x}.
\end{aligned}
$$

Hence the second condition of the Lumer-Phillips theorem holds for every $\mathbf{u} \in H^1(\mathbb{R}^m)$, and we can take limits to show it holds for every $\mathbf{u} \in \mathcal{D}(\mathbf{A})$.

The main part of the work is in verifying the third condition, namely, that $\mathbf{A} - \lambda\mathbf{I}$ is onto for sufficiently large $\lambda \in \mathbb{R}$. From the proof of the Lumer-Phillips theorem it is clear that it suffices to show that the range of $\mathbf{A} - \lambda\mathbf{I}$ is dense. For this purpose, we first consider a new operator $\tilde{\mathbf{A}}$, which agrees with $\mathbf{A}$ except for the domain. Namely, we take the domain of $\tilde{\mathbf{A}}$ to be the set of all $\mathbf{u} \in L^2$ such that $\mathbf{A}^i \partial\mathbf{u}/\partial x_i$ (interpreted as a element of $H^{-1}(\mathbb{R}^m)$) is also in $L^2(\mathbb{R}^m)$. It is clear that $\tilde{\mathbf{A}}$ is an extension of $\mathbf{A}$. We next use a Galerkin argument to show that $\tilde{\mathbf{A}} - \lambda\mathbf{I}$ is onto for λ sufficiently large. For this, let $\{\phi_n\}$, $n \in \mathbb{N}$, be a complete orthogonal system in $H^1(\mathbb{R}^m)$. Let X_n be the span of $\{\phi_1, \ldots, \phi_n\}$, and let P_n be the orthogonal projection from $L^2(\mathbb{R}^m)$ onto X_n. We now consider the equation

$$(\phi, (\mathbf{A} - \lambda)\mathbf{u}_n) = (\phi, \mathbf{f}) \quad \forall \phi \in X_n, \tag{12.49}$$

where we seek a solution $\mathbf{u}_n \in X_n$. Using the fact that $\mathbf{A}$ is quasidissipative, it is easy to see that such a solution $\mathbf{u}_n$ exists and that the L^2-norm of $\mathbf{u}_n$ is bounded independently of n. Hence a subsequence of the $\mathbf{u}_n$ converges weakly, and the limit is easily shown to be a solution of $(\tilde{\mathbf{A}} - \lambda)\mathbf{u} = \mathbf{f}$.

It remains to be shown that $\tilde{\mathbf{A}}$ is indeed $\mathbf{A}$. That is, given $\mathbf{u}, \mathbf{f} \in L^2(\mathbb{R}^m)$ such that $\tilde{\mathbf{A}}\mathbf{u} = \mathbf{f}$, we have to find $\mathbf{u}_n \in H^1(\mathbb{R}^m)$, $\mathbf{f}_n \in L^2(\mathbb{R}^m)$ such that $\mathbf{A}\mathbf{u}_n = \mathbf{f}_n$ and $\mathbf{u}_n \to \mathbf{u}$, $\mathbf{f}_n \to \mathbf{f}$ in $L^2(\mathbb{R}^m)$. For this, we shall need the following result, known as Friedrichs' lemma.

Lemma 12.27 (Friedrichs). *For $\epsilon > 0$, there exists a mapping Q_ϵ from $L^2(\mathbb{R}^m)$ into $H^1(\mathbb{R}^m)$ such that the following properties hold:*

1. *For every $\mathbf{f} \in L^2(\mathbb{R}^m)$, $Q_\epsilon\mathbf{f} \to \mathbf{f}$ in $L^2(\mathbb{R}^m)$ as $\epsilon \to 0$.*

2. *The operator $\mathbf{A}Q_\epsilon - Q_\epsilon\mathbf{A}$ can be extended to a bounded operator in $L^2(\mathbb{R}^m)$. Moreover, for any fixed $\mathbf{u} \in \mathcal{D}(\tilde{\mathbf{A}})$, $\mathbf{A}Q_\epsilon\mathbf{u} - Q_\epsilon\tilde{\mathbf{A}}\mathbf{u} \to \mathbf{0}$ in $L^2(\mathbb{R}^m)$.*

With this lemma, the completion of the proof is immediate. We simply set $\mathbf{u}_\epsilon = Q_\epsilon\mathbf{u}$ and $\mathbf{f}_\epsilon = \mathbf{A}\mathbf{u}_\epsilon$. Then

$$\mathbf{f}_\epsilon = Q_\epsilon\tilde{\mathbf{A}}\mathbf{u} + (\mathbf{A}Q_\epsilon - Q_\epsilon\tilde{\mathbf{A}})\mathbf{u}. \tag{12.50}$$

As $\epsilon \to 0$, the first term on the right converges to $\tilde{\mathbf{A}}\mathbf{u} = \mathbf{f}$, and the second term converges to zero.

It thus remains to prove Friedrichs' lemma. Let $k(\mathbf{x})$ be a non-negative test function supported on the ball $|\mathbf{x}| \leq 1$ and having unit integral. We define $k_\epsilon(\mathbf{x}) = \epsilon^{-m}k(\mathbf{x}/\epsilon)$, and we define Q_ϵ to be convolution with k_ϵ. Then, for every $\epsilon > 0$, Q_ϵ takes $L^2(\mathbb{R}^m)$ to $H^1(\mathbb{R}^m)$ (in fact to $C^\infty(\mathbb{R}^m)$); moreover, Lemma 8.19 implies that Q_ϵ has norm less than or equal to 1 as an operator from $L^2(\mathbb{R}^m)$ to itself. It is easy to show that $Q_\epsilon\mathbf{u} \to \mathbf{u}$ for a dense subset of $L^2(\mathbb{R}^m)$ (e.g., for $\mathbf{u} \in \mathcal{D}(\mathbb{R}^m)$), and together with the uniform boundedness of Q_ϵ, we conclude that Q_ϵ converges to the identity strongly; i.e., property 1 of the lemma holds.

To verify property 2, one calculates

$$
\begin{aligned}
\mathbf{A}Q_\epsilon\mathbf{u} - Q_\epsilon\tilde{\mathbf{A}}\mathbf{u} &= \int_{\mathbb{R}^m} k_\epsilon(\mathbf{x} - \mathbf{y})(\mathbf{A}^i(\mathbf{x}) - \mathbf{A}^i(\mathbf{y}))\frac{\partial \mathbf{u}}{\partial y_i}(\mathbf{y})\, d\mathbf{y} \\
&= \int_{\mathbb{R}^m} k_\epsilon(\mathbf{x} - \mathbf{y})\frac{\partial \mathbf{A}^i}{\partial y_i}(\mathbf{y})\mathbf{u}(\mathbf{y})\, d\mathbf{y} \qquad (12.51) \\
&\quad + \int_{\mathbb{R}^m} \frac{\partial}{\partial x_i}(k_\epsilon(\mathbf{x} - \mathbf{y}))(\mathbf{A}^i(\mathbf{x}) - \mathbf{A}^i(\mathbf{y}))\mathbf{u}(\mathbf{y})\, d\mathbf{y}.
\end{aligned}
$$

As $\epsilon \to 0$, the first expression on the right-hand side clearly tends to zero for smooth enough $\mathbf{u}$ (e.g., $\mathbf{u} \in \mathcal{D}(\mathbb{R}^m)$). Hence it suffices to show that $\mathbf{A}Q_\epsilon - Q_\epsilon\tilde{\mathbf{A}}$ is uniformly bounded for $\epsilon \to 0$ as an operator from $L^2(\mathbb{R}^m)$ to itself. For this purpose, we use the second expression on the right-hand side of (12.51). We can see from the proof of Lemma 8.19 that the first term in this expression has an operator norm bounded by

$$
\sup_{\mathbf{y}\in\mathbb{R}^m}\left|\frac{\partial \mathbf{A}^i}{\partial y_i}(\mathbf{y})\right| \times \int_{\mathbb{R}^m} |k_\epsilon(\mathbf{x})|\, d\mathbf{x}, \qquad (12.52)
$$

and in this expression the first factor is independent of ϵ and the second is 1. The operator norm for the second term can, again by Lemma 8.19, be estimated by

$$
\sup_{\mathbf{x}\in\mathbb{R}^m}\int_{\mathbb{R}^m}\left|\frac{\partial}{\partial x_i}k_\epsilon(\mathbf{x} - \mathbf{y})\right|\,|\mathbf{A}^i(\mathbf{x}) - \mathbf{A}^i(\mathbf{y})|\, d\mathbf{y}. \qquad (12.53)
$$

Since the support of k_ϵ is contained in $|\mathbf{x}| \leq \epsilon$, we can estimate $|\mathbf{A}^i(\mathbf{x}) - \mathbf{A}^i(\mathbf{y})|$ by a constant times ϵ. On the other hand, the L^1-norm of $\partial k_\epsilon/\partial x_i$ is a constant times $1/\epsilon$. Hence (12.53) is bounded by a constant independent of ϵ.   □

12.3.2   The Wave Equation

We consider the second-order equation

$$
u_{tt} = \frac{\partial}{\partial x_i}\left(a_{ij}(\mathbf{x})\frac{\partial u}{\partial x_j}\right), \quad \mathbf{x} \in \Omega,\ t > 0, \qquad (12.54)
$$

with boundary condition

$$
u(\mathbf{x}, t) = 0, \quad \mathbf{x} \in \partial\Omega, \qquad (12.55)
$$

and initial conditions

$$
u(x, 0) = u_0(x), \quad u_t(x, 0) = u_1(x). \qquad (12.56)
$$

Here we assume that Ω is a bounded domain in $\mathbb{R}^m$ with smooth boundary and that the matrix a_{ij} is symmetric, strictly positive definite and of class $C^1(\overline{\Omega})$.

We denote by A the operator represented by the right-hand side of (12.54). We view A as an operator in $L^2(\Omega)$ with domain $H^2(\Omega) \cap H_0^1(\Omega)$. In order to apply semigroup theory, we must first transform the second-order equation (12.54) to a first-order system. For this purpose, we set $v = u_t$, and we rewrite (12.54) as

$$\begin{pmatrix} \dot{u} \\ \dot{v} \end{pmatrix} = \mathcal{A} \begin{pmatrix} u \\ v \end{pmatrix} := \begin{pmatrix} 0 & I \\ A & 0 \end{pmatrix} \begin{pmatrix} u \\ v \end{pmatrix}. \tag{12.57}$$

We regard $\mathcal{A}$ as an operator in the space $X = H_0^1(\Omega) \times L^2(\Omega)$ with domain $\mathcal{D}(\mathcal{A}) = (H_0^1(\Omega) \cap H^2(\Omega)) \times H_0^1(\Omega)$. We claim

Theorem 12.28. *The operator $\mathcal{A}$ generates a C_0-semigroup.*

Proof. We again use the Lumer-Phillips theorem. For this purpose, we consider the following inner product on X, which is equivalent to the usual one:

$$\left[\begin{pmatrix} u \\ v \end{pmatrix}, \begin{pmatrix} f \\ g \end{pmatrix} \right] = \int_\Omega a_{ij}(\mathbf{x}) \frac{\partial u}{\partial x_i} \frac{\partial f}{\partial x_j} + vg \, d\mathbf{x}. \tag{12.58}$$

It is easy to see that

$$\left[\begin{pmatrix} u \\ v \end{pmatrix}, \mathcal{A} \begin{pmatrix} u \\ v \end{pmatrix} \right] = 0 \tag{12.59}$$

for $(u, v) \in \mathcal{D}(\mathcal{A})$. Moreover, it is clear that $\mathcal{D}(\mathcal{A})$ is dense. We claim that every positive λ is in the resolvent set of $\mathcal{A}$. For this, we need to consider the problem

$$v - \lambda u = f, \; Au - \lambda v = g, \; f \in H_0^1(\Omega), \; g \in L^2(\Omega). \tag{12.60}$$

These two equations can be combined into the one equation

$$Au - \lambda^2 u = g + \lambda f. \tag{12.61}$$

The solvability of this equation follows from our results on elliptic boundary-value problems in Chapter 9. □

12.3.3 The Schrödinger Equation

The Schrödinger equation is the fundamental equation of quantum mechanics. It reads

$$u_t = i(\Delta u - V(\mathbf{x})u), \; \mathbf{x} \in \mathbb{R}^m. \tag{12.62}$$

Here i is the imaginary unit, i.e., u is complex-valued. However, the function $V(\mathbf{x})$ is real-valued. In a formal sense, the operator on the right-hand side of (12.62) is skew-symmetric, and we expect it to be skew-adjoint if the domain is chosen correctly. In this case, it follows immediately that the operator generates a C_0-semigroup.

A major part of the effort in mathematical quantum mechanics goes into proving that this operator is skew-adjoint. How difficult this is depends, obviously, on how badly V is allowed to behave. The case where V is bounded is basically trivial, but of little physical interest. Somewhat more interesting are potentials with point singularities. We prove here the following result.

Theorem 12.29. *Assume that $m \leq 3$ and $V \in L^2(\mathbb{R}^m)$. Then the operator A given by $Au = i(\Delta u - V(\mathbf{x})u)$ with domain $\mathcal{D}(A) = H^2(\mathbb{R}^m)$ is skew-adjoint.*

Proof. We note that

$$\|Vu\|_2 \leq \|V\|_2 \|u\|_\infty \leq C\|V\|_2 \|u\|_{m/2+\delta,2} \leq C\|V\|_2 \|u\|_{2,2}. \qquad (12.63)$$

Here δ is any sufficiently small positive number. This shows that A is indeed defined on $H^2(\mathbb{R}^m)$. Moreover, a straightforward calculation involving Fourier transforms shows that we can improve on (12.63) as follows. For every $\epsilon > 0$, there is a constant $C(\epsilon)$ such that

$$\|u\|_{m/2+\delta,2} \leq \epsilon \|u\|_{2,2} + C(\epsilon)\|u\|_2. \qquad (12.64)$$

We can use this to show that A is closed; see Problem 12.19. It is clear that A is skew-symmetric. To show it is skew-adjoint, we have to prove that $A \pm I$ is surjective. This follows by considering the family of operators $i(\Delta - tV) \pm I$ for $t \in [0,1]$ and using the homotopy invariance of the Fredholm index. $\qquad \square$

Clearly, we can more generally consider a V which is the sum of an L^2-function and a bounded function. This includes, for example, a Coulomb potential ($V = -1/r$) in $\mathbb{R}^3$.

Problems

12.15. State an existence theorem for the equation $u_t = a(x,y)u_x + b(x,y)u_y + c(x,y)u + f(x,y,t)$, $(x,y) \in \mathbb{R}^2$, $t > 0$, with initial condition $u(x,y,0) = u_0(x,y)$. Similarly, state an existence theorem for the wave equation.

12.16. Consider the problem $\mathbf{u}_t = \mathbf{A}(x)\mathbf{u}_x$, where the matrix $\mathbf{A}$ has real and distinct eigenvalues. Show that this problem can always be transformed to one with a symmetric matrix to which the results of Section 12.3.1 are applicable. What difficulty arises in several space dimensions?

12.17. In $L^2(\mathbb{R})$, let $Au = u_{xxx} + a(x)u_x$ with domain $\mathcal{D}(A) = H^3(\mathbb{R})$. Assume that $a \in C_b^1(\mathbb{R})$. Prove that A generates a C_0-semigroup.

12.18. Discuss the wave equation with Neumann conditions in a manner similar to Section refWVHYP.

12.19. Use (12.64) to show that the operator A in Theorem 12.29 is closed.

12.4   Analytic Semigroups

12.4.1   Analytic Semigroups and Their Generators

We shall now discuss a class of semigroups which in many respects allows for much stronger results than strongly continuous semigroups. In particular we shall obtain:

1. Better regularity of solutions to initial-value problems.

2. Better results concerning perturbations of the infinitesimal generator.

3. A relationship between the type of the semigroup and the spectrum of the infinitesimal generator.

We begin with the definition of an analytic semigroup.

Definition 12.30. *A strongly continuous semigroup* $\exp(At)$ *is called an* **analytic semigroup** *if the following conditions hold:*

1. *For some* $\theta \in (0, \pi/2)$, $\exp(At) \in \mathcal{L}(X)$ *can be extended to* $t \in \Delta_\theta = \{0\} \cup \{t \in \mathbb{C} \mid |\arg t| < \theta\}$ *and the conditions of Definition 12.1 hold for* $t \in \Delta_\theta$.

2. *For* $t \in \Delta_\theta \backslash \{0\}$, $\exp(At)$ *is analytic in* t *(in the sense of the uniform operator topology).*

The most important question now is how to characterize the infinitesimal generators of analytic semigroups. This is addressed by the following theorem.

Theorem 12.31. *A closed, densely defined operator* A *in* X *is the generator of an analytic semigroup if and only if there exists* $\omega \in \mathbb{R}$ *such that the half-plane* $\operatorname{Re} \lambda > \omega$ *is contained in the resolvent set of* A *and, moreover, there is a constant* C *such that*

$$\|R_\lambda(A)\| \leq C/|\lambda - \omega| \qquad (12.65)$$

for $\operatorname{Re} \lambda > \omega$. *If this is the case, then actually the resolvent set contains a sector* $|\arg (\lambda - \omega)| < \frac{\pi}{2} + \delta$ *for some* $\delta > 0$, *and an analogous resolvent estimate holds in this sector. Moreover, the semigroup is represented by*

$$e^{At} = \frac{1}{2\pi i} \int_\Gamma e^{\lambda t} (\lambda I - A)^{-1} \, d\lambda, \qquad (12.66)$$

where Γ *is any curve from* $e^{-i\phi}\infty$ *to* $e^{i\phi}\infty$ *such that* Γ *lies entirely in the set* $\{|\arg (\lambda - \omega)| \leq \phi\}$. *Here* ϕ *is any angle such that* $\frac{\pi}{2} < \phi < \frac{\pi}{2} + \delta$.

Proof. The necessity of the condition is not hard to see. Assume $\exp(At)$ is an analytic semigroup defined in Δ_θ. It is easy to show that for any $\psi < \theta$, there are constants M and ω such that $\| \exp(At)\| \leq M \exp(\omega|t|)$

for $t \in \Delta_\psi$; the proof proceeds along the lines of Theorem 12.7. From the proof of the Hille-Yosida theorem, we know that

$$-R_\lambda(A) = \int_0^\infty e^{-\lambda t} T(t)\, dt \qquad (12.67)$$

for Re $\lambda > \omega$. Without loss of generality, let Im $\lambda > 0$. Then we shift the contour to the ray arg $t = -\delta$, $\delta = \max(\psi, \arg \lambda)$ and obtain

$$-R_\lambda(A) = \int_0^\infty \exp(-\lambda e^{-i\delta} t) T(e^{-i\delta} t)\, e^{-i\delta}\, dt, \qquad (12.68)$$

and hence

$$\|R_\lambda(A)\| \leq \frac{M}{\text{Re}(\lambda \exp(-i\delta)) - \omega} \leq \frac{C}{|\lambda - \omega|}. \qquad (12.69)$$

Assume now that (12.65) holds for Re $\lambda > \omega$. If $\lambda - \omega = Re^{i\theta}$, $\theta > \pi/2$, and ϕ is any angle less than $\pi/2$, we can write

$$A - \lambda I = A - (\omega + Re^{i\phi})I - R(e^{i\theta} - e^{i\phi})I, \qquad (12.70)$$

and the inverse can be represented as a geometric series as long as $|e^{i\theta} - e^{i\phi}| \leq C$. This shows that the resolvent set extends to a sector beyond the half-plane Re $\lambda > \omega$, and a resolvent estimate is obtained by term-by-term estimation of the geometric series. In particular, this makes (12.66) well defined.

We next show that (12.66) does indeed represent an analytic semigroup generated by A. Clearly, (12.66) is defined for $|\arg t| < \phi - \frac{\pi}{2}$ and it is analytic in t. We next check the semigroup property. We denote by $T(t)$ the right-hand side of (12.66) and we take Γ' to be a path with the same properties as Γ, but lying entirely to the right of Γ. Then we have

$$
\begin{aligned}
T(s)T(t) &= \left(\frac{1}{2\pi i}\right)^2 \int_{\Gamma'} \int_{\Gamma} e^{\lambda' s + \lambda t} (\lambda' I - A)^{-1} (\lambda I - A)^{-1}\, d\lambda\, d\lambda' \\
&= \left(\frac{1}{2\pi i}\right)^2 \left[\int_{\Gamma'} e^{\lambda' s} (\lambda' I - A)^{-1} \int_{\Gamma} e^{\lambda t} (\lambda - \lambda')^{-1}\, d\lambda\, d\lambda' \right. \\
&\qquad \left. - \int_{\Gamma} e^{\lambda t} (\lambda I - A)^{-1} \int_{\Gamma'} e^{\lambda' s} (\lambda - \lambda')^{-1}\, d\lambda'\, d\lambda \right] \\
&= \frac{1}{2\pi i} \int_{\Gamma} e^{\lambda(t+s)} (\lambda I - A)^{-1}\, d\lambda = T(t+s).
\end{aligned}
$$

Here we have used the identity $(\lambda I - A)^{-1} (\lambda' I - A)^{-1} = (\lambda - \lambda')^{-1} [(\lambda' I - A)^{-1} - (\lambda I - A)^{-1}]$ and the fact that

$$
\begin{aligned}
\int_{\Gamma} e^{\lambda t} (\lambda - \lambda')^{-1}\, d\lambda &= 0, \\
\int_{\Gamma'} e^{\lambda' s} (\lambda - \lambda')^{-1}\, d\lambda' &= -2\pi i e^{\lambda s}.
\end{aligned}
$$

By shifting the contour to the right if necessary, we may assume that Γ in (12.66) lies entirely in the set $\{|\arg \lambda| < \phi\}$. For sufficiently small t, we may then make the substitution $\lambda' = \lambda|t|$ and then deform the contour back to Γ. In this fashion, we obtain

$$T(t) = \frac{1}{2\pi i |t|} \int_\Gamma e^{\lambda' t/|t|} \left(\frac{\lambda'}{|t|}I - A\right)^{-1} d\lambda'. \tag{12.71}$$

From this, we find

$$\|T(t)\| \le \frac{C}{2\pi} \int_\Gamma |e^{\lambda' t/|t|}| |\lambda'|^{-1} d\lambda'. \tag{12.72}$$

Clearly, this expression remains bounded as $t \to 0$. Moreover, for $u \in \mathcal{D}(A)$, we find, if we choose Γ lying to the right of the origin,

$$\begin{aligned}
T(t)u - u &= \frac{1}{2\pi i} \int_\Gamma e^{\lambda t}((\lambda I - A)^{-1} - \lambda^{-1})u \, d\lambda \\
&= -\frac{1}{2\pi i} \int_\Gamma e^{\lambda t}\lambda^{-1}(\lambda I - A)^{-1} \, d\lambda \, Au.
\end{aligned}$$

As $t \to 0$, the integral converges to

$$-\frac{1}{2\pi i} \int_\Gamma \lambda^{-1}(\lambda I - A)^{-1} \, d\lambda = 0, \tag{12.73}$$

as can be seen by closing the contour by a circle on the right. This shows that $T(t)$ converges strongly to the identity as $t \to 0$.

For $t \ne 0$, we find from (12.66):

$$\begin{aligned}
\frac{d}{dt}T(t) &= \frac{1}{2\pi i} \int_\Gamma \lambda e^{\lambda t}(\lambda I - A)^{-1} \, d\lambda \\
&= \frac{1}{2\pi i} \int_\Gamma e^{\lambda t} A(\lambda I - A)^{-1} \, d\lambda + \frac{1}{2\pi i} \int_\Gamma e^{\lambda t} I \, d\lambda \\
&= AT(t).
\end{aligned}$$

Here the closedness of A is used to justify taking it outside the integral. Hence we have

$$\begin{aligned}
\frac{T(h) - I}{h}u &= \lim_{\epsilon \to 0} \frac{T(h) - T(\epsilon)}{h}u \\
&= \lim_{\epsilon \to 0} \frac{1}{h}\int_\epsilon^h AT(t)u \, dt = \frac{1}{h}\int_0^h T(t)Au
\end{aligned}$$

for $u \in \mathcal{D}(A)$. As $h \to 0$, we obtain Au in the limit; hence the infinitesimal generator of the semigroup is an extension of A. Since A has a resolvent in the half-plane $\mathrm{Re}\, \lambda > \omega$, this extension cannot be proper. $\square$

In the course of the proof above, we proved that $\exp(At)$ is differentiable for $t > 0$ and that $\frac{d}{dt}\exp(At) = A\exp(At)$. Using in essence the same argument, we can show that $\exp(At)$ is actually infinitely often differentiable and $\frac{d^n}{dt^n}\exp(At) = A^n \exp(At)$. In particular, this implies that the range of $\exp(At)$ is contained in $\mathcal{D}(A^n)$ for every n. In PDE applications, this usually translates into a smoothing property. Even if the initial conditions have singularities, the solution is smooth (of class C^∞) for positive t. For future use, we note the following bound on the norm of $A\exp(At)$.

Lemma 12.32. *Let A be the infinitesimal generator of an analytic semigroup. Then there are constants C and ω such that, for every $t > 0$, we have*

$$\|A\exp(At)\| \le C\exp(\omega t)/t. \tag{12.74}$$

The proof is based on a calculation similar to that leading to (12.72); we leave the details as an exercise (Problem 12.20).

In ODE courses we learn that the zero solution of the equation $\dot u = Au$ is stable if all eigenvalues of A have negative real parts. Indeed, this is the most widely applied practical criterium for evaluating stability. Naturally, one would like to apply the same type of criterium for physical systems described by PDEs. This raises the following question: If A is the generator of a semigroup and the spectrum of A is entirely in the left half-plane at a positive distance from the imaginary axis, does it follow that $\|\exp(At)\|$ is bounded as $t \to \infty$? Or equivalently, is there a connection between an upper bound on the real part of the spectrum of A and the type of the semigroup? Unfortunately, in the general context of C_0-semigroups, the answer is negative. Although the counterexamples are not the kind that would arise "naturally" in physical applications, no practical conditions are known that would lead to a positive result. The state of current knowledge is clearly unsatisfactory. For analytic semigroups, however, the problem is quite easy. We have the following result.

Theorem 12.33. *Let A be the infinitesimal generator of an analytic semigroup and assume that the spectrum of A is entirely to the left of the line $\mathrm{Re}\,\lambda = \omega$. Then there exists a constant M such that $\|\exp(At)\| \le M\exp(\omega t)$.*

The idea of the proof is to shift the contour in (12.66) so that it lies entirely to the left of $\mathrm{Re}\,\lambda = \omega$. We leave the details of the argument as an exercise (Problem 12.21).

12.4.2 Fractional Powers

Throughout this subsection, we assume that A is the infinitesimal generator of an analytic semigroup and that the spectrum of A lies entirely in the (open) left half-plane. In this case, we shall define fractional powers of $-A$. If the spectrum of A lies in the half-plane $\mathrm{Re}\,\lambda < \omega$, then we can of course apply the same considerations to define fractional powers of $\omega I - A$. If the spectrum is in the open left half-plane, we can choose a $\delta > 0$ such that the spectrum of A is in the half-plane Re $\lambda < -\delta$, and from the previous subsection we have bounds of the form $\| \exp(At) \| \leq M \exp(-\delta t)$, $\| A \exp(At) \| \leq M_1 \exp(-\delta t)/t$. Moreover, by noting that $A^n \exp(At) = (A \exp(A\frac{t}{n}))^n$, we find $\| A^n \exp(At) \| \leq M_n \exp(-\delta t)/t^n$.

We define

$$(-A)^{-\alpha} = -\frac{1}{2\pi i} \int_\Gamma \lambda^{-\alpha}(\lambda I + A)^{-1} \, d\lambda, \tag{12.75}$$

where Γ is a curve from $\exp(-i\theta)\infty$ to $\exp(i\theta)\infty$ such that the spectrum of $-A$ lies to the right and the origin lies to the left of Γ. Here $\pi/2-\delta < \theta < \pi$, where δ is as in Theorem 12.31 and $\lambda^{-\alpha}$ denotes the branch of the function which takes positive values on the positive real axis. It follows from (12.65) that the integral in (12.75) is absolutely convergent for any $\alpha > 0$. If α is an integer, we can deform the contour in (12.75) to a circle around the origin and use residues to evaluate the integral; we find that indeed $(-A)^{-n} = (-A^{-1})^n$; see Problem 12.22. If $0 < \alpha < 1$, we can deform Γ into the upper and lower sides of the negative real axis. This leads to the expression

$$(-A)^{-\alpha} = \frac{\sin(\pi\alpha)}{\pi} \int_0^\infty \lambda^{-\alpha}(\lambda I - A)^{-1} \, d\lambda, \ 0 < \alpha < 1. \tag{12.76}$$

Recall from the proof of the Hille-Yosida theorem that

$$(\lambda I - A)^{-1} = \int_0^\infty e^{-\lambda t} e^{At} \, dt. \tag{12.77}$$

We insert this expression into (12.76) and then exchange the order of integrations. This yields

$$
\begin{aligned}
(-A)^{-\alpha} &= \frac{\sin(\pi\alpha)}{\pi} \int_0^\infty \exp(At) \int_0^\infty \lambda^{-\alpha} e^{-\lambda t} \, d\lambda \, dt \\
&= \frac{\sin(\pi\alpha)}{\pi} \left(\int_0^\infty u^{-\alpha} e^{-u} \, du \right) \left(\int_0^\infty t^{\alpha-1} e^{At} \, dt \right).
\end{aligned}
$$

We have

$$\int_0^\infty u^{-\alpha} e^{-u} \, du = \frac{\pi}{\sin(\pi\alpha)\Gamma(\alpha)}, \tag{12.78}$$

and hence

$$(-A)^{-\alpha} = \frac{1}{\Gamma(\alpha)} \int_0^\infty t^{\alpha-1} e^{At} \, dt. \tag{12.79}$$

The argument we gave here applies only if $0 < \alpha < 1$. However, both (12.75) and (12.79) are defined for any $\alpha > 0$ and analytic in α, hence by the uniqueness of analytic continuation they agree for every $\alpha > 0$. Using (12.79) and the bound $\| \exp(At) \| \leq M \exp(-\delta t)$, one easily establishes the following result.

Lemma 12.34. *There is a constant C such that $\|(-A)^{-\alpha}\| \le C$ for $0 < \alpha \le 1$.*

Theorem 12.35. *The following hold.*

1. $(-A)^{-\alpha}(-A)^{-\beta} = (-A)^{-(\alpha+\beta)}$.

2. $\lim_{\alpha \to 0}(-A)^{-\alpha} = I$ *in the strong operator topology.*

With the obvious convention $(-A)^0 = I$, this theorem asserts that the operators $(-A)^{-\alpha}$, $\alpha \ge 0$, form a C_0-semigroup. It is natural to call the infinitesimal generator $-\log(-A)$, but we shall not pursue this point further.

Proof. We have

$$(-A)^{-\alpha}(-A)^{-\beta}$$
$$= \frac{1}{\Gamma(\alpha)\Gamma(\beta)} \int_0^\infty \int_0^\infty t^{\alpha-1} s^{\beta-1} e^{At} e^{As} \, dt \, ds$$
$$= \frac{1}{\Gamma(\alpha)\Gamma(\beta)} \int_0^\infty t^{\alpha-1} \int_t^\infty (u-t)^{\beta-1} e^{Au} \, du \, dt$$
$$= \frac{1}{\Gamma(\alpha)\Gamma(\beta)} \int_0^\infty \int_0^u t^{\alpha-1}(u-t)^{\beta-1} \, dt \, e^{Au} \, du$$
$$= \frac{1}{\Gamma(\alpha)\Gamma(\beta)} \int_0^1 v^{\alpha-1}(1-v)^{\beta-1} \, dv \int_0^\infty u^{\alpha+\beta-1} e^{Au} \, du$$
$$= \frac{1}{\Gamma(\alpha+\beta)} \int_0^\infty u^{\alpha+\beta-1} e^{Au} \, du$$
$$= (-A)^{-\alpha-\beta}.$$

Since we already know that $\|(-A)^{-\alpha}\|$ is bounded as $\alpha \to 0$, it suffices to show that $(-A)^{-\alpha} u \to u$ for u in a dense subset of X. Choose $u \in \mathcal{D}(A)$, then $u = -A^{-1}y$ for some $y \in X$. We then have

$$(-A)^{-\alpha} u - u = (-A)^{-1-\alpha}y - (-A)^{-1}y, \qquad (12.80)$$

and it is clear from either (12.75) or (12.79) that for $\alpha > 0$, $(-A)^{-\alpha}$ is actually continuous (indeed analytic) in the uniform operator topology. $\square$

Since $(-A)^{-n}$ is one-to-one for $n \in \mathbb{N}$ and $(-A)^{-n} = (-A)^{-n+\alpha}(-A)^{-\alpha}$ for $n > \alpha$, it follows that $(-A)^{-\alpha}$ is one-to-one. Hence it has an inverse, which naturally we denote by $(-A)^\alpha$. It is clear that $(-A)^\alpha$ is closed with domain $\mathcal{D}((-A)^\alpha) = \mathcal{R}((-A)^{-\alpha})$; since $\mathcal{R}((-A)^{-n}) = \mathcal{D}(A^n) \subset \mathcal{R}((-A)^{-\alpha})$ for $n > \alpha$, it follows that the domain of $(-A)^\alpha$ is dense. Moreover, it is easy to check that $(-A)^{\alpha+\beta} u = (-A)^\alpha(-A)^\beta u$ for any $\alpha, \beta \in \mathbb{R}$ and any $u \in \mathcal{D}((-A)^\gamma)$, where $\gamma = \max(\alpha, \beta, \alpha+\beta)$.

We conclude this subsection with a result relating $(-A)^\alpha$ to the semigroup.

Lemma 12.36. *Let $\alpha > 0$. For every $u \in \mathcal{D}((-A)^\alpha)$, we have*

$$\exp(At)(-A)^\alpha u = (-A)^\alpha \exp(At)u.$$

Moreover, the operator $(-A)^\alpha \exp(At)$ is bounded, with a bound of the form

$$\|(-A)^\alpha \exp(At)\| \leq M_\alpha t^{-\alpha} e^{-\delta t}. \tag{12.81}$$

If $0 < \alpha \leq 1$ and $u \in \mathcal{D}((-A)^\alpha)$, we have a bound of the form

$$\|\exp(At)u - u\| \leq C_\alpha t^\alpha \|(-A)^\alpha u\|. \tag{12.82}$$

Proof. If $u \in \mathcal{D}((-A)^\alpha)$, then $u = (-A)^{-\alpha} v$ for some v, and we find

$$
\begin{aligned}
\exp(At)u &= \exp(At)(-A)^{-\alpha} v = \frac{1}{\Gamma(\alpha)} \int_0^\infty s^{\alpha-1} \exp(As) \exp(At) v \, ds \\
&= (-A)^{-\alpha} \exp(At)v = (-A)^{-\alpha} \exp(At)(-A)^\alpha u.
\end{aligned}
$$

The first claim of the lemma follows by applying $(-A)^\alpha$ to both sides.

Let $n - 1 < \alpha < n$, then

$$
\begin{aligned}
\|(-A)^\alpha \exp(At)\| &= \|(-A)^{\alpha-n} A^n \exp(At)\| \\
&\leq \frac{1}{\Gamma(n-\alpha)} \int_0^\infty s^{n-\alpha-1} \|A^n \exp(A(t+s))\| \, ds \\
&\leq \frac{M_n}{\Gamma(n-\alpha)} \int_0^\infty s^{n-\alpha-1} (t+s)^{-n} e^{-\delta(t+s)} \, ds \\
&\leq \frac{M_n e^{-\delta t}}{\Gamma(n-\alpha)t^\alpha} \int_0^\infty r^{n-\alpha-1} (1+r)^{-n} \, dr.
\end{aligned}
$$

Finally, we have

$$
\begin{aligned}
\|e^{At}u - u\| &= \left\| \int_0^t Ae^{As}u \, ds \right\| = \left\| \int_0^t (-A)^{1-\alpha} e^{As} (-A)^\alpha u \, ds \right\| \\
&\leq C \int_0^t s^{\alpha-1} \|(-A)^\alpha u\| \, ds \leq C_\alpha t^\alpha \|(-A)^\alpha u\|.
\end{aligned}
$$

This completes the proof. $\square$

12.4.3 Perturbations of Analytic Semigroups

Suppose A is the infinitesimal generator of a semigroup. In what sense must B be "small" to be sure that $A + B$ also generates a semigroup? In the context of C_0-semigroups, we saw in Problem 12.11 that it suffices if B is bounded. Indeed, unless conditions other than smallness are imposed, e.g., that both A and B are dissipative, then this is basically the best we can do. In particular, it does not help if B is "of lower order." The reader may verify that although the operator $Au = u_{xxx}$ generates a C_0-semigroup in $L^2(\mathbb{R})$, the operator $(A + B)u = u_{xxx} - u_{xx}$ does not. For analytic semigroups, however, the situation is much better. The main result is the following theorem.

Theorem 12.37. *Let A be the infinitesimal generator of an analytic semigroup. Then there exists a positive number δ such that, if B is any operator satisfying*

1. *B is closed and $\mathcal{D}(B) \supseteq \mathcal{D}(A)$,*

2. *$\|Bu\| \leq a\|Au\| + b\|u\|$ for $u \in \mathcal{D}(A)$, where $a \leq \delta$,*

then $A + B$ is also the infinitesimal generator of an analytic semigroup.

Proof. Since A generates an analytic semigroup, there exists $\omega \in \mathbb{R}$ and $M > 0$ such that $\|R_\lambda(A)\| \leq M/|\lambda - \omega|$ for $\operatorname{Re} \lambda > \omega$. The operator $BR_\lambda(A)$ is bounded, and we find

$$
\begin{aligned}
\|BR_\lambda(A)u\| &\leq a\|AR_\lambda(A)u\| + b\|R_\lambda(A)u\| \\
&\leq a\left(1 + \frac{M|\lambda|}{|\lambda - \omega|}\right)\|u\| + \frac{bM}{|\lambda - \omega|}\|u\|.
\end{aligned}
$$

For any $\epsilon > 0$, we can find ω' such that $\|BR_\lambda(A)\| \leq a(1 + M + \epsilon)$ for $\operatorname{Re} \lambda > \omega'$. If moreover $a < (1 + M)^{-1}$, then we can choose ϵ such that $\|BR_\lambda(A)\| < 1$ for $\operatorname{Re} \lambda > \omega'$. The rest follows from the identity

$$
R_\lambda(A + B) = R_\lambda(A)(I + BR_\lambda(A))^{-1}, \tag{12.83}
$$

from which we find $\|R_\lambda(A + B)\| \leq M'/|\lambda - \omega'|$ for $\operatorname{Re} \lambda > \omega'$. $\square$

In applications, B is often "of lower order" than A, and a in the last theorem can be taken arbitrarily small. The abstract form of the notion of "lower order" can be phrased in term of fractional powers. We have the following lemma.

Lemma 12.38. *Let A be the infinitesimal generator of an analytic semigroup and assume that B is closed and $\mathcal{D}(B) \supseteq \mathcal{D}((\omega I - A)^\alpha)$ for some $\alpha \in (0, 1)$. Then there is a constant C such that*

$$
\|Bu\| \leq C(\rho^\alpha \|u\| + \rho^{\alpha-1}\|(A - \omega I)u\|) \tag{12.84}
$$

for every $u \in \mathcal{D}(A)$ and every $\rho > 0$.

By choosing ρ sufficiently large and applying the last theorem, we conclude that $A + B$ generates an analytic semigroup.

Proof. Without loss of generality, we may assume $\omega = 0$. If $\mathcal{D}(B) \supseteq \mathcal{D}((-A)^\alpha)$, then $B(-A)^{-\alpha}$ is bounded; i.e., there is a constant C such that $\|Bu\| \leq C\|(-A)^\alpha u\|$. Hence it suffices to show (12.84) for $B = (-A)^\alpha$. We have, for $u \in \mathcal{D}(A)$,

$$
\begin{aligned}
\|(-A)^\alpha u\| &\leq \left|\frac{\sin(\pi\alpha)}{\pi}\right| \left|\int_0^\rho \lambda^{\alpha-1}\|AR_\lambda(A)\|\|u\|\, d\lambda\right. \\
&\quad + \left|\frac{\sin(\pi\alpha)}{\pi}\right| \left|\int_\rho^\infty \lambda^{\alpha-1}\|R_\lambda(A)\|\|Au\|\, d\lambda\right.
\end{aligned}
$$

We now use the fact that $\|R_\lambda(A)\| \leq M/\lambda$ and $\|AR_\lambda(A)\| \leq 1 + M$ to complete the proof of the lemma. $\qquad\square$

In applications, it is often difficult to precisely characterize the domains of fractional powers. Instead of checking that $\mathcal{D}(B) \supseteq \mathcal{D}((\omega I - A)^\alpha)$, one usually checks (12.84) directly. In this context, the following result is of interest.

Lemma 12.39. *Let A be the generator of an analytic semigroup and let B be a closed linear operator such that $\mathcal{D}(B) \supseteq \mathcal{D}(A)$ and, for some $\gamma \in (0,1)$ and every $\rho \geq \rho_0 > 0$, we have*

$$\|Bu\| \leq C(\rho^\gamma \|u\| + \rho^{\gamma-1}\|Au\|) \tag{12.85}$$

for every $u \in \mathcal{D}(A)$. Then $\mathcal{D}(B) \supseteq \mathcal{D}((\omega - A)^\alpha)$ for every $\alpha > \gamma$.

Proof. Again we assume without loss of generality that $\omega = 0$. Let $u \in \mathcal{D}((-A)^{1-\alpha})$ so that $(-A)^{-\alpha}u \in \mathcal{D}(A)$. We have

$$B(-A)^{-\alpha}u = \frac{1}{\Gamma(\alpha)} \int_0^\infty t^{\alpha-1} B\exp(At)\, dt, \tag{12.86}$$

provided that the integral is convergent. We split the integral as

$$\int_0^\delta t^{\alpha-1} B\exp(At)\, dt + \int_\delta^\infty t^{\alpha-1} B\exp(At)\, dt. \tag{12.87}$$

We set $\delta = 1/\rho_0$ and use (12.85) with $\rho = \rho_0$ in the second integral and $\rho = 1/t$ in the first integral. The result is that $B(-A)^{-\alpha}$ is bounded for $\alpha > \gamma$, which implies the lemma. $\qquad\square$

We now present an application to parabolic PDEs. Let Ω be a bounded domain in $\mathbb{R}^m$ with smooth boundary, let $a_{ij}(\mathbf{x})$ be of class $C^1(\overline{\Omega})$ be such that the matrix a_{ij} is symmetric and strictly positive definite and let $b_i(\mathbf{x})$, $c(\mathbf{x})$ be of class $C(\overline{\Omega})$. In $L^2(\Omega)$, we consider the operator

$$Au = \frac{\partial}{\partial x_i}\left(a_{ij}(\mathbf{x})\frac{\partial u}{\partial x_j}\right) + b_i(\mathbf{x})\frac{\partial u}{\partial x_i} + c(\mathbf{x})u \tag{12.88}$$

with domain $H^2(\Omega) \cap H_0^1(\Omega)$. We claim

Theorem 12.40. *A generates an analytic semigroup.*

Proof. Let

$$A_0 u = \frac{\partial}{\partial x_i}\left(a_{ij}(\mathbf{x})\frac{\partial u}{\partial x_j}\right). \tag{12.89}$$

Then A_0 is self-adjoint with negative spectrum; hence it clearly generates an analytic semigroup. Moreover, we find

$$\|(A - A_0)u\|_2 \leq C\|u\|_{1,2} \leq C\|u\|_2^{1/2}\|u\|_{2,2}^{1/2}$$
$$\leq C\|u\|_2^{1/2}\|A_0 u\|_2^{1/2} \leq C(\rho^{1/2}\|u\|_2 + \rho^{-1/2}\|A_0 u\|_2).$$

Hence $\mathcal{D}(A - A_0)$ contains $\mathcal{D}((-A_0)^\alpha)$ for any $\alpha > 1/2$. $\square$

Remark 12.41. The intelligent reader may suspect that $\mathcal{D}((-A_0)^{1/2})$ is actually $H_0^1(\Omega)$. Indeed, this suspicion is well founded. A proof, however, would be significantly more involved than the discussion given above.

12.4.4 Regularity of Mild Solutions

We now turn our attention to the inhomogeneous initial-value problem

$$\dot{u}(t) = Au(t) + f(t), \ u(0) = u_0. \tag{12.90}$$

The mild solution is given by

$$u(t) = e^{At}u_0 + \int_0^t e^{A(t-s)} f(s) \ ds. \tag{12.91}$$

If A generates an analytic semigroup, we already know that the term $e^{At}u_0$ is analytic in t for $t > 0$; moreover, $e^{At}u_0$ is in $\mathcal{D}(A^n)$ for every n. Moreover, we know that $\|A^n e^{At}u_0\| \le C\|u_0\|/t^n$ as $t \to 0$. We can hence focus attention on the term

$$v(t) := \int_0^t e^{A(t-s)} f(s) \ ds. \tag{12.92}$$

We need the following definition:

Definition 12.42. *We say that $f \in C^\theta([0,T]; X)$, $0 < \theta < 1$, if there is a constant L such that*

$$\|f(t) - f(s)\| \le L|t - s|^\theta \quad \forall s, t \in [0,T]. \tag{12.93}$$

Lemma 12.43. *Let A be the infinitesimal generator of an analytic semigroup in X and assume that $f \in C^\theta([0,T]; X)$ for some $\theta \in (0,1)$. Let*

$$w(t) = \int_0^t e^{A(t-s)}(f(s) - f(t)) \ ds. \tag{12.94}$$

Then $w(t) \in \mathcal{D}(A)$ for every $t \in [0,T]$ and $Aw \in C^\theta([0,T]; X)$.

Proof. Let us assume that $\|\exp(At)\| \le M$ and $\|A\exp(At)\| \le C/t$ for $t \in (0,T]$. The fact that $w(t) \in \mathcal{D}(A)$ follows from the estimate

$$\left\| \int_0^t A e^{A(t-s)}(f(s) - f(t)) \ ds \right\| \le \int_0^t \frac{C}{t-s} L(t-s)^\theta \ ds = \frac{CLt^\theta}{\theta}. \tag{12.95}$$

It remains to prove the Hölder continuity. We first note that

$$\|A\exp(At) - A\exp(As)\| = \left\|\int_s^t A^2\exp(A\tau)\,d\tau\right\|$$

$$\leq \int_s^t \|A^2\exp(A\tau)\|\,d\tau \tag{12.96}$$

$$\leq 4C\int_s^t \tau^{-2}\,d\tau = 4Ct^{-1}s^{-1}(t-s).$$

We next write

$$Aw(t+h) - Aw(t)$$
$$= A\int_0^t [\exp(A(t+h-s)) - \exp(A(t-s))][f(s) - f(t)]\,ds$$
$$+ A\int_0^t \exp(A(t+h-s))(f(t) - f(t+h))\,ds$$
$$+ A\int_t^{t+h} \exp(A(t+h-s))(f(s) - f(t+h))\,ds$$
$$=: I_1 + I_2 + I_3.$$

We now use (12.96) to obtain

$$\|I_1\| \leq \int_0^t 4C(t+h-s)^{-1}(t-s)^{-1}hL(t-s)^\theta\,ds \leq C_1 h^\theta. \tag{12.97}$$

(Use the substitution $s = t - h\tau$.) I_2 can be rewritten as $(\exp(A(t+h)) - \exp(At))(f(t) - f(t+h))$, and the Hölder estimate simply follows from that for the second factor. For the last term, we have

$$\|I_3\| \leq \int_t^{t+h} \frac{C}{t+h-s}L(t+h-s)^\theta\,ds \leq C_2 h^\theta. \tag{12.98}$$

□

The following regularity result says that, except for a neighborhood of $t = 0$, $\dot{u}$ and Au are as smooth as f is. Note that there is no comparable result for C_0-semigroups.

Theorem 12.44. *Let A be the infinitesimal generator of an analytic semigroup and let u be the solution of (12.90) as given by (12.91). Moreover, let $f \in C^\theta([0,T];X)$. Then:*

1. *For every $\delta > 0$, Au and $\dot{u}$ are in $C^\theta([\delta,T];X)$.*

2. *If $u_0 \in \mathcal{D}(A)$, then Au and $\dot{u}$ are in $C([0,T];X)$.*

3. *If $u_0 = 0$ and $f(0) = 0$, then Au and $\dot{u}$ are in $C^\theta([0,T];X)$.*

Proof. Let $v(t)$ be as given by (12.92). Recall from the proof of Theorem 12.16 that if $Av(t)$ exists, then $\dot{v}$ also exists and $\dot{v} = Av + f$.

Hence, it suffices to consider Au in verifying the theorem. For this purpose, we write

$$u(t) = e^{At}u_0 + \int_0^t e^{A(t-s)}(f(s) - f(t))\, ds + \int_0^t e^{A(t-s)}f(t)\, ds. \quad (12.99)$$

In view of the previous lemma, it suffices to consider the last term. We have

$$A\int_0^t e^{A(t-s)}f(t)\, ds = (e^{At} - I)f(t). \quad (12.100)$$

Since f was assumed in $C^\theta([0,T]; X)$, we need only consider $\exp(At)f(t)$. We have, for $t \geq \delta$ and $h > 0$,

$$\begin{aligned}
&\| \exp(A(t+h))f(t+h) - \exp(At)f(t)\| \\
&\leq \quad \| \exp(A(t+h))\|\ \|f(t+h) - f(t)\| \\
&\quad + \| \exp(A(t+h)) - \exp(At)\|\ \|f(t)\| \quad (12.101)\\
&\leq \quad C_1 h^\theta + C_2 \frac{h}{\delta}.
\end{aligned}$$

This implies part 1. Next we note that

$$\| \exp(At)f(t) - f(0)\| \leq \| \exp(At)f(0) - f(0)\| + \| \exp(At)\|\|f(t) - f(0)\|; \quad (12.102)$$

the strong continuity of the semigroup now implies part 2. To show part 3, we first proceed as in (12.101), but then we estimate

$$\begin{aligned}
\|(e^{A(t+h)} - e^{At})f(t)\| &= \left\| \int_t^{t+h} Ae^{A\tau}f(t)\, d\tau \right\| \\
&\leq \int_t^{t+h} \left\| Ae^{A\tau}(f(t) - f(0)) \right\|\, d\tau \\
&\leq C\int_t^{t+h} \tau^{-1}t^\theta\, d\tau \\
&\leq C\int_t^{t+h} \tau^{\theta-1}\, d\tau \leq Ch^\theta.
\end{aligned}$$

This completes the proof $\qquad\qquad\qquad\qquad\qquad\qquad\qquad\qquad\square$

Problems

12.20. Prove Lemma 12.32.

12.21. Prove Theorem 12.33.

12.22. Verify that for $n \in \mathbb{N}$, we have $(-A)^{-n} = (-A^{-1})^n$, where $(-A)^{-n}$ is defined by (12.75).

12.23. Prove part 3 of Theorem 12.44, assuming that u_0 and $f(0)$ lie in the domain of appropriate fractional powers of $\omega - A$.

12.24. Let A be the infinitesimal generator of an analytic semigroup on X and let B be a closed operator with $\mathcal{D}(B) \supseteq \mathcal{D}(A)$. Show that the operator defined by $\mathcal{A}(u,v) = (Bv, Av)$ generates an analytic semigroup on $X \times X$.

12.25. Discuss how analytic semigroups can be applied to the equation $u_{tt} = \Delta u_t + \Delta u$ with Dirichlet boundary conditions.

AppendixA
References

A.1 Elementary Texts

[Bar] R.G. Bartle, *The Elements of Real Analysis*, 2nd ed., Wiley, New York, 1976.

[BC] D. Bleecker and G. Csordas, *Basic Partial Differential Equations*, Van Nostrand Reinhold, New York, 1992.

[BD] W.E. Boyce and R.C. DiPrima, *Elementary Differential Equations and Boundary Value Problems*, 4th ed., Wiley, New York, 1986.

[Bu] R.C. Buck, *Advanced Calculus*, 3rd ed., McGraw-Hill, New York, 1978.

[Kr] E. Kreysig, *Introductory Functional Analysis with Applications*, Wiley, New York, 1978.

[MH] J.E. Marsden and M.J. Hoffman, *Basic Complex Analysis*, W.H. Freeman, New York, 3rd ed., 1999.

[Rud] W. Rudin, *Principles of Mathematical Analysis*, 3rd ed. McGraw Hill, New York, 1976.

[Stak] I. Stakgold, *Boundary Value Problems of Mathematical Physics*, Vol. 1/2, Macmillan, New York, 1967.

[ZT] E.C. Zachmanoglou and D.W. Thoe, *Introduction to Partial Differential Equations with Applications*, Dover, New York, 1986.

A.2   Basic Graduate Texts

[CH1] R. Courant and D. Hilbert, *Methods of Mathematical Physics I*, Wiley, New York, 1962.

[CH2] R. Courant and D. Hilbert, *Methods of Mathematical Physics II*, Wiley, New York, 1962.

[DiB] E. DiBenedetto, *Partial Differential Equations*, Birkhäuser, Boston, 1995.

[[Eva] L.C. Evans, *Partial Differential Equations*, American Mathematical Society, Providence, 1998.

[GS] I.M. Gelfand and G.E. Shilov, *Generalized Functions*, Vol. 1, Academic Press, New York, 1964.

[Ha] P.R. Halmos, *A Hilbert Space Problem Book*, 2nd ed., Springer-Verlag, New York, 1982.

[In] E.L. Ince, *Ordinary Differential Equations*, Dover, New York, 1956.

[Jo] F. John, *Partial Differential Equations*, 4th ed., Springer-Verlag, New York, 1982.

[La] O.A. Ladyzhenskaya, *The Boundary Value Problems of Mathematical Physics* (English Edition), Springer-Verlag, New York, 1985.

[Rau] J. Rauch, *Partial Differential Equations*, Springer-Verlag, New York, 1992.

[RS] M. Reed and B. Simon, *Methods of Modern Mathematical Physics I: Functional Analysis*, Academic Press, New York, 1972.

[Sc] L. Schwartz, *Mathematics for the Physical Sciences*, Addison-Wesley, Reading, MA, 1966.

[Wlok] J. Wloka, *Partial Differential Equations*, Cambridge University Press, New York, 1987

A.3   Specialized or Advanced Texts

[Adam] R.A. Adams, *Sobolev Spaces*, Academic Press, New York, 1975.

[Dac] B. Dacorogna, *Direct Methods in the Calculus of Variations*, Springer-Verlag, Berlin, 1989.

[DS] N. Dunford and J.T. Schwartz, *Linear Operators I*, Wiley, New York, 1958.

[ET] I. Ekeland and R. Temam, *Convex Analysis and Variational Problems,* North-Holland, Amsterdam, 1976.

[EN] K.J. Engel and R. Nagel, One-parameter semigroups for linear evolution equations, Springer-Verlag, New York, 2000.

[Fri1] A. Friedman, Partial Differential Equations, Holt, Rinehart and Winston, New York, 1969.

[Fri2] A. Friedman, Partial Differential Equations of Parabolic Type, Prentice Hall, Englewood Cliffs, 1964.

[GT] D. Gilbarg and N.S. Trudinger, *Elliptic Partial Differential Equations of Second Order*, Springer-Verlag, New York, 1983.

[Go] J.A. Goldstein, *Semigroups of Linear Operators and Applications*, Oxford University Press, New York, 1985.

[GR] I.S. Gradshteyn and I.M. Ryshik, *Table of Integrals, Series and Products*, Academic Press, New York, 1980.

[He] G. Hellwig, *Differential Operators of Mathematical Physics*, Addison-Wesley, Reading, MA, 1964.

[Ka] T. Kato, *Perturbation Theory for Linear Operators*, 2nd ed., Springer-Verlag, New York, 1976.

[Ke] O.D. Kellogg, *Foundations of Potential Theory*, Dover, New York, 1953.

[KJF] A. Kufner, O. John, and S. Fucik, *Function Spaces*, Noordhoff International Publishers, Leyden, 1977.

[LSU] O.A. Ladyzhenskaya, V.A. Solonnikov and N.N. Uraltseva, Linear and Quasilinear Equations of Parabolic Type, American Mathematical Society, Providence, 1968.

[LU] O.A. Ladyzhenskaya and N.N. Uraltseva, Linear and Quasilinear Elliptic Equations, Academic Press, New York, 1968.

[LM] J.L. Lions and E. Magenes, *Non-Homogeneous Boundary Value Problems and Applications I*, Springer-Verlag, New York, 1972.

[Li] J.L. Lions, *Quelques Méthodes de Résolution des Problèmes aux Limites non Linéaires*, Dunod, Paris, 1969.

[Mor] C.B. Morrey, Jr., *Multiple Integrals in the Calculus of Variations*, Springer-Verlag, Berlin, 1966.

[Pa] A. Pazy, *Semigroups of Linear Operators and Applications to Partial Differential Equations*, Springer-Verlag, New York, 1983.

[PW] M.H. Protter and H.F. Weinberger, *Maximum Principles in Differential Equations*, Prentice-Hall, Englewood Cliffs, 1967.

[Sm] J. Smoller, *Shock Waves and Reaction-Diffusion Equations*, Springer-Verlag, New York, 1983.

[Ze] E. Zeidler, *Nonlinear Functional Analysis and its Applications II/B*, Springer-Verlag, New York, 1990.

A.4 Multivolume or Encyclopedic Works

[DL] R. Dautray and J.L. Lions, *Mathematical Analysis and Numerical Methods for Science and Technology*, 6 vol., Springer-Verlag, Berlin, 1990-1993.

[ESFA] Y.V. Egorov, M.A. Shubin, M.V. Fedoryuk, M.S. Agranovich (eds.), Partial Differential Equations I-IX, in: *Encyclopedia of Mathematical Sciences*, Vols. 30-34, 63-65, 79, Springer-Verlag, New York, from 1993.

[Hor] L. Hörmander, *The Analysis of Linear Partial Differential Operators*, 4 vol., Springer-Verlag, Berlin, 1990-1994.

[Tay] M.E. Taylor, *Partial Differential Equations*, 3 vol. Springer-Verlag, New York, 1996.

A.5 Other References

[Ab] E.A. Abbott, *Flatland*, Harper & Row, New York, 1983.

[ADN1] A. Douglis and L. Nirenberg, Interior estimates for elliptic systems of partial differential equations, *Comm. Pure Appl. Math.* **8** (1955), 503-538.

[ADN2] S. Agmon, A. Douglis and L. Nirenberg, Estimates near the boundary for solutions of elliptic partial differential equations satisfying general boundary conditions, *Comm. Pure Appl. Math.* **12** (1959), 623-727 and **17** (1964), 35-92.

[Ba] J. Ball, Convexity conditions and existence theorems in nonlinear elasiticy, *Arch. Rational Mechan. Anal.*, **63** (1977), 335-403.

[Fra] L.E. Fraenkel, On regularity of the boundary in the theory of Sobolev spaces, *Proc. London Math. Soc.* **39** (1979), No. 3, 385-427.

[Fri] K.O. Friedrichs, The identity of weak and strong extensions of differential operators, *Trans. Amer. Math. Soc.* **55** (1944), 132-151.

[GNN] B. Gidas, W.M. Ni and L. Nirenberg, Symmetry and related properties via the maximum principle, *Comm. Math. Phys.* **68** (1980), 209-243.

[La] P.D. Lax, Hyperbolic systems of conservation laws II, *Comm. Pure Appl. Math.* **10** (1957), 537-566.

[Max] J.C. Maxwell, Science and free will, in: L. Campbell and W. Garnett (eds.), *The Life of James Clerk Maxwell*, Macmillan, London, 1882.

[Mas] W.S. Massey, *Singular Homology Theory*, Springer-Verlag, New York, 1980, p. 218ff.

[Mo] T. Morley, A simple proof that the world is three-dimensional, *SIAM Rev.* **27** (1985), 69-71

[Se] M. Sever, Uniqueness failure for entropy solutions of hyperbolic systems of conservation laws, *Comm. Pure Appl. Math.* **42** (1989), 173-183.

[Vo] L.R. Volevich, A problem of linear programming arising in differential equations, *Uspekhi Mat. Nauk* **18** (1963), No. 3, 155-162 (Russian).

Index

Texts in Applied Mathematics

(continued from page ii)